RODENTS IN DESERT ENVIRONMENTS

MONOGRAPHIAE BIOLOGICAE

Editor

J. ILLIES

Schlitz

VOLUME 28

SPRINGER-SCIENCE+BUSINESS MEDIA, B.V. 1975

RODENTS IN DESERT ENVIRONMENTS

Edited by

I. PRAKASH & P. K. GHOSH

SPRINGER-SCIENCE+BUSINESS MEDIA, B.V. 1975

ISBN 978-94-010-1946-0   ISBN 978-94-010-1944-6 (eBook)
DOI 10.1007/978-94-010-1944-6

CONTENTS

CHAPTERS' CONTENTS

PREFACE

Ever since BUXTON published his Animal Life in Deserts in 1923, much individual, collective and organisational effort, both at national and international levels, has gone into the unravelling of the mystique of desert living. Man's interest in the desert must have been a primordial element in his cultural evolution as it was in the womb of deserts that human civilization had dawned and has since been thriving for many millennia. The vision of vast expanses of sun-baked, wind-swept, water-less, forbidding wasteland that seems to occur to most of us the moment we pronounce the word 'desert' is, on the whole, a faithful projection of the reality. But, an important aberration in our thinking that often mars this projection is that the deserts of the world are devoid of any trace of animal life. The situation is, to be sure, quite different and the desert tracts are, if not actually teeming with surface-active animal types, the home of quite a few specialised animal forms and, probably, the last haven of many a vanishing species.

Of all animal types, the rodent would unquestionably be the most numerically heavy tenant of desert lands around the globe. Nature's master plan has guaranteed rodents a place in the sun and they are, apparently, making the most of this arrangement. A study of the biology of deserts would, therefore be, in a large measure, a study of the most important component of the desert biomass, viz. the rodent fauna. It is surprising, however, that with all the recent spate of interest in desert biology, accentuated in no mean measure by the patronage extended by the UNESCO and other world bodies, so little effort has been expended so far in collating the available information on the biology of rodents in desert environments. In our view this has been a major lapse and hence this volume.

Since aridity, rather than great heat or shifting sand masses, truly characterises a desert, both 'hot' and 'cold' (the icy wastes in the Arctic and the Antarctic and at high altitudes on mountains in all latitudes) deserts co-exist on the earth. The scope of this volume has, however, been restricted to 'hot' desert regions. Although the forces causing deserts, or rather aridity, may be several, the present trend of thinking would tend to consider deserts basically as climatic phenomena. But there can also be no two opinion regarding man's role as the principal biotic element in further extending the limits of desert areas. The 'hot' deserts of the world are distributed in two discontinuous belts, one in the Northern Hemisphere and the other in the Southern, more or less centred along the Tropic of Cancer and the Tropic of Capricorn, with neither strip deviating by less than 15 degrees or more than 40 degrees from the equator. There are, however, a few patches of desert tract that do not fall into this pattern of distribution.

Although the exact definition of a desert has long been a moot point, the KÖPPEN classification of 1918 is generally accepted as the most convenient means of differentiating desert regions from non-desert ones. Temperature and precipitation figures are combined mathematically in the KÖPPEN system to establish the boundaries of 'vegetative distributions' for various geographical purposes. KÖPPEN defined deserts as having generally high temperatures and under 255 mm (10 in.) of rain annually. According to this estimate, 14 per cent of the earth's 145.6 million sq. km of land are classed as desert. 'KÖPPEN steppes', with about 254 to 508 mm (10 to 20 in.) of annual rainfall, and high daily and annual temperature ranges, comprise an additional 14 per cent. Thus the combined desert and steppe areas, all of the arid and semi-arid regions, add up to 41.6 million sq. km. The land areas of the principal deserts of the world are as follows*:

Region	Square kilometer (millions)
Sahara Desert	9.1
Australian Desert	3.4
Arabian Desert	2.6
Turkestan Desert	1.9
Great American Desert (includes the Majave, Great Basin, Sonoran, Colorado, Great Salt Lake, Gila and Chihuahuan Deserts of Southwestern North America)	1.3
Patagonian Desert (Argentina)	0.67
Thar Desert (India & Pakistan)	0.60
Kalahari and Namib Deserts (South west Africa)	0.57
Takla Makan Desert including the Gobi (western China to Mongolia)	0.52
Iranian Desert	0.39
Atacama Desert (Peru and Chile)	0.36

* Adapted from CLOUDSLEY-THOMPSON, Chapter I in this book.

In each of these major deserts of the world, a surprising variety of rodent species has evolved and thrives. Although derived from unrelated stocks, there is often a clearly recognisable resemblance in the morphology, physiology and behaviour of many of these rodent species occupying their respective niches in deserts around the world. Of the two major problems of the desert, heat and water, the first can be actively dealt with only at the expense of the second and unavailable commodity. It seems clear now that the rodent's overall strategy of desert survival encompasses both behavioural and physiological adaptations. Much of what follows in this volume is an elaboration of this theme.

As we have emphasised earlier, any worthwhile study of a desert biome must necessarily include a study of its rodent fauna. There seems clearly a case for a proper assessment of the rodent's place in any desert eco-system. For example, we ought to have quantitative information on the rodent's contributions to the maintenance and aggravation of desertic conditions, its role as a pest of food grains, grasslands and other vegetation and as a carrier of diseases, its relationships with its predator fauna, viz. birds, reptiles and small carnivores of the desert and its susceptibility to chemical and biological control measures. Hopefully, the present volume will provide sufficient background information to generate research interest in these applied aspects of desert rodent biology. We are aware of the deficiencies and limitations of the present volume but such lacunae are, perhaps, inherent in the production of any volume having as multifaceted a scope as this one. Inspite of our best efforts, many species, geographical regions and aspects of research have remained uncovered in this book. Since time was running out fast we decided to plunge into publication without waiting for the manuscripts of several committed authors. Perhaps it will be possible to fill up these gaps in a subsequent edition of the book.

We are thankful to all the contributors to this volume for their excellent cooperation, understanding and patience. We must also thank the Publishers – Dr. W. Junk b.v. and Prof. Dr. J. ILLIES, Editor-in-Chief of the Series for sparing no pains in the production of this volume. Finally, we would like to record our sincere gratitude to Mrs. LAKSHMI I. PRAKASH for her help at all levels of production of this book.

January 31, 1975   ISHWAR PRAKASH
Jodhpur, India   PULAK K. GHOSH

AUTHORS' ADDRESSES

CLOUDSLEY-THOMPSON, J. L., Department of Zoology, Birkbeck College, London, England.

CORBETT, L. K., Division of Wildlife Research, C.S.I.R.O., Alice Springs, N.T., Australia.

DE VOS, A., Food and Agriculture Organisation of the United Nations, Rome.

EISENBERG, J. F., National Zoological Park, Smithsonian Institution, Washington D.C., U.S.A.

FRENCH, N. R., Environmental Resources Centre, Colorado State University, Fort Collins, Colorado, U.S.A.

GAISLER, J., Institute of Zoology, Purkyně University, Brno, Czechoslovakia.

GHOBRIAL, L. I., Middlesex Hospital Medical School, Cleveland Street, London W1, United Kingdom.

GHOSH, P. K., Central Arid Zone Research Institute, Jodhpur, India.

HAPPOLD, D. C. D., Department of Zoology, University of Ibadan, Ibadan, Nigeria.

HARRISON, D. L., Bowerhood House, Sevenoaks, Kent, England.

HAWBECKER, A. C., Fresno State College, Fresno, California, U.S.A.

JOHNSON, S., Department of Zoology, University of Jodhpur, Jodhpur, India.

JORGENSEN, C. D., Department of Zoology, Brigham Young University, Provo, Utah, U.S.A.

KRAFT, A., Department of Biology, University of New Mexico, Albuquerque, U.S.A.

LOBACHEV, V. S., Moscow University, Moscow, U.S.S.R.

MACFARLANE, W. V., Waite Agricultural Research Institute, Adelaide, Australia.

MARES, M. A., Biology Department, University of Pittsburgh, Pittsburgh Pennsylvania 15260, U.S.A.

MISONNE, X., Institut Royal Des Sciences Naturelles de Belgique, Brussels, Belgique.

NAUMOV, N., Biological Faculty, Moscow University, Moscow, U.S.S.R.

NEWSOME, A. E., Division Of Wildlife Research, CSIRO, Canberra, Australia.

NOUR, T. A., Department of Chemistry, University of Alger, Algerie.

PETTER, F., Muséum National d'Histoire Naturelle, Laboratoire de Zoologie, Mammifères et Oiseaux, 55 rue de Buffon, Paris 5e, France.

PRAKASH, I., Central Arid Zone Research Institute, Jodhpur, India.

ROSENZWEIG, M. L., Department of Biology, University of New Mexico, Albuquerque, New Mexico 87106, U.S.A.

SCHMIDT-NIELSEN, K., Department of Zoology, Duke University, North Carolina, U.S.A.

SMIGEL, BARBARA, Department of Biology, University of New Mexico, Albuquerque, New Mexico 87106, U.S.A.

SMITH, H. D., Department of Zoology, Brigham Young University, Provo, Utah 84601, U.S.A.

TCHERNOV, E., Department of Zoology, The Hebrew University of Israel, Jerusalem, Israel.

TURNBULL, PRISCILLA F., Field Museum of Natural History, Chicago, Ill. U.S.A.

I. THE DESERT AS A HABITAT

by

J. L. CLOUDSLEY-THOMPSON

Introduction

Desert regions are not necessarily characterised by great heat, nor do they always consist of vast expanses of shifting sand dunes. The one characteristic common to them all is their aridity throughout most or all of the year. When the air is humid, not only does less solar heat penetrate to the ground during the day, but less is lost from the earth by radiation at night. Consequently, humid climates tend to show daily or seasonal stability, while deserts are characterised by extremes of temperature and humidity. The most adverse conditions for life consist of a combination of aridity and high temperature, and it is the effects of these two factors that have been studied most extensively. The combination presents an unusually acute thermoregulatory challenge to homeothermic animals, because it poses the singularly intractable problem of losing heat to a hot environment while simultaneously keeping water loss at a minimum.

Weather or climate exist as part of a global system of air movements and desert areas are often small in relation to the wind systems that dominate them. A detailed bibliographic review of the state of knowledge about desert climates is given by REITAN & GREEN (1968).

Distribution

The deserts of the world can be divided into five types on a climatic basis. These are as follows: sub-tropical deserts, cool coastal deserts, rain-shadow deserts, interior continental deserts and polar deserts. Only the first four of these will be discussed in the present volume, but all of them are arid. In the case of polar deserts, water is present in the form of ice and is therefore not available to plants and animals. In the other types of desert, water is deficient throughout most of the year because the amount of evaporation greatly exceeds the annual precipitation.

The distribution of deserts throughout the world is due mainly to the way in which the atmosphere circulates, particularly in its lower layers. Sub-tropical deserts are the result of semi-permanent belts of high pressure in tropical regions within which the air has a tendency to descend from high altitudes towards the surface of the land. At the beginning of its descent this air is cold and dry but it becomes warmed by compressional heating at the adiabatic rate of 10 degrees C. per 1,000 m. Consequently it reaches ground level very hot and with an extremely low

1

Table 1. Land Areas of Principal Deserts

Region	Square kilometres (millions)	Square miles (millions)
Sahara Desert	9.1	3.5
Australian Desert	3.4	1.3
Arabian Desert	2.6	1.0
Turkestan Desert	1.9	0.75
Great American Desert (includes the Mojave, Great Basin, Sonoran, Colorado, Great Salt Lake, Gila and Chihuahuan Deserts of south-western North America)	1.3	0.5
Patagonian Desert (Argentina)	0.67	0.26
Thar Desert (India)	0.60	0.23
Kalahari and Namib Deserts (South West Africa)	0.57	0.22
Takla Makan Desert, including the Gobi (Western China to Mongolia)	0.52	0.20
Iranian Desert (Persia)	0.39	0.15
Atacama Desert (Peru and Chile)	0.36	0.14

(From BROWN, 1968, adapted from CLOUDSLEY-THOMPSON, 1965)

relative humidity so that it is completely incapable of producing any precipitation.

Cool coastal deserts are almost always rainless, yet drenched with chilly moisture. They include the Namib, Atacama and the coastal desert of Baja California. Rainlessness results from descending air masses, the high humidity and cold from nearby cool ocean currents, respectively the Benguela, Humboldt and Californian Currents. In each case these currents have originated in polar regions, but they pull up even colder water from the depths which lie near the shore in those parts of the world.

Rain-shadow deserts are situated on the lee sides of mountains which cause the prevailing wind to rise and drop its moisture in the form of orographic precipitation. An example is afforded by the Mojave Desert which owes its winter aridity to the Sierra Nevada and Transverse Ranges, whilst its aridity in summer is caused by the presence at that time of the sub-tropical high pressure cell which dominates the Sonoran Desert, just to the south, throughout the year. The Great Basin Desert is likewise sheltered by the Sierra and Cascade Ranges to the west and by the Rocky Mountains to the east. The deserts of Patagonia are in rain-shadow as far as the prevailing westerly winds from the Pacific are concerned, while air masses moving from the south Atlantic are cooled from the Falkland Islands Current and carry little moisture.

2

The Australian desert, too, is to some extent in the rain shadow of the Great Dividing Range. Like the Mojave and Great Basin Deserts, however, it also falls into the category of the continental interior deserts, which are arid through the lack of marine influence and other factors related to the massive bulk of the land surrounding them. Distance from water is the final factor in the creation of the deserts of Central Asia (LOGAN, 1968; WALLÉN, 1966). The deserts of the world (excluding polar deserts) are summarised in Table 1.

Soils

The ground surface of the desert is typically barren and, where relief is appreciable, bare rock stands out. Soils have restricted topographic distribution and occur mainly on flat or gently sloping surfaces. Elsewhere, bare rock predominates. Weathering is largely a mechanical process, with chemical effects playing an important secondary rôle. Exposed rock experiences wide and rapid temperature variations which were long thought to cause erosion by splitting and fragmentation. Serious doubt has been cast upon this interpretation, however, and it is now believed that moisture may play a greater part through hydration of constituent mineral grains with their resulting expansion. Moisture also supplies saline solutions to pores and cracks in rocks. When these evaporate, the salts crystallize and produce a wedging effect (see discussion in SMITH, 1968). The most conspicuous effect of weathering is to form *screes* or *bajadas* and aprons of rock waste that, as their tops erode, gradually bury the bases of rocks, cliffs and even hills, in the products of their own decay. In time these too disappear and the land surface may become completely flat.

Although rainstorms are infrequent in deserts, they exert a profound effect on bare rock and soil unprotected by vegetation. Gully and sheet erosion often occur, while water courses are quickly choked with sand and silt. When desert *wadis* terminate in alluvial basins, the sediments they carry become stratified and gradually extend until they fill the whole internal drainage basin, forming an immense level plain. Such plains include stony desert with a mosaic of gravel, the *reg* of the western Sahara, or of pebbles, the *serir* of Libya and Egypt. In western North America, alluvial plains are called *playas* and temporary lakes often form in them. When dry, however, such areas are usually covered with glistening salt. The effect of sub-surface water is partly mechanical, through sapping and piping, and partly chemical, by transporting and depositing soluble salts.

The abrasive effects of wind erosion involve the lossening and removal of particles from the bedrock. These materials, along with those already loosened by other processes, are removed by deflation. Sand is rolled along the ground; dust and silt are carried aloft in suspension, sometimes

3

Fig. 1. Effect of wind erosion, southern Algeria.

for many hundreds of kilometres. On settling, a thin but widespread layer is formed, which may accumulate in the steppe lands bordering the desert to form loess. The sand travels more slowly and accumulates in dunes nearer to its source.

Rocky desert, known as *hammada* in the Sahara, is composed of denuded rock plateaux, smoothed and polished by wind abrasion and flattened by deflation. Deposited soils, composed of accumulations of medium or fine wind-blown sand, often take the form of *ergs* – vast, sandy wastes occupied by great masses of dunes. These may grow to enormous size, reaching a height of 200 m or more in the central Sahara and Rub'al Khali of Arabia.

In most deserts, dunes cover less than half the land surface, and usually very much less than this. They are of many types and occur in assemblages of endless variation. According to BAGNOLD (1941), there are two main types of sand dune: crescent-shaped, moving *barchan* dunes and *seif* or sword-shaped dunes. The latter vary in form according to the wind. The four most easily distinguished types are therefore *barchan*, *seif*, *transverse* and *stellar* dunes (CLOUDSLEY-THOMPSON, 1965).

The winnowing effect of the wind, sorting out particles of different sizes and depositing them elsewhere, results in the three main desert types described above, viz. *hammada*, *reg* and *serir*, and *erg*. In semi-arid regions where rainfall varies between about 120–250 mm per year, the soil types usually formed are brown or grey semi-desert soils: the latter are sometimes called sierozems. True desert soils contain no humus and are little more than fragmented rock.

Sierozems, too, may contain less than one per cent of organic matter in the surface horizon with calcium carbonate deposited on the surface. But they often support low desert-scrub vegetation. Between them and the black earths or chernozems of the steppe may be found a type of soil known as chestnut earth, covered with low grass steppe vegetation and occasionally some scrub. Thus there is often a complete gradation from the bare desert sands containing virtually no humus at all, to the rich black loess of the steppes (CLOUDSLEY-THOMPSON and CHADWICK, 1964; JEWITT, 1966).

From the above it will be seen that the desert soil must influence its fauna not only indirectly through the vegetation it supports, but also directly – by providing deposits into which animals can burrow and rocky crevices into which they can escape from the heat of the day and the cold of the night. Shelter from wind and sand, and elevation that prevents the risk of temporary flooding, are also factors important to the survival of desert animals. LUSTIG (1968) gives an encyclopaedic report on the geomorphology and surface hydrology of desert environments; DREGNE (1968) an appraisal of research on surface materials in which desert soils and weathering are discussed.

Climate

Desert climates are subject to extremes. High temperatures and low humidity during the day are followed by comparatively cold nights. Long periods of drought are broken by torrential rainfall and flooding. Strong winds and sandstorms are characteristic, especially of the summer. Desert areas are best defined, however, in terms of their rainfall characteristics and either evaporation, evapotranspiration or some feature of the climate related to these (CLOUDSLEY-THOMPSON & CHADWICK, 1964).

Perhaps the best index of aridity so far devised is that of THORNTH-WAITE (1948) as adapted by MEIGS (1953). The classification of climate does not depend on reducing all the parameters to the same thermal terms. THORNTHWAITE's calculations, like those of KÖPPEN (1923) both show that a desert area may be one that is cold and has a very low rainfall, or hotter and with relatively more precipitation. PENMAN's (1948) formula is recommended for more careful analysis. (See discussion by WALLÉN, 1966).

RAINFALL

Desert rainfall tends to be seasonal, but it is most erratic as the total annual precipitation varies considerably from year to year. The presence

5

Fig. 2. Desert rainstorm, New Mexico.

of *wadis* and dry saline lake beds, such as the *chotts* of North Africa, show that torrential rain may sometimes fall. Most of this runs off so rapidly, however, that it tends to be wasted.

The erratic nature of desert rainfall can best be illustrated by some typical figures. During the 3 years following September 1933, 2 mm, 3 mm, and 5 mm respectively fell at Helwan, Egypt, but no less than 125 mm fell in the year 1945–46. Again, 679 mm were recorded at Erkowit in the Red Sea Hills during 1951, but only 40 mm fell there in the following year. The monthly rainfall at Khartoum for July 1946 was 129.5 mm while Shambat, about 8 km away, received only 38.7 mm. Rainfall figures for the year in the two localities were 247.7 mm and 143.1 mm respectively. The previous year, however, the situation was reversed, for Khartoum recorded 90.6 mm and Shambat 224.8 mm. Between 1900 and 1957 the annual rainfall at Khartoum varied from 48 to 380 mm, but only in 5 years did it exceed 250 mm.

Cairo had only 18 rainfalls of more than 100 mm between 1890 and 1919, and in 17 years out of the 30 rain was entirely lacking. By contrast, 430 mm fell in a single storm on 17 January 1919 after which there were boats in the streets of the city, trams were sunk in the mud up to the level of their windows and houses of unbaked brick in the suburbs melted away like lumps of sugar. There was no rain at all at Baghdad, California, over a period of 32 months between 1909 and 1912, and again during the 3 years following February 1917. Prolonged droughts have been known to persist for up to 10 years or more in the central Sahara, and form a measure of the variability of desert rainfall.

The biological significance of meteorological data in deserts is therefore limited. Furthermore, winter and spring rains which fall in North Africa are probably more beneficial to plants and animals than the summer precipitation that occurs south of the Sahara because they do not evaporate so rapidly (CLOUDSLEY-THOMPSON, 1965). More important is the mean period between storms of sufficient magnitude for some water to remain stored and available in favoured localities once surface evaporation has ceased. The occurrence of such storms enables the seeds that have lain dormant through years of drought to germinate, grow into mature plants and then re-seed themselves from a single shower. It is upon such plants and their seeds that rodents and larger nomadic mammals depend for their food (CLOUDSLEY-THOMPSON & CHADWICK, 1964).

TEMPERATURE

When the air is laden with water vapour, either diffused or in the form of clouds, not only does less of the sun's heat penetrate to the ground during the day, but less is lost by radiation at night. Thus, whereas in clear desert air only about 10 per cent of solar radiation is deflected by dust particles and cloud, in humid regions some 20 per cent may be deflected by clouds, 10 per cent by dust and 30 per cent by water surfaces and vegetation. At night, on the other hand, up to 90 per cent of accumulated heat escapes from the desert surface to the upper air while, in humid countries, only 50 per cent escapes, 30 per cent being deflected by clouds and dust, and the remainder being retained by the land cover and water. Consequently, humid equatorial climates tend to show diurnal and seasonal stability whereas deserts are characterised by extremes of temperature and humidity.

Maximum temperatures are high in deserts and semi-arid areas, especially in summer. But, even in winter, temperatures during daytime are very high compared with other regions. The mean maximum temperature may reach 43.6 °C in Baghdad, 41.7 °C in Biskra and 40.0 °C in Phoenix (Table 2). In arid regions, influenced by cool ocean currents, mean maximum temperatures in summer are considerably lower – for example, 24.4 °C in Antofagasta and 29.4 °C in Windhoek. Winter mean maxima are low in the cold interior continental deserts of Asia. In Kashgar the mean maximum is only 0.6 °C in January but it is as high as about 30 °C in Khartoum and Bilma (WALLÉN, 1966).

Mean figures are of little biological significance in deserts, however,

Table 2. Climatic data at various arid and semi-arid stations (From WALLÉN, 1966)

Station	Lat.	Long.	January Mean	January Mean max.	January Mean min.	January Rain-fall	January Inc. rad.	July Mean	July Mean max.	July Mean min.	July Rain-fall	July Inc. rad.	Year Rain-fall	Year Period
1. Sacramento, N. Am.	38°31′N	121°30′W	+ 7.5	+11.1	+ 3.9	96.5	190	+23.3	+32.2	+14.4	< 2.5	650	472	1850–1949
2. Phoenix, N. Am.	33 26 N	112 01 W	+11.1	+18.3	+ 3.9	20.3	290	+32.5	+40.0	+25.0	25.4	610	191	1896–1949
3. Zaragoza, Eur.	41 39 N	00 53 W	+ 5.7	+ 9.7	+ 1.8	14.0	170	+23.5	+30.5	+16.6	20.0	680	377	1951–60
4. Marrakesh, Afr.	31 37 N	08 02 W	+11.5	+17.3	+ 5.7	28.0	290	+28.2	+36.9	+19.5	2.0	670	242	1951–60
5. Biskra, Afr.	34 48 N	05 44 E	+11.4	+16.1	+ 6.7	17.8	290	+34.2	+41.7	+26.7	2.5	690	158	1913–50
6. Kashgar, As.	39 24 N	76 07 E	− 5.3	+ 0.6	−11.1	15.2	190	+26.7	+33.3	+20.0	10.2	650	81	1934–44
7. Palmyra, As.	34 33 N	38 17 E	+ 7.9	+13.1	+ 2.6	24.0	240	+29.8	+38.3	+21.2	0	740	153	1951–60
8. Baghdad, As.	33 20 N	44 24 E	+10.6	+16.7	+ 4.5	25.0	300	+34.6	+43.6	+25.6	0	730	151	1951–60
9. Teheran, As.	35 41 N	51 19 E	+ 4.0	+ 8.9	+ 1.0	37.0	230	+28.6	+35.5	+21.6	0	700	213	1951–60
10. Jodhpur, As.	26 18 N	73 01 E	+16.7	+24.4	+ 8.9	2.5	370	+31.4	+36.1	+26.7	101.6	490	356	1891–1943
11. Cairo/Helwan, Afr.	29 52 N	31 20 E	+13.3	+18.3	+ 8.3	5.1	300	+28.4	+35.6	+21.1	0	730	28	1904–45
12. La Paz, N. Am.	24 10 N	110 18 W	+18.0	+22.2	+13.9	5.0	380	+19.4	+35.0	+23.9	10.1	560	145	1917–42
13. Monterrey, N. Am.	25 40 N	100 18 W	+14.4	+20.0	+ 8.9	15.2	250	+26.9	+32.2	+21.7	58.4	600	579	1910–45
14. Bilma, Afr.	18 41 N	12 55 E	+17.2	+27.2	+ 7.2	0	480	+33.0	+42.2	+23.9	2.5	630	23	1931–40, 1949–55
15. Khartoum, Afr.	15 36 N	32 33 E	+23.6	+32.2	+15.0	<2.5	520	+31.7	+38.3	+25.0	53.3	600	157	1900–45
16. Antofagasta, S. Am	23 28 S	70 26 W	+20.8	+24.4	+17.2	0	700	+13.8	+17.2	+10.5	5.8	340	13	1904–42
17. Windhoek, S. Afr.	22 34 S	17 06 E	+23.3	+29.4	+17.2	76.2	420	+13.0	+20.0	+ 6.1	<2.5	340	363	1891–1950
18. Alice Springs, Aus.	23 38 S	133 53 E	+28.6	+36.1	+21.1	43.1	690	+11.7	+19.4	+ 3.9	7.6	390	251	1921–50
19. Merredin, Aus.	30 46 S	121 27 E	+26.0	+34.3	+17.6	18.0	690	+10.3	+15.7	+ 4.9	49.0	250	300	1951–60
20. Mendoza, S. Am.	32 53 S	68 49 W	+23.9	+32.2	+15.5	22.8	680	+ 8.4	+15.0	+ 1.7	5.0	250	191	1921–50

Temperatures in °C. Rainfall in mm. Radiation in cal/cm²/day.

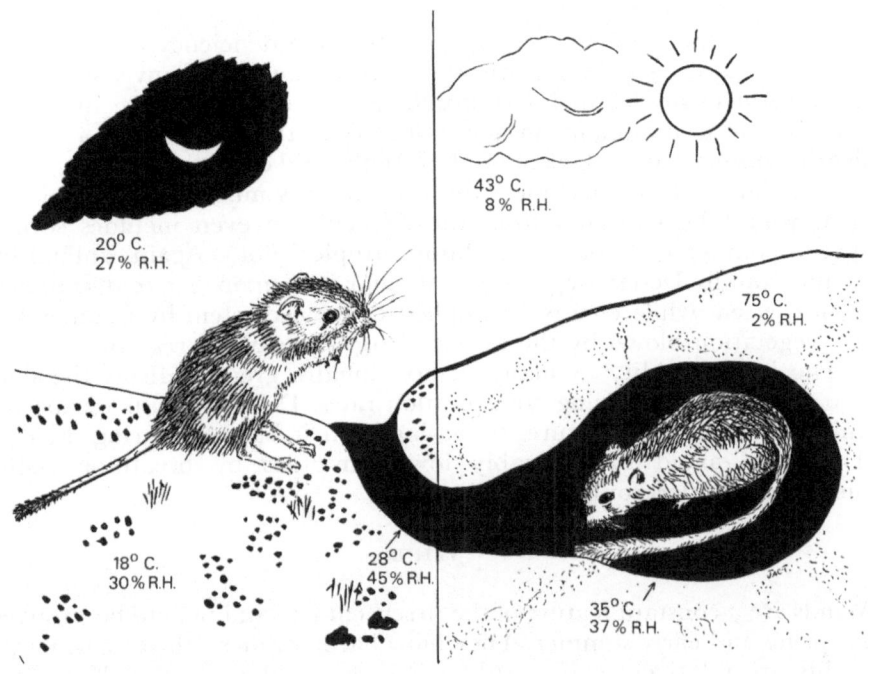

20° C.
27% R.H.

43° C.
8% R.H.

75° C.
2% R.H.

18° C.
30% R.H.

28° C.
45% R.H.

35° C.
37% R.H.

Fig. 3. Typical microclimatic readings above ground and in the burrow of a jerboa at night and during the day. (Drawing by Mrs. J. A. CLOUDSLEY-THOMPSON).

because fluctuations are so large. It is the incidence of high temperatures liable to cause heat damage and of low temperatures causing frost damage, and limiting the growing season of the vegetation, that are important to plants and animals. Shade temperatures as high as 56.5 °C have been registered in Death Valley, California, and even higher in the Sahara. As annual range of shade temperature from —2 °C to 52.5 °C has been recorded from Wadi Halfa and a daily range in summer of 29 °C at In Salah. The record appears to be a fluctuation of 38 °C (—0.5 °C to 37.5 °C) within 24 hours at Bir Mighla in southern Tripolitania in December. During 1910 there were 14 days of frost at Tamanrasset at an altitude of 1400 m, and absolute minima of —7 °C and —2 °C were recorded in January and February of that year (CLOUDSLEY-THOMPSON & CHADWICK, 1954).

HUMIDITY

Not only is lack of moisture in the form of rain the chief factor causing desert conditions, and the absence of clouds responsible for extremes of temperature, but low humidity itself has an adverse effect upon plant

9

and animal life. This is because the saturation deficiency of the atmosphere increases as temperatures soar during the day. Conversely, there are several examples in low latitudes where a high relative humidity, usually due to dominant onshore winds from the sea, may to a certain degree compensate for lack of rainfall. In the arid or semi-arid zones along the Persian Gulf, for instance, the vegetation is much better developed than would be expected from the rainfall and even includes several species characteristic of a more humid tropical flora. Again, animal life in the Namib Desert depends for its moisture upon fog that comes in from the sea, while energy is supplied to the ecosystem by fragments of dry vegetation blown by the easterly berg winds of the region.

The low humidity of many desert climates greatly affects the flora and fauna by influencing transpiration rates. Desert rodents escape not only from high temperatures by hiding in their burrows during the day. They also avoid unfavourably low humidities by breathing cooler moister air underground (Fig. 3).

Wind

Winds are a constant feature of the desert climate and tend to be strongest in spring and early summer. They blow hardest during the day while the nights are relatively calm. Although speeds seldom exceed 80 km per hour and average only about 16 km per hour annually, the effect of wind is enhanced by drought, high temperature and lack of vegetation.

Strong, hot winds are frequently associated with sand storms. Well-known examples related to large scale cyclonic disturbances include the *khamsin* of North Africa and the Near East, the *harmattan* on the southern border of the Sahara and the *sumum* in Iran and Pakistan. The *haboobs* of the northern Sudan are of a more local character, as are the *chubaseos* of the North American deserts. Another familiar phenomenon, the whirlwind, 'dust devil' or *tornillo* results from a sudden irregular upward rush of heated air on a still day. Such columnar vortices carry sand and other objects from the soil surface often to a great height. No doubt many small animals and plants are killed in this way. Whenever possible, small animals shelter from the wind in holes, burrows and other retreats.

Microclimates

For rodents and other small animals, standardised meteorological measurements have only indirect relevance. Their environment is one of holes and crevices, tunnels and nests, in which distances are measured in metres rather than kilometres and where the differences between sun and shadow, day and night, or winter and summer may have lethal significance. Although tremendous daily and seasonal temperature fluctuations are characteristic of the desert surface, conditions beneath

10

are more equable. Thus, a diurnal range of 56.5 °C has been recorded at a depth of 0.4 cm in Arizona. This diurnal range is very similar to the yearly range but the effects are not transmitted much below the surface. A relatively moderate and constant temperature is reached at a depth of 100–200 cm, which is well within the range of burrowing animals.

Sand surface temperatures often reach an extremely high level. The world record is 84 °C, registered on the Loango Coast near the equator and at Wadi Halfa. Below 50 cm there is hardly any diurnal temperature variation in the sands of the Sahara and, at a depth of 100 cm, the annual variation is not more than 10 °C. According to PIERRE (1958) an outstanding feature of the Saharan sands is high humidity. Even during summer, the air surrounding loose, dry grains at a depth of only 50 cm has a relative humidity of 50 per cent. The moisture rises from extra-Saharan water which underlies most of the dunes in the Great Western Erg. Often, too, the uniformity of the internal climate of the sand is broken up into distinct microclimates. Immediately underneath a stone the temperature may be the same as at a depth of 10 cm below the bare surface, and shows less variation. In contrast, microclimatic measurements in the sands of the Red Sea Hills and coastal plain in autumn have revealed relative humidities seldom in excess of 35 per cent, and the distribution of the fauna here seems to depend entirely upon the availability of shade (CLOUDSLEY-THOMPSON, 1962).

Water descends very quickly through wind-blown sand since its anti-wetting properties are low and its permeability high. Owing to capillary tension, a given charge of water applied at the surface of dry sand will sink to a certain depth and no more. This depth is about eight times the immediate precipitation. Water that has reached a depth of 20–30 cm keeps the soil in a moist, unsaturated condition for several years because the temperature is relatively constant and there is no ventilation. The sand above and below is dry.

By burrowing deeply or, to a lesser extent, by entering caves and rock fissures, desert animals can avoid the extreme heat and drought of the day and, by leaving their retreats at night, they can even avoid the peak temperature below, for there is a considerable time-lag before heat begins to penetrate deeply into the sand (Fig. 3). Nocturnal habits associated with the selection of favourable microhabitats play an important role in the survival of rodents in the desert.

Vegetation

Most desert and semi-desert regions support some degree of vegetation, although their climate and soils are so dry. The least amount of plant life is found in areas of *hammada* and clay, apart from where they are traversed by wadis. Except on dunes, which are usually quite bare, sandy desert usually has a less scanty flora.

11

Beyond the edges of true desert are shrub-steppe lands where the rainfall is very scanty. Such is the *Acacia* desert scrub which lies south of the Sahara extending from Senegal to the Red Sea and the mountains of Ethiopia. Where rain is moderate in amount, even if it falls only in a few days of the year, grass-steppe is found.

The fauna of any region is largely dependent upon the vegetation. Some animals are able, however, to live in areas of desert, such as *ergs*, where there are no plants growing. Their food chains are based upon dried vegetation and grass seeds, blown often from a considerable distance. As BRINCK (1956) points out, dried vegetable matter is continually being transported by the wind into the more arid desert regions of the world. Part of this is eaten or destroyed immediately; the remainder becomes buried and, in the absence of bacteria, does not decompose but comes to the surface, often years later, when it supports Lepismidae, the larvae of Tenebrionidae, and occasionally other forms. Even vegetationless desert, therefore, may support a sparse fauna provided that a sufficient concentration of dried plant material is achieved.

REFERENCES

BAGNOLD, R. A. (1941). The Physics of Blown Sand and Desert Dunes. London: Methuen.

BROWN, G. W. jr. (Ed.) (1968). Desert Biology Vol. 1. New York: Academic Press.

BRINCK, P. (1956). The food factor in animal desert life. In W. G. WINGSTRAND (Ed.). Bertil Hanstrom: Zoological Papers in Honour of his 65th Birthday, November 20th, 1956, pp. 120–137. Lund: Zool. Inst.

CLOUDSLEY-THOMPSON, J. L. (1962). Bioclimatic observations in the Red Sea Hills and coastal plain, a major habitat of the desert locust. *Proc. R. Ent. Soc. Lond. (A)* 37: 27–34.

CLOUDSLEY-THOMPSON, J. L. (1965). Desert Life. Oxford: Pergamon Press.

CLOUDSLEY-THOMPSON, J. L. & CHADWICK, M. J. (1964). Life in Deserts. London: Foulis.

DREGNE, H. E. (1968). Appraisal of research on surface materials of desert environments. In McGINNIES, W. G., GOLDMAN, B. J. & PAYLORE, P. (Eds.). Deserts of the world. An appraisal of research into their physical and biological environments. 285–377. University of Arizona Press.

JEWITT, T. N. (1966). Soils of arid lands. In E. S. HILLS (Ed.). Arid Lands. A Geographical Appraisal. 103–126. London: Methuen.

KÖPPEN, W. (1923). Die Klimate der Erde. Berlin: De Gruyter.

LOGAN, R. F. (1968). Causes, climates, and distribution of deserts. In G. W. BROWN, jr. (Ed.). Desert Biology, 1: 21–50. New York: Academic Press.

LUSTIG, L. K. (1968). Geomorphology and surface hydrology of desert environments. In McGINNIES, W. G., GOLDMAN, B. J. & PAYLORE, P. (Eds.). Deserts of the world.

12

An appraisal of research into their physical and biological environments. 93–283. University of Arizona Press.

MEIGS, P. (1953). World distribution of arid and semi-arid homoclimates. Reviews of Research on Arid Zone Hydrology, 203–209. Paris: U.N.E.S.C.O.

PENMAN, H. L. (1948). Natural evaporation from open water, bare soil and grass. *Proc. Roy. Soc. (A)* 193: 120–145.

PIERRE, F. (1958). Écologie et Peuplement Entomologique des Sables vifs du Sahara Nord-Occidental. Paris: Centre national de la Recherche Scientifique.

REITAN, C. R. & GREEN, C. R. (1968). Weather and climate of desert environments. In McGINNIES, W. G., GOLDMAN, B. J. & PAYLORE, P. (Eds.). Deserts of the world. An appraisal of research into their physical and biological environments. 19–92. University of Arizona Press.

SMITH, H. T. U. (1968). Geologic and geomorphic aspects of deserts. In G. W. BROWN, jr. (Ed.). Desert Biology, 1: 51–100.

THORNTHWAITE, C. W. (1948). An approach towards a rational classification of climate. *Geograph. Ref.*, 21: 633–655.

WALLÉN, C. C. (1966). Arid zone meteorology. In E. S. HILLS (Ed.). Arid Lands. A Geographical Appraisal, 31–52. London: Methuen.

II. THE ECOLOGY OF RODENTS IN THE NORTHERN SUDAN

by

D. C. D. HAPPOLD

Introduction

The northern Sudan extends from approximately 22 °E to 37 °E, and from 12 °N to 22 ° (Fig. 1). In the north near the Egyptian border it is desert with practically no vegetation. South of the desert, at about the 25 mm isohyet, there is semidesert with a sparse vegetation and this merges gradually into a dry woodland savanna between the 200 and 400 mm isohyets. The rainfall increases from north to south (Fig. 2). At Wadi Halfa the annual rainfall is about 4 mm, and in some years there is no rain. At Khartoum the annual rainfall is about 160 mm (OLIVER, 1965) and south of latitude 17 °N there is a large increase in the rainfall spread over a longer period of the year. The Nile valley flows northwards through the Sudan; along its banks and floods plains is a moister environment than in the surrounding desert and semidesert.

Rocky hills, or jebels, project from the flat desert and semidesert plains. These jebels are single and isolated, or in groups. Some are the eroded remains of large hills and are now covered with small stones, others are of granite or syenite, or are the remains of extinct volcanos. The northern jebels have only a few grasses growing on them.

In the semidesert, there is considerable variation in the annual rainfall from year to year (Table 1). The majority of the rain falls from July to September, and during these months several large storms account for most of the annual rainfall. After heavy storms the desert is flooded, especially in low lying areas, and it takes several days for the water to evaporate or to percolate into the sand. Within seven to ten days of the first rains, the desert is covered with short grasses. In a wet year, e.g. 1964, these grasses are dense and may grow to a height of two to three feet high; in a dry year, the grass is patchy and only a few inches high. Two to three weeks after the last rains, the grass withers and is gradually broken up and blown away. Even within a small area, the rainfall is very varied, and so the distribution of grass and seeds is patchy.

BARBOUR (1961) gives a general account of the geography of the Sudan, and details of the vegetation are given by HARRISON & JACKSON (1958).

In the northern Sudan, there are four main habitats for rodents:
1. The desert and semidesert: there is little vegetation except after the

15

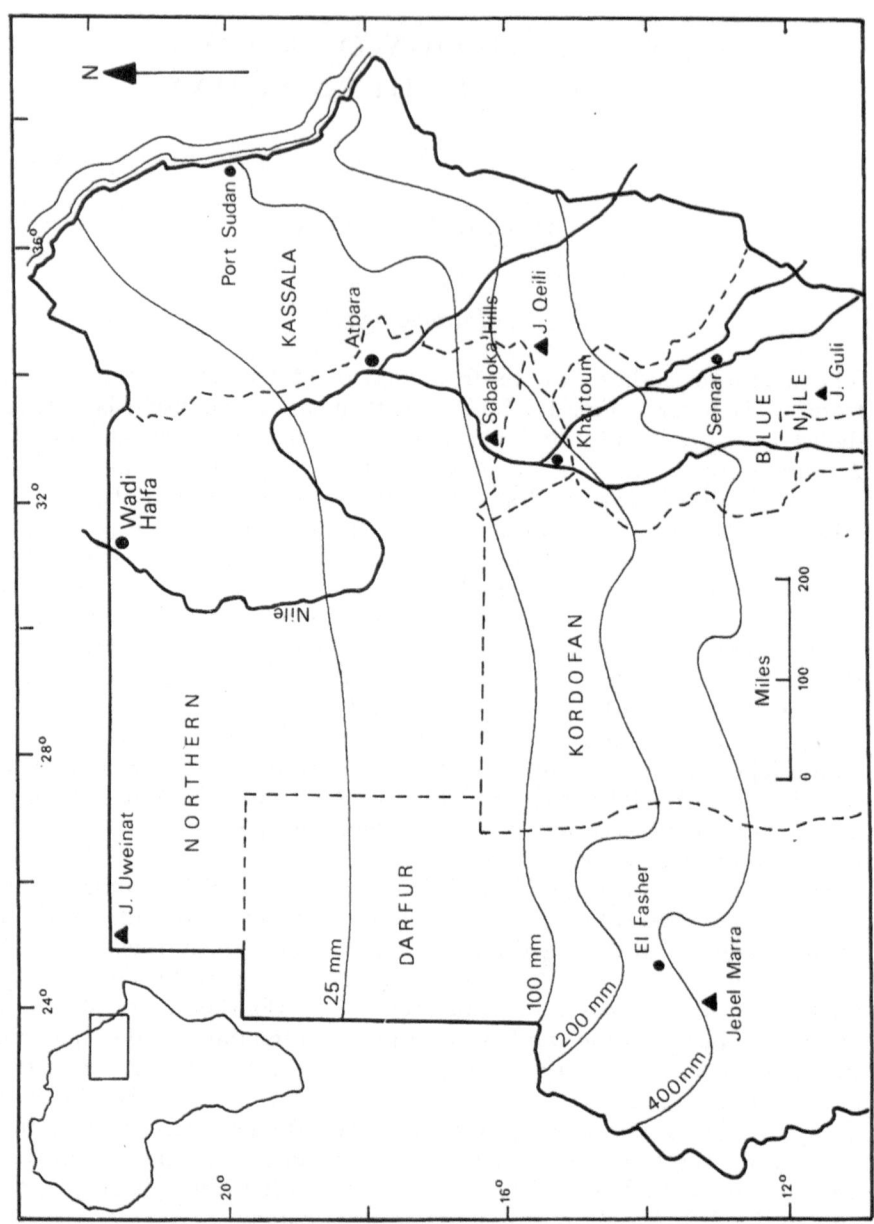

Fig. 1. Map of the northern Sudan to show the 25, 100, 200 and 400 mm isohyets and localities mentioned in the text.

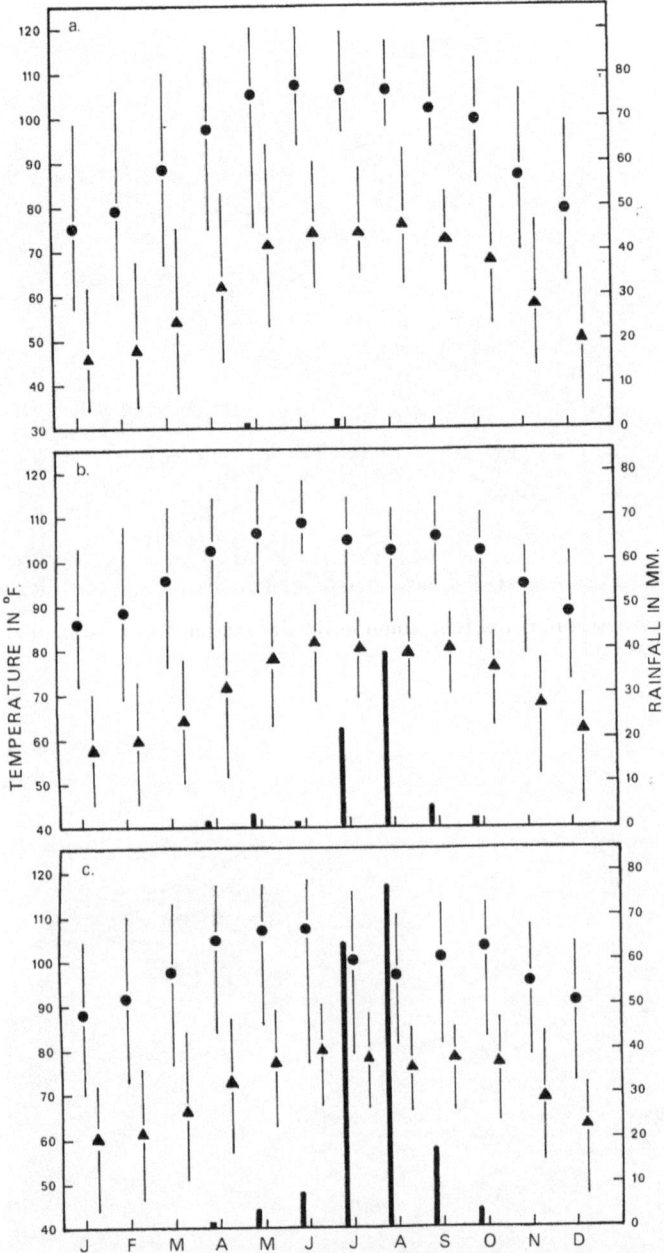

Fig. 2. Climatic data for a. Wadi Halfa, b. Atbara, and c. Khartoum. ●, monthly mean maximum temperature; ▲, monthly mean minimum temperature; range of temperature is indicated by thin lines. Monthly mean rainfall is shown by the thick lines. (Original numerical data from the Sudan Almanac, 1963).

Fig. 3. Semi-desert north of Khartoum in the dry season. Sandbathing patches are seen in the foreground.

Fig. 4. Semi-desert near Khartoum. The grass tussocks in the foreground are partially covered by sand. Distant jebels are visible on the horizon.

18

Table 1. Rainfall in mm at Khartoum Airport during 1963 to 1965

	Jan.	Feb.	Mar.	April	May	June	July	Aug.	Sept.	Oct.	Nov.	Dec.	Total
1963	0	0	0	tr.	19.0	2.9	82.5	17.6	19.9	tr.	0	0	141.9
1964	0	0	0	0	1.3	6.2	82.8	191.7	11.5	0.4	0	0	293.9
1965	0	0	0	0	tr.	6.7	90.6	22.1	5.9	9.0	0	0	134.3

Heaviest rainfall in 1963: 36.2 mm (10 July); in 1964: 24.4 (10 July) and 83.5 (14 August); in 1965: 46.6 (24 July).
tr = trace of rain too small to measure.

19

Table 2. The rodents of the northern Sudan

Species found during the study period 1963–1966

Sciuridae
  Euxerus erythropus   Ground squirrel
Cricetidae
  Tatera robusta
  Gerbillus pyramidum   Greater Egyptian gerbil
  Gerbillus watersi   Waters' gerbil
  Gerbillus campestris venustus   Sundevall's gerbil
  Meriones crassus pallidus   Sand rat
Muridae
  Arvicanthis niloticus   Nile rat
  Rattus rattus frugivorous   Cream bellied roof rat
  Mus musculus gentilis   White bellied house mouse
  Acomys cahirinus   Spiny mouse
Dipodidae
  Jaculus jaculus butleri   Sudan jerboa

Species recorded from the northern Sudan (SETZER 1956)
but not found during 1963–1966. Most are probably very rare.

Cricetidae   Recorded from
  Gerbillus stigmonyx   Khartoum region
  Gerbillus gerbillus agag   El Fasher region
  Gerbillus gerbillus sudanensis   North east Sudan
  Gerbillus nancillus   North of El Fasher
  Gerbillus (Dipodillus) principulus   Jebel Meidob
  Gerbillus (Dipodillus) mackilligini   Eastern Egyptian desert
  Gerbillus (Dipodillus) bottai   Sennar
  Taterillus emini perluteus   El Fasher region
  Desmodilliscus braueri   El Fasher region
  Psammomys obsesus   Port Sudan

---

rains, and the substrate is sandy or stony. Burrows in the sand are necessary for protection, and rodents have to be independent of exogenous water. 2. The jebels: these are also characterised by lack of vegetation except for grasses growing in crevices. Cracks and spaces between rocks provide cover and protection. As in the sandy desert, no exogenous water is available.
3. Riverine habitat: along the Nile valley and irrigated areas close to the river. Logs and human buildings are available for protection, although some individuals use burrows. Water is available throughout the year.
4. Human habitations: houses and stores in the principal towns and villages. This habitat is protected from the harshness of the desert environment, and there is adequate food and water throughout the year.

My studies on rodents in the northern Sudan in 1963–66 were made in the region of Khartoum, northwards towards Atbara, and southwards along the Blue and White Niles to about 12 °N. Eleven species of rodents

Table 3. Measurements of Sudan Cricetidae, Muridae and Dipodidae (from HAPPOLD 1967c). Lengths in mm, weights in gms.

Species	n	Head and body	Tail	Hindfoot	Ear	Weight	Greatest length of skull	Condylo-incisive length	Greatest width across zygomatic arch	Greatest width of braincase	Crown length of molars	Length of auditory bulla
Talera robusta	1♀	162	180	38	21	—	38.4	35.0	18.2	17.3	6.0	10.7
Gerbillus pyramidum	5♂	104	148	29	15	44	32.1	28.2	16.4	15.6	4.4	—
Gerbillus watersi	5♂	77	126	22	12	18	25.1	22.3	13.5	13.6	3.2	9.6
Gerbillus campestris venustus	3♂ 3♀+	90	139	25	13	27	29.3	26.0	14.8	13.9	4.1	9.6
Meriones crassus	1♀	130	115	29	18	77	39.9	34.9	19.7	20.7	4.7	15.9
Arvicanthis niloticus	1♀ 5♀+	146	144	31	17	107	34.5	32.5	17.3	14.6	6.4	6.8
Rattus rattus frugivorus	2♀+	147	190	34	22	99	38.1	35.5	19.1	16.2	6.4	6.7
Mus musculus gentilis	4♂ 2♀+	76	83	18	13	13	21.3	20.4	11.1	9.5	3.4	3.8
Acomys cahirinus	3♂ 2♀+	98	98	15	15	30	27.9	25.4	13.1	—	4.2	—
Jaculus jaculus butleri	10♂	101	182	56	19	49	31.9	28.1	21.9	22.0	4.9	13.6

Note: Measurements are not available for Euxerus erythropus.

21

Table 4. Ecological separation of the rodents of the northern Sudan

Desert	Jebel	Riverine	Commensal
Euxerus erythropus	*Acomys cahirinus*	*Arvicanthis niloticus*	*Rattus rattus frugivorus*
Jaculus jaculus	*Gerbillus campestris*		*Mus musculus gentilis*
Gerbillus pyramidum			
Gerbillus watersi			
Meriones crassus			

were found in the desert and semidesert although SETZER (1956) has recorded a further ten species which are also known from this area (Table 2).

Standard measurements for the common species of small rodents are given in Table 3.

The small rodents may be arranged according to their habitat characteristics (Table 4).

DESERT SPECIES

Jaculus jaculus

The Sudan jerboa *Jaculus jaculus butleri* is one of the commonest rodents in the Sudan deserts. Its burrows tend to be on sandy ridges, or on any land that is slightly higher than the surrounding plains. Although it was not possible to watch jerboas emerging from their burrows in the desert, this was observed in captivity. The jerboa pushes the plug of sand blocking the entrance with the flattened part of its nose. This is usually done slowly and the animal pulls itself out of the hole with its long hindlegs and tail trailing behind the body. There may be a brief period of grooming at the entrance of the burrow, and then the sand is scrapped back with the hindlegs. Finally the animal turns around and the sand is patted flat with the nose so that the entrance becomes invisible.

In the desert, jerboas were seen frequently sitting in the sandy tracks made by vehicles; often they were feeding, presumably on seeds and pieces of grass uncovered by the wheels of vehicles. Usually jerboas were seen singly, but occasionally they were in groups of up to four animals (Table 5). After being approached, the animal(s) try to run away using their 'medium speed gait' (HAPPOLD, 1967 Fig. 5b), or if being chased they ran at 'fast speed' when the two hind legs appeared to move together to give a strong propulsive thrust. At this fast speed, jerboas can run at about 14 m.p.h. When chased, jerboas can often escape by virtue of their speed and manoeverability, and not because they went into burrows.

The evidence from watching jerboas in the desert at night suggests

Table 5. The numbers of jerboas seen in the desert at night (from HAPPOLD 1967a)

Single animals	Pairs	Trios	Quartets	Total
231	21	5	2	296

that apparently they are not social animals, but in captivity they show considerable amicable behaviour once they have settled down. A number of these animals were kept in cages 8 ft long by 3 ft wide, with sand and grass on the floor and six nest boxes so that they could nest independently if they wanted to. In cool weather (December to February) they nested together, up to six in a nest box, each animal curled into a ball and close to each other. In the hot weather, they slept singly or in pairs, usually lying on their sides with legs stretched out. It would be interesting to know if they do this in the wild as well, or whether this change in nesting habits was due to the lack of insulation (and hence a different micro-climate) in the nest boxes compared with a burrow in the sand.

There was a lot of contact between individuals in captivity when they were moving about at night. Nose-nose contact with the eyes partially or completely closed was frequently observed; occasionally there was nose-anus contact. Chasing or fighting was rare except when a new jerboa was placed in a cage with an already established group.

There is a wide range of body weights in jerboas (Fig. 5): females (average weight 60 g) are heavier than males (average weight 50 g), and all animals weighing over 60 g were females. Body weight can not be used as an index to determine whether a female is pregnant or not since animals weighed from 53 to 71 g. Small animals, of less than 40 g body weight, were only found during September to November, and again in February to March. The relative ages of jerboas were determined from the wear of the upper left cheek teeth (M1, M2 and M3). Absolute age could not be known since skulls from animals of known age were not available, but since jerboas can live for up to six years in captivity (FLOWER, 1931), it is likely that the oldest animals with the greatest tooth wear are perhaps three to four years old. Each tooth was given an arbitrary grading based on the pattern of enamel and dentine (HAPPOLD, 1967 Fig. 3). The lowest grading for the three cheek teeth was 3 and the highest was 17. In the period September to December (Fig. 6) there are probably three age groups. First, there were those animals that weighed less than 40 g and with a tooth wear grading of less than 8 which were born during the previous June to July (sector a). Second, there were animals greater than 40 g with a tooth wear grading of less than 8 which were born during the previous September to December and were about a year old when captured; some jerboas more than a year old might also have been included in this age group (sector b). The third group comprised of those animals greater than 40 g and with a tooth wear

23

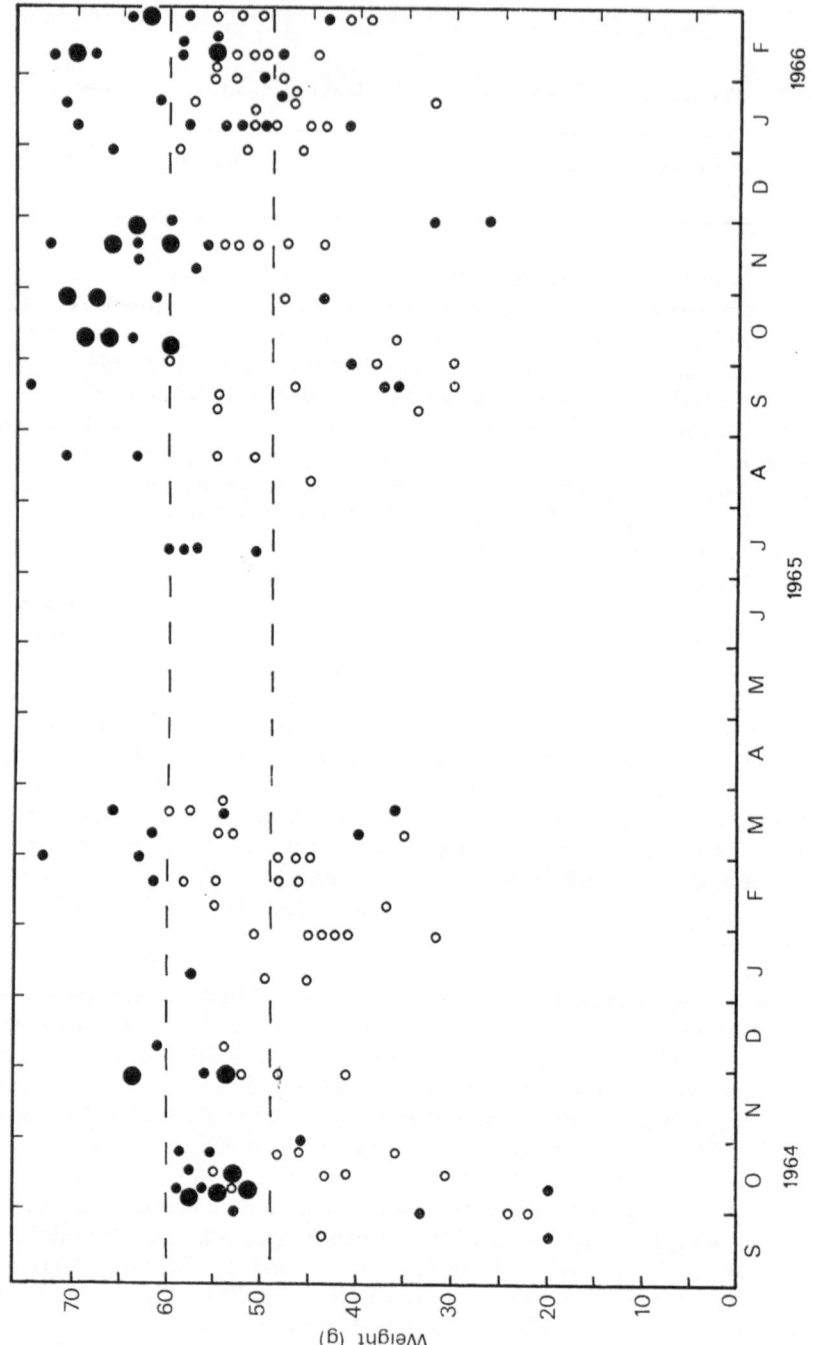

Fig. 5. The weights of jerboas when caught in the desert. ○, males; ● non-pregnant females; ● pregnant females. The broken line at 60 g indicates the average weight of females, that at 49 g the average weight of males. (From HAPPOLD 1967a, reprinted by permission).

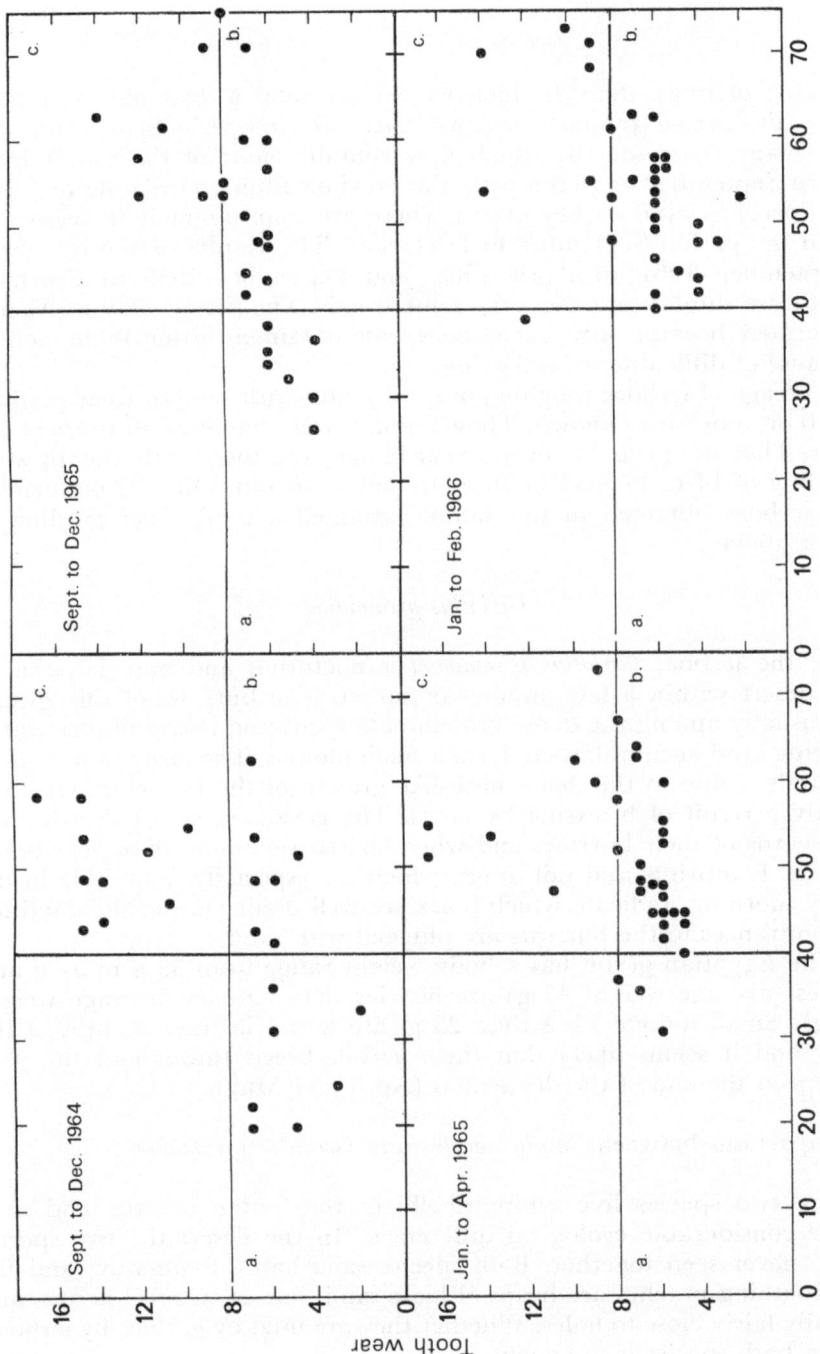

Fig. 6. The amount of tooth wear in relation to weight in *Jaculus jaculus butleri* at Khartoum, plotted in four-monthly intervals. Description in the text. (From HAPPOLD 1967a, reprinted by permission).

grading of more than 8 which were more than a year old (sector c).

By the period January to April, there are few jerboas in sector (a), but many in sector (b) which now contains most of the recent litter (born September to December), the previous litter (born June to July), and perhaps even earlier litters. There are some animals in sector (c) as in the period September to December. The results of the two years (September 1964 to April 1965, and September 1965 to February 1966) are similar and show the same trends. The period May to August is omitted because only four jerboas were obtained during these months because of difficulties of collecting.

The age of jerboas weighing over 40 g and with a tooth wear grading of 10 or above is unknown. These animals were found at all times of the year. They are probably over a year in age, and those with a tooth wear grading of 14 to 17 may be three to four years old. Only 25 per cent of the jerboas obtained in this sample attained a tooth wear grading of 10 or above.

Gerbillus pyramidum

Like the jerboa, *Gerbillus pyramidum* is nocturnal and may be seen in the desert within a few minutes of sunset. The burrows of this species are usually among the dense growth of *Capparis* and *Acacia* bushes where drifting sand accumulates to form a small mound. The formation of these mounds is due to the dense bush-like growth of the branches, which is partly a result of browsing by goats. The gerbils usually stay within a few yards of their burrows and when chased they soon disappear down a hole. Footprints and tail drags, which are especially noticeable in the early morning, indicate which holes are well used. During the daytime, the entrances to the burrows are plugged with sand.

The Egyptian gerbil has a body weight range from 33 g to 51 g and males (average weight 45 g) are heavier than females (average weight 37 g). Small animals, less than 25 g, are found in most months of the year and it seems likely that these gerbils breed throughout the year except at the end of the dry season (April and May).

Comparisons between *Jaculus jaculus* and *Gerbillus pyramidum*

These two species live sympatrically in the Sudan deserts, and they show considerable ecological differences. In the desert the two species were never seen together. Both species sand-bathe frequently, and the sandbathing patches are heaps of loose sand, one to two feet across, and usually fairly close to holes. Whether they are used by gerbils, by jerboas, or by both species is unknown.

Since jerboas and gerbils live in the same area, it is likely that they may have occasional encounters. In captivity both species live peacefully

Fig. 7. A track in the semi-desert near Khartoum. Scattered *Acacia* trees have a bush like form because of overgrazing by goats.

Fig. 8. The base of an *Acacia* bush which is partially submerged by windblown sand. These sand mounds are used frequently for burrowing by *Gerbillus pyramidum*.

27

Table 6. Some ecological and behavioral comparisons between the jerboa *Jaculus jaculus butleri* and the gerbil *Gerbillus pyramidum* at Khartoum (from HAPPOLD 1967a)

Jaculus	*Gerbillus*
Often moves well away from burrow at night	Tends to remain near burrow entrance at night
Relies on speed and manoeverability to escape from predators at night	Relies on hiding in bushes and running into burrow to escape predators at night
Individuals tend to be solitary in desert	Individuals tend to be in colonies
Rarely bites when handled	Frequently bites when handled
Does not store food (?)	Makes large food stores
Relatively 'tame' when caught	Wild and jumpy when caught
Does not eat hard seeds or fruits	Can eat hard seeds, and other seeds not eaten by *Jaculus*
Principal breeding period Oct–Nov and sometimes to Feb.	Breeding period from June to Feb–March
Average of 3 young per litter (range 2–5)	Average 4 young per litter (range 2–7)
Young not found in desert at night until $\frac{1}{2}$ to $\frac{2}{3}$ adult weight	Young found in desert when $\frac{1}{5}$ to $\frac{1}{4}$ adult weight

Table 7. Comparison between the three species of *Gerbillus* found near Khartoum (from HAPPOLD 1967c)

	G.pyramidum	*G.watersi*	*G.campestris venustus*
General colour	Greyish-Orange	Clay-Bronze Brown	Brownish-Orange
Tuft on tail	small, 40–50mm	small, c.40mm	bushy, 50–60mm
Hair	close fitting, dense, slightly coarse, 7–8mm	close fitting, dense, 5–8mm	long, soft, 10–12mm
Hindfeet	hairy	not hairy	not hairy
Habitat	sandy desert	sandy desert	hills and jebels among rocks
Abundance	common	rare	rare, but locally common.

together in the same cage but usually they sleep separately. No fighting has been observed between the members of two species, although after capture gerbils are very active and jumpy; consequently jerboas in the same cage start to jump as well and make frightened wheezing noises.

The observed ecological comparisons between *Jaculus jaculus* and *Gerbillus pyramidum* are summarised in Table 6.

Gerbillus watersi

This gerbil is found in the same sort of habitat as *Gerbillus pyramidum*

Fig. 9. A *Capparis* bush, with windblown grasses and seeds around its base which will be buried gradually by windblown sand.

Fig. 10. A pool of water in the semi-desert near Khartoum at the beginning of the rains.

29

but it is in much rarer and appears to prefer regions where bushes are fairly common. A plantation of mesquite trees near Khartoum contained a colony of these gerbils but no *Gerbillus pyramidum*. No doubt there are considerable ecological differences between the two species, but since *G. watersi* was so rare, no comparison is possible.

The main differences between the three species of gerbils occurring near Khartoum are shown in Table 7.

Meriones crassus

The sand rat is a well known species from Egypt (HOOGSTRAAL, 1963), but is rare in the northern Sudan, and only one specimen was found in the semidesert near Khartoum.

All the other Sudan specimens of this species (SETZER, 1956) come from localities along the Nile valley, although the exact habitat is not known.

JEBEL SPECIES

Acomys cahirinus

The spiny mouse is common in suitable habitats in northern Africa and the Middle East. In the arid areas of the Sudan, the spiny mouse is found on jebels where there are large crevices for them to hide in during the day. Consequently only the larger jebels are suitable habitats for this species. Where the conditions are suitable, *Acomys* are found in dense populations; for example, 48 specimens were caught in 198 trap-nights on Jebel Qeili. On most desert jebels it is the only species of rodent; on two jebels it was found along with *Gerbillus campestris*. *Acomys* appear to be completely nocturnal in the Sudan. They do not leave the jebels to feed on the sandy parts of the desert.

Further south in the *Acacia* savanna *Acomys* are found where fallen trees and dead grass occur around the bases of jebels, which suggests that these animals require cover of some sort for their existence. In this habitat, *Acomys* is associated with *Arvicanthis niloticus*. *Acomys* was a very rare species on Jebel Marra (HAPPOLD, 1965) where other species were the dominant small mammals.

Gerbillus campestris venustus

This species is widespread throughout northern Africa, but the subspecies *venustus* is only known from the Sudan (HAPPOLD, 1967b); here, it has been found only on two jebels, but it is likely it occurs on most of the larger desert jebels. Like *Acomys*, this species is nocturnal and hides in crevices during the daytime. It is not as common as *Acomys* since trapping records for 78 rodents on two jebels show a ratio of 5 *Acomys*:1 *Gerbillus*.

Arvicanthis niloticus

In desert areas the Nile rat is limited to the Nile valley and to irrigated farms which are situated near the river. Further south in the savanna, it is found where grass and cover are abundant but not necessarily where the soil and vegetation are moist. Near Khartoum, there were large populations of *Arvicanthis* and they were pests in fields and food stores. There is no evidence that this species can survive desert conditions, and its survival and distribution in the northern Sudan depends on the Nile river and on irrigated cultivation.

COMMENSAL SPECIES

Rattus rattus frugivorus

The cream-bellied roof rat is found occasionally in buildings in Khartoum, and at the Zoological Gardens. It probably occurs in other large towns in the northern Sudan. It is completely dependent on human habitations for survival.

Mus musculus gentilis

The white bellied house mouse occurs in buildings in Khartoum, and probably in other towns in the Nile valley. ANDERSON (1902) recorded it from Wadi Halfa on the Sudan-Egypt border. It has not been found in any villages in the desert. Like the roof rat, it is dependent on human habitations for survival.

Reproduction

The reproductive period for desert rodents in the northern Sudan is towards the end of the rains and in the few months succeeding the rains when desert grasses and seeds are abundant. October is the month of greatest breeding activity. Species living along the Nile and the commensal species probably breed throughout the year (Fig. 12).

Most pregnances in *Jaculus jaculus* were in October and November, 1–2 months after the end of the rains. In 1964, when there was above average rain, pregnancies were recorded in October to November; in 1965, a year of below average rain, pregnancies were spread out from October to February. It is likely that some females are pregnant during

Fig. 11. Grasses in the semi-desert after the end of the rains.

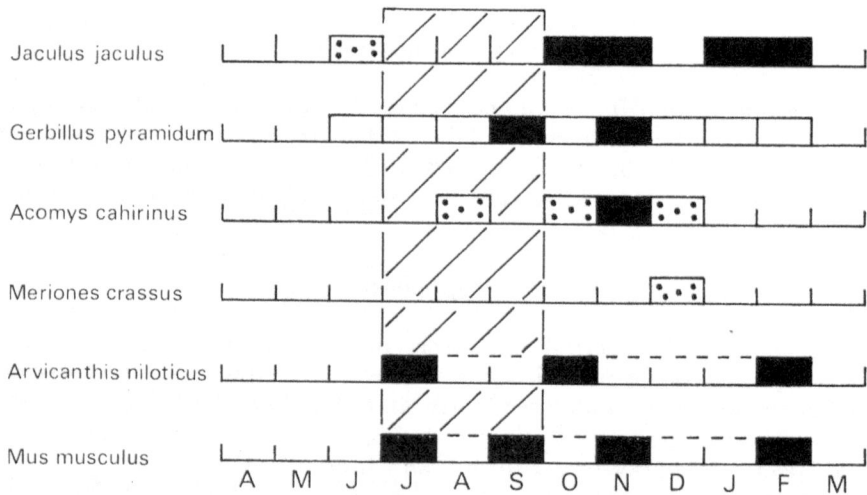

Fig. 12. The breeding periods of six species of rodents in the northern Sudan. Unshaded areas – young animals; dotted areas – placental scars; black areas – pregnancies and births. The rains occur at Khartoum in July, August and September and are indicated by diagonal lines.

Table 8. Comparative data on reproduction and development of *Jaculus jaculus* and *Gerbillus pyramidum*. Figures refer to days unless stated otherwise

	Gerbillus	Jaculus
Gestation period	22	27
Usual number in litter	4	3
Movement by rolling	2	?
Crawling using front feet	6	12
Crawling on all fours	8	?
Hair growing on dorsal side	8–10	22–27
Grooming	15	38 approx.
Ears open	16–18	?
Incisor teeth erupting	16–18	27
Eyes open	19–20	38
Weight when first found in desert (as fraction of adult weight)	⅕ to ¼	½ to ⅔
Weight of smallest animal found in desert (g)	11	20

the rains (when collecting was difficult) since young jerboas are found in the desert in September and October. (Figs. 5,6).

The gestation period of *Jaculus* is recorded as 27 days (DIETERLEN, 1963). The jerboa usually produces three young, but there may be from two to five (HAPPOLD, 1967a). At birth, the naked young resemble young rats of the same age; there is no enlargement of the head nor elongation of the hindlegs. The hairs on the back appear at 22 days and the eyes open at about 38 days. Until the eyes open, the young remain in the nest and crawl only short distances using the forefeet. Compared with the rest of the body, the hindfeet develop quickly; by 35 days they measure about 50 mm which is nearly adult size. Consequently the hindfeet are extremely long in relation to the rest of the body at this age, and they are un-coordinated and unable to support the weight of the body. The process of development of the hindlimbs and the bones of the head have been described by HAPPOLD (1970).

The other common desert species, *Gerbillus pyramidum*, has a more prolonged breeding season. For this species, the births of young animals caught in the desert have been calculated from information on laboratory reared animals (Fig. 15). This shows that pregnancies occur in every month from June to February. The gestation period of *G. pyramidum* is about 22 days, and the development of the young to maturity is rapid. The hair appears on the back at 15–18 days and the eyes open at 19–20 days. The young gerbil leaves the nest when 19–23 days old, and weighs about 11 g. Sexual maturity is attained at 75–80 days. A comparison of the stages of development of *Jaculus* and *Gerbillus* is given in Table 8. The larger litters and the quicker speed of development of *Gerbillus* suggests that populations of *Gerbillus* can increase more rapidly than

Fig. 13. The entrance of a burrow showing the spill-heap and tail-drag marks.

Fig. 14. The Sabaloka Hill, north of Khartoum. *Acomys cahirinus* and *Gerbillus campestris venustus* live in the rocky area on the hills (left), and *Jaculus jaculus butleri* and *Gerbillus pyramidum* live in the sandy semi-desert at the base of the hills (right).

34

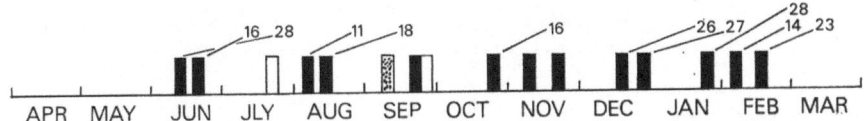

Fig. 15. The breeding season of *Gerbillus pyramidum* at Khartoum. Black bars – births; white bars – placental scars; dotted bar – pregnancy. The positions of the black bars (births) are calculated from the weights of young animals; the weights of these young animals are shown at the date when each was caught.

those of the jerboa; this may be important to the species in years of heavy rain and abundant food supplies.

No pregnant females of *Gerbillus watersi* were found, but immature animals (7 g) in early February were probably born at the end of December. Reproductively active males were found in September, October and February.

The jebel species, *Acomys cahirinus* and *Gerbillus campestris*, appear to have a breeding season similar to that of the sandy desert species. Samples of these species were only obtained in September, November and December, but young animals caught in November suggest that breeding began at the end of the rains (Fig. 19). None of the *Gerbillus campestris* were pregnant, but several immature animals (19–22 g) suggest that this species also began breeding at the end of the rains.

Arvicanthis niloticus, *Rattus rattus frugivorus* and *Mus musculus gentilis* probably breed throughout the year, except possibly during the hot season (April to June) before the rains begin.

Other Adaptations for Life in the Desert

TEMPERATURE REGULATION

During the day *Jaculus jaculus* and *Gerbillus pyramidum* block the entrances of their burrows; this presumably helps to maintain a cooler temperature and a higher humidity in the burrow compared with the outside air. No measurements were made in the Sudan on burrow microclimate, but it is likely that it shows similar characteristics to desert burrows which have been studied elsewhere. During the 'cold' weather, *Jaculus jaculus*, *Gerbillus pyramidum* and *G. watersi* sleep curled up with the head tucked under the abdomen, and the hairs erected. Fewer animals appeared to be out in the desert during the cold weather, but the occasional jerboa that was found was extremely active. The lower lethal temperature for the desert species is unknown, but several *G. watersi* trapped when the overnight low was 53 °F were dead in the traps in the morning.

In the 'hot' weather, *Jaculus* and *Gerbillus pyramidum* sleep on their sides with the legs stretched out, and they continually wake up and move

35

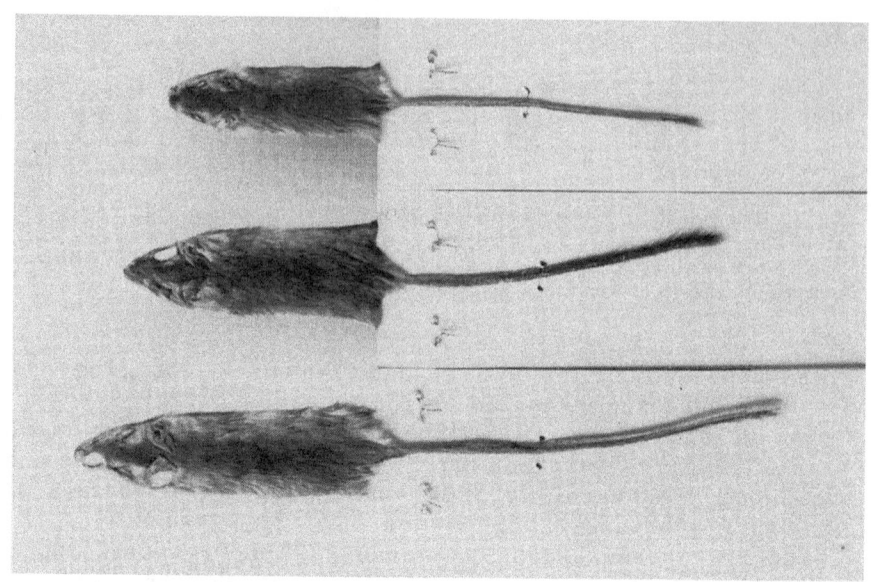

Fig. 16. The three species of *Gerbillus* at Khartoum: *Gerbillus pyramidum* (bottom), *Gerbillus campestris venustus* (centre), *Gerbillus watersi* (top).

Fig. 17. Gerbillus pyramidum at Khartoum.

36

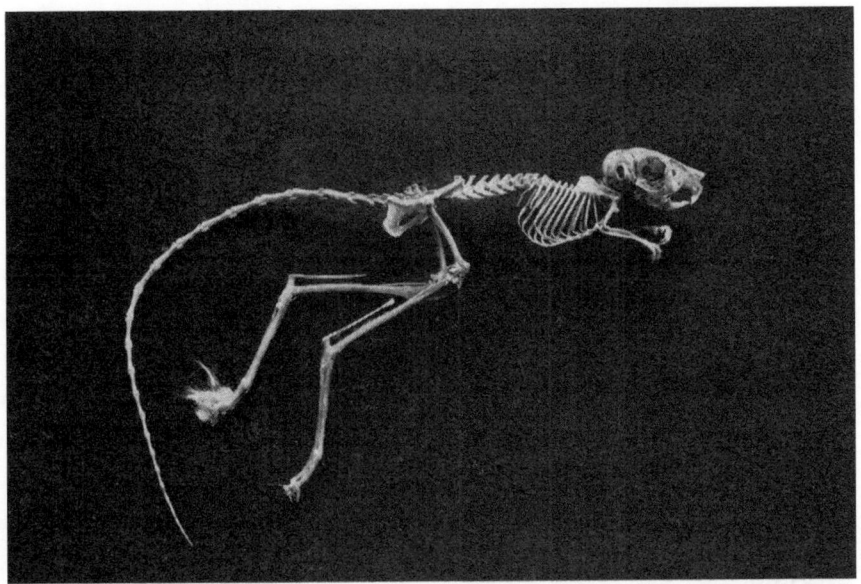

Fig. 18. The skeleton of adult *Jaculus jaculus butleri*. Special adaptations include the long hindlimbs, long tail, hair tufts on the hindtoes, and a broad skull with large tympanic bullae and zygomatic plates.

positions. When they are overheated, jerboas lick the fur so that it becomes wet and this undoubtedly results in cooling by evaporation; similarly some jerboas salivate and lick themselves after being chased for a long time in the desert.

Since the jerboa does not have sweat glands, it can cool itself only by losing heat through expired air, or by diffusion of heat through the general body surface. The skin of jerboas is very thin; the epidermis is 0.016–0.025 mm thick, and the dermis is 0.15–0.52 mm thick (GHOBRIAL, 1970). For an animal as active as the jerboa, a thin skin may be advantageous despite the possible disadvantage of loss of water by diffusion as suggested by RIAD (1960). It may be that the clustering of hairs to form dense patches surrounded by naked areas (GHOBRIAL, 1970) has a double function since it allows heat retention during cold weather, yet also allows heat transfer away from the body when the animal is hot.

FOOD REQUIREMENTS

All the species kept in captivity were fed on sunflower seeds and millet, and cucumber was given occasionally to provide additional water; all the animals thrived on this diet. Most of them put on weight and became fat, as do wild-living animals after the rains when there is plenty of food

37

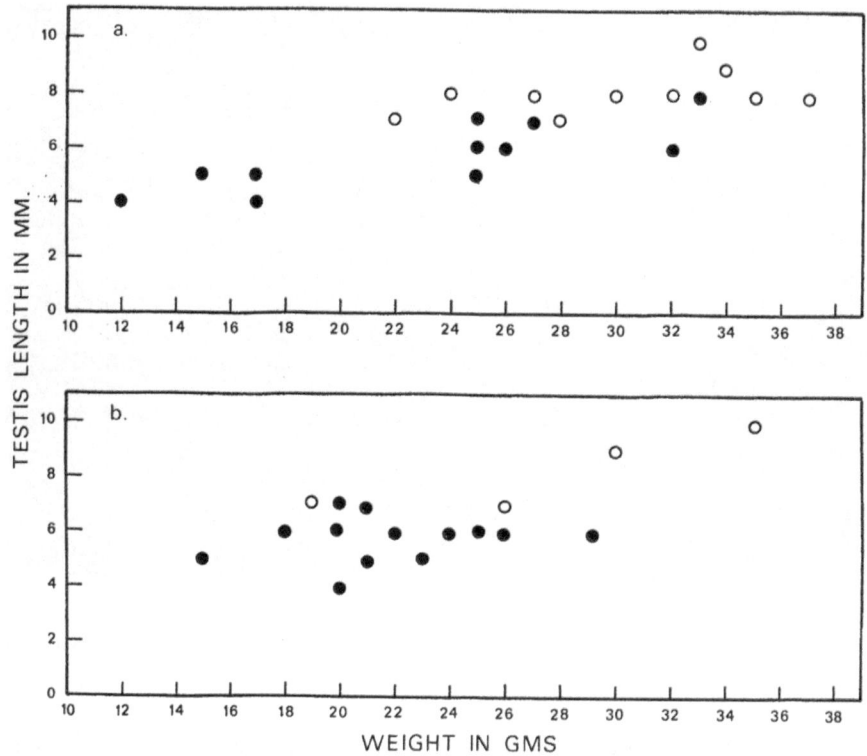

Fig. 19. The weights of male *Acomys cahirinus* in relation to testis length at two localities in the northern Sudan. a. Jebel Qeili in November; b. Sabaloka Hills and Jebel Guli in January.

in the desert. *Gerbillus pyramidum* hoards seeds and small pieces of chopped grass, but there is no evidence that *Jaculus jaculus* does this. I have the impression that *Jaculus* is more 'opportunistic' and consequently not so desert adapted as *Gerbillus pyramidum*. This is suggested also by the experiments of SHKOLNIK (quoted by SCHMIDT-NIELSEN, 1964) who showed that after being kept for three weeks at 30 °C and 30% R.H. on a diet of whole barley, *Gerbillus pyramidum* lost 5% of its original weight and *Jaculus jaculus* lost 34%. *Acomys cahirinus*, which is found only on jebels but including some of those in the most arid areas, lost 38%, and *Gerbillus gerbillus*, found only in the sandy areas in the extreme north of the Sudan, gained 10%.

WATER REQUIREMENTS

Jaculus jaculus and *Gerbillus pyramidum* can live on a diet that does not

38

include exogenous water. In captivity, where conditions in the nest-boxes were warmer and less humid than in natural burrows, both species ate cucumber as a source of water. *Jaculus* partly solves the problem of water conservation by a special arrangement between the cortical and juxtamedullary glomeruli which gives a large filtration area (MUNKASCI & PALKOVITS, 1964), and results in the secretion of highly concentrated urine. The long renal papillae of *Jaculus jaculus* and *Gerbillus gerbillus* project into the ureter, and this also helps to produce a concentrated urine (KHALIL & TAWFIC, 1963). Captive animals produced concentrated urine infrequently. The urine of *Jaculus jaculus* contains 4320 mm urea/litre urine, and is more concentrated than any other African desert rodent (SCHMIDT-NIELSEN, 1964). After being deprived of water for a month *Jaculus jaculus* drank 4.3% of its body weight of water, *Acomys cahirinus* drank 11.38%, and *Gerbillus pyramidum* did not drink any water (SHKOLNIK, in SCHMIDT-NIELSEN, 1964). With my specimens, *Gerbillus pyramidum* lapped water when given the chance, but *Jaculus jaculus* did not.

In the Sudan, *Gerbillus gerbillus* appears to be the most 'desert adapted', followed by *Gerbillus pyramidum*, *Jaculus jaculus*, and *Acomys cahirinus*. This sequence is the same as that suggested by the habits and distribution of these species. The least desert adapted species, *Acomys*, is able to live only on jebels where it is protected from the desert environment by the special conditions on the jebels.

Discussion

The limited information on the rodents of the northern Sudan suggests that their way of life and adaptations are similar to those of other rodents in warm semideserts and deserts. In the Sudan semidesert, the life histories and reproduction are regulated by the annual rainfall. Although the numbers of animals seen in the desert varied during the year, partly due to daily climatic changes and partly to births and deaths, there was no large population changes. There is no evidence for cyclical periods of great abundance, even though a very wet year, like 1964, could have resulted in high population numbers. The renewal of food supplies occurs infrequently for those species living near the Egyptian-Libyan border since there is no regular annual rainfall. Presumably there are large areas of sandy desert where no vegetation occurs, and consequently there are no rodents. However, isolated jebels and the surrounding dried wadis can maintain a limited rodent fauna. The rocky regions of Jebel Uweinat, at the point where Egypt, Libya and the Sudan join, contain *Acomys cahirinus* and *Gerbillus campestris;* and in the sandy 'river beds' at the base of the jebel among the clumps of *Panicum* grass, there are *Jaculus jaculus* and *Gerbillus gerbillus* (MISONNE, 1969, OSBORN & KROMBEIN, 1969). MISONNE reports that in some areas it is possible to find densities of up to 100 *Jaculus* and *G. gerbillus* per hectare; similar densities

39

may also be found in some valleys where *Citrullus colocynthis* grows since the animals are fond of the seeds of this plant.

There are some interesting comparisons between the ecology of rodents living in the northern Sudan with those in Egypt (HOOGSTRAAL, 1963) and Libya (RANCK, 1969). In Egypt *Acomys cahirinus* is mostly associated with human habitations and it is not as common on jebels as in the Sudan. In Libya it is found in areas of loose sand near the bases of palm trees at oases; it does not appear to be common on jebels, nor is it commensal. *Gerbillus campestris* lives in fig groves and fallow fields in Egypt, and in Libya it is found in mesic habitats in palm groves or among tamarisk and acacia trees; it is not found in similar damp environments in the Sudan. In Libya, as in the Sudan, *Gerbillus campestris* is found on jebels. *Gerbullus pyramidum* also shows differences in habitat preferences; in Egypt it is associated with human cultivations and domestic animals and does not live away from cultivations (HOOGSTRAAL, 1963). In Libya it is the commonest species of Saharan oases but it also lives in open sand regions and sand dunes covered with vegetation. These three species have not been recorded in mesic habitats along the Nile valley although similar to those in Egypt and Libya. These differences might be explained if the Sudan semidesert is not as arid as the desert regions of Libya and Egypt, or that the presence of *Arvicanthis*, *Rattus* and *Mus* in mesic areas of the Sudan has resulted in the exclusion of *Acomys* and *Gerbillus*. The inter-relationships of these species and their exact requirements deserve further study.

It will be realised from this account that little is known about the ecology of the rodents of the Northern Sudan. As a result of three years work in the Sudan, I think the following studies would be worthwhile. They have been chosen since they deal with the most easily obtained species, and they will help to clarify the relationships between the species and the environment, and between sympatric species.

1. A detailed study of burrows of each species, and the variations in microclimate throughout the year. Is the burrow depth significant in sheltering the animal from the outside environment? Does this depth vary according to the soil or sand type? What temperatures and humidities are the animals subjected to during the 24 hour cycle?

2. A study of the ecological conditions on jebels. What is the interspecific relationship between *Acomys* and *Gerbillus campestris?* Why are these species only found on jebels in the desert? Why is the ratio of abundance 5 *Acomys*:1 *Gerbillus?* Because of the known physiological differences of these species compared with more desert adapted species, how do these differences affect the ecology of the jebel species?

3. A more detailed study of the life history of *Jaculus jaculus*, especially growth rates and life tables. Studies on natural populations would be valuable (but difficult).

4. A population study on *Gerbillus pyramidum*. This should be fairly easy

Fig. 20. Postures in *Jaculus jaculus butleri* (terminology from Happold, 1967a) *a*. 'Sitting' with metatarsal bones resting on the ground.

Fig. 20b. 'Standing' with metatarsal bones raised above the ground.

41

Fig. 20c. Movement on all fours.

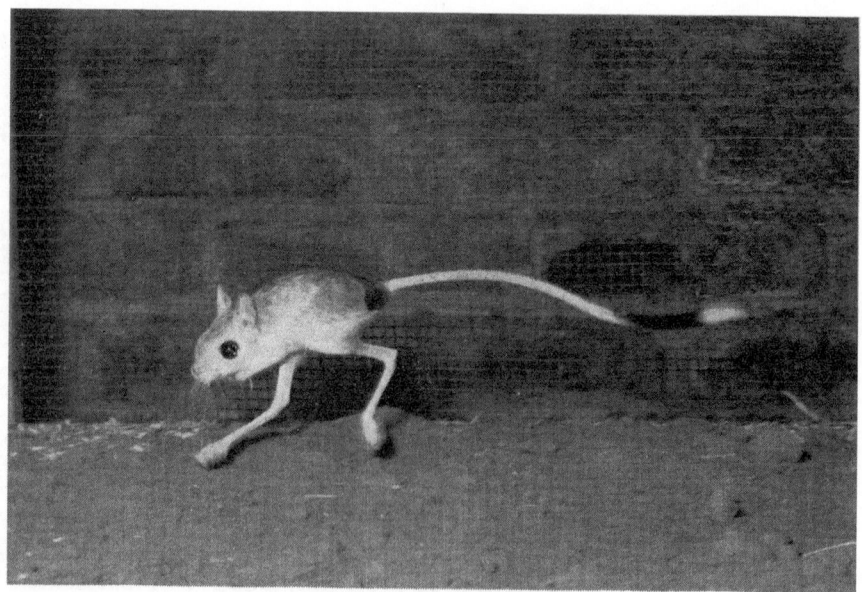

Fig. 20d. Medium speed running using a bipedal gait.

42

Fig. 20e. A faster version of medium speed running with the body and legs raised well above the ground.

Fig. 21. Nose-to-nose contact between three individuals of *Jaculus jaculus butleri*. The eyelids are almost closed.

since colonies grouped round several *Capparis* and *Acacia* trees can be found easily.

5. Comparative studies and interactions between all the common small rodents; in particular water and food requirements, aestivation, and population regulation.

43

6. Behavioural patterns and social interactions of the different species of cricetids.

Acknowledgements

I am grateful to the Editor of Journal of Zoology for permission to reprint Figs 3 and 4, and Tables 5 and 6; and to the Editor of Mammalia for permission to reprint Fig. 6.

REFERENCES

ANDERSON, J. 1902. Zoology of Egypt: Mammalia. Hugh Rees, London.

BARBOUR, K. M. 1961. The Republic of the Sudan. Univ. London Press.

DIETERLEN, F. 1963. Vergleichende Untersuchungen zur Ontogenese von Stachelmaus *(Acomys)* und Wanderratte *(Rattus norvegicus)*. Beitrage zum Nesthoker – Nestfluchter-Problem bei Nagetieren. *Z. Saugetierk.* 28: 193–227.

FLOWER, S. S. 1931. Contributions to our knowledge of the duration of life in vertebrate animals. 5. Mammals. *Proc. Zool. Soc. Lond.* 1931: 145–234.

GHOBRIAL, L. I. 1970. A comparative study of the integument of the camel, Dorcas gazelle and jerboa in relation to desert life. *J. Zool. Lond.* 160: 509–521.

HAPPOLD, D. C. D. 1965. The mammals of Jebel Marra. *J. Zool. Lond.* 149: 126–136.

HAPPOLD, D. C. D. 1966. Breeding periods of rodents in the northern Sudan. *Rev. zool. bot. Afr.* 74: 357–363.

HAPPOLD, D. C. D. 1967a. Biology of the jerboa, *Jaculus jaculus butleri* (Rodentia, Dipodidae), in the Sudan. *J. Zool., Lond.* 151: 257–275.

HAPPOLD, D. C. D. 1967b. *Gerbillus (Dipodillus) campestris* (Gerbillinae, Rodentia) from the Sudan. *J. nat. Hist.* 1: 315–317.

HAPPOLD, D. C. D. 1967c. Guide to the natural History of Khartoum Province. 3. Mammals. *Sudan Notes Rec.* 48: 111–132.

HAPPOLD, D. C. D. 1968. Observations on *Gerbillus pyramidum* (Gerbillinae, Rodentia) at Khartoum, Sudan. *Mammalia* 32: 44–53.

HAPPOLD, D. C. D. 1970. Reproduction and development of the Sudanese jerboa, *Jaculus jaculus butleri* (Rodentia, Dipodidae). *J. Zool. Lond.* 162: 505–515.

HARRISON, M. N. & JACKSON, J. K. 1958. Ecological Classification of the vegetation of the Sudan. *Bull. Forests Dept. Khartoum N.S.* 2.

HOOGSTRAAL, H. 1963. A review of the contemporary land mammals of Egypt (including Sinai). 2. Lagomorpha and Rodentia. *J. Egypt. publ. Hlth Ass.* 38: 1–35.

KHALIL, F. & TAWFIC, J. 1963. Some observations on the kidney of the desert *J. jaculus* and *G. gerbillus* and their possible bearing on water economy of these animals. *J. exp. Zool.* 154: 259–271.

MISONNE, X. 1969. La faune – Expedition scientifique Belge dans le desert de Libye: Jebel Uweinat 1968–1969. *Africa-Tervuren* 15: 117–119.

MUNKACSI, I & PALKOVITS, M. 1965. Volumetric analysis of glomerular size in kidneys of mammals living in desert, semidesert or water-rich environment in the Sudan. *Circ. Res.* 17: 303–311.

OLIVER, J. 1965. The climate of Khartoum Province. *Sudan Notes Rec.* 46: 90–129.

OSBORN, D. J. & KROMBEIN, K. V. 1969. Habitats, flora, mammals, and wasps of Gebel 'Uweinat, Libyan desert. *Smithsonian Contrib. Zool.* 11: 1–18.

RANCK, G. L. 1968. The Rodents of Libya. *U.S. Nat. Mus. Bull.* 275: 1–264.

RIAD, Z. M. 1960. Integumentary peculiarities related to life in the desert. *Proc. Egypt. Acad. Sci.* 15: 37–42.

SCHMIDT-NIELSEN, K. 1964. Desert Animals. Oxford Univ. Press.

SETZER, H. W. 1956. Mammals of the Anglo-Egyptian Sudan. *Proc. U.S. Nat. Mus.* 106: 447–587.

III. THE RODENTS
OF THE IRANIAN DESERTS

by

XAVIER MISONNE

Introduction

The complexity of the structure of Iran makes it necessary to examine first briefly the different types of deserts existing in this huge country composed of a variety of desert, subdesert, plateau and mountain landscapes.

A first factor is the position of the country extending from 25 °N to 39 °N, and, for this reason, the temperatures may vary widely between the north and the south, at least during winter and spring.

A second factor is the strong relief of most of Iran: range after range extend over most of the country, and the mean altitude of the plateau is lying around 1,500 metres; many mountains are above 3,000 metres, while in the south-east, two large depressions, the Dasht-i-Lut and the Jaz Murian, are at a much lower altitude, with only 250–400 metres of elevation. For this reason again, winter and spring temperatures may be low even far south and close to the Gulf coast. This explains why a large part of the country, though showing fairly high summer and fall temperatures, may be cold in winter and early spring.

A third factor is the amount of rainfall and its periodicity. Most of the rains (and snow) are to be observed in winter and spring, at most from October to May. Summer and autumn are usually extremely dry. These characteristics allow the growth of a fairly dense spring vegetation with many annuals; these soon dry up in June and then most of the country looks rather arid. In order to exemplify the variety of the country, it may be said that the wheat is ripe already in the third week of April in Baluchistan, while in Azerbaidjan and Kurdistan, the plateaus are still covered with snow at that moment, and the wheat ripens only in August.

The amount of rain is not even upon the whole country; the mountains receive more water, and during a longer period, than the lower parts; accordingly, the vegetation grows higher there, on the slopes and along the river beds, even far down in the plains. In opposition to accepted opinions, most of the central desert receives some rain in the spring, usually in the form of light showers, and even the central, arid part of the southern Dasht-i-Lut receives casual showers from thunderstorms in April or May. Because of the more abundant rains of the mountains, which also last longer in the season, the vegetation is also more abundant and longer lasting than in the plains. Above 1,000 metres it becomes more abundant than in the plains, and it may be fairly rich above 2,000 metres.

When considering the whole year, it appears that spring is the best season for the rodents, with much green vegetation, and also early summer, with many seeds. In late summer and early fall, the temperature is high, the earth is very dry; there is no, or very little green vegetation, but still a number of seeds. In winter, the temperature falls, not so much in the south and in the depressions, but strongly in the north, which may be covered with heavy snow, as also on a large part of the plateau. The worst season for most of the rodents is the end of the winter, with very little available food under cold temperature conditions.

The Iranians make a clear distinction between what they call 'garmsir', the regions with hot summer and mild winter, and 'sardsir', the regions presenting warm summer but cold to very cold winter. The limit between garmsir and sardsir may very well be defined as being the northern limit of the date palm. This distinction is important in Iran; it is in point of fact the limit of occasional frosts.

The difficulty of an adequate definition of the word 'desert' is well known; in its general sense it includes not only the true desert, but very often also the subdesert and even the short steppes.

The term of true desert is used here under the meaning of parts of the country completely devoid of vegetation. In Iran, only the central and western parts of the southern Dasht-i-Lut make a true desert of appreciable size. However, though this desert was considered until very recently as the very type of an 'abiotic' desert, it seems that it is not the case; it is true that there is no vegetation there; nevertheless, this desert is crossed over by a number of insects, probably born by the strong winds, and lizards have been found right in its centre, which live on these insects, together with a number of birds and even jerboas *(Jaculus blanfordi)* which have been found crossing it. It seems that there is nothing in Iran to be compared with such a desert as most of the Libyan Desert.

Another definition of the desert, much in use and less restricted, is a region with less than 50 mm of rainfall per year; actual limits of such regions are not yet accurately known in Iran, but they include large parts of the northern, central and southern Dasht-i-Lut, and other smaller areas. There some rains occur in winter and spring in the form of light showers, and thunderstorms occur as late as the end of May. There is some vegetation, sometimes very scarce, richer along the virtual river-beds. These winter and spring rains allow the persistance of vegetation during the whole year, and accordingly we should use for such regions the denomination of subdesert; then the limit between desert and sub-desert is made clear: the persistent vegetation stops and even if there are some annuals to be seen in the spring, nothing remains in summer, while in subdesert, there is at least some vegetation throughout the year.

48

On the whole, there is little sand in the desertic and subdesertic regions of Iran; the only large sand dune areas are those lying on the eastern side of the southern Dasht-i-Lut, then the Rig-i-Djinn, south of the great kavir, and the dune areas south and south-east of the Jaz Murian. Most of the northern and central Dasht-i-Lut is covered with gravel or stones, with clay.

An important characteristic of the deserts and subdeserts of Iran is the presence of huge salt flats. When salt reaches a high concentration on lower ground, it makes salt marshes called 'kavir'; these kavir are sometimes very large and very difficult to cross; around these kavir there is usually a line of vegetation characterized by a high salt tolerance, this vegetation includes many *Chenopodiacea* and different *Tamarix;* smaller kavirs may be too dry to support any vegetation, or do not receive enough water in the spring to allow the growth of permanent vegetation. This happens for instance in the kavirs south of Naiband.

When the ground contains less salt, other plants appear but the typical plants are still the *Chenopodiacea,* mainly *Calligonum* on clay, and *Haloxylon* on sand. In the region near Baluch Ab appears another type of association with *Prosopis* and *Calligonum* which may cover wide areas such as for instance the Jaz-Murian. It is only when the ground contains very little sand that *Artemisia* appears on higher grounds, or *Acacia* along the Gulf coast and in the plain of Minab. The *Artemisia* steppe is an important association in Iran and it covers extensive areas. On the foothills and on the slopes of the mountains appears still another type of vegetation with many annual and bulb plants such as *Gagea, Tulipa, Eremurus,* etc.; there grows a rich vegetation in the spring.

As a first conclusion, it appears that the subdeserts of Iran are not of a single type, but they make a mosaic of small units, much diversified according to ground conditions (salt, clay, sand, stones). The diversity of the plants is surprising, and this is caused as much by the diversity of the association as by the number of different plant species within each association.

The presence and the density of the different plant species is also ruled by the rains, and altitude (modifying both temperatures and amount of rain). On higher ground and on the plateau, the vegetation is usually more abundant, though also xerophytic. In summer and autumn, most of Iran may appear as a desert in the broad sense, but in the spring, only a very little part of it is really very dry.

The Rodents: northern and southern Types

As already said above, the winter conditions prevailing in the north of the country and at high altitude, are different from those to be found in the south and in the depressions. The limit is that of the occasional frosts. All the species living in the northern part of the country are able to resist cold or very cold winters, while most of the species living in the south, which never endure temperatures below freezing point, seem unable to resist northern winters. This will lead to a subdivision into northern, winter-resistant rodents, and southern rodents with more tropical and subtropical affinities, unable to resist the northern winters. There are some intermediate cases which will be examined below.

The ecological needs of the northern species are not yet completely known; it is to be hoped that the study of these needs will make further progress and such a knowledge would readily explain the distribution of the rodents in Iran, this opens an interesting field for future field research. Incomplete as they are, the characteristics of the different northern species may be summarized as being the following:

Citellus fulvus: large colonies in alluvial plains, with abundant green vegetation in the spring; the vegetation must last at least three months; summer and winter temperature do not seem to be important.

Allactaga williamsi: a solitary species apparently restricted to the coolest parts of Iran: the north-west and the Alborz mountains; probably a mountain species able to live in regions with little vegetation in summer; hibernating species.

Allactaga elater: this species is close to *A. williamsi* but lives in more varied habitats, mainly on the plateau and the foothills; it is not found in very dry areas. This is probably typically a short steppe species; its distribution extends far in the south but apparently then at higher elevation. It is also an hibernating species, at least in the north.

Ellobius fuscocapillus: this is not an inhabitant of the subdeserts, but of the steppe. Its underground activities are observed only during the wet months of the year; it needs good winter and spring rains; it lives only in the north and on the central plateau.

Meriones vinogradovi: among the numerous *Meriones* species, this is the 'coolest'; it resists very well winter cold. In the melting snow, it is the first species to come outside, to clean its burrows and to have youngs. It seems to need much green vegetation; its shallow burrows perhaps prevent its establishing farther south, being too hot in summer.

Meriones tristrami comes close to *M. vinogradovi* in regard to vegetation needs; the distribution of both species is very similar in Iran, though *M. tristrami* is less restricted.

Meriones persicus: this species lives in a variety of habitats on the Iranian plateau; it is clearly a species living on the slopes of hills and mountains. It makes deep burrows. In autumn when everything is extremely dry, it may be easily captured with a bait of juicy vegetables; this demonstrates its need for green food. *Allactaga*, for instance, is not attracted by such baits.

Meriones libycus: its needs are very similar to those of *M. persicus* and it may be found at many places close to it, but then while *M. persicus* is

50

always on the hills, *M. libycus* makes its burrows on flat ground. Its distribution extends far south and even the Jaz Murian depression.

Meriones meridianus: little is known of this species in Iran; its habitat seems to be more or less the sandy areas of Khorassan.

Meriones crassus is, in North Africa, a species rather well adapted to severe desertic conditions and it is often found there in association with *Gerbillus gerbillus*, another species of very dry areas. However, it seems that in Iran *M. crassus* does not present exactly similar characteristics. It is not found in the north of the country and is then not a 'cold' species; it is no more an inhabitant of the warm south, and this is more surprising, since in the Sahara it may be found under extremely hot conditions. In Iran, it is found in the central half of the country, and not under very arid conditions, mostly in the plains, and east and west of the great depressions.

Rhombomys opimus: this is an interesting species living in a variety of conditions, though not in very warm areas; its habitat is restricted to the eastern half of Iran, with a smaller colony in the Araxe valley, Azerbaidjan. It is a northern species extending as far south as to reach the limits of Baluchistan. A good account of this species has been given recently by NERONOV & FAHRANG AZAD (1972) who show that the distribution of *Rhombomys* in Iran is a mosaic of rather isolated populations. This species is diurnal.

Calomyscus bailwardi: this might hardly be considered as a subdesert species, though it is often found in the middle of very dry areas; it lives in the rocks in the mountains and seems to need green vegetation throughout the year. Little is known of its habits.

The southern species are less abundant:

Jaculus blanfordi: this jerboa is found in the south and the east of Iran, and it is the only species commonly found in the driest parts and in true desert. It seems to be nowhere abundant but it may be found scattered in most of the very dry places. Unlike the Gerbillidae, *Jaculus* (and *Allactaga*) have a very wide territory and a stronger tendency to be erratic, and this particularity gives the jerboas a larger capability for survival in very dry areas.

Acomys cahirinus is, among the southern species, the 'warmest' and it does not extend very far to the north. This is rather surprising since in North Africa it is commonly found in areas which are rather cold in winter. Few captures are known from Iran; this species seems to be more linked to rocks and big stones.

Tatera indica is also a species of warmer areas but it extends much farther north than *Acomys*. It needs green vegetation throughout the year and, accordingly, it is found along perennial rivers and around cultivated fields irrigated by qanat, or again in places where the water is not deep in the soil, with green vegetation during the whole year such as the natural growths of the native palm-tree *Nanorrhops*. *Tatera* does not

51

present commonly large populations in Iran; in India, this species is very common in the cultivations as for instance in Uttar Pradesh. In Iran, it is scarcer and restricted to green areas, while still farther west, in Iraq and Syria, it seems to be still more linked to man and lives around the villages.

Gerbillus gerbillus is a common species much linked to sand in southern Iran. It is clearly more psammophilous than *G. nanus* which is also found on clay, and *G. gerbillus* is able to live under more arid conditions. Both species live in zones with mild winters, with temperature minima much above freezing point.

Meriones hurrianae is an Indian *Meriones* extending westwards to Iranian Baluchistan and as far as Bandar Abbas. It seems unable to endure the cold and prefers habitats with *Acacia*.

Two conclusions may be reached at this point; the first is that winter temperature is an important factor in the distribution of the rodents in Iran: while several species are able to endure cold winters in the north, others do not have such a capability and are thus restricted to the warmest areas. Southern species are clearly limited in the north by winter temperatures, but northern species are possibly limited in the south not only by the warmth but by the differences in the vegetation.

The second conclusion is that the northern species are not present in very dry areas, while the southern species seem, at least many of them, to venture far more into the subdesert. One species only, *Jaculus blanfordi*, may be found in true desert, as for instance right in the middle of the southern Lut. Densities are usually low in subdeserts; however some areas apparently belonging to very arid zones may show a high density of rodents, and this occurs for example all along the foot of the large dunes in the eastern Lut. There large populations of *Gerbillus gerbillus* are found in areas where the only possible food is *Haloxylon* and some insects (together with some annuals in spring, but possibly not every year).

A preliminary classification of the rodents according to their needs for green vegetation may be as follows: first comes *Jaculus blanfordi* with no need for green food; then the two *Gerbillus* (*G. gerbillus, G. nanus*) which live in very dry areas. The other Gerbillidae apparently need far more green vegetation and are not truly subdesertic, and this character varies for each of them: *Meriones crassus* and *M. libycus* have comparatively small needs for green food; then come *M. persicus* and *M. meridianus*, then *M. hurrianae, M. tristrami, M. vinogradovi* and *Tatera indica*. The needs of *Rhombomys* are not yet evident.

Annual Rhythm

A short examination of the different seasons of the year will give indications of the needs and adaptations of the different species to variable conditions.

52

At the end of the winter in the north, the ground is still covered with snow; melting occurs very fast and is usually accompanied by rains. This is the most difficult period of the year: the nights are still very cold, there is little food available, and there is so much water on the ground that many burrows are drowned; the grass-litters of the others are completely wet, giving little protection against the cold, while the winter provisions of the *Meriones* are completely exhausted. In the south, the conditions are less severe but food is not more abundant. For all of them, the end of the winter is a severe period. However, many species breed during that period, in order to have the first young born as soon as food is again available. The beginning of the spring is sudden, and the season develops rapidly. A few days after snow melting appear the first annual and bulb plants and within a week, there is a lot of food available.

At this moment, the rodents quickly come out of their burrows; the first to do so is *Meriones vinogradovi*, which is also the first to clean up its burrow and the first to have young. The other species follow a week later, and within three weeks there are a great number of young everywhere in the burrows. The vegetation develops completely and this period of vegetation lasts for 6 to 8 weeks, then gradually begins to dry out, sometimes very quickly. It may last longer along favourable river beds, but after 10 to 12 weeks everything becomes dry. Then there is no more available green food except from the bushes or the other permanent hard plants. However many seeds become now available and this may last for another two months, sometimes more. The Chenopodiacea for instance have seeds late in the season. So the spring and most of the summer are not too difficult for the rodents.

Autumn is usually the driest season; the earth is completely dry and seeds are much less abundant. The different *Meriones* species soon begin making their winter provisions, sometimes as early as the end of July; they do not utilize them and continue to feed outside. At this season (September and October), trapping with juicy baits such as carrots, beets, potatoes, etc. is very successful for *Meriones*, less for *Allactaga*. Later in October the first rains may occur, with colder winds, and soon snow in November. During the bad weather, the rodents do not move out of their burrows, and if this lasts a few days, then they are forced to begin to eat their winter provisions. Thus it appears that the length of the winter, differing from year to year, is a very important factor; when rains, snow and strong winds make an early appearance, then the rodents are forced to make early a first cut into their winter provisions, and when this is cumulated with a long winter, lasting longer than usual, then their provisions are at an end before the end of the winter, and they must die.

In November, the temperatures are much cooler; in the regions in which live wild boars, these animals begin then to search systematically for the burrows containing winter provisions, and they plunder them one after the other. This causes serious destructions among the *Meriones*, as it is then too late to reconstitute these provisions. During the same months, the nights become very cold, and *Allactaga* begin their hibernation in their shallow burrows; the *Meriones* are still active but they become more and more diurnal. Accordingly, the foxes also become diurnal, and the strong pressure made by these carnivores on the rodents then becomes more visible. By sitting on a hill for a whole day in a region with many burrows, one may see the *Meriones* running along their tracks in search for food, but one sees also many foxes patiently hunting them. The difficulties for survival for both *Meriones* and foxes is then clearly apparent.

Population Densities

The densities of the populations of rodents in Iran vary considerably according to the season, and also from year to year. The variations within a year change the population levels from one to ten at its peak, followed by a decrease leading to a level which is normally not very far from the initial level.

At the end of the winter, the populations are at their minimum. With the very first appearance of spring vegetation, all the species have then their first litters; this occurs usually one or two days after the first growth of the vegetation, but sometimes already a few days before it, so it must be agreed that breeding occurs before the appearance of spring and under the snow (at least in the north). For the different *Meriones* species, the mean number of young per litter exceed 6 young. A second and a third litter soon follow the first with intervals of 5–6 weeks. There is still a normal mortality among all the individuals, together with an increased pressure from the predators, such as foxes and owls, which have their own young to feed during the same period. This leads to a population reaching, in the middle of July, a density of about 9–10 times the density of the beginning of the spring.

The youngs stay with their mother for about 6 weeks, and it occurs very often that some females have with them half-grown young together with babies; these half-grown young soon leave their mother and begin to dig their own small burrows in the vicinity of the familial burrow. The last litter occurs in August, and this allows the last young to be fully grown up for the wintering. This last litter normally stays with the parents during the winter, and so they find food for the long winter with the reserves made by the parents; otherwise, they should not be able to winter since they are too small at the time the winter reserves are gathered.

The density of the population reaches its peak with the birth of the last young in August; from that time on, the population begins to decrease under normal mortality and predator pressure, and the climatic factors now begin to act more severely. The populations reach the next end of the winter at a level not very different from the level of the previous spring.

However, population densities are not stable from year to year; the reasons for this are complex, but the main factors are climatic. When the rains last longer in the spring, this increases the duration of the green vegetation, which is favourable; but if these rains are too heavy, many litters are drowned by flooding of the burrows, a negative factor. Summer thunderstorms are not frequent and have only casual action on the populations. On the other hand, predator pressure also vary from year to year, usually in the same direction as that of the rodent populations. Successful nesting for the owls, or large litters for the foxes increase the pressure on the rodents and tend to diminish the favourable action of a good spring.

In autumn, the moment of the beginning of the rains, strong winds and cold with snow, is very important; when it occurs early, the Gerbillidae must begin to eat their winter reserves, and when this is combined with a long duration of the winter, a catastrophic situation develops which may lead to the destruction of a good part of the remaining individuals. When this occurs, the population level may drop down to a quarter or less of that of the previous spring. On the opposite, when winter begins late in November or December, and spring is early, then the population level may begin the new season at a high level. When there is a succession of mild winters, then summer populations may reach extremely high levels.

So it appears that there is no 'normal' population level, but only constant variations and oscillations from year to year with limits to be observed between minimal populations, and this may be very low with few surviving individuals, and maximal populations which may reach a level 100 times higher than the minimal, or even more. This may have some practical importance since the outbreaks of plague, for instance, are observed when the populations reach very high levels.

It appears clearly from all this that the length of the winter and the importance of spring rains are the main factors affecting population levels. For *Allactaga* which does not make winter reserves but hibernates, it is the spring rains which make the main factor and not winter severity. The fluctuations for both *Meriones* and *Allactaga* follow similar curves, but while for *Meriones* wide variations are observed, the fluctuations of *Allactaga* populations are much less important, and this species keeps a more constant level.

Winter is usually severe in the North, cold and snow make the wind-swept plateaux wholly uninhabitable; the larger mammals such as moufflons, wild boars, foxes and hares take refuge in more hospitable valleys, while the owls migrate locally. *Allactaga* sleep in their burrows and *Meriones* are still active in their deep burrows in the depth of which they find appropriate temperature; when the weather is fine, they come outside and even dig galleries under the snow; in early spring however, the cold rains soon drench their grass litters completely, and this is another difficulty in keeping supportable temperatures inside the burrows.

Nothing is known so far for the wintering of the southern species, which do not endure such a severe winter; it may only be assumed that for them the regulating factors are winter and spring rains.

Some Peculiarities

To live in a desert of a dry steppe is a challenge for all mammals; the scarcity of food and the lack of water at least during a large part of the year make it difficult to survive. Few species have developed a capability to live normally without juicy plants, and these species live mainly on seeds and insects. *Allactaga*, *Jaculus* and some *Gerbillus* for instance may live on dry seeds only; others do not reach such a specialized level and need some water at least during some periods. For example, different species of *Meriones* do not resist very long when fed on dry seeds only; others, like *Meriones crassus* or *M. hurrianae* may resist for a longer period. However, it seems that these species are not found in very dry areas; thus it appears that though they are able experimentally to live on dry seeds only, they do not do it actually in nature. The reason for this is not known and it would be extremely interesting to investigate why they do not venture into very dry areas, while *Jaculus* and *Gerbillus* are able to maintain normal populations in the same areas.

The capability to resist the scarcity of water for more or less long periods of time is not the only possible way to survive in dry steppes. Two other interesting cases may be examined: *Citellus fulvus* and *Ellobius fuscocapillus*. These are certainly not desert animals, and yet they venture far south into regions which are almost desertic during more than half of the year.

Citellus fulvus is found in Iran in Khorassan, where it lives in large colonies, and in some points of South Azerbaidjan and eastern Kurdistan. The colonies are usually located along temporary river beds. *C. fulvus* has an imperative need for juicy plants, and such plants do not last long in these regions. As soon as they dry out, and this occurs suddenly, *C. fulvus* goes back to its burrow, falls into lethargy and spends the summer estivating. This continues uninterruptedly during the fall and, with the coming of the winter, estivating becomes hibernating. This extraordinary aptitude makes *C. fulvus* active during only three or four months of the year. A study of the southern limits of its distribution in Iran make clear the minimum needs for its survival.

At the end of the winter, the individuals come out of their burrows and when the vegetation becomes fully developped, the numerous individuals of the colony feed endlessly day and night, in order first to recover from their long sleep, then to prepare for the next period of sleep, which requires large fat supplies. The estivation soon begins with the drying up of the green vegetation. When captive individuals are given fresh food, they do not estivate and continue their activity until the temperature drops to around $+10\,°C$; this shows that estivation and hibernation have different causes though showing the same effects.

The last young are born about 30 days after the beginning of the spring and they are fully adult some 60 days later. When the vegetation dries up early, as it happens in some years, the young which are not yet fully adult are apparently unable to fall into estivation and soon die since they do not find any more food outside. So it appears that the length of the spring is the essential factor limiting the presence of *Citellus fulvus* in the south. When the spring lasts less than 60 days, *C. fulvus* cannot survive, and so it does not live in regions which have shorter springs. Its southern limit is indicated by the limit of a 90 days spring.

Ellobius fuscocapillus is another steppe species of northern Iran; it lives completely underground. In summer, it seems to be completely inactive in its deep burrow; its activity becomes apparent with the return of the autumn rains, as soon as these have given back to the soil some moisture to a depth of 40–50 cm. Then *Ellobius* begins endless galleries night after night and does not stop with the coming of cold and snow. Its activities under the snow are difficult to control. However, it seems that breeding must occur during the coldest months of the year, since no sexually active individual has ever been found between April and December. Its gallery-digging continues during the spring as long as the earth is not dry, then its ceases all its activities. Thus, though living in dry areas, *Ellobius fuscocapillus* completely avoids the effects of drought by being active during the rainy months. This is an entirely negative position, with underground nocturnal life, and activity limited to the least favourable months of the year. However, this species is common and successful in its very specialization.

This very short introduction to Iranian rodents is not an ecological study of all the facets of the numerous and interesting aspects of this huge country. But through the complexity of the country and its geographical position close to the great deserts of the north Palearctic and to Arabia and the Sahara, the study of the rodents of Iran should make clear many problems and open the gate for further desert studies. It is clear, for example, that the presence in this single country of seven species of *Meriones* constitutes in itself an interesting problem. The reasons for the limitations of the distribution of a species are not always easy to find out. We still lack a comparative study of the ecological needs of the different rodents of Iran. Another aspect still poorly understood is the difference between the high diversity of plant associations in Iran, and the apparent lack of diversity among the rodents; there seems to be no relation between the numerous vegetal units and the rodents. This is unusual and it is even more remarkable because the vegetation of Iran is unusually diversified. Other aspects need better investigation; different species show, under experimental conditions a high capability to resist well to drought, and in spite of this, they do not venture very far into dry areas. Other species show some adaptation to salty food and, yet the large areas covered with salt vegetation seem almost devoid of rodents. The ecology of the rodents of Iran is still at its very beginning.

REFERENCES

(Extensive bibliographies will be found in Misonne (1959) and Lay (1967))

Agrawal, V. C. 1965. Field observations on the biology and ecology of the Desert Gerbil *(Meriones hurrianae)* in Western India. *J. Zool. Soc. India*, 17, 125–134.
Lay, D. M. 1967. A study of the mammals of Iran, resulting from the Street Expedition of 1962–1963. *Fieldiana Zool.*, 54, 1–282.
Misonne, X. 1959. Analyse zoogéographique des mammifères de l'Iran. *Mem. Inst. R. Sci. Nat. Belgique*, 2°Ser., 59, 1–164.
Mobayen, S. & Tregubov, V. 1970. Carte de la végétation naturelle de l'Iran (échelle 1:2,500,000) avec guide explicatif, Tehran University.
Neronov, V. M. & Fahrang Azad, A. 1972. On distribution of the Great Gerbil *(Rhombomys opimus)* and spatial structure of its range in Iran. *Zool. Journ. Moscow*, LI, 715–723, in Russian.
Petter, F. 1961. Répartition géographique et écologie des rongeurs désertiques du Sahara occidental à l'Iran oriental. *Mammalia*, Paris, 25, n° special, 1–222.
Popov, G., Zeller, W. & Cochemé, J. 1965. Prospection écologique. Rapport sur les études faites en Inde, au Pakistan et en Iran en 1963–1964. FAO, Proj. Criquet pélerin, rapport UNSF/DL/ES/7, Rome.

IV. COMPARATIVE ECOLOGICAL NOTES ON AFGHAN RODENTS

by

JIŘÍ GAISLER

Introduction

Although I do not yet know the contents of the other contributions, I presume that the present communication somewhat deviates from the framework of the monograph. Lack of information does not allow us for the time being to draw from the available data general conclusions on the ecology of desert rodents. Therefore I hesitated to accept the kind offer of the editors to contribute about the desert rodents of Afghanistan. In the end various circumstances made me accept it. Above all it was the ecological peculiarity of Afghan mammalian habitats due to an interesting geographic situation, climate and phytogeocoenoses, and, furthermore, the fact that most of our material from Afghanistan came from a small area, the vicinity of the town Jalalabad. This has its negative consequences (see below), but its positive feature is a relatively detailed knowledge of the local rodentofauna, which is rarely achieved in this part of the world. Most of our material has been acquired by a uniform method of trapping which enabled us to obtain comparative data also applicable beyond the territory of Afghanistan. Finally, since the rodent material collected especially during the recent years in Afghanistan by various foreign zoologists is by no means negligible, it is necessary to summarize published ecological data.

It has already been said that this contribution is gravely limited. The material of Czechoslovak expeditions, on which my study is based, could be collected but in a few regions and cannot be considered representative of the entire territory of the state. This material also offers one-sided seasonal aspect (February–May). Only some of the results of other foreign teams have been published, and in particular the systematic position of many forms remains unexplained. Even the published ecological data are incomplete. The data available so far including Czechoslovak ones enable us only the basic, 'traditional' ecological evaluation. They do not suffice for a more profound analysis from modern points of view, e.g. of questions concerning fecundity, population dynamics, etc.

The original material used for the present study comprises 510 specimens of 12 species of rodents. They are not species exclusively adapted to 'dry' habitats in the narrower sense of the word. With regard to the generally dry climate of the regions where we worked I think that all species found there should be discussed. Our activities in Afghanistan have been reported by POVOLNÝ (1967) and GAISLER et al. (1967). In the latter contribution our material is examined according to species, with localities and exact numbers of individuals stated. These faunistical data will not be repeated here.

The heart of this contribution is the comparison of colonization of various habitats by rodents as it appears in the results of trapping. We used break-back traps of a larger type (18 × 8 cm), usually spaced 5 m. The traps were baited with wick soaked in lard; carrots; native non-leavened bread; and raisins. In our experience, local big raisins were the most attractive bait for all species of small terrestrial mammals. Their advantage was both easy handling and, particularly, the fact that this bait was not dried by the scorching sun. The big raisins were used for trapping in all habitats in the environs of Jalalabad. A minor part of the material was obtained by shooting, net-catching, or by purchase. In the latter case the material was also obtained by trapping according to our instructions by natives in their houses and around them.

The material was weighed, measured and dissected. These results have not yet been published. Some of them are used in the present contribution, mainly for distinguishing sexually active individuals from sexually inactive ones. The development of testes, epididymis and, especially, of seminal vesicles was considered a sign of male sexual activity; in females it was the presence of corpora lutea, embryos, fresh placental scars, vaginal plug or lactation. Body weight was also taken into consideration, similarly as in European mammals.

Results

TRAPPING IN AND NEAR HUMAN SETTLEMENTS

At three places in Afghanistan we could trap rodents in buildings, either ourselves or with the aid of natives: in Kabul, the capital, in Jalalabad and a few neighbouring villages, and in Baghlan (northern Afghanistan). However, the bag in Baghlan consisted of mere two specimens of *Mus musculus* captured with ten traps in the building of an agricultural school. The bag from other places is shown in Table 1. Although the material from Kabul is not numerous, the prevalence of *M. musculus* is obvious. In contrast, in buildings inside settlements in eastern Afghanistan *Rattus rattus* is markedly predominant. It seems that this species does not live in towns like Kabul situated at a high altitude; it is replaced there by the related *Rattus rattoides* (cf. NIETHAMMER, 1965). On the contrary, we never obtained *R. rattoides* from buildings in eastern Afghanistan. *R. rattoides* lives in the vicinity of Jalalabad in very peculiar habitats beyond human settlements, which will be described in the following chapter.

60

Table 1. Results of trapping in buildings. T = number of trap-nights. N = total catch. n = catch of individual species.

Species	n	active males	active females	inactive	n % T	n % N
Kabul; town: T = 39, N = 5						
Mus musculus	3	—	2	1	7.7	60.0
Rattus rattoides	1	1	—	—	2.6	20.0
Cricetulus migratorius	1	—	—	1	2.6	20.0
Jalalabad, Bisut, Laghman, Dar-i-Nur; town and villages: T = 332, N = 144						
Mus musculus	45	20	15	10	13.5	31.3
Rattus rattus	99	30	26	43	29.8	68.7
Jalalabad environs; isolated houses on the edge of semi-deserts; T = 275, N = 60						
Mus musculus	12	6	4	2	4.4	20.0
Tatera indica	41	21	11	9	14.9	68.3
Nesokia indica	7	3	2	2	2.5	11.7

Table 2. Results of trapping in the vicinity of houses, in the gardens and parks of the town Jalalabad. T = 327, N = 39 (for explanation, see Table 1).

Species	n	active males	active females	inactive	n % T	n % N
Mus musculus	23	9	9	5	7.0	59.0
Nesokia indica	16	3	4	9	4.9	41.0

Special attention should be given to bags from isolated tea-houses and dwellings on the fringe of semideserts in eastern Afghanistan. *Tatera indica* lives there hemisynanthropically (cf. PRAKASH's 1962 observations from Rajasthan desert), prevailing in number over *M. musculus*. With the third species, *Nesokia indica*, it was impossible to ascertain whether the individuals had been captured inside buildings. It is more probable that they were caught in the immediate surroundings of the buildings, in closed courtyards with vegetation, etc. Anyway, *N. indica* lives there in the immediate neighbourhood of man.

Two species of rodents, *M. musculus* and *N. indica* (Table 2), were commonly captured in gardens and parks, near buildings and walls in

Jalalabad, as well as one insectivore, *Suncus murinus;* the bag of this species, expressed in percentage of the number of traps (n % T), amounted to 4.6%. Comparing total bags of mammals in anthropogenic habitats in eastern Afghanistan we find maximum concentration inside buildings in continuous agglomerations (n % T = 43.3%), in isolated buildings on the fringe of semideserts (n % T = 21.8%), and the lowest concentration near buildings (n % T = 11.9%; 16.5% including *S. murinus*). *R. rattus* (40.7%) and *M. musculus* (32.9%) are the most numerous species in the rodent material from anthropogenic habitats in eastern Afghanistan, followed by *T. indica* (16.9%) and *N. indica* (9.5%). All the species captured, excepting *Cricetulus migratorius* represented by one specimen, were intensively reproducing at the time of our trapping. The highest percentage of sexually active females (37.5%) was ascertained in the species *M. musculus*.

It should be pointed out that the values given in this chapter show the actual abundance of the studied species in their respective habitats only to a certain degree. It is well known that individual species of rodents as well as different population components of the individual species vary in their trappability. However, I think it premature to speculate in this respect, especially because as far as I know the numbers given above are the first factual data on the relative abundance of the Afghan rodents.

Trapping in eastern semideserts

The investigated region and places of trapping are shown in Fig. 1. According to Hassinger (1968) the Jalalabad Vale is one of the natural areas of Afghanistan. Although it is situated in the monsoon region, its climate is rather dry because the surrounding high mountains extract most of the rainfall from the high clouds. Most of the localities where we trapped mammals are of a quite semidesert character. The Kabul river, with tributaries rather richly supplied with water from the surrounding mountains, runs in the middle of the valley. This enables a thriving agriculture along the rivers and irrigation ditches. Because the field cultures usually adjoin semidesert biotopes, I shall also discuss in this chapter the results of trapping in artificially irrigated areas beyond human agglomerations.

Rocky slopes and ridges with sparse xerothermic vegetation are, generally speaking, one of the commonest biotopes in Afghanistan. The investigated localities of this type were at the altitude of 500–1500 m. As shown in Table 3, the petrophilous *Calomyscus bailwardi* is the commonest species in this habitat. It was not caught in any of the other biotopes. *Meriones persicus* is another species inhabiting this biotope. Both were caught at night only, mostly under overhanging rocks.

In dry river beds with scattered boulders the bag contained a greater variety of species. As shown in Table 3, *Tatera indica* predominated there.

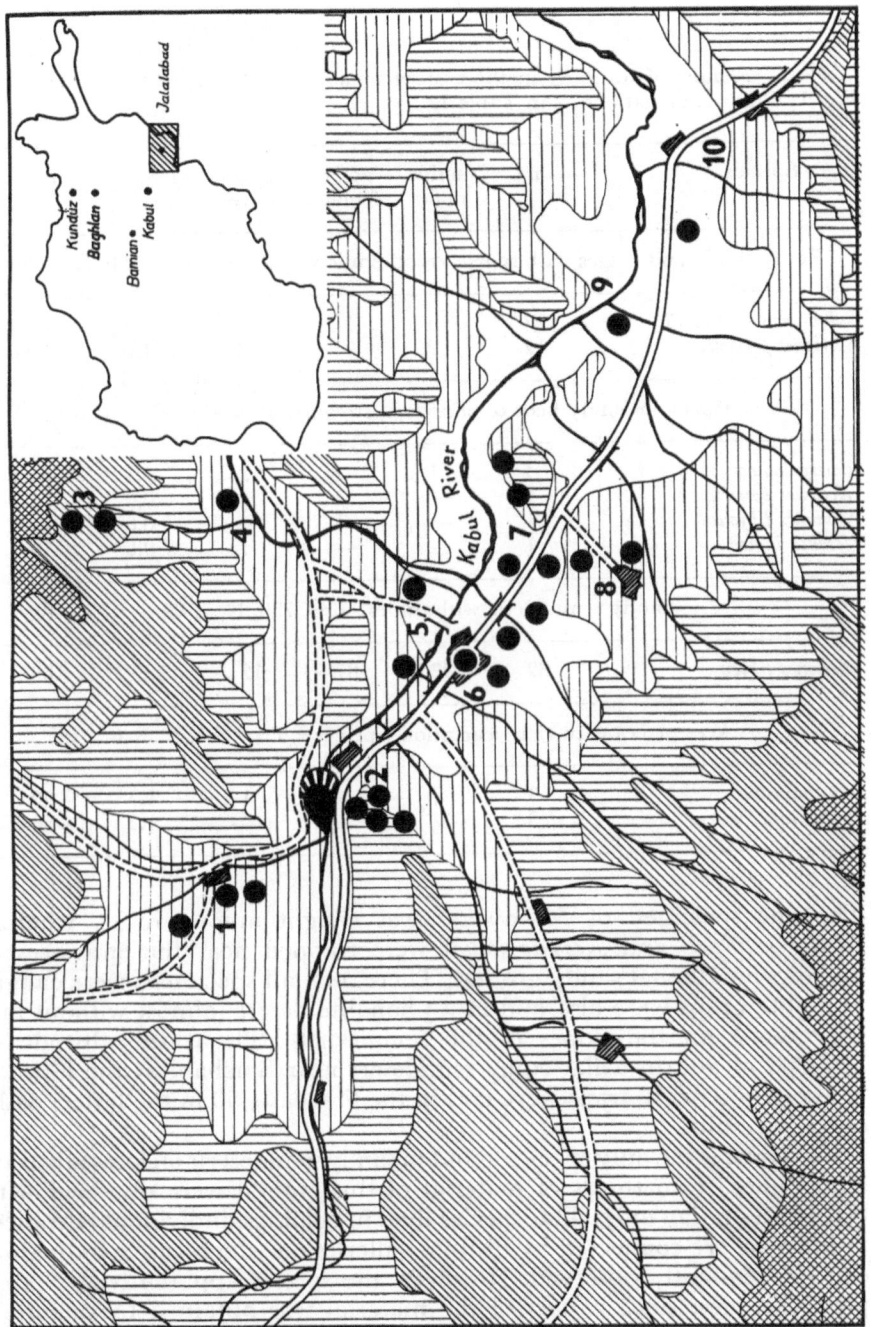

Fig. 1. A very approximate map of the Jalalabad Vale showing the main study area.
Elevations: white, 300–600 m; horizontal strips, 600–900 m; vertical strips, 900–1500 m;
oblique strips, 1500–2000 m; reticulation, over 2000 m. Localities: 1, Laghman; 2,
Darunta (village, basin and Darunta Hills); 3, Dar-i-Nur; 4, Abdukil; 5, Bisut; 6,
Jalalabad; 7, Somarkhel; 8, Sarshahi; 9, Chak-Naur; 10, Tor Khama.

Table 3. Results of trapping in the eastern semideserts. The localities are shown in Fig. 1. Other explanations as in Table 1.

Species	n	active males	active females	inac- tive	n % T	n % N
Rocky slopes and ridges with sparse xerothermic vegetation: T = 750, N = 39						
Calomyscus bailwardi	27	15	11	1	3.6	69.2
Meriones persicus	12	9	1	2	1.6	30.8
Rocky or sandy beds of periodical streams: T = 175, N = 7						
Tatera indica	4	1	1	2	2.3	57.1
Meriones persicus	1	—	1	—	0.6	14.3
Meriones libycus	1	—	1	—	0.6	14.3
Nesokia indica	1	—	—	1	0.6	14.3
Sandy semideserts with dunes: T = 750, N = 82						
Meriones libycus	82	33	37	12	10.9	100.0
Wet places with rich vegetation (springs): T = 225, N = 6						
Mus musculus	3	2	—	1	1.3	50.0
Rattus rattoides	3	2	1	—	1.3	50.0

The one specimen of *M. persicus* was caught among rocks, whereas *Meriones libycus* and *Nesokia indica* were trapped near the entrances to burrows on alluvial sandy ground. There, too, all specimens were caught at night. Traps were also laid in a phonolite desert (almost without any vegetation) on the slopes of the Somarkhel Hills at the altitude of 500–800 m and in a flat stony semidesert near Sarshahi at the altitude of about 600 m. 72 traps were laid in the first case, 50 in the latter, without any results.

M. libycus is the only species found in sandy semideserts on loess substrate with dunes. Two thirds of material of this animal were obtained with traps laid on day only. This species lives in colonies, individual entrances to burrows are connected by distinct galleries on which active individual can be frequently observed. However, the species apparently has polyphase activity, since a part of the material was obtained by definitely nocturnal trapping. A high percentage of the bag (Table 3) was certainly affected by our preference for trapping in sandy semideserts at places with visible burrows of rodents.

Table 4. Results of trapping in field cultures and on banks of irrigation canals in the Jalalabad Vale. Explanations as in Table 3. T = 470, N = 59.

Species	n	active males	active females	inac- tive	n % T	n % N
Nesokia indica	37	11	9	17	7.9	62.7
Mus musculus	13	10	3	—	2.8	22.0
Meriones libycus	9	3	3	3	1.9	15.3

Quite a special biotope was a not very large locality on the slope of the Darunta Hills with lush subtropical vegetation growing at sulphurous springs, consisting in particular of the strata of herbs and shrubs. The locality was situated at about 3 km from the nearest human settlement, or rather from a camp of quarry workers. During two trap-nights there were caught three specimens of *Mus musculus* and three individuals of *Rattus rattoides* which was not found anywhere else near Jalalabad. *C. bailwardi* was common in the immediate neighbourhood of this locality, in a rocky habitat without the shrub stratum.

The results of trapping in fields and on the banks of irrigation ditches are shown in Table 4. There the most abundant species was *N. indica* whose activity – according to the results of trapping – is polyphase, diurnal as well as nocturnal. The abundance of this rodent is apparently greater than the trapping has shown. This species lives largely underground and often buries traps laid at the exits from burrows. *Mus musculus* is another species frequently occurring in cultivated plots, whereas *M. libycus* is found but on their borders adjoining sandy semideserts. However, colonies of *M. libycus* were often found in the immediate neighbourhood of human settlements or cultivated land, i.e. on the very outskirts of the town of Jalalabad; similar observations are mentioned in literature (e.g. BOBRINSKIJ *et al.*, 1965).

Comparing the relative sizes of bag in individual habitats (n % T) we arrive at the following order: field cultures (12.6%), sandy semideserts (10.9%), rocky slopes (5.2%), beds of periodical desert streams (4.1%), vicinity of springs with thick vegetation (2.6%) and stony semideserts (0.0%). Owing to great differences in the ecological conditions of individual habitats we cannot sum up the material and calculate the representation of individual species. *M. libycus*, *N. indica* and *C. bailwardi* can be considered abundant, of course in suitable habitats only. A truly rare rodent among the species found near Jalalabad is only *R. rattoides*.

All the species of rodents were intensively reproducing at the time of our investigations, as our stay in Afghanistan partly coincided with the period of the rapidest growth of vegetation. The percentage of reproducing females in the bag amounted to 42.1% in *M. libycus*, 23.7% in *N. indica* and 40.7% in *C. bailwardi*. A surprisingly high number of inactive individuals was found among *N. indica*, mostly in lower weight classes (young or subadult). The causes of this relatively high trappability of immature individuals are not quite clear.

COLLECTING IN CENTRAL SEMIDESERTS

In mountain valleys and on plateaux around Kabul we could trap only twice. Our results thus only supplement a much more numerous material of German and American expeditions, which, unfortunately, has not yet been published in full. The first trapping was done in a semidesert habitat on a plateau near Istalif at the altitude of 1700 m. 16 traps were laid and 2 immature females of *Meriones libycus* were caught between 10 a.m. and 4 p.m. At the other locality, Bamiyan Vale, 20 traps were laid overnight on a rocky slope at the altitude of 2700 m. Three sexually active males of *Calomyscus bailwardi* and one male and one female (both active) *Meriones persicus* were captured. Four other specimens of *M. persicus*, three active males and one inactive female, were shot at night in the Shebar Kotal valley at 2500 m.

Two specimens (male and female, both active) of *Cricetulus migratorius* caught at night on a road near Charikar at the altitude of about 1800 m also belong among the rodent material from the vicinity of Kabul. Finally, several times we had an opportunity to join in nocturnal hunts for porcupines, *Hystrix indica*. All specimens obtained (for localities see GAISLER et al., 1967) were shot at night when foraging in fields and along streams in alpine valleys, about 2000–2500 m above sea level. Finds of spines indicate that on day the porcupines hide in caverns and various cavities under boulders on rocky slopes.

COLLECTING IN NORTHERN STEPPES

In northern Afghanistan we trapped successfully at one place only, on the bank of an irrigation ditch and at the margin of a field near the town of Kunduz. One active male and one active female of *Nesokia indica* and one active male of *Mus musculus* were captured overnight with 40 traps laid. Trapping on a rocky slope and on the fringe of steppe did not yield any results.

A night hunt in the steppe 25 km NW of Kunduz to Sherchan-Port was far more interesting. The material collected by shooting and net-catching between 9.30 p.m. and 1.00 a.m. consisted of 19 rodents and 9 insectivores (all hedgehogs *Hemiechinus auritus*). The rodents were 6 active and 1 inactive males and 2 active and 1 inactive females of *Meriones libycus;* 1 active male *Meriones meridianus;* 3 active males, 3 active and 2 inactive but mature females of *Allactaga elater*. No active rodents except for one suslik (not captured) were observed at the same locality on day.

TENTATIVE ECOLOGICAL REVIEW OF AFGHAN RODENTS

Previous information on the rodents of Afghanistan (ELLERMAN & MORRISON-SCOTT 1951, ZIMMERMANN, 1955, and others) has been summarized by NIETHAMMER (1965) in a paper which is a preliminary evaluation of a quite extensive material collected by German zoologists working at the Institute of Zoology of the university in Kabul. This material is being gradually examined in more detail – so far there have been published studies on *Petauristinae* (NIETHAMMER, 1967), *Apodemus* (NIETHAMMER, 1969) and *Microtinae* (NIETHAMMER, 1970). Unfortunately, the groups of rodents adapted to life in the desert have not yet been properly evaluated. Another large collection of Afghan rodents was acquired by the American '1965 Street Expedition to Afghanistan'. As far as I know, the material collected by the American zoologists has not yet been studied in detail. There is only the 'Introduction to the mammal survey' by HASSINGER (1968), where in Table 1 on pp. 74–75 there is a survey of the species obtained, with numbers of individuals given according to fifteen major collecting localities. Considering information from the sources mentioned above including our own study (GAISLER *et al.*, 1967) we find that so far 34 species of rodents have been reported from Afghanistan.

An ecological classification of the Afghan rodents was made by NIETHAMMER (1965) who divided 27 species known at that time into eight groups according to their habitats. Later, HASSINGER (1968) analysed in detail the physiography, climate and phytogeography of Afghanistan and outlined his own classification of mammalian biotopes. HASSINGER's description of individual biotopes is very detailed, with good photographs. However, an account of mammals confined to single biotopes is lacking. Although the delimitation of some categories (e.g. as far as the anthropogenic habitat is concerned) is controversial, in the following text I shall stick to HASSINGER's classification. I shall limit myself to a mere account of individual species, presuming that the American zoologists will follow HASSINGER in publishing information on their material; only then it will be possible to work out a more detailed ecological classification. The survey given below is based on data from the papers referring to Afghanistan as mentioned above, as well as on information on the bionomy of the same species in adjacent regions (BOBRINSKIJ *et al.*, 1965, LAY, 1967, and others). My own findings described in the previous chapter are also taken into consideration.

I. Rodents preferring or confined to the biotopes of dry habitat.
    Dry habitat sensu HASSINGER (1968) includes 97.5 per cent of the whole Afghan territory.

A. MOUNTAINS AND ASSOCIATED TERRAIN.
   1. Alpine habitat above 3000 m: *Marmota caudata, Microtus afghanus, Microtus juldaschi* and *Microtus arvalis* live on high montane meadows; *Alticola roylei* is typical of high rocky slopes; also *Cricetulus migratorius, Calomyscus bailwardi* and *Allactaga williamsi* penetrate above 3000 m.
   2. Slopes and plateau below 3000 m with a clay loess substrate and a covering of small stones: *Marmota caudata* lives along the upper border of this habitat; *Cricetulus migratorius, Hystrix indica* and *Allactaga williamsi* occur there, too.
   3. Rock-covered slopes and plateau below 3000 m: *Calomyscus bailwardi* and *Meriones persicus* are typical of this biotope, above 2400 m also *Alticola roylei*; various other species including *Hystrix inaica, Apodemus sylvaticus* and *Dryomys nitedula* may live here, too.
   4. Montane watercourses: *Apodemus sylvaticus, Microtus arvalis* and *Dryomys nitedula* live in strips of vegetation along mountain streams; if the stream runs through the rock biotope, *Alticola roylei, Calomyscus bailwardi* and *Meriones persicus* may be found there.

B. STEPPES AND SEMIDESERTS.
   1. Clay and loess, including stony deserts: *Spermophilopsis leptodactylus, Rhombomys opimus, Meriones libycus, Meriones meridianus, Meriones zarudnyi* and *Allactaga elater* are typical especially of the loess steppes in northern Afghanistan; also the subterranean rodent *Ellobius talpinus* is confined to this habitat, whereas the related *Ellobius fuscocapillus* lives in similar dry soils at a higher altitude (above 1000 m) in central Afghanistan; stony deserts and even semideserts sometimes seem to lack any rodent life; however, *Meriones* or *Gerbillus* species, *Cricetulus migratorius, Tatera indica* and perhaps others may be found in places.
   2. Sand: *Meriones libycus, Meriones crassus, Allactaga elater, Allactaga williamsi, Allactaga hotsoni, Jaculus blanfordi, Gerbillus cheesmani, Gerbillus nanus* and *Tatera indica* are considered typical species of sandy steppes and semideserts; this biotope is also frequented by *Cricetulus migratorius* which seems to be relatively little specialized ecologically. *Citellus fulvus* lives on sandy ground at the altitude of 1000–3000 m.
   3. Watercourses, including ephemeral ones: various species living in the biotope through which the stream flows can be found near it; *Nesokia indica* and *Mus musculus* live on alluvia with a relatively lush vegetation, occasionally perhaps also *Rattus rattoides*; dry beds are inhabited e.g. by *Tatera indica* and *Meriones persicus*.

C. ANTHROPOGENIC HABITAT.
   According to my opinion, the definition of this habitat (HASSINGER 1968, p. 28) is somewhat controversial and its subdivision (HASSINGER 1968, pp. 44–51) is incomplete. Therefore I suggest the following division of this habitat into two basic types of biotopes:
   1. Interiors and immediate vicinity of houses: *Mus musculus, Rattus rattoides* or *Rattus rattus* (these two species seem to exclude each other) are synanthropes; *Cricetulus migratorius, Tatera indica, Nesokia indica, Meriones persicus* and perhaps others may be considered hemisynanthropes in places.
   2. Gardens, orchards, parks, field cultures, banks of irrigation ditches, etc.: *Nesokia indica* and *Mus musculus* are the most typical species, especially at lower altitudes; both species of *Ellobius* locally occur in dry fields, and above 1000 m *Microtus afghanus* is often abundant in fields (its ecology was studied in detail by NIETHAMMER 1970); a number of other species, e.g. *Cricetulus migratorius, Meriones libycus, Meriones persicus* and *Tatera indica*

are found on the fringes of fields, near irrigation ditches and buildings beyond settlements.

II. Rodents preferring or confined to the biotopes of wet habitat. Wet habitat sensu HASSINGER (1968) includes only 2.5 per cent of the entire territory of Afghanistan – parts of Nuristan and Paktia.

A. MOUNTAINS AND ASSOCIATED TERRAIN.

I cannot follow HASSINGER's subdivision, as too little is known of the rodent ecology in the monsoon region of Afghanistan. There are only two true forest species: *Hylopetes fimbriatus* and *Petaurista petaurista*; according to NIETHAMMER (1967) they live in Nuristan in a mixed *Quercus baloot* – *Cedrus deodara* forest. Another species living in forests in this part of Afghanistan is *Apodemus sylvaticus*. Besides, *Calomyscus bailwardi* and *Cricetulus migratorius* were found there at similar localities as elsewhere. Apparently no collecting was done above 3000 m.

B. ANTHROPOGENIC HABITAT.

In the reports published so far there is no reference to species living inside buildings in this part of the state. HASSINGER (1968) mentions *Mus musculus* as the most abundant species at Kamdesh; of the other synanthropes or hemisynanthropes he found there *Rattus rattoides* and *Nesokia indica*.

The rodents of Afghanistan could certainly be ecologically classified according to other criteria as well, such as vertical distribution, adaptation to a certain way of life (subterranean, terrestrial, petricolous, arboricole), rhythm of activity, type of reproduction, etc. Such a classification will probably be fully justified when all information on the collected material will be published. One species could not be included even in the basic classification given above. It is *Salpingotus thomasi*, known so far from a single specimen coming 'probably from some part of Afghanistan' (ELLERMAN & MORRISON-SCOTT, 1951). However, our present knowledge is sufficient to allow us to draw a conclusion which is the most important one from our point of view: the species adapted to life in the arid zone constitute 64.7%, i.e. majority of the Afghan rodentofauna.

Discussion

Various morphological characters (besides physiological ones) are usually interpreted as a manifestation of the adaptation of rodents to life in dry areas without forests – e.g. colouring, structure of extremities, elongation of tail and auricles, or the growing size of bullae tympani (OGNEV, 1951, HARRISON, 1964, CLOUDSLEY-THOMPSON & CHADWICK, 1964, etc.). However, to a mammalogist from Central Europe the most conspicuous phenomenon is a larger size of rodents in Afghanistan. And not only in areas at a high altitude, where it might be considered a manifestation of Bergmann's rule, e.g. in the Jalalabad Vale, at 300–1000 m, we captured 5 species of rodents: *Meriones libycus*, *Meriones persicus*, *Tatera indica*, *Calomyscus bailwardi* and *Nesokia indica;* I do not take into consideration synanthropes of the genera *Rattus* and *Mus*. Of these five species, apparently autochthonous, only *C. bailwardi* is a 'small' mammal in the European

69

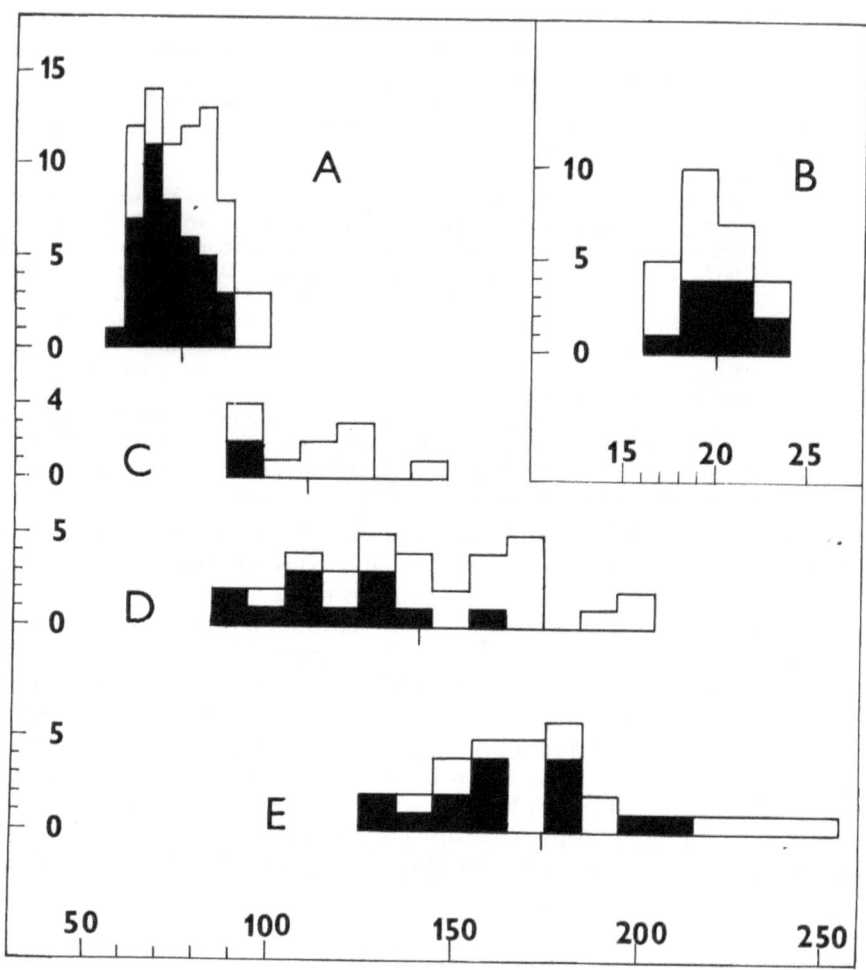

Fig. 2. Graphical representation of weights of sexually active individuals of five species, collected in the Jalalabad Vale. Ordinate, number of specimens; abscissa, weight classes in g; A, *Meriones libycus*; B, *Calomyscus bailwardi*; C, *Meriones persicus*; D, *Tatera indica*; E, *Nesokia indica*; black, females; white, males; short abscissae, average weight for both sexes.

sense of the word. Actual values of the body weight of sexually active individuals of these species are shown graphically in Fig. 2. Obviously, adult specimens of all the species excepting *C. bailwardi* weigh over 50 g. In contrast, a great majority of European rodents weigh less than 40 g. Considering only species inhabiting comparable biotopes, i.e. relatively dry land without forests (the so-called cultivated steppe), in Central Europe they are mainly members of the genera *Microtus*, *Apodemus* and

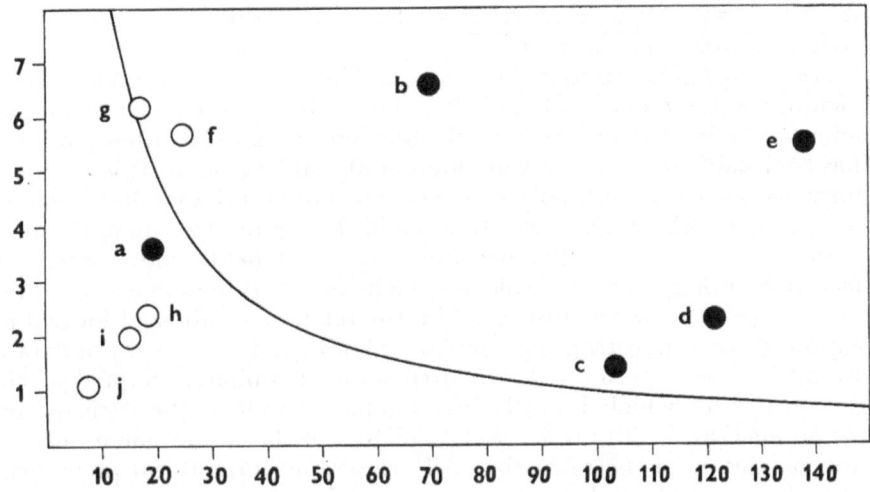

Fig. 3. An example of a simple comparison of relative abundance combined with average weights in Afghan and European rodents. The Afghan material (dots) collected in semideserts and fields in the Jalalabad Vale: a, *Calomyscus bailwardi*; b, *Meriones libycus*; c, *Meriones persicus*; d, *Tatera indica*; e, *Nesokia indica*. The European material (circles) collected in fields and other woodless biotopes in Czechoslovakia (the data by GAISLER *et al.* 1967): f, *Microtus arvalis*; g, *Apodemus agrarius*; h, *Apodemus sylvaticus*; i, *Apodemus microps*; j, *Micromys minutus*. *Mus musculus* is excluded from both samples. The weights were calculated from all specimens, including the sexually inactive. The curve denotes all combinations equal to the average weight of 20 g and a relative abundance of 5%. – Ordinate, number of individuals captured in per cent of traps laid (n % T); abscissa, average weight in g.

Micromys. Only in two species of rodents living in such environment – *Cricetus cricetus* and *Citellus citellus* – the body weight is higher and comparable with the weight of most of the Afghan rodents.

Although this observation might be extended to the rodentofauna of the entire large arid zone from northern Africa to central Asia, I do not maintain that the increase in body size is an indication of adaptation of the rodents to life in arid regions. I am pointing out the body weight of the Afghan rodents with regard to possible later evaluations of productivity. Surprisingly, the relative size of bag from snap-traps in dry biotopes in Afghanistan is similar to the relative size of bag in analogous but more humid biotopes in Central Europe (except for periods of overcrowding). An example of such comparison made with my own material appears in Fig. 3. Undoubtedly, this comparison has a number of drawbacks. More precise density estimations are necessary, such as those made in investigations of desert rodent communities e.g. in the USA (ROSENZWEIG & WINAKUR, 1969, and others). Nevertheless, at the time being we know enough to consider the biomass of rodents in the dry biotopes investigated in Afghanistan relatively large. This is due particularly to the weight parameters discussed.

Nesokia indica has been included among the 5 rodent species found in the vicinity of Jalalabad, although it is not a desert rodent. Its ecological adaptation is of a different kind and deserves special attention. As it has been said, *N. indica* lives predominantly underground. It is a strongly burrowing animal with polyphase activity. Often it lives in large colonies with a network of burrows. It inhabits banks of streams and canals, various earthworks and, especially, cultivated fields and gardens. In fact, it occupies the same ecological niche as – and lives in a way similar to – *Arvicola terrestris* in Europe. Also the relative numbers of individuals captured with snap-traps are similar in both species (cf. e.g. the data by PELIKÁN et al., 1971a for *A. terrestris*). Another similarly adapted species, the ecology of which I partly investigated as well, is the African *Arvicanthis niloticus* (cf. PELIKÁN et al., 1971b). All the species mentioned are approximately of the same size. All are serious agricultural pests, being also of some epidemiological importance. The comparative study of their ecology might be useful in providing data for control measures.

Summary

Material consisting of 510 individuals of 12 rodent species collected in February–May 1966 and 1967 in Afghanistan by Czechoslovak zoologists is ecologically evaluated. Most of the material was obtained by trapping in the Jalalabad Vale, eastern Afghanistan (Tables 1–4). In and near human settlements there were found *R. rattus*, *M. musculus*, *T. indica* and *N. indica*, the first two species being the most abundant. *T. indica* lives hemisynanthropically in isolated dwellings on the fringe of semideserts. In fields and on the banks of irrigation ditches *N. indica* is the most abundant species; *M. musculus* is also commonly found there. Rocky slopes are inhabited by *C. bailwardi* and *M. persicus*, sandy semideserts by *M. libycus*. *R. rattoides* was found in the Jalalabad Vale at one locality only, beyond human settlements. The relative size of bag from individual biotopes decreases in the following order: buildings in settlements (43.3%), isolated buildings (21.8%), field cultures (12.6%), vicinity of buildings (11.9%), sandy semideserts (10.9%), rocky slopes (5.2%), beds of periodical streams (4.1%), vicinity of springs (2.6%), stony semideserts (0.0%). Notes on reproduction and activity are given for individual species. Occasional collecting and observations of rodents in central and northern Afghanistan are mentioned.

With the aid of literary sources, in particular papers by NIETHAMMER (1965, 1967, 1969, 1970), a provisional ecological classification of all 34 rodent species recorded so far in Afghanistan is presented. According to HASSINGER's (1968) classification of mammalian habitats the individual species are placed in one of eleven biotopes delimited. Published information and my own observations show that the species adapted to life in desert environment represent the majority (about 60%) of the Afghan rodentofauna. The values of the body weight of some Afghan rodents are discussed; they must be taken into consideration in the evaluation of productivity. The results of trapping indicate that the biomass of rodents is relatively large even in some of the dry biotopes. The ecological adaptation of *Nesokia indica*, with serious economical consequences, is pointed out. An almost identical ecological niche is occupied by other species in other regions, e.g. *Arvicola terrestris* in Europe and *Arvicanthis niloticus* in Africa.

REFERENCES

BOBRINSKIJ, N. A., KUZNECOV, B. A. & KUZJAKIN, A. P. 1965. Opredelitel mlekopitajuščich SSSR (Review of mammals of the USSR). Moscow (in Russian).

CLOUDSLEY-THOMPSON, J. L. & CHADWICK, M. J. 1964. Life in deserts. London.

ELLERMAN, J. R. & MORRISON-SCOTT, T. C. S. 1951. Checklist of Palearctic and Indian mammals 1758 to 1946. London.

GAISLER, J., ZAPLETAL, M. & HOLIŠOVÁ, V. 1967. Mammals of ricks in Czechoslovakia. *Acta Sc. Nat. Acad. Sc. Brno*, 1: 299–348.

GAISLER, J., POVOLNÝ, D., ŠEBEK, Z. & TENORA, F. 1967. Faunal and ecological review of mammals occurring in the environs of Jalal-Abad, with notes on further discoveries of mammals in Afghanistan. I. Insectivora, Rodentia. *Zool. listy*, 16: 355–364.

HARRISON, D. L. 1964. The mammals of Arabia. Vol. 1. London.

HASSINGER, J. D. 1968. Introduction to the mammal survey of the 1965 Street expedition to Afghanistan. *Fieldiana: Zoology*, 55: 1–81.

LAY, D. M. 1967. A study of the mammals of Iran resulting from the Street expedition of 1962–1963. *Fieldiana: Zoology*, 54: 1–282.

NIETHAMMER, J. 1965. Die Saugetiere Afghanistans. II. Insectivora, Lagomorpha, Rodentia. Science, Kabul, 1965: 18–41.

NIETHAMMER, J. 1967. Die Flughörnchen (Petauristinae) Afghanistans. *Bonner Zool. Beiträge*, 18: 2–14.

NIETHAMMER, J. 1969. Die Waldmaus, *Apodemus sylvaticus* (LinnF, 1758), in Afghanistan. *Säug. Mitteilungen*, 17: 121–128.

NIETHAMMER, J. 1970. Die Wühlmäuse (Microtinae) Afghanistans. *Bonner Zool. Beiträge*, 21: 1–24.

OGNEV, S. I. 1951. Očerki ekologii mlekopitajuščich (Ecology of mammals). Moscow (in Russian).

PELIKÁN, J., ZEJDA, J. & HOLIŠOVÁ, V. 1971. Catch curve and analysis of the catch of *Arvicola terrestris* on trap lines. *Zool. listy*, 20: 215–228.

PELIKÁN, J., MADKOUR, G. & GAISLER, J. 1971. On some mammals from Egypt (Insectivora, Lagomorpha, Rodentia). *Zool. listy*, 20: 307–318.

POVOLNÝ, D. 1967. Beiträge zur Kenntnis der Fauna Afghanistans. Reisebericht und Charakteristik des Sammelgebietes. *Acta Mus. Morav. Brno, Sc. nat.*, 52 (Suppl.): 12–14, 21–29.

PRAKASH, I. 1962. Ecology of the gerbils of the Rajasthan desert, India. *Mammalia*, 26: 311–331.

ROSENZWEIG, M. L. & WINAKUR, J. 1969. Population ecology of desert rodent communities: Habitats and environment complexity. *Ecology*, 50: 558–572.

ZIMMERMANN, K. 1955. Zur Fauna Afghanistans. *Zeitschr. f. Säugetierkunde*, 20: 189–191.

APPENDIX

After sending this paper to press I was given valuable information on the material of rodents, which had been gained by the participants to the Czechoslovak Hindu Kush expedition in July and August 1965. The information has been granted by the kindness of Dr. M. DANIEL, Prague, who was the zoologist of the expedition. The material is kept in the collections of the National Museum in Prague. Below I state a brief extract from the information concerning rodents. Besides, also Insectivora *(Crocidura russula)* and Lagomorpha *(Ochotona rufescens* and *O. roylei)* were collected. The results quoted here have not been published so far.

1. Surroundings of Kabul, height above sea level not given. Rodents were trapped (number of trap-nights not stated) on the banks of irrigation canals among grain fields and on the stubble fields, not irrigated after the harvest. Catch: *Microtus afghanus* 10, *Apodemus sylvaticus* 6, *Mus musculus* 4 specimens.

2. Surroundings of Faisabad, height above sea level 1500 m. Rodents were trapped (number of trap-nights not stated) on fallows among parched grain fields on hills behind the river Koktcha. Catch: *Meriones meridianus* 2, *Meriones persicus* 1, *Mus musculus* 1 specimen; 1 specimen of *Nesokia indica* was caught in a stone wall between mulberry orchards.

3. Surroundings of the village Ishmurkh, Wakhan, height above sea level 2750 m. Rodents were trapped (T = 278) on the banks of irrigation canals with dense vegetation, constituted above all by *Calamagrostis pseudophragmites, Agrostis stolonifera*, and *Carex turkestanica*. Catch: *Cricetulus migratorius* 6, *Apodemus sylvaticus* 2 specimens.

4. The Ishmurkh Darrah valley, connecting the main chain with the vale of the river Ab-i-Panj, height above sea level 2850 m. Rodents were trapped (T = 440) in a thicket of willow bushes on the confluence of mountain torrents, with an undergrowth of thick grass and moist moss. Catch: *Alticola roylei* 24, *Apodemus sylvaticus* 14 specimens.

5. The Chap Darrah valley, branching from the Ishmurkh Darrah valley, height above sea level 3700 m. Rodents were trapped (T = 668) with the exception of marmots which were shot. The locality is a slowly rising plain, the dimensions of which are approximately 1.5 by 0.5 km, covered by a steppe of the association *Artemisia leucotricha* + *Stipa himalaica*. The burrows of marmots were scattered all over the plain, the catch of voles was successful solely in the surroundings of small rivulets where *Lonicera asperfolia* together with *Pentaphylloides dryadanthoides* were growing. Catch: *Alticola roylei* 15, *Marmota caudata* 7 specimens.

6. Surroundings of the glacier front in the Ishmurkh Darrah valley, height above sea level 3800 to 4000 m. Rodents were trapped (T = 530) on minor grass-covered spots under steep rocks. Catch: *Alticola roylei* 13 specimens.

7. Upper part of the Ishmurkh Darrah valley between the eastern and the western arm of the glacier, height above sea level 4550 m. Rodents were trapped (T = 730) on a small island, not covered with ice, between the moraines of the glacier flowing from the Anushah saddle, in the surroundings of a small rivulet. Catch: *Alticola roylei* 13 specimens.

74

V. THE POPULATION ECOLOGY OF THE RODENTS OF THE RAJASTHAN DESERT, INDIA*

by

ISHWAR PRAKASH

Introduction

Rodents constitute one of the largest mammalian groups in the Rajasthan desert, India, both in total number and in the number of species represented. They exhibit a great plasticity in respect of their choice of a wide spectrum of desert habitats and their impact on the desert ecosystems is all too prominent. The rodents' role in the intensification of desertic conditions may be comprehended from the fact that a single species, viz. *Meriones hurrianae*, is able to excavate about 61,400 kg of stabilised soil per km² per day during summer and deposit it outside its burrow openings in a loose formation. This dug-up soil is easily blown away by the strong desert winds. The rodents, thus, are a prime biotic factor for soil erosion. Their impact on the desert vegetation as a result of their gnawing, debarking, cutting and feeding propensities is easily discernible throughout the tract. In the desert region, orchards, crop fields and rangelands cannot attain their optimum productivity without a rodent control programme. The rodents are, therefore, regarded to be the most potent enemies of man in the desert in his endeavours for maximizing output from this inhospitable terrain. In view of the considerable economic importance of a large number of desert vertebrates, and especially rodents, a study was initiated in the Rajasthan desert in 1953 with financial assistance of the UNESCO. Later on, in 1959, the Central Arid Zone Research Institute at Jodhpur provided a small cell of workers to study the ecology of the rodents and to devise means of their control in the Rajasthan desert. A brief review of our work on the population structure of rodents, their ecological distribution, home ranges, food, predators and reproduction aspects has been presented here. The control aspects of our work have been deliberately left out of this review.

The Rajasthan desert, where most of these studies have been carried out, is situated almost in the centre of the Great Indian or the Thar desert. The Indian desert is spread over four Indian States: Punjab, Haryana, Rajasthan and Gujarat – all situated in the west of India,

---

* This communication is dedicated to my parents.

*Table 1.** Mean maximum and minimum temperature (°C), rainfall (mm) and relative humidity (per cent) at 17 hrs. (IST) in some of the localities in the Rajasthan desert.

Stations	Jan.	Feb.	March	April	May	June	July	Aug.	Sept.	Oct.	Nov.	Dec.	Yearly average
Sri Ganganagar													
I	19.8	23.1	28.3	35.2	42.1	41.6	38.9	37.7	37.8	35.1	29.1	23.1	32.7
II	3.4	7.6	11.2	16.7	23.9	27.4	28.1	26.9	23.4	15.4	8.7	4.0	16.7
III	6.6	11.2	5.3	5.1	6.1	31.0	68.3	70.6	6.9	1.5	0.0	5.0	217.9
IV	42	41	31	23	20	31	43	45	47	28	32	39	34
Jodhpur													
I	24.6	27.0	32.5	37.4	40.8	39.8	36.1	33.2	34.6	35.3	30.9	26.1	33.2
II	9.2	11.4	16.4	21.6	26.3	27.9	26.8	25.0	23.8	18.6	13.0	10.3	19.2
III	3.8	6.1	2.8	3.3	10.4	36.1	100.8	122.9	62.0	8.2	2.8	2.8	360.9
IV	22	21	17	11	15	34	46	50	40	16	18	22	27
Jaisalmer													
I	23.8	25.4	31.7	38.5	42.8	40.4	37.7	34.7	37.7	35.6	30.4	24.8	33.6
II	8.9	8.9	16.9	21.4	26.8	26.8	26.8	24.9	25.1	19.5	11.3	6.2	18.7
III	4.1	4.8	3.6	2.8	7.9	14.7	52.3	62.5	21.8	1.3	1.3	2.0	179.1
IV	45	25	35	37	32	44	65	73	64	29	23	21	41
Barmer													
I	24.7	27.4	32.3	37.9	41.3	39.7	35.8	33.6	35.1	36.2	31.9	27.3	33.6
II	10.0	13.1	17.9	23.5	26.7	26.8	26.1	24.9	24.3	21.3	16.1	11.9	20.2
III	2.5	3.6	3.1	1.5	9.4	24.6	89.4	86.6	37.1	2.5	1.3	1.8	263.4
IV	28	28	30	25	37	40	51	51	44	29	26	28	34
Sikar													
I	22.7	24.8	31.0	37.6	42.2	40.6	36.2	33.5	34.2	34.2	29.2	23.9	32.5
II	6.9	7.5	14.2	20.3	26.7	28.6	26.4	24.6	23.1	17.6	8.9	5.3	17.5
III	9.1	8.4	7.6	2.8	16.3	46.7	126.7	147.3	49.3	6.1	3.6	6.9	430.5
IV	34	24	21	15	20	25	74	60	50	28	30	38	35

* After **PRAMANIK & HARIHARAN** (1952). – I = Mean maximum temperature; II = Mean minimum temperature; III = Mean rainfall; IV = Mean relative humidity.

Fig. 1. Sandy plain with hummocks in the background.

Fig. 2. Sandy plain with vegetation.

Fig. 3. Bare sand dunes.

Fig. 4. Bare sand dune. Trees, *Prosopis cineraria*, are visible in the background (Photo by B. L. TAK).

bordering Pakistan. The Rajasthan desert, encompassing about 196,150 km², lies between 25 °N and 30 °N and 69.5 °E and 76 °E. Its eastern limit is bordered by the Aravalli ranges. The climatic conditions are typical of a desert. The rainfall is low and erratic. Ninety per cent of the rain falls during the monsoon season, from July to September. The average annual precipitation ranges from 80 to 425 mm in various parts of this arid region (Table 1). Air temperature exhibits great extremes. The winter is quite cold and at many places the temperature falls below freezing point and frost occurs. The average minimum temperatures, during January, vary from 3.4 °C to 10 °C. During the summer, the heat is intense and the mean maximum temperature ranges between 40.8 and 42.8 °C (Table 1). The highest maximum temperature has been recorded at 50 °C. The mean relative humidity at 17.00 hours have been given in Table 1. The relative humidity remains minimum during the hot weather, and maximum during the months of July, August and September.

The Indian desert has been subjected to marine transgressions during Jurassic, Cretaceous and Eocene periods. The sea receded during Miocene and Pliocene (KRISHNAN, 1952). WADIA (1960), however, believed that this region began to dry gradually after the Pleistocene and the last glacial period. On the basis of their findings of Chaloclithic period on drainage systems, kilnburt bricks and other features of the Indus civilisation, archaeologists are of the view that this desert has a rather recent origin (GHOSH, 1952). ROY & PANDEY (1971) and PRAKASH (1974) have, however, supported the view of AHMED (1969) that this desert is actually much older.

The Indian desert is an undulating vast plain of sand. In certain localities out-crops of hills and large gravel plains form a part of its topography. From the point of view of rodent distribution, the major habitats have been classified into sandy, gravel plains, rocky and ruderal (PRAKASH, 1962–64). The sandy habitat occupies by far the largest proportion of the total area and is interspersed with wind blown and stabilised sand dunes. The gravel plains are usually situated on the foothills of hillocks, except in the Jaisalmer region where extensive gravel plains occur. Rocky out-crops are found all over the desert region but cover comparatively larger areas in the northeastern desert. The rocks are a mixture of rhyolite and sandstone. The ruderal habitat (the village complex) is scattered almost in all parts of the desert depending upon the availability of drinking water. Rainfed crops are grown in the vicinity of villages while in some areas cropping under irrigation is now feasible. The various habitats and the vegetation types of this desert have been discussed in detail by PRAKASH (1974). Rodents are found in almost all these habitats and, curiously enough, they are relatively more abundant in the 100 mm rainfall zone in the western most sector of the Rajasthan desert.

Fig. 5. Gravel plain.

Fig. 6. A close view of the gravelly habitat.

Fig. 7. Rocky habitat (Photo by **B. L. Tak**).

Population Structure

Habitat selection

Certain species of rodents show marked habitat specificity while others occur over diversified forms of desert landscapes. This became evident from our trapping records from four major habitats: sandy, gravel plains, rocky and ruderal, at 12 representative localities within the Indian desert (Prakash *et al.*, 1971). Additional trapping on a prolonged basis was also carried out in a sandy plain at Jodhpur (Prakash & Rana, 1970), over gravel slopes (Prakash & Rana, 1972) and in sand dunes in a 100 mm rainfall zone (Prakash & Rana, 1973). These studies revealed that a few species occur exclusively in a particular habitat such as *Gerbillus n. indus* in the sandy habitat, *Rattus c. cutchicus* and *Mus cervicolor phillipsi* in the rocky habitat and *Rattus rattus*, *Mus musculus* and *Mus booduga* in the ruderal habitat. Other rodents inhabit more than one habitat type but based on the frequency of their occurrence in large numbers in a certain habitat, these rodents may be assigned a particular niche (Table 2). For example, 60 per cent of *Meriones hurrianae* were collected from the sandy habitat, 17 per cent from gravel plains and the

81

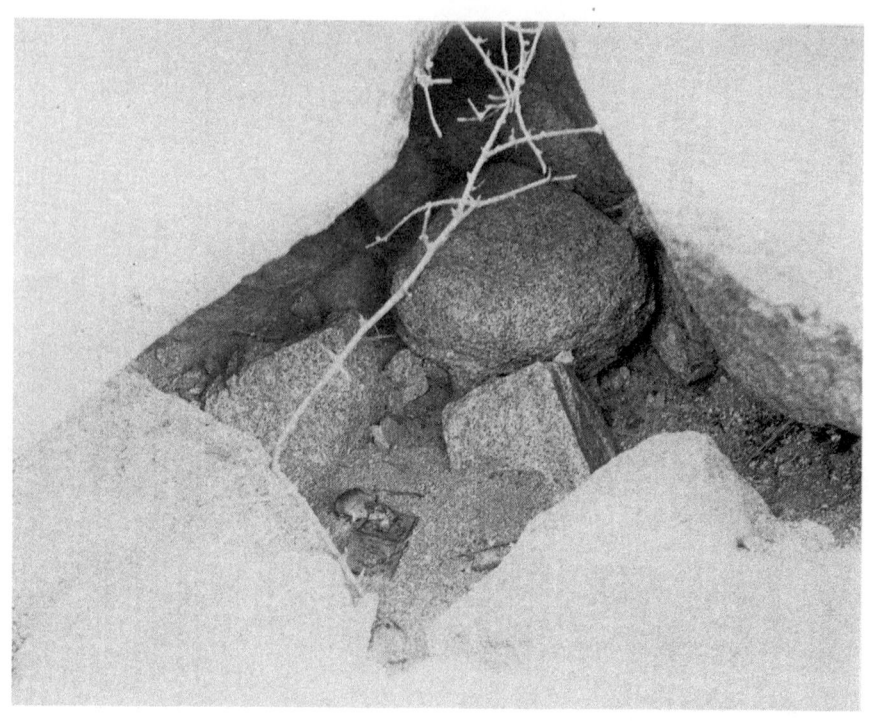

Fig. 8. Typical habitat of *Rattus cutchicus.*

Table 2. Habitat preference exhibited by rodents in the Indian desert.

| Sandy | | Rocky | Ruderal | |
Sand dunes	Sandy plains		Residential buildings/huts	Surrounding crop fields
Gerbillus gleadowi	*Gerbillus nanus indus*	*Hystrix i. indica*	*Funambulus pennanti*	*Tatera i. india*
	Meriones hurrianae	*Rattus c. cutchicus*	*Rattus rattus rufescens*	*Rattus meltada pallidior*
	Rattus gleadowi	*Mus cervicolor phillipsi*	*Mus musculus bactrianus*	*Mus booduga*
		Mus platythrix sadhu		*Golunda ellioti gujerati*
				Nesokia i. indica
				Bandicota bengalensis

Fig. 9. The most abundant rodent of the Rajasthan desert, *Meriones hurrianae.*

rest from the ruderal habitat. *M. hurrianae* is considered as an animal of the sandy habitat in view of its preference for this habitat.

The habitat preferences of the rodents appear to be related to their modes of living. The squirrel, *F. pennanti* is arboreal and constructs nests of cloth rags, leaves, human hair etc. There are not many trees in the desert far away from human habitations. It has, therefore, taken to living near the village complexes, and, to some extent, on hillocks where trees are found in some numbers. The hairy-footed Gerbil, *Gerbillus gleadowi* prefers sand dunes and it is fairly abundant in the 100 mm rainfall zone of the desert, probably due to its higher tolerance of salt in its food, and its ability to live in shallow, simple burrows from which it can find its way out even when these are buried under the blowing sand. The merion Gerbil, *M. hurrianae* and *Tatera indica* are not found over loose, shifting sand dunes but prefer sandy plains and inter-dunal regions

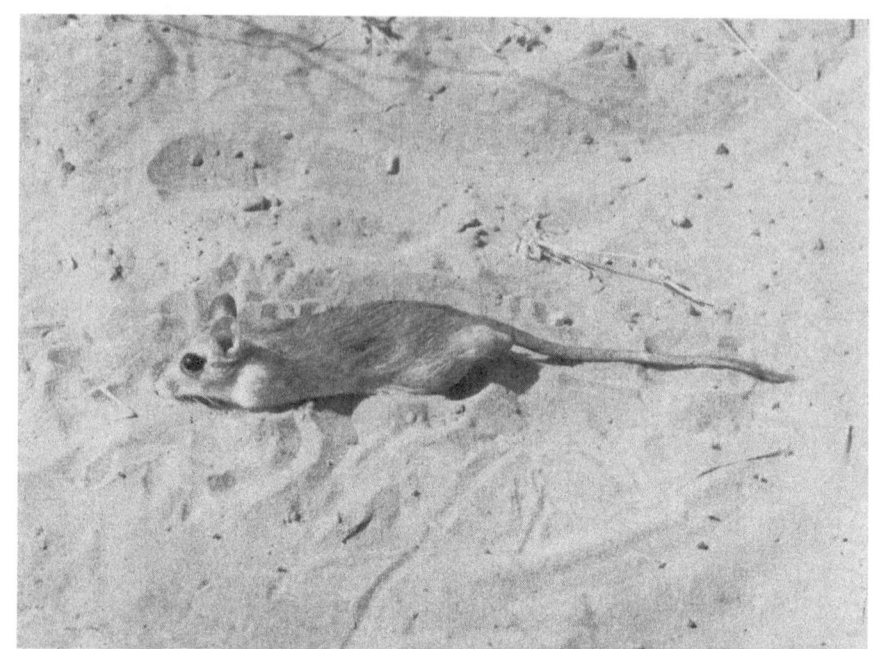

Fig. 10. The Indian Gerbil, *Tatera indica.*

Fig. 11. Burrow openings of *M. hurrianae* and *T. indica* in the vicinity of crop fields.

84

Table 3. Rodent associations in various habitats of the Indian desert.

Habitat types	Rodent associations[1]	Source
Sand dunes	*Gerbillus gleadowi-Meriones hurrianae*	PRAKASH & RANA, 1973
Sandy plain[2]	*Tatera i. indica-Meriones hurrianae-Gerbillus n. indus*	PRAKASH & RANA, 1970
Sandy plain[3]	*Meriones hurrianae-Tatera i. indica-G. gleadowi*	PRAKASH et al. 1971
Rocky[4]	*Rattus c. cutchicus-Mus cervicolor phillipsi*	PRAKASH & RANA, 1972
Rocky[3]	*Rattus c. cutchicus-Mus platythrix sadhu-M. c. phillipsi*	PRAKASH et al. 1971
Piedmont zone	*Rattus meltada pallidior-Mus p. sadhu-Golunda e. gujerati*	PRAKASH & RANA, 1972
Gravelly plain[3]	*Meriones hurrianae-Tatera indica*	PRAKASH et al. 1971
Ruderal[2]	*Tatera i. indica-Rattus m. pallidior-M. hurrianae*	PRAKASH et al. 1971
Crop fields		
unirrigated	*Meriones hurrianae-Tatera indica*	PRAKASH, 1973
Irrigated[5]	*Tatera indica-M. hurrianae*	PRAKASH, 1973
Irrigated[6]	*Rattus m. pallidior-Mus* spp.[7]	PRAKASH, 1973
Orchards	*Tatera indica-Funambulus pennanti*	PRAKASH, 1973

[1] The first named species is the most prolific, the second less so and so on.
[2] Results of a sustained study at Jodhpur in protected grasslands with comparatively dense vegetation cover.
[3] Results of 72 hours trapping, each at 12 localities in various habitats.
[4] Results of a sustained study at Erinpura and Jalore.
[5] Irrigated by water from dug wells, soil not very clayey.
[6] Irrigated by canal water, soil very clayey due to excessive water and gradual alluvial deposits, mostly in northern Rajasthan, Punjab and Haryana.
[7] *Mus b. booduga* and *Mus musculus bactrianus.*

where they can dig extensive and deep burrow systems. *Tatera indica* is also not found in the very low rainfall zone of the Indian desert probably due to its higher water requirements as compared to *M. hurrianae.* *Tatera indica* has a marked preference for the ruderal habitat and crop fields where the soil moisture regime is much superior compared to the bare sandy plains. This may assist them in thermo-regulation as well. The non-desert elements like *Rattus meltada pallidior* and *Golunda ellioti* inhabit either crop fields or densely vegetated plains. The latter species stays in thickets of bushes. *Nesokia indica* and *Bandicota bengalensis* occur only in irrigated crop lands where the sub-soil moisture is very high throughout the year. On the other hand, merion gerbils have been driven away to more arid parts due to the incoming of irrigation channels (PRAKASH, 1958; TABER et al., 1967), creating higher humidity conditions within

85

their burrow systems and replacing the sand cover with very compact clayey soils. PRAKASH *et al.* (1971b) have found a relationship between the number of desert gerbils and the clay per cent of the soil in any particular region. The gerbils shun clayey soils presumably as such a habitat would require arduous digging of their complex burrow systems. It may also be speculated that such habitat preferences may serve another important function viz. of minimising interactions between species. In the hilly terrain, *R. cutchicus* is only found in crevices, whereas *Mus c. phillipsi* was invariably collected in association of the shrub, *Euphorbia caducifolia*. The crevice dwelling rat rarely comes in contact with the smaller mice. *Mus platythrix sadhu* prefers gravelly patches in the rocky and sandy habitats as it is in the habit of plugging its burrow openings with small pebbles and also arranges them in a circular fashion around the burrow openings (PRAKASH & RANA, 1972). Due to its habit of doing so it has selected the rocky habitat where such pebbles are available and not purely sandy plains which are devoid of pebbles.

SPECIES COMPOSITION

As a consequence of habitat specificity, the species composition of rodents varies from one habitat type to another within the Indian desert. Very often the same species may be found in two distinct habitats, but their relative abundance will not be the same in the two locations. Table 3 sums up the abundant species as found in the various sub-habitats within this desert.

RELATIVE ABUNDANCE

In the sandy habitat, *Meriones hurrianae* was found to be the most common rodent (Fig. 12). *Tatera indica* and *Gerbillus gleadowi* were found to be the next in the order of abundance in this habitat (PRAKASH *et al.*, 1971). In the gravel plains *Meriones hurrianae* and *Tatera i. indica* were present in relatively high numbers while the other species were found in comparatively low numbers (Fig. 13). On the rocky habitats, 66.6 per cent of the rodent population has been found to be composed of *Rattus c. cutchicus* *i.e.* it was found in appreciable numbers (Fig. 14). In the ruderal area, where trapping was done on the outskirts of the desert villages, *Tatera i. indica* and *Rattus meltada pallidior* were found to be most common rodent species (Fig. 15). *Meriones hurrianae* and *Gerbillus gleadowi* were also found in this habitat in fair numbers but they were usually present when the village was situated on the top of a sand dune. Inside the houses, which are inhabited by two species only, *Rattus rattus rufescens* was found to be somewhat more common than *Mus musculus bactrianus*.

Disregarding habitat diversity and considering the overall picture of all the rodent species inhabiting the Indian desert biome, it may be seen

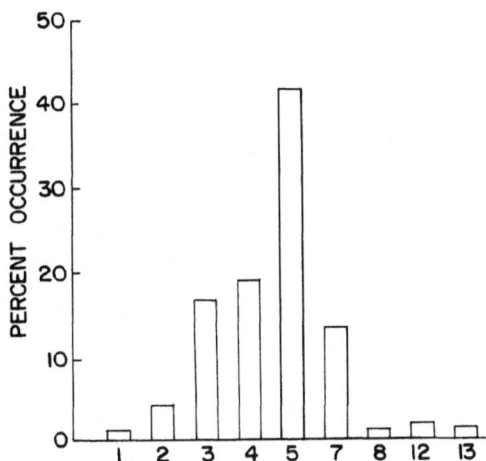

SANDY (throughout the desert)

Fig. 12. Frequency distribution of various rodent species in the sandy habitat. 1. *Funambulus pennanti*, 2. *Gerbillus nanus indus*, 3. *Gerbillus gleadowi*, 4. *Tatera indica indica*, 5. *Meriones hurrianae*, 7. *Rattus meltada pallidior*, 8. *Rattus gleadowi*, 12. *Mus platythrix sadhu*, 13. *Golunda ellioti gujerati*.

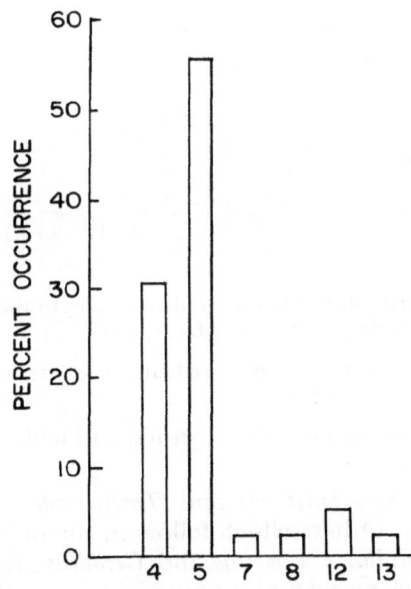

Fig. 13. Frequency distribution of various rodent species in the gravel plains.

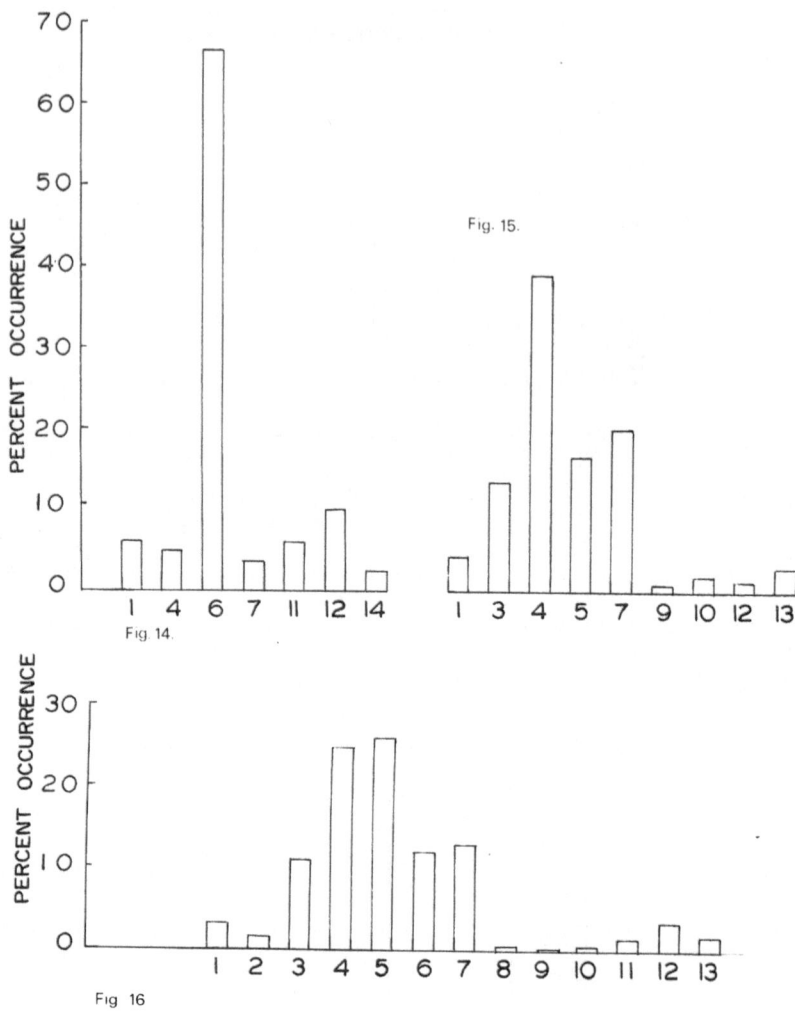

Fig. 14. Frequency distribution of various rodent species in the rocky habitat. 6. *Rattus c. cutchicus,* 11. *Mus cervicolor phillipsi,* 14. *M. c.* sp.

Fig. 15. Frequency distribution of various rodent species in the ruderal habitat. 9. *Mus musculus bactrianus,* 10. *Mus b. booduga.*

Fig. 16. Frequency distribution of various rodent species in the Rajasthan desert biome.

(Fig. 16) that *Meriones hurrianae* and *Tatera indica indica* are the most abundant ones here. Others which follow in the order of abundance are *Rattus m. pallidior, Rattus c. cutchicus* and *Gerbillus gleadowi, Mus platythrix sadhu* and *Funambulus pennanti* also occur in fair numbers throughout the desert (PRAKASH *et al.,* 1971).

Table 4. Association between predominant rodent species and certain grass and crop types in the Indian desert*

Grassland/Crop types	Predominant rodent species
Grassland types	
Sehima nervosum-Aristida spp. type	*Rattus cutchicus, Mus cervicolor phillipsi*
Eleusine compressa type	*Rattus meltada pallidior, Rattus gleadowi, Mus platythrix sadhu.*
Cenchrus spp.-*Lasiurus sindicus* type	*Gerbillus gleadowi, G. nanus indus, Meriones hurrianae*
Dichanthium type	*Tatera indica, Rattus m. pallidior, Golunda ellioti.*
Cenchrus spp.-*Aristida* spp. type	*Meriones hurrianae, Tatera indica, Gerbillus gleadowi.*
Panicum turgidum type	*Gerbillus gleadowi*
Sporobolus-Dichanthium type	*Meriones hurrianae, Tatera indica.*
Crops	
Millet-Sorghum-Sesame	*Meriones hurrianae, Tatera indica, Rattus meltada pallidior*
Wheat-Barley-Gram	*Tatera indica, R. m. pallidior, Nesokia indica, Bandicota bengalensis, Mus booduga.*
Sugarcane	*Nesokia indica, Mus booduga, Mus musculus bactrianus*

* Modified from PRAKASH *et al.* (1971).

RODENT POPULATIONS IN RELATION TO VEGETATION COMPOSITION

The rodents of the Indian desert are fairly versatile in respect of associating with various vegetation communities and it appears that none of the rodent species is associated with a particular vegetation type or a plant species, except for *Mus cervicolor phillipsi* which was invariably collected in association of the shrub, *Euphorbia caducifolia* on the rocky habitat. An attempt has, however, been made to group dominant rodent species in accordance with their association with certain grass and crop types as observed in the Indian desert (Table 4).

No correlation between basal cover of the vegetation in the sandy and rocky habitats and the trap indices in respect of various rodent species could also be established (PRAKASH *et al.*, 1971). However, in an earlier study (PRAKASH *et al.*, 1971b), an inverse relationship between the basal cover of the vegetation and the number of desert gerbil, *Meriones hurrianae* was found to exist. This was attributed to the fact that grasses have anastomosing, fibrous roots and, as such, burrowing in grassy tracts is very arduous for the rodent. On the contrary, the desert gerbil is usually found burrowing in hummocks around the shrubs, *Capparis decidua, Calligonum polygonoides, Zizyphus nummularia,* etc. The roots of these shrubs are not fibrous and penetrate deep into the soil to tap the sub-soil water and hence do not hinder the subterranean digging activity of rodents. *Gerbillus gleadowi* populations, when inhabiting stabilised sandy plains,

Table 5. Rodent biomass and the vegetation cover during winter (after PRAKASH *et al.*, 1971)

Administrative districts in the Rajasthan desert	Biomass (g) per hectare			Percent basal cover of vegetation	
	Habitats			Habitats	
	Sandy	Rocky	Ruderal	Sandy	Rocky
Barmer	811	955	1266	0.1	0.1
Jaisalmer	271	222	610	6.9	4.7
Bikaner	460	—	490	3.2	—
Jhunjhunu	2580	933	2930	1.0	—
Churu	2160	—	3122	1.3	—
Jodhpur	182	805	1652	1.0	1.2
Nagaur	2555	—	333	9.0	—
Pali	1100	1142	1423	6.2	1.5
Jalore	233	1093	1644	0.8	0.7
Sirohi	1350	433	2522	7.3	1.1
	Cotton wheat field	Sugarcane field	Date palm field		
Shri Ganganagar	1111	2600	2711	—	—

also prefer to stay in the hummocks of these bushes and that of *Calotropis procera*.

We also found that the number of merion gerbils was higher in plots of natural pastures with high densities of *Dactyloctenium sindicum, Aristida adscensionis, Lasiurus sindicus, Perotis hordeiformis* and *Digitaria marginata* and very low in plots in which the frequency of occurrence of *Cenchrus biflorus* was high. The latter grass which has a high seeding rate, bears hard awns over the seeds, which stick to the bodies of the desert gerbils. These awns are difficult to remove and thus this grass is somewhat repulsive to the desert gerbil (PRAKASH *et al.*, 1971b).

RODENT BIOMASS AND THE VEGETATION COVER

It is normally expected that the rodent biomass should be higher in localities supporting a dense vegetation cover where the availability of food and shelter is abundant than in the habitats with poor vegetation cover. During an intensive ecological reconnaissance survey of rodent populations in the Indian desert (PRAKASH *et al.*, 1971) the situation did not quite fit in with the above conjecture. In the sandy habitat at Jaisalmer (Table 5), though the basal cover is good (6.9 per cent), the rodent biomass is very low viz. 271 g/ha, whereas at Churu and Jhunjhunu, where the vegetation cover is 1.26 and 1.00 per cent respectively, the rodent biomass is fairly high, 2160 and 2580 g per ha. Similarly, in the rocky habitat, at Jaisalmer, the vegetation cover is 4.7 per cent but the rodent biomass is only 222 g/ha while at Jalore it is 0.7 per cent and the rodent biomass is 1093 g/ha. The collection data from Jodhpur, Jalor, Nagaur and Pali, however, indicate a dependence of rodent biomass on the vegetation cover. In the Shri Ganganagar district (Table 5) where dense green crops were standing at the time of the survey, the rodent biomass was not particularly high. The data also point out that the carrying capacity of the land does not always determine the rodent biomass which could be sustained by it. It would, therefore, appear that rodent biomass may not be directly dependent on the density of vegetation. It may be that physiologic adaptations of rodents to withstand the various stresses of the xeric environment, the availability of shelter, and other edaphic and micro-climatic conditions play important roles in the maintenance of rodent biomass in the desert.

FLUCTUATIONS IN NUMBERS

We have undertaken studies on the monthly or seasonal variations in numbers of the predominant rodent species found in the Indian desert. Unfortunately, however, we have not been able to study their fluctuations from year to year on a long-term basis. Available information on eight rodent species is presented here.

Funambulus pennanti: The population of this squirrel was studied by capture-recapture method in a part of a garden at Jodhpur, during 1963 and 1964. The population trend during both the years was found to be almost identical. The capture data for 1963 (Table 6) indicated a gradual decrease in their number from April to October.

Tatera indica indica: The Indian gerbils were live trapped in Sherman traps on two consecutive nights every month for 12 months and on every occasion about 50 traps were used. The collections were made around households in Bikaner town. The trapping data as per this schedule indicate only a broad pattern of fluctuations during the year. Two minor peaks are observed, the first one during March-April, and the other during August–September. The two peaks coincide with the increase in the number of subadults in the samples as is clearly indicated from the data collected by JAIN (1970). This correlation suggests that the build up in the population is mainly brought about by the breeding activity of the rodents as these peaks in their numbers fall soon after their maximum reproductive activity which occur during February and in July–August (JAIN, 1970).

Meriones hurrianae: The method of censussing the desert gerbils was an indirect one. It was observed that soon after venturing out of their burrows in the morning, the desert gerbils feed continuously for 30–45 minutes when they do not indulge in burrow digging. The census technique was based on this habit of the gerbils. The technique may be described in a nutshell as follows: all gerbil burrow openings in each plot

91

Table 6. Observed monthly fluctuations in the number of various desert rodent species

Species	Fluctuations in numbers												Source of data
	Jan.	Feb.	Mar.	Apr.	May	June	July	Aug.	Sept.	Oct.	Nov.	Dec.	
F. pemanti	—	—	—	102	54	61	36	47	28	30	10	6	PRAKASH & KAMETKAR, 1969
T.i. indica	30	39	51	53	63	69	55	74	78	42	40	43	JAIN, 1970
M. hurrianae	—	—	300	—	—	131	—	170	—	—	—	385	PRAKASH et al. 1971.
R.c. cutchicus	—	—	49	—	26	—	47	15	—	32	—	23	PRAKASH & RANA, 1972.
R.m. pallidior	—	—	12	—	10	—	12	0	—	10	—	—	PRAKASH & RANA, 1972.
M.c. phillipsi	—	—	15	—	11	—	21	0	—	2	—	15	PRAKASH & RANA, 1972.
M.p. sadhu	—	—	21	—	9	—	11	1	—	1	—	—	PRAKASH & RANA, 1972.
G.e. gujerati	—	—	4	—	4	—	5	1	—	1	—	—	PRAKASH & RANA, 1972.

measuring 95×95 m at six localities in three bio-climatic zones of the Indian desert were counted and plugged with soil in the evening after the cessation of all surface activities. This was followed by counting of the freshly opened burrows in the next morning (PRAKASH *et al.*, 1971b). The pooled mean counts made at the six localities in respect of the four major seasons of the desert (Table 6) indicate a population build up during winter which continues till spring and then their number declines during summer. It appears that the population explosion during winter is due to the higher rate of breeding of merion gerbils during and after monsoon (PRAKASH, 1964) which is influenced by the availability of green food at that time. The data also indicate that there is no correlation between gerbil population and the density of grass and other vegetation cover in the native grasslands.

Rattus c. cutchicus: The Cutch Rock-rat, and the further ones, were all collected by trapping at two-monthly intervals in the rocky habitat and piedmont zones at Erinpura, situated in the south-eastern sector of the Rajasthan desert. Two trap lines, each having 30 snap traps each fixed at 10 m intervals, were fixed in each locality. The data presented (Table 6) pertain to 72 hours of trapping.

The number of Rock-rats trapped do not show a wide fluctuation during the year but relatively more rats were collected in the months of March, July and October (Table 6). These periods coincide with the peaks in their reproductive activity (PRAKASH *et al.*, 1973) and also with the 'comfortable' climatic conditions prevailing in the desert during a year.

Rattus meltada pallidior: Although population data for the whole year are not available, an almost similar number of these rodents were captured in March, May, July and October.

Mus cervicolor phillipsi: The data for six months (Table 6) indicate a very low density of population during the monsoon and post-monsoon seasons. It is quite possible that due to the availability of natural green food, during these periods, the mice are not attracted towards the baits in the traps or possibly again, many of them may be perished due to rains. They are mostly found on the hill slopes under *Euphorbia caducifolia* shrubs and only occasionally they would occupy a burrow. The mice were, however, collected in larger numbers during the month of March and during the period from July to December. Pregnant females were collected only during July and December, which appear to be their major breeding periods.

Mus platythrix sadhu: Like *Mus c. phillipsi* the density of this spiny mouse was also observed to be very low during the monsoon and post-monsoon seasons and highest numbers were collected in March (Table 6).

Golunda ellioti gujerati: These Bush rats were collected in almost similar numbers in the earlier half of the year but during the later period of the year, very few rodents were captured in the trap lines.

Table 7. Ranges of movements of some Indian desert rodents.

Rodent species	Mean home range ± S.E.	Mean maximum distance travelled	Source of data
Funambulus pennanti	♂ 0.21 ± 0.073 ha ♀ 0.15 ± 0.034 ha	65.61 ± 4.80 m 46.87 ± 5.40 m	PRAKASH *et al.*, 1968
Tatera i. indica	♂ 1875 m² ♀ 1912.5 m²	36.87 ± 12.2 m 60.38 ± 9.35 m	PRAKASH & RANA, 1970
Meriones hurrianae	♂ 88.7 ± 44.3 m² ♀ 154.7 ± 24.6 m²	16.03 ± 0.98 m 18.46 ± 1.5 m	FITZWATER & PRAKASH, 1969
Rattus meltada pallidior	1217 m²	31.0 m 53.9 m	MANN, 1969 SAGAR, 1972
Mus musculus bactrianus	675 ± 390 m²	25.0 m 39.0 m	MANN, 1969 SAGAR, 1972
Mus b. booduga	1275 ± 52 m²	31.0 m 62.1 m	MANN, 1969 SAGAR, 1972
Mus platythrix	—	22.5 m	SAGAR, 1972
Golunda ellioti	—	22.5 m	SAGAR, 1972
Bandicota bengalensis	945 ± 515.8 m²	—	MANN, 1969

Table 8. Range length in relation to sex and maturity in squirrels.

Range length categories	Per cent Adult	Male Subadult	Per cent Adult	Female Subadult
1–50 m	45.1	83.3	71.4	100
50.1–100 m	37.1	16.6	25.0	0
100.1–150 m	17.7	0.0	3.5	0

The data presented above for eight species of rodents were collected by different methods and hence it is not worthwhile to compare the monthly fluctuation trends in their numbers. It is, however, observed that, in general, the number of most of the species is lowest during summer and shows peaks during the spring and soon after the monsoon. These two periods coincide with that of maximum reproductive activity among these desert rodents.

Ranges of Movements

Funambulus pennanti: The study was carried out in a part of a large garden and we found that the home ranges (Table 7) of animals of the two sexes do not differ significantly from each other. In fact, the home ranges of the male and female animals showed considerable overlapping and no single range was found to be occupied solely by an individual. Out of about 98 squirrels studied, only one had shifted its sub-adult range to a distance of 128.3 m when it attained sexual maturity. The observed range length data suggest that the adult male squirrels wander more ($P < 0.05$) than both the adult females and the sub-adults of the two sexes. This fact is further substantiated by the data presented in Table 8.
Tatera i. indica: The ranges of the Indian gerbil were found to be overlapping in every case. No significant difference between the home ranges of the male and female gerbils was observed. The observed range lengths of male and female, adult and subadult, animals also did not differ significantly from each other (PRAKASH & RANA, 1970). However, a larger number of female *T. i. indica* moved longer distance as compared to the males of the species. It will be apparent from Table 9 that about 50 per cent male Indian gerbils move to a maximum distance of 30 m whereas more than 50 per cent females move from 46 to 105 m.
Meriones hurrianae: Like the Indian gerbil, this one, too, does not seem to defend its entire range and almost all the ranges of both male and female desert gerbils were found to be overlapping (FITZWATER & PRAKASH, 1969). The ranges remained similar during March–April when no mating was observed. During June–July, one of the males in the natural observation paddock, extended its range considerably, for

Table 9. Percentage distribution of male and female *Tatera i. indica* in relation to various range lengths.

Range length categories (m)	Percent male	Percent female
0–15	25.00	17.64
16–30	33.33	11.64
31–45	8.33	17.64
46–60	8.33	11.64
61–75	8.33	11.64
76–90	8.33	0.0
91–105	8.33	29.41

the purpose of mating with a number of females. No significant difference was found either in the home range area or in the observed range lengths between the sexes.

The home ranges and observed range lengths of six more species have been presented in the Table 7. Most of the data have been derived from observations made on the rodents found in the crop fields in the Punjab (SAGAR, 1972; MANN, 1969).

Food

Out of the various rodent species inhabiting the Indian desert, we have detailed information about the food of only a few of them.

Funambulus pennanti: This commensal rodent depends on the kitchen refuse in the towns but the squirrel populations living away from human habitations feed upon seeds and fruits of trees. They particularly prefer grapes, guava, berries, pomegranate and vegetable crops. Squirrels also consume insects, particularly locust. KRISHNASWAMI & CHOUHAN (1957) reported that *F. pennanti* is quite injurious to lac plantations.

Tatera i. indica: The stomach contents of freshly captured Indian gerbils were examined every month during a year (PRAKASH, 1962). The contents were sorted out into: seeds, stems and rhizomes, leaves and flowers, and insects. The stomach contents of all the gerbils collected during a month were sorted into the above categories and the volume of each food item comprising the mixture of contents, was measured (Fig. 17). The food items, stems and rhizomes and leaves and flowers, do not show any marked monthly fluctuation in their occurrence in the gerbil stomach. These fluctuations vary between 15 to 30 per cent of the total food taken. The seed item shows the greatest fluctuation, its consumption being maximum during winter and minimum during monsoon. This decline in seed consumption is, however, made good of by the increased intake of insects, the volume of which increases from 10 per

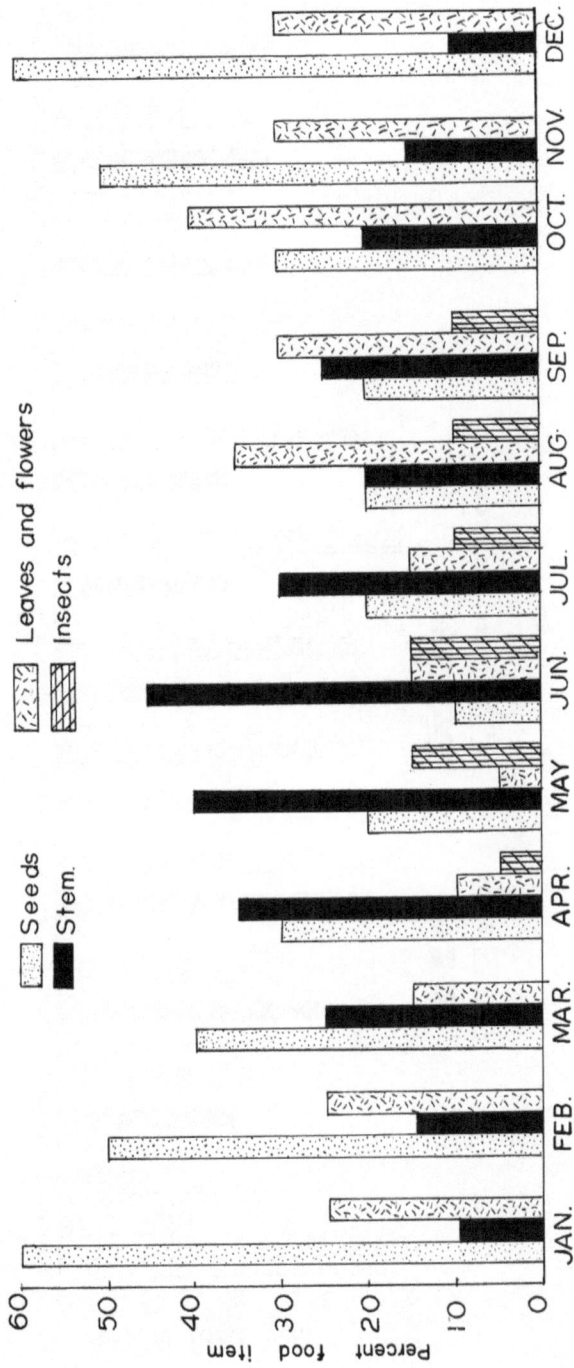

Fig. 17. Monthly distribution of different food items in the stomach contents of *Tatera indica.*

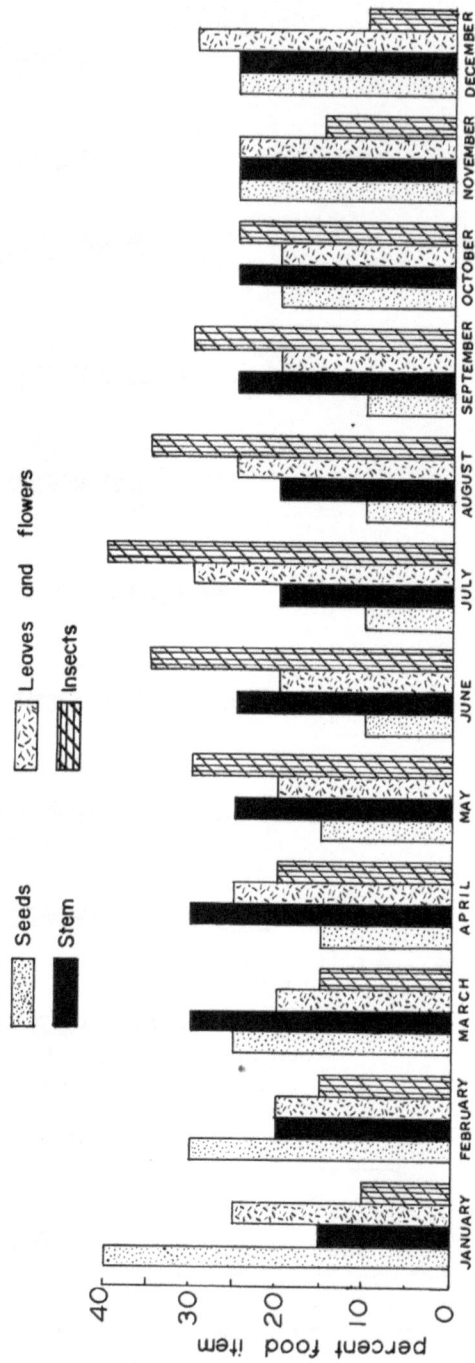

Fig. 18. Monthly distribution of different food items in the stomach contents of *Meriones hurrianae*.

cent in January to 40 per cent in July and, thereafter, steadily declines to low levels around 10 per cent during December. These fluctuations in proportions of various food items of the Indian gerbil are apparently influenced by their availability in the nature. Insects are fed upon in larger quantities during and after the monsoon when there is an abundance of them. Almost all seeds sprout after the showers when these are not available and consequently, their consumption declines. Leaves and flowers have been found to be consumed in larger quantities during the season presumably due to their profuse availability during this period.

In the desert, *Tatera indica* is dreaded by the farmers. It inhabits, sometimes in very large numbers, the crop fields, both rain-fed and irrigated. It inflicts severe losses to the standing crops, by feeding upon the sown seeds, sprouting saplings, cobs and ears, and sometimes by felling the whole plant for reaching the seeds in milk (PRAKASH, 1973). *Meriones hurrianae:* The desert gerbil shows spectacular fluctuations in the occurrence of various food items in its stomach contents over a year. The volumetric analysis of their stomach contents (Fig. 18) indicates that the desert gerbil chiefly thrives on seeds during winter, on stems and rhizomes during summer and on leaves and flowers during monsoon and post-monsoon seasons. The desert gerbil, curiously, starts feeding on insects during summer. The physiological implications of the inclusion of insects in its dietary during summer are explained by Dr. P. K. GHOSH in his chapter in this book. The desert gerbil shows a great versatility in its food habit, moulding it according to the availability of different items during various seasons of the year (PRAKASH, 1959; 1962).

In nature, during the monsoon, when green vegetation comprising, principally, the grasses, is available to the desert gerbil in abundance, it shows a fair amount of selectivity and prefers the grass blades and inflorescence of *Cenchrus ciliaris* (PRAKASH, 1969). I studied its food preferences in the field during the monsoon. Plant remains were identified and the frequencies of occurrence of unconsumed plant species lying near their burrows were compared with the frequencies of occurrence of various species in four plant communities. The data revealed that so far as the gerbil is concerned the palatability index of the grasses, which are the chief fodder for the livestock, is the highest. It has been estimated that in a moderately gerbil infested natural grassland, the gerbil can consume over a year the entire production of edible fodder species, leaving almost nothing for the livestock (PRAKASH, 1969). Thus, these rodents are serious competitors of sheep and other livestock as far as fodder is concerned. The desert gerbils also devastate large quantities of sown seeds of various rain-fed crops and their sprouted saplings. They have a special preference for sorghum and millet inflorescence, particularly when these are in 'milk'. They colonise the areas where the ears and cobs of harvested crops are hoarded for threshing. The desert gerbils are also very partial to groundnut crops.

Fig. 19. The seeds of wild, *Citrullus colocynthis* are scooped from the fruits by desert gerbils.

The gerbil population being high in this desert and their demand for food being insatiable, these rodents maintain an appreciable pressure on the desert grasslands in particular, and on the plant communities in general. The rodents thus become an important component of the desert ecosystem. They comprise perhaps the largest group of primary consumers in the xeric environment and at times become secondary consumers also.

The rodent populations influence the critical ecological balance in the desert by their changing modes of feeding in keeping with the seasonal availability pattern of various food items in nature. Due to their very large numbers the rodents play havoc with the vegetation wherever they are. At Bikaner, gerbils devastated a 16 ha rain-fed range of *Lasiurus sindicus* within a month (PRAKASH, 1973). In a range-land at Maulasar, in north-eastern Rajasthan, the total food requirement of the resident rodent population during monsoon was estimated to be 1044 kg/ha (PRAKASH, 1969). The annual production of edible (for sheep) species in this fenced patch of land was estimated to be, on an average, 865 kg/ha and the total forage production to be of the order of 1100 kg/ha. Grazing pressure was minimal here. This shows that the desert maintains a much higher load of rodent population than its carrying capacity would allow. The above estimate of feeding does not include the wastage of green vegetation that the rodents indulge in, which would be about eight times the amount required for feeding. The severity of the desert rodent problem may thus be easily appreciated.

In the highly rodent-infested areas, almost no perennial grasses exist. These have been replaced by less productive annuals, like *Aristida* spp., *Tephrosia purpurea*, *Cyperus arenarius*, *C. rotundus*, and *Cenchrus biflorus*. These are the characteristic vegetation of the degenerated ranges. On the rocky habitats also the palatable grasses have been devoured by dense populations of *Rattus cutchicus* and *Mus cervicolor phillipsi*. What we see for most part of the year is the association of *Euphorbia caducifolia*, *Barleria acanthoides*, *Lepidagathis trinervis*, *Fagonia cretica* etc. In both the habitats, the vegetation composition has changed due to rodent depradations, furthered by over-grazing, from highly-productive and nutritive perennial grasses to unpalatable, thorny, and more or less useless weed species. Undoubtedly, over-grazing of livestock has exerted tremendous pressure on the vegetation of the Indian desert and attempts are being made now to separate the grazing factor quantitatively, so that an estimate of the rodents' devastating role in the desert ecosystem may be obtained. Our present rough estimates indicate that rodents are causing twice as much damage to vegetation as is being caused due to overgrazing in the Indian desert.

In addition to being consumers and destroyers of vegetation, the rodents are also serious inhibitors of the process of plant generation. Due to their seedivorous nature, they do not leave enough seeds of the palatable grasses and other vegetation to sprout again. In the hilly terrains within the desert region, the densities of *Rattus cutchicus* and *Mus c. phillipsi* are very high (PRAKASH & RANA, 1972). They mostly feed on the flowers, seeds and fruits of grass and tree species, leaving very few to regenerate.

Due to the rodents' preferences for the more nutritive and highly palatable grasses and other edible plants, the nature of the original plant communities in this desert has totally changed to non-productive, degenerated vegetation type. Admittedly, the process has been greatly helped by man through grazing of his animals.

Predator-Prey Relationship

Predators play an important role in regulating the population numbers of small mammals, especially rodents. The impact of a predator on its prey is, however, directly proportional to the number and breeding intensity of the former as the latter is usually in abundance, spread over a variety of habitats and is a prolific breeder. In the Indian desert rodent predation is chiefly effected by reptiles, birds and mammals (Fig. 20). *Reptiles:* Amongst the desert lizards, only two species are capable of capturing rodents. The monitor lizards, *Varanus bengalensis* and *V. griseus koniecznyi* are large animals powerful enough to tackle a rodent. Both are, however, diurnal and, as such, are left with only a few diurnal rodents,

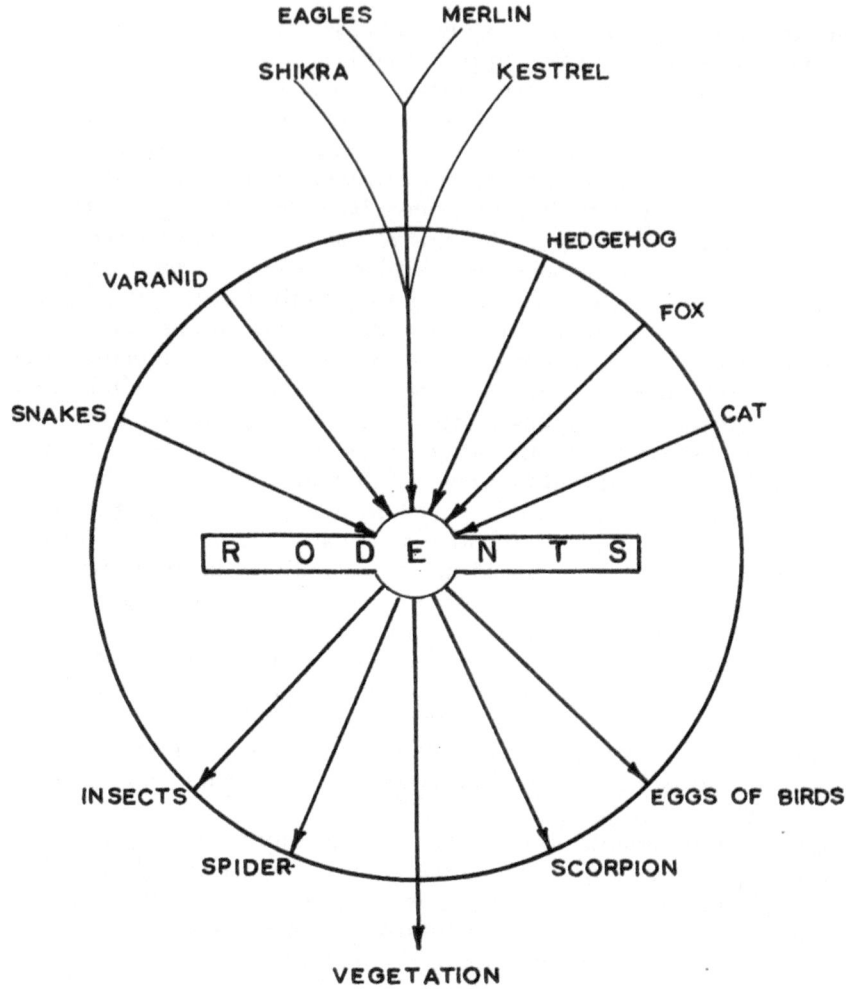

Fig. 20. Predators and foods of Rajasthan desert rodents.

viz. *Funambulus pennanti, Meriones hurrianae* and the crepascular *Golunda ellioti.* All other species of rodents are nocturnal. MINTON (1966) found the squirrel, *F. pennanti* in the stomach of *V. bengalensis* but the monitors were also feeding on shrew, snake, bird, lizard, fish, beetle, crab, crayfish and locust. It appears that these lizards have a varied dietary and are not solely dependent on rodents.

I have earlier listed a number of snake species which feed upon rodents (PRAKASH, 1962). However, MINTON (1966) inferred from his studies on the stomach contents of a number of snake species from the desert region of Pakistan that snakes prefer to feed upon lizards and inverte-brates. Rodents were found in stomach of only a few species (cf. *Sphale-rosphis arenarius, Echis carinatus*). It is, however commonly believed in this part of the desert that the main food of snakes is the rodent.

Birds: In the Indian desert, the predatory birds are poorly represented. The shikra, *Accipiter badius* and the Tawny eagle, *Aquila rapax*, are, prob-ably, the most common ones. Merlin, *Falco chicquera;* and Kestrel, *Falco tinnunculus*, are only occasionally seen. Not only that there are only a few species of predatory birds, their numbers are also very low in the desert region. In the extreme western extremity of the arid region, which is a tree-less zone, the raptores are still rarer, probably due to the paucity of roosting sites in that locale.

Mammals: The long-eared hedgehog, *Hemiechinus auritus collaris*, some-times feed upon rodent youngs or on ailing adults (KRISHNA & PRAKASH, 1960). The other mammals which are predatory on the rodents are the carnivores. The Asiatic jackal, *Canis aureus aureus* occasionally feeds upon gerbils but it prefers hares and other larger mammals. The two foxes, *Vulpes v. pusilla*, and *Vulpes bengalensis* do feed on rodents but most of the stomachs of these animals which were examined were full with scorpions, centipede, beetles, leguminous seeds and berries of *Z. nummularia* suggesting that the foxes tend to feed on readily obtainable food. The desert cat, *Felis libyca ornata* and the Jungle cat, *F. chaus prateri* mostly feed on beetles which are readily available above ground but feed upon rodents also (PRAKASH, 1959).

From the foregoing account it is clearly indicated that most of the predators of rodents feed upon a variety of food. Contrary to the common belief, probably originating with observations made on captive snakes provided rats and mice as food, that snakes feed on rodents; it has, how-ever, been clearly evidenced by the examination of stomach contents of snakes (MINTON, 1966) that in nature, lizards, frogs and insects, and not rodents, are their main dietary items. It would, therefore, be in-teresting to study the role played by snakes in regulating the field popula-tions of rodents.

Breeding Season and Litter Size

Indian Crested Porcupine, *Hystrix indica* Kerr: Very little is known about the breeding season of the porcupine in nature. PRATER (1965) observed that it bears young in the month of March. In West Pakistan, TABER *et al.* (1967) had dug out a pair accompanied by two small young in April, our information is restricted to the animals in the zoo. In the Jodhpur and Bikaner zoos the porcupines have been found to litter from March to December. It appears that peak littering occurs during the monsoon followed by another peak during December.

The litter size in the zoos in the desert regions varied from 1 to 3, the average being 1.45 (PRAKASH, 1974). A female porcupine had littered on June 3, 1963 and again on September 20, 1963, the interval between the two deliveries being of 109 days. Although it may not be proper to call this period as the period of gestation in the absence of any specific information on super-foetation and post-parturition oestrous in the species, yet interestingly, it is near about the reported gestation of 112 days period for the African species of porcupine (ZUCKERMAN, 1953).

Northern Palm Squirrel, *Funambulus pennanti* Wroughton: BANERJI (1957) observed that in central north India *F. pennanti* breeds every fourth month in a year. PUROHIT et al., (1966) reported that it breeds from March to September in the arid and semi-arid areas of Rajasthan. AGRAWAL (1965), however, reported a case of pregnancy with five young ones in the month of December. I have observed two litters of 5 young each, delivered early in May and in July, 1971. My observations on the percentage distribution of 13 litters in various months at Jodhpur during 1963 indicate that there are two peaks in the breeding activity of this squirrel, one during March–April and the other during the period July to September. SETH & PRASAD (1969) confirmed these observations and found that, at New Delhi, pregnant females occurred in the population from February until September. They also found two peaks, during March and July, when 51 and 38 per cent of the females were pregnant, respectively.

From the data on litters born in the laboratory and from the captured pregnant females, it was found that the number of young born at a time varied from 1 to 4, the average being 2.9 (PUROHIT et al., 1966). AGRAWAL (1965) and SETH & PRASAD (1969) found litter sizes ranging from 2–5 and 1–5 respectively. I also observed two litters of 5 young each delivered during May and July, 1971 (PRAKASH, 1971).

WAGNER's Gerbil, *Gerbillus nanus indus* (Thomas): Pregnant females were collected during April, June and December (PRAKASH & JAIN, 1971) indicating that this gerbil breeds twice in a year – during summer and winter seasons.

Although we do not have records of a sufficiently large number of pregnant females but it may be a safe guess that the number of young ones born to a female each time is rather small from 2 to 3, the average being 2.33 (PRAKASH & JAIN, 1971).

Indian Hairy-footed Gerbil, *Gerbillus gleadowi* Murray: This psammophile gerbil too, it appears, has two breeding seasons, one in the summer and the other in the winter. Pregnant females were observed during May–June and October–January.

A rather interesting observation has been made in respect of its litter size during the two seasons. While, during the summer, the number of young per litter varied from 2 to 4, the average being 2.75 (PRAKASH & PUROHIT, 1967), the litter size was considerably bigger (5 to 6, average 5.5) in the pregnant females collected during the winter. These observa-

Fig. 21. Gerbillus gleadowi.

tions are based on a small sample but if it is a generality, the larger litter size in winter may be ascribed to better climatic and food availability factors associated with this season.

Indian Gerbil, *Tatera indica indica* Hardwicke: It breeds all the year round and the percentage of females found pregnant varied from 9.7 to 61.0, the annual average being 29.7 (Table 10). Peaks in breeding activity were observed during February, July–August and November. The prevalence of pregnancy runs parallel to the rate of fecundity in males throughout the year. On an average, the male fecundity remains at a 50 per cent level all the year round but peaks occur during February, August and November, corresponding to the peaks in the number of pregnant females (JAIN, 1970). JAIN's specimens were collected from within the Bikaner town where fluctuations in the availability of food and drinking water are minimal and the impact of climatic parameters not felt as much as it may be felt in the open field. Collections of *T. i. indica* from natural desert grasslands, however, do not suggest a year-long breeding period in this species. This study is at present inconclusive.

BLANFORD (1888–91) mentioned a rather large litter size (from 8 to 12) for this Gerbil. We, however, found that anything from 1 to 9 young may be born at a time (PRAKASH *et al.*, 1971a; JAIN, 1970). In another subspecies of *Tatera indica* found in South India, namely *cuvieri*, the litter size has been found to vary from 1 to 10 with that of 6 occurring at maximum frequency (PRASAD, 1961) and in the subspecies *hardwickei*, from 2 to 9, with that of 5 at maximum frequency (CHANDRAHAS, Personal Communication).

GHOSH & TANEJA (1968) found the duration of the oestrous cycle in *Tatera i. indica* to be of 4.82 days. The gestation period varies from 26 to 30 days, the young are weaned after 30 days of birth, and sexual maturity is attained around the age of 16 weeks (PRAKASH *et al.*, 1971a).

Indian Desert Gerbil, *Meriones hurrianae* (Jerdon): Like *T. indica*, the desert gerbil, too, breeds all through the year (PRAKASH, 1964; KAUL & RAMASWAMY, 1969). The data on the prevalence of pregnancy indicate that the reproductive rate of females has three peaks in a year, viz. during the period February to April, in July, and during September to November, while the rate is low during December and January, the severe winter months in this desert. These findings are in general agreement with a previous report (PRAKASH, 1964), wherein, however, peaks in littering were observed during February and July. The distribution of litters through the year as reported by KAUL & RAMASWAMY (1969), however, shows a clear peak during the month of February only.

My studies have shown that, on an average, 4.4 young are born to a female at a time, the range being from 1 to 9 (PRAKASH, 1964). At Jaipur, a city situated at close proximity to the main Rajasthan desert, the productivity of the desert gerbil was found to be slightly lower, the average number born to a female at a time being 4.0 with a range from 2 to 7 (KAUL & RAMASWAMY, 1969).

The average period of oestrous cycle has been found to be of 6.22 days (GHOSH & TANEJA, 1968) and 6.1 days (KAUL & RAMASWAMY, 1969). The gestation period varies from 28 to 30 days and the young attain sexual maturity at the age of 15 weeks (PRAKASH, 1964).

House Rat, *Rattus rattus rufescens* (Gray): Our observations in this desert indicate that the house rat breeds from April to September but TABER *et al.* (1967) found evidence of breeding in the post monsoon period and in winter at Lyallpur in West Pakistan.

In the desert region we have encountered pregnant house rats carrying 1 to 9 embryos each. From Lyallpur, TABER *et al.* (1967) reported a litter size of 4 to 9.

Cutch Rock Rat, *Rattus cutchicus cutchicus* (Wroughton): The rock rats

Table 10. Percent adult female rodents found pregnant in the Rajasthan desert through the year.

Species	Jan.	Feb.	March	April	May	June	July	Aug.	Sept.	Oct.	Nov.	Dec.	Source
F. pennanti	0.0	38.7	70.0	31.2	32.5	16.2	42.5	17.5	10.0	0.0	0.0	0.0	SETH & PRASAD, 1969
T. indica	26.6	47.4	30.0	12.5	16.6	41.1	46.6	61.0	9.7	10.5	38.8	15.7	JAIN, 1970
M. hurrianae	7.6	24.2	20.8	20.0	12.0	11.4	21.0	16.6	18.2	26.6	20.0	9.3	KAUL & RAMASWAMI, 1969
R. cutchicus	—	—	50.0	—	45.5	—	5.9	88.9	—	77.0	—	0.0	PRAKASH *et al.*, 1973b
R.m. pallidior	—	—	14.3	—	50.0	—	75.0	—	—	100.0	—	—	Unpublished data

were collected in seven administrative districts of the Rajasthan desert, every month from November, 1968 to April, 1969. Out of the 19 females, captured during these six months, only one was found to be pregnant in the month of April. We conducted another study during 1970–1971 and found pregnant females from March to October. 1968 was a drought year whereas during 1970 the rains in the desert exceeded the normal level. It may be possible that moisture scarcity had adversely affected the breeding potential of these rodents during 1968–69. During year long study conducted in 1970–71, the prevalence of pregnancy was the maximum during August and it gradually decreased to zero in the month of December (Table 10). After a rise in March and another again in May, the pregnancy rate declined to 5.9 per cent during July which is rather surprising as in all other desert rodents studied, breeding activity has been found to be generally very high during this month (PRAKASH, 1971). The litter size varied from 2 to 8, the average being 4.0 (PRAKASH et al., 1973).

Soft-furred Field Rat, *Rattus meltada pallidior* Ryley: A collection of 25 adult females from various localities in the Rajasthan desert, made from November, 1968 to April, 1969, revealed only one pregnant and one lactating female, both in the month of December. In another series of collections made from October, 1970 to July, 1971 pregnant females were encountered throughout the year. The peak prevalence of pregnancy (100 per cent) was found during October. In the State of Punjab, situated towards north of the Rajasthan desert, BINDRA & SAGAR (1968), however, found that in captivity littering was restricted to two periods in a year, viz. from March to May, and from July to October.

From the Rajasthan desert collections, examination of pregnant females revealed that, on an average, 5.9 young are born to a female at a time, the range varying from 4 to 10. BINDRA & SAGAR (1968) found a wide diversity in the litter sizes of freshly-captured and laboratory-bred rodents of this species. In the wild ones, the number of young per litter varied from 5 to 8, the average being 6.0 while that in the laboratory bred ones was from 1 to 8, the average being 3.4.

Sand-coloured Rat, *Rattus gleadowi* (Murray): Sufficient information is not available regarding this species. In the Rajasthan desert pregnant specimens were collected from August to October. The litter size varied from 2 to 3, with an average of 2.3.

House Mouse, *Mus musculus bactrianus* Blyth: It reportedly breeds throughout the year (MANN, 1969). TABER et al. (1967) found an evidence of continuous breeding from November to April, the period of their study in West Pakistan. Among laboratory bred *Mus*, the litter size varied from 3 to 6, with an average of 4.6 but in the field trapped ones the range was from 1 to 8 (MANN, 1969).

Little Indian Field Mouse, *Mus booduga booduga* (Gray): SRIVASTAVA (1968) mentioned that the little field mouse breeds in September,

108

October, February and June. From the Rajasthan desert, where this mouse is not so common, we have only one record of littering in the month of October. The male mice collected in January, however, had scrotal testes.

While SRIVASTAVA (1968) observed litter sizes varying from 6 to 13, MANN (1969) found that, in the Punjab, the range may be from 1 to 5 with an average of 3.5 young per litter.

Fawn-coloured Mouse, *Mus cervicolor phillipsi* (Wroughton): In a year round study conducted in the Rajasthan desert, pregnancy was recorded during December (prevalence of pregnancy 25 per cent), and during July (50 per cent). On an average, 4.4 embryos were observed per pregnant female and the embryo number ranged from 2 to 6 (PRAKASH, 1971).

Brown spiny Mouse, *Mus platythrix sadhu* (Wroughton): The females collected in January, 1969 and in March, 1971 exhibited evidence of breeding but pregnant females could be collected only during August and October. These observations made over a period of several years may point to a year-long breeding season in this species. However, since the breeding season is very likely to change from year to year, chiefly depending on the amount of precipitation received, the idea of a year-long breeding season may not hold true for all years. 3 to 10 embryos have been recorded per pregnant female in the Rajasthan desert.

Indian Bush Rat, *Golunda ellioti gujerati* Thomas: Pregnancy was observed during the period from March to August and the embryo numbers varied from 5 to 10, the average being 6.6.

Short-tailed Mole Rat, *Nesokia indica indica* (Gray): In the Punjab it litters during winter. LAY (1967) reported their breeding in Iran from October to January but the laboratory maintained ones litter all the year round. The litter size varied from 2 to 7. SOUTHWICK (1966), however, mentioned that the average size is 8 to 10.

Lesser Bandicoot Rat, *Bandicota bengalensis* (Gray): The bandicoot rat breeds all the year round (SPILLET, 1968; GOKHALE, 1956). Very little information is, however, available about its reproductive activity in the desert region. Litter size varies from 2 to 14.

BODENHEIMER (1957) observed that the birth season of mammals is in late winter in the Sahara–Sindian region, in late summer in Sudano–Deccanian, and in spring in the Irano–Turanian regions. In the Indian desert, however, most of the rodents breed from March to September, although a few breed all through the year. The minimum number of rodent species breeding is during the winter, quite contrary to what BODENHEIMER *(loc. cit.)* had stated. The peak number of other mammals also breed during monsoon (PRAKASH, 1960), just like the large herbivores of northeast Africa. It appears, therefore, that the breeding pattern of rodents occurring in the Indian desert is an admixture of breeding patterns observed in northeast Africa and in the Sudano–Deccanian and Irano–Turanian regions.

Table 11. Number of rodent species breeding, day length and mean monthly rainfall in the Rajasthan desert.

	Jan.	Feb.	March	April	May	June	July	Aug.	Sept.	Oct.	Nov.	Dec.
No. of rodents species breeding	5	4	6	12	11	13	13	17	14	7	6	4
Average sunshine hours	10.8	11.3	12.5	12.8	13.6	13.8	13.7	13.2	12.2	11.6	10.9	10.6
Mean monthly rainfall (mm) at Jaisalmer	4.1	4.8	3.6	2.8	7.9	14.7	52.3	62.5	21.8	1.3	1.3	2.0

Table 12. Monthly prevalence of pregnancy of three rodents, day length and mean monthly rainfall in the Rajasthan desert.

	Jan.	Feb.	March	April	May	June	July	Aug.	Sept.	Oct.	Nov.	Dec.
					Percent adult females pregnant							
F. pennanti	0.0	38.7	70.0	31.2	32.5	16.2	42.5	17.5	10.0	0.0	0.0	0.0
T. indica	26.6	47.4	30.0	12.5	16.6	41.1	46.6	61.0	9.7	10.5	38.8	15.7
M. hurrianae	7.6	24.2	20.8	20.0	12.0	11.4	21.0	16.6	18.2	26.6	20.0	9.3
Average sunshine hours	10.8	11.3	12.5	12.8	13.6	13.8	13.7	13.2	12.2	11.6	10.9	10.6
Mean monthly rainfall (mm) at Jaisalmer	4.1	4.8	3.6	2.8	7.9	14.7	52.3	62.5	21.8	1.3	1.3	2.0

111

With a winter drop in breeding activity, the Indian desert rodents exhibit peak activity in spring and the activity decreases during June, the hottest month of the year. A major breeding peak then occurs during the monsoon season (Table 11). During the first peak, corresponding to spring, many desert plants flower and several of them produce fresh shoots and leaves, thus making fresh, green food available to the rodents which accelerates their reproductive activity. Secondly, after passing through a partial quiescent period during the preceding winter, their inherent, internal physiologic activity is, probably, enhanced during the spring and thus a large number of rodent species are found to be participating in reproductive activity in the spring season. Although this activity generally declines during June yet surprisingly, quite a few rodent species continue to breed throughout the summer season when the air temperature is high, the relative humidity of the air is low and the food supply scarce. The maintenance of a comparatively high level of reproduction during summer is presumably an indication of the animals' efficient eco-physiological adaptive mechanisms suitable for arid conditions. Most of these rodents are fossorial. The temperature inside their burrows does not vary by more than 1.1 °C during summer months and is on an average 20.8 °C lower than the soil surface temperature (PRAKASH et al., 1965). The rodents, in a way, stay in 'air conditioned' chambers and are not actually exposed to hostile temperatures and low relative humidity conditions of the outside environment. Moreover, most of the Indian desert rodents are nocturnal. The diurnal Desert Gerbil, Meriones hurrianae, is also behaviourally adapted to unload the hyperthermia developed by exposure to day-time temperatures by intermittently visiting the cooler environment of the burrows (FITZWATER & PRAKASH, 1969), and by shifting its daily surface activity to early mornings and late evenings during the summer (PRAKASH, 1962). The scarcity of vegetable food and water during summer is met with by switching over to insect food which has a relatively high water content (PRAKASH, 1962). Due to these desert adaptations, probably, some rodents are able to participate in breeding activity during the summer, except in June, which is the hottest month in the Indian desert. It is also probable that in addition to the nutritive green food, the day length also plays an important role by influencing the reproductive cycle of Indian desert rodents, as has been observed in the case of the Indian desert Hare, Lepus nigricollis dayanus (PRAKASH & TANEJA, 1969). In the Indian desert, the day length starts increasing from March and reaches a peak in late summer after which it decreases till winter (Table 11). The number of rodent species breeding in a month is broadly correlated to the mean monthly day length. This relationship also holds true when day length and the results of detailed study of breeding activity of three rodent species are considered (Table 12). Only a few rodent species breed at a very low rate during winter when day length is the shortest. SETH & PRASAD

Table 13. Mean monthly litter size of four desert rodents through the year.

Rodent species	Jan.	Feb.	March	April	May	June	July	Aug.	Sept.	Oct.	Nov.	Dec.
F. pennanti	—	—	3.0	2.6	4.0	3.0	3.0	4.0	3.0	—	—	—
Tatera i. indica	4.25	4.33	5.83	4.66	4.0	3.85	4.71	5.30	4.50	4.0	5.42	3.66
M. hurrianae	3.3	4.2	4.3	4.3	4.5	3.6	4.3	4.3	4.5	5.0	4.3	4.0
R.c. cutchicus	—	—	3.0	—	3.0	—	3.0	4.4	—	5.1	—	—

(1969) also observed that this is apparently due to ovarian refractoriness to photoperiods. The short days during winter may terminate the refractoriness of the gonads and lead to recrudescence.

The breeding activity of three species of rodents, detailed studies on which have been done by us, also show a parallel pattern with the mean monthly precipitation (Table 12) in the Indian desert. The rainfall directly influences the availability of nutritive green food which in turn, probably, affects the breeding rate of desert rodents.

In the Indian desert, not only the largest number of rodent species breed during monsoon, but their litter size is also largest during the period of abundant food and comfortable climate as is observed in *T. i. indica* (JAIN, 1970), *M. hurrianae* (PRAKASH, 1964) and *Funambulus pennanti* (PRAKASH, 1968), *Rattus c. cutchicus* (PRAKASH *et al.*, 1973, Table 13) and in the lagomorph, *L. n. dayanus* (PRAKASH & TANEJA, 1969). Larger litters are produced by the female rodents during the optimum (rainy) season in the desert, perhaps due to their good health which is maintained by the abundant availability of green food. The mothers are also in a position to meet lactation needs of the larger litters. The litters after weaning, too, get a supply of good food and the survival rate of these litters is much superior to those produced earlier during the summer.

It appears that during the drought years, the breeding potential of rodents is decreased considerably. This has been clearly observed in the case of *Meriones hurrianae* (PRAKASH, 1968), which breeds only during monsoon in drought years, whereas in normal rainfall years it breeds all the year round (PRAKASH, 1964). During the 1968–69 rodent survey of the entire Indian desert, we found that only 12 per cent females of *M. hurrianae* were pregnant. Rains during 1968 were meagre in this desert, and it was a year of severe drought (PRAKASH, 1971).

REFERENCES

AGRAWAL, V. C. 1965. Observations on habits of five-striped squirrel, *Funambulus pennanti*, in Rajasthan. *J. Bengal nat. Hist. Soc.*, 34: 76–83.
AHMED, E. 1969. Origin and geomorphology of the Thar desert. *Ann. Arid Zone*, 8: 171–180.
BANERJI, A. 1957. Further observations on the family life of the five-striped squirrel, *Funambulus pennanti* Wroughton. *J. Bombay nat. Hist. Soc.*, 54: 336–343.
BINDRA, O. S. & SAGAR, P. 1968. Breeding habits of the field rat, *Millardia meltada* (Gray). *J. Bombay nat. Hist. Soc.*, 65: 477–481.
BLANFORD, W. T. 1888–1891. The fauna of British India, including Ceylon and Burma, Mammalia (Vols 1 & 2), London, Taylor & Francis.

BODENHEIMER, F. S. 1957. The ecology of mammals in arid zones. In Human and Animal Ecology. *Reviews of Research, Arid Zone Research*, 8: 100–137, Paris. UNESCO.

FITZWATER, W. D. & PRAKASH, I. 1969. Burrows, behaviour and home range of the Indian desert gerbil, *Meriones hurrianae* Jerdon. *Mammalia*, 33: 598–606.

GHOSH, A. 1952. The Rajputana desert – its archaeological aspects. in Symp. on Rajputana desert. *Bull. Natl. Instt. Sci.*, India, 1: 37–42.

GHOSH, P. K. & TANEJA, G. C. 1968. Oestrous cycle in the desert rodents, *Tatera indica* and *Meriones hurrianae*. *Indian J. Exptl. Biol.*, 6: 54–55.

GOKHALE, M. S. 1956. Studies on Bombay rats and their flea parasites. Ph.D. Dissertation, Univ. of Bombay.

JAIN, A. P. 1970. Body weights, sex ratio, age structure, and some aspects of reproduction in the Indian gerbil, *Tatera indica indica* Hardwicke, in the Rajasthan desert, India. *Mammalia*, 34: 415–432.

KAUL, D. K. & RAMASWAMY, L. S. 1969. Reproduction in the Indian desert gerbil, *Meriones hurrianae* Jerdon. *Acta Zoologica*, 50: 233–248.

KRISHNA, D. & PRAKASH, I. 1960. Hedgehogs of the desert of Rajasthan. Pt. III. Food in nature. *Proc. Raj. Acad. Sci.*, 7: 60–62.

KRISHNAN, M. S. 1952. Geological history of Rajasthan and its relation to present day conditions. in Symp. on Rajputana desert. *Bull. Natl. Instt. Sci.*, India, 1: 19–31.

KRISHNASWAMI, S. & CHOUHAN, N. S. 1957. A note on insects consumed as food by squirrels and birds at Kundi Forest, Palamau District, Bihar. *J. Bombay nat. Hist. Soc.*, 54: 457–459.

LAY, D. M. 1967. A study of the mammals of Iran. *Fieldiana: Zoology*, 54: 1–282.

MANN, G. S. 1969. Studies on the biology and control of field mice and analysis of rodent population around Ludhiana. M.Sc. Thesis, Punjab Agri. Univ., Ludhiana.

MINTON, S. A. 1966. A contribution to the herpetology of west Pakistan. *Bull. Amer. Mus. nat. Hist.*, 134(2): 27–184.

PRAKASH, I. 1958. Extinct and vanishing mammals from the desert of Rajasthan and the problem of their preservation. *Indian For.*, 84: 642–645.

PRAKASH, I. 1959. Foods of Indian desert mammals. *J. Biol. Sci.*, 2: 100–109.

PRAKASH, I. 1960. Breeding of mammals in Rajasthan desert, India. *J. Mamm.*, 42: 386–389.

PRAKASH, I. 1962. Ecology of gerbils of the Rajasthan desert India. *Mammalia*, 26: 311–331.

PRAKASH, I. 1962–1964. Taxonomical and ecological account of the mammals of Rajasthan desert. *Ann. Arid. Zone*, 1: 143–162, 2(2): 150–161.

PRAKASH, I. 1964. Ecotoxicology and control of Indian desert gerbille, *Meriones hurrianae* Jerdon Pt. 2. Breeding season, litter size, and post-natal development. *J. Bombay nat. Hist. Soc.*, 61: 142–149.

PRAKASH, I. 1968. Biology of the rodents of Rajasthan desert. Proc. Symp. 'Natural Resources of Rajasthan'. 337–352.

PRAKASH, I. 1969. Eco-toxicology and control of Indian desert Gerbille, *Meriones hurrianae* Jerdon, V. Food preference in the field during monsoon. *J. Bombay Nat. Hist. Soc.*, 65: (3) 581–589.

PRAKASH, I. 1971. Breeding season and litter size of Indian desert rodents. *Zeit. f. angewandte Zool.*, 58(4): 442–454.

PRAKASH, I. 1973. Rodent control in the Desert. *Indian Fmg.* 23(3): 41–43 & 47.

PRAKASH, I. 1974. The ecology of vertebrates of the Indian desert. In Biogeography and ecology in India. The Hague, Junk: 369–420.

PRAKASH, I., GUPTA, R. K., JAIN, A. P., RANA, B. D. & DUTTA, B. K. 1971. Ecological evaluation of rodent populations in the desert biome of Rajasthan. *Mammalia*, 35: 384–423.

PRAKASH, I. & JAIN, A. P. 1971. Some observations on the WAGNER's Gerbil, *Gerbillus nanus indus* Thomas in the Indian desert. *Mammalia*, 35(4): 614–628.

PRAKASH, I., JAIN, A. P. & PUROHIT, K. G. 1971a. Breeding and post-natal development of the Indian Gerbil, *Tatera indica* Hardwicke in Rajasthan desert. *Saugetier. Mittei-lungen*, 19(4): 375–380.

PRAKASH, I. & KAMETKAR, L. R. 1969. Body weight, sex and age factors in population of the Northern Palm squirrel, *Funambulus pennanti* Wroughton. *J. Bombay nat. Hist. Soc.*, 66(1): 99–115.

PRAKASH, I., KAMETKAR, L. R. & PUROHIT, K. G. 1968. Home range and territoriality of the Northern Palm Squirrel, *Funambulus pennanti* Wroughton. *Mammalia*, 32: 604–611.

PRAKASH, I., KUMBKARNI, C. G. & KRISHNAN, A. 1965. Ecotoxicology and control of the Indian desert gerbil, *Meriones hurrianae* (Jerdon), III. Burrow temperature. *J. Bombay nat. Hist. Soc.*, 62: 237–244.

PRAKASH, I & PUROHIT, K. G. 1966. Some observations on the hairy footed gerbille, *Gerbillus gleadowi* Murray, in the Rajasthan desert. *J. Bombay nat. Hist. Soc.* 63: 431–434.

PRAKASH, I. & RANA, B. D. 1970. A study of field population of rodents in the Indian desert. *Zeit. f. Angewandte Zool.*, 57: 129–136.

PRAKASH, I. & RANA, B. D. 1972. A study of field population of rodents in the Indian desert. II. Rocky and piedmont zones. *Zeit. f. angewandte Zool.*, 59(4): 129–139.

PRAKASH, I. & RANA, B. D. 1973. A study of field population of rodents in the Indian desert. III. Sand dunes in 100 mm rainfall zone. *Zeit. f. angewandte Zool.*, 60: 31–41.

PRAKASH, I., RANA, B. D. & JAIN, A. P. 1973. Reproduction in the Cutch rock-rat, *Rattus cutchicus cutchicus*, in the Indian desert. *Mammalia*, 37(3): 457–467.

PRAKASH, I. & TANEJA, G. C. 1969. Reproduction biology of the Indian desert hare, *Lepus nigricollis dayanus* Blanford. *Mammalia*, 33: 102–117.

PRAKASH, I., TANEJA, G. C. & PUROHIT, K. G. 1971b. Ecotoxicology and control of Indian desert gerbil, *Meriones hurrianae* Jerdon. VII. Relative numbers in relation to ecological factors. *J. Bombay nat. Hist. Soc.* 68(1): 86–93.

PRAMANIK, S. K. & HARIHARAN, P. S. 1952. The climate of Rajasthan. in Symp. on Rajputana desert. *Bull. Natl. Instt. Sci.*, India, 1: 167–178.

PRASAD, M. R. N. 1961. Reproduction in the female Indian gerbille, *Tatera indica cuvieri* (Waterhouse). *Acta Zool. Stockh.*, 42: 245–256.

PRATER, S. H. 1965. The book of Indian animals. Bombay, Bombay Natural History Society.

PUROHIT, K. G., KAMETKAR, L. R. & PRAKASH I. 1966. Reproduction biology and post-natal development in the Northern palm squirrel, *Funambulus pennanti* Wroughton. *Mammalia*, 30: 538–546.

ROY, B. B. & PANDEY, S. 1971. Expansion or contraction of the great Indian desert. *Proc. Indian Natl. Sci. Acad.*, 36(B): 331–344.

SAGAR, P. 1972. Studies on the biology of the lesser bandicoot rat, *Bandicota bengalensis* (Gray) in the Punjab, Ph.D. Dissertation, Punjab Agri. Univ. Ludhiana.

SETH, P. & PRASAD, M. R. N. 1969. Reproduction cycle of the female five-striped Indian palm squirrel, *Funambulus pennanti* Wroughton. *J. Reprod. Fert.* 20: 211–222.

SOUTHWICK, C. H. 1966. Reproduction, growth and mortality of murid rodent populations. Proc. Indian Rodent Symp. New Delhi, Johns Hopkins Univ. & USAID: 152–176.

SPILLET, J. J. 1968. The ecology of the Lesser bandicoot rat in Calcutta. Bombay, Bombay Natural History Society & The Johns Hopkins Univ. CMRT.

SRIVASTAVA, A. S. 1968. Rodent Control for increased food production. Kanpur, Rotary Club (West).

TABER, R. D., SHERI, A. N. & AHMED, M. S. 1967. Mammals of the Lyallpur region, West Pakistan. *J. Mamm.*, 48: 392–407.

WADIA, D. N. 1960. The post-glacial desiccation of central Asia: Evolution of the arid zone of Asia. *Natl. Instt. Sci.*, India Monogr., 10: 1–25.

ZUCKERMAN, S. 1953. The breeding seasons of mammals in captivity. *Proc. Zool. Soc. London*, 122: 859.

116

VI. OUTBREAKS OF RODENTS IN SEMI-ARID AND ARID AUSTRALIA: CAUSES, PREVENTIONS, AND EVOLUTIONARY CONSIDERATIONS

by

A. E. NEWSOME & L. K. CORBETT

Introduction

The native rodents of Australia belong to the Family Muridae. There are 47 species all told (RIDE, 1970) compared with 173 in continental United States of America (HALL & KELSON, 1959) which is about the same size. Seventeen of the Australian species are endemic to the arid zone (defined below) which covers about 5.2 million sq. km (70%) of the Australian land-mass. Mostly they are uncommon or rare. An alien Murid, the house-mouse *(Mus musculus)*, is now a resident, even in the arid zone where it forms occasional outbreaks along with some of the native species (FINLAYSON, 1939; this study).

Here, then, are four major, possibly inter-related ecological problems associated with the rodent fauna of the Australian arid zone, which can be stated as follows:
(i)   Rarity of most species most of the time;
(ii) Sporadic widespread irruptions of some species;
(iii) Low species diversity;
(iv) The super-position of *Mus*.

Of these four problems, it has been the irruptions, or plagues as they are termed in Australia, that have excited most curiosity (CLELAND, 1918, FINLAYSON, 1939, DUNNET, 1956, CARSTAIRS, 1971, PLOMLEY, 1972). Their cause has been a puzzle. Plagues arrive unexpectedly, stay varying times, and depart suddenly. Sometimes more than one species is involved. That several species can irrupt simultaneously over vast areas of land indicates a general cause. FINLAYSON (1939) was first to be explicit, linking simultaneous plagues of three species (including *Mus musculus*) around Lake Eyre, the driest region in Australia (see Figs. 1, 2), to vegetative flushes caused by heavy rains and floods. The trouble is, not all such rains or floods are so efficaceous.

Between 1966 and 1971, data have been gathered on a large central Australian study-area (delineated in Figs. 1-4) to study predator/prey relationships between the dingo *(Canis familiaris dingo)* and the rodents (NEWSOME & CORBETT, unpublished). As luck would have it, we trapped

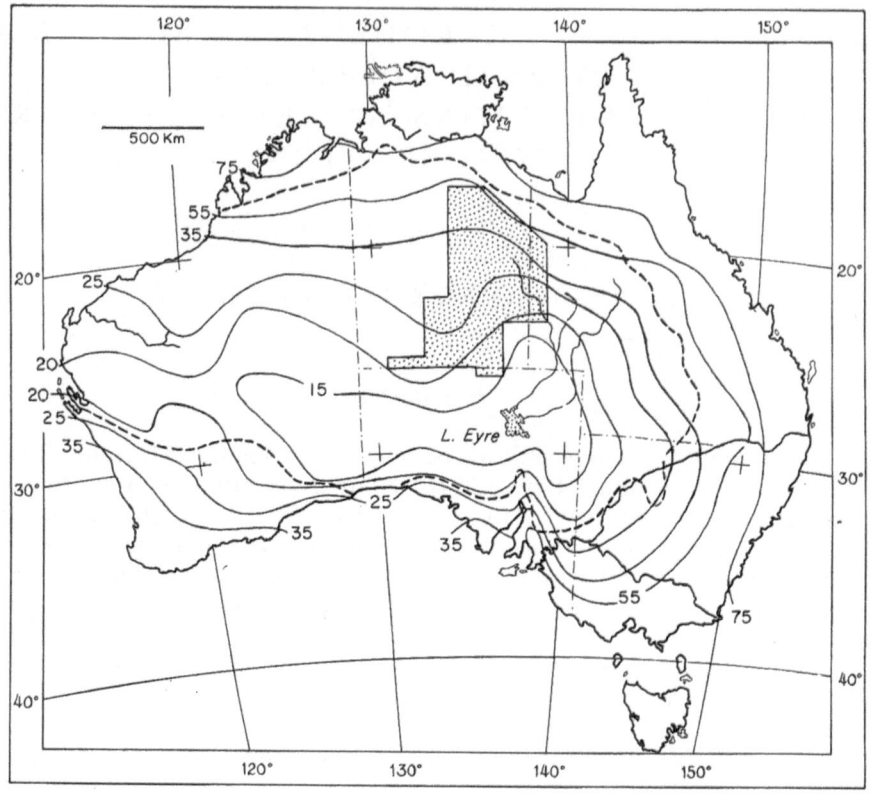

Fig. 1. Isohyets (cm) within and around the Australian arid-zone (– – – –), defined in the text, and the location and size of the study-area in central Australia.

clean through plagues of three species of rodent, two natives, *Rattus villosissimus* and *Notomys alexis*, and, almost inevitably, the alien *Mus musculus*.

There is presently no body of ecological information on any rodent inhabiting the Australian arid zone comparable to that for *Mus* (NEW-SOME, 1969a; b; 1970; unpublished). So, as a first approach to the faunal problems mentioned above, these recent data are analysed here in relation to the hypothesis advanced for the formation of *Mus* plagues in wheatlands just south of the arid zone. Data for *Mus* in central Australia are first presented, then data for the two native species. The major themes of this chapter are, therefore, irruption and rarity; but the super-position of *Mus* and the low species diversity also feature.

Since the initial irruption hypothesis for plague formation rests on unusually wet weather ameliorating the aridity of the usually hot dry summer of southern Australia, weather in central Australia is examined

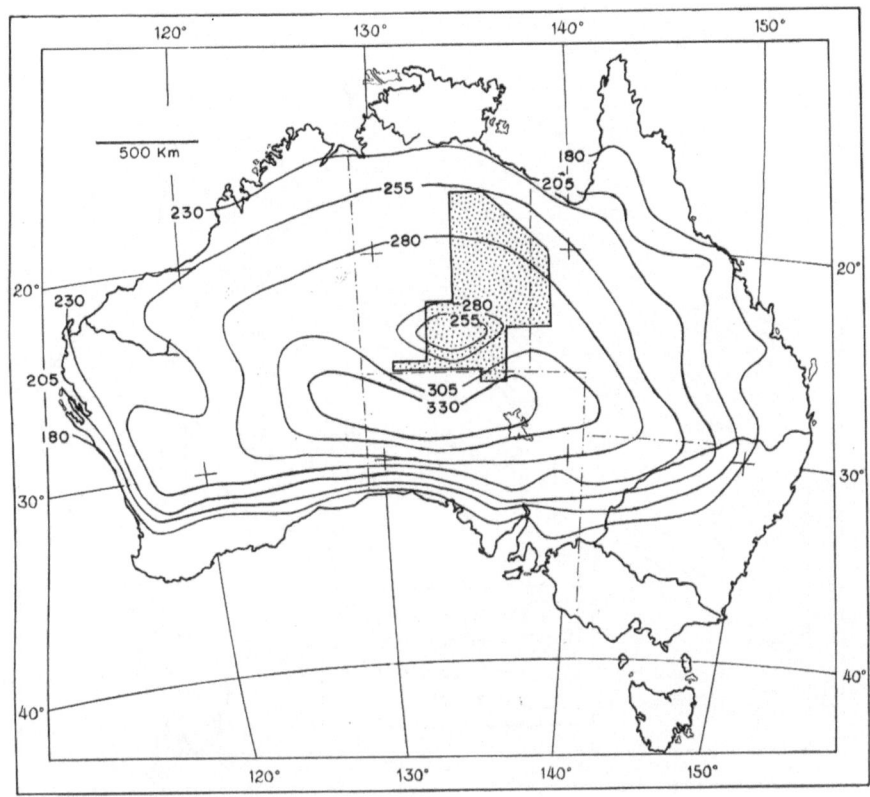

Fig. 2. Mean annual evaporation (cm) in and around the arid-zone (cf. Fig. 1) with the study-area super-imposed.

for unusually rainy periods. Details of the general weather patterns follow below.

Climates in Central Australia

Australia is the world's driest continent. The cause is increasing aridity with distance from the coast (see Figs. 1, 2) such that most of the land-mass is arid or semi-arid (MEIGS, 1953) producing deserts, or, in the better watered regions, sparse woodlands and grassland (see below).

The extent of the arid zone* is delimited in Fig. 1. Its perimetre

---

* The arid zone as defined here is more conservative than that of PERRY (1970). Sectors of *Eucalyptus* woodland and shrubland about 185 to 370 km wide have been deleted to the north and east of this zone because that genus characterises regions of high rainfall (in contrast to *Acacia* in the arid zone), and because large tracts of eucalypts have been cleared just to the east and south of it to farm cereal crops. The boundaries of the arid zone in this chapter are based on vegetation patterns given by MOORE & PERRY (1970).

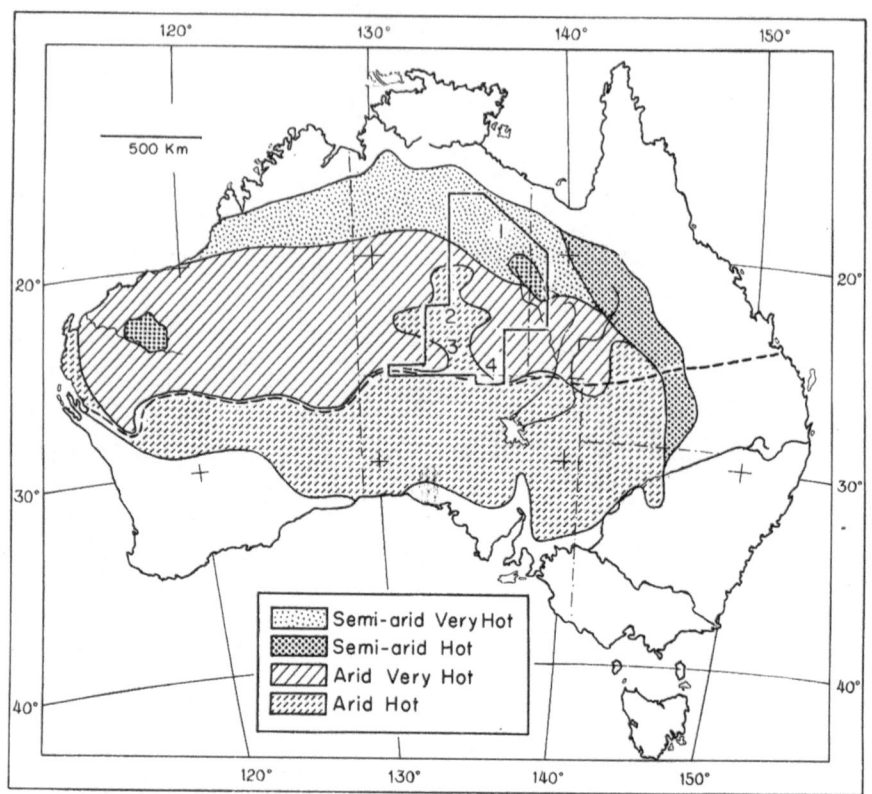

Fig. 3. Climates within the Australian arid zone and study-area (after Meigs 1953). The four Habitats are numbered. *Arid:* Monthly evapo-transpiration > rainfall by 40 cm. *Semi-arid:* Monthly evapo-transpiration > rainfall by 20 to 40 cm. *Very Hot:* Mean temperature in hottest month > 30 °C. *Hot:* Mean temperature in hottest month > 20 °C. The transverse dotted line separates regions of predominantly summer and winter rainfalls.

roughly co-incides with the 60 cm isohyet in the north, the 35 cm isohyet in the east, and the 25 cm isohyet in the south, the asymmetry being created by the hotter summer to the north, the cooler winter to the south, and the great potential evaporation throughout (see Fig. 2). The driest regions surround Lake Eyre (mean annual rainfall of 12 cm and potential evaporation of 330 cm (see Figs. 1, 2). Summer temperatures throughout this arid zone are high, often reaching 35 to 40 °C (e.g., see Table 1). The general aridity is often exacerbated by droughts, for there is no guarantee of rain.

A gradient in rainfall exists in the study-area. Rainfall in the northern sub-tropical portion comes almost entirely in summer (90%, see Table 1). Though it comes with some certainty, the amount, however, is variable.

120

Table 1. Climatic Gradients within Study-Area

Station	Latitude	Maximum and Minimum Mean Temp. (°C)						Rainfall (cm)	
		Hottest month		Coolest month		Year		Annual	Summer
Tennant Creek	19°S	37.7	24.4	24.1	10.6	32.0	18.4	35.2	31.8
Alice Springs	23°S	35.2	21.0	19.4	3.8	28.3	12.9	25.2	19.1
Charlotte Waters	26°S	37.1	22.3	19.4	4.9	29.1	14.1	17.9	11.7

121

Rain in the south is far less certain and even more variable making droughts common and often severe. Summer rains still predominate but the winter component increases from 10% in the north to 25% half way down, and to 35% in the south (see Table 1).

About one year in four may bring a drought with only two or fewer months receiving enough rain for grasses to grow (see Table 2). Sometimes, runs of dry years occur producing very severe drought. The longest drought on record ended just before the beginning of the study reported here, and lasted 6.5 years (Commonwealth of Australia, 1965).

The prevailing aridity is due to central Australia being in latitudes where warm dry tropical air descends from the upper atmosphere, absorbing whatever moisture there is. For there to be any chance of rain, the stability of the prevailing high pressure system has to be disrupted by invading moist air, an unusual event. Unlike large arid regions elsewhere, no large obstructive land-masses or mountains lie to north or south of the Australian arid region. Consequently, the northern and southern low pressure systems rarely deviate from their paths tangential to the arid zone.

The uncertain rainfall comes from a tiered series of four chance events. Two of them, summer and winter components, are byproducts from the edges of the sub-tropical monsoons that bring rain to northern Australia in summer, and the temperate low pressure systems that bring rain to the south in winter. Sometimes, by an even lower chance, warm moist air invading from the tropics collides with a cold front emanating from a low pressure system over the ocean to the south, bringing widespread heavy rains. Each low pressure system providing this confluence may be 2000 km away from central Australia in either direction at the time.

Super-imposed on these three rainy events is the most unusual though possibly the most biologically significant of them all: the tropical cyclone. Usually tangential to the arid zone also, cyclones (hurricanes) sometimes change direction and sweep inland, right into the arid interior drenching great swathes of country. The average annual rainfall may come in a matter of a few days. Water lies everywhere, and the countryside responds with lush and seeding herbage for many months. One year in about ten to fifteen receives enough rain from these chance events to keep herbage flourishing for most of the year (see Table 2), the rainfall being about double the average (see Table 3). Sometimes there are runs of wet years, as in 1921–23, 1935–36, 1947–48, and most recently from 1966–68 (see below).

To illustrate the way in which chance climatic events bring rain to central Australia, I should like to describe one that occurred in March 1967 when amazing amounts of rain fell in a few days over hundreds of thousands of square kilometres providing the land with one of its biggest drenchings on record. The effect of this rain, compared with the severity of the previous drought (that had lasted 6.5 years) illustrates best the

Table 2. Number of Years with stated Number of Months receiving Rainfall adequate for full Growth of Herbage*

No. of months	0	1	2	3	4	5	6	7	8
No. of years	1	1	7	9	7	8	4	1	2

* Monthly Rainfalls for 1921–1960 classified by SLATYER's (1962) estimates of adequate rainfall.

Table 3. The probability of annual rainfall exceeding a stated amount*

Amount (cm)	2	5	10	20	40	60
Probability (%)	100	99	95	65	10	1.5

* At Alice Springs (see Fig. 5) where mean annual rainfall is 25 cm. From GIBBS (1972).

aptness of the description of central Australia as a land of feast or famine.

The great rains resulted from straying tropical air and a southern cold front. In only two days, Alice Springs received 14.4 cm of rain, half its annual rainfall, and the desert to the southwest, 17.2 cm, more than the annual average (W. HARE, pers. comm.). The Finke River (see Fig. 5 below) ran almost a kilometre wide at its highest, and flowed for six months that year, emptying itself into the Simpson Desert. Together with other streams it filled clay-pans and inundated swales between sand-dunes over great areas. To fly over the country at 10,000 m altitude was to see temporary ponds and lakes extending to the horizon on all sides all the way down to Lake Eyre. Water lay about for months, and the land had one of the flushest seasons in memory.

The importance of such a rain to the well-being of the biota of central Australia cannot be sufficiently emphasized. When such extraordinary rains invade central Australia, the land blooms. Many animals usually rare breed up. Swarms of birds especially budgerigahs *(Melopsittacus undulatus)* and zebra finches *(Taeniopygia castanotis)* appear. And sometimes the inland rats plague.

The Study-Area: Habitats and Rodents

The size and location of the study-area are shown on Figs. 1–4 in relation to mean rainfall, potential evaporation, climatic regions, and vegetation, respectively. I have broken the area into four basic Habitats numbered 1 to 4 on Figs. 3 and 4 to locate them. From north to south they range along a line of increasing aridity, and are, respectively, a grassland,

123

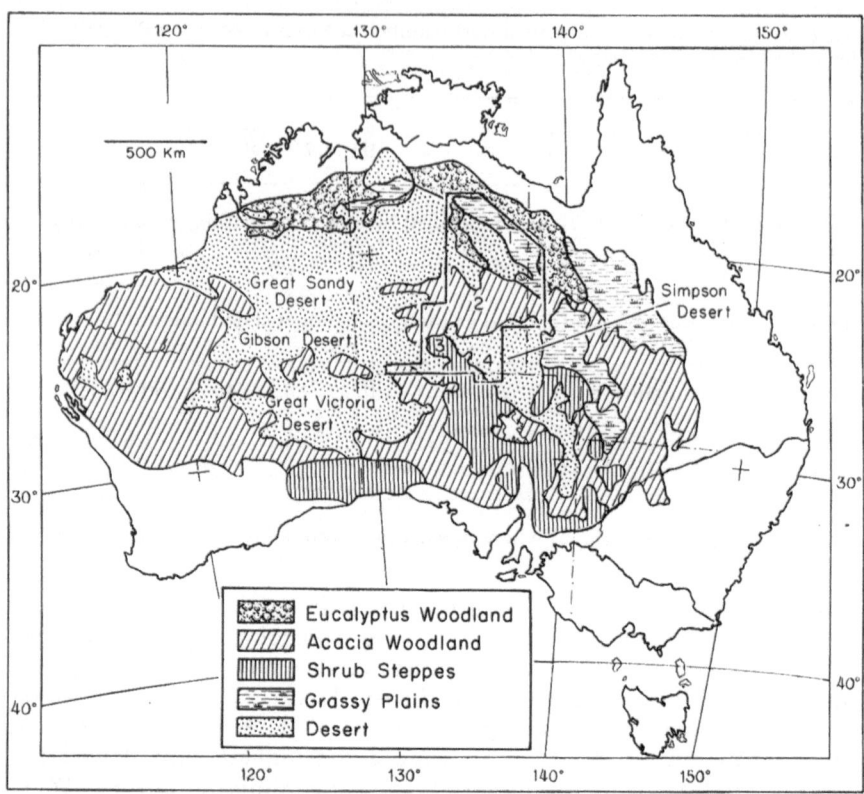

Fig. 4. Vegetation patterns within the Australian arid zone and study-area (after MOORE & PERRY, 1970).

woodland, shrubland, and desert. Pertinent features of each Habitat are now described. Detailed descriptions exist in PERRY & CHRISTIAN (1954) and PERRY *et al.*, 1962.

The species of rodent caught in this study are given in Table 4, and the 13 known species (PARKER, 1973) in Table 8, according to Habitat. The taxonomy used is by J. CALABY, J. MAHONEY, and W. RIDE (J. MAHONEY, pers. comm.). This list and some natural history notes were presented by RIDE (1970). For some details of distribution, TROUGHTON (1947) was consulted. Several matters of detail were clarified by Mr. P. AITKEN (1968; in litt.) and Mr. J. MAHONEY (pers. comm.). PARKER's (1973) recent check-list proved invaluable.

HABITAT I: THE BARKLY TABLELAND (see Fig. 5)

The vast, flat, almost treeless, grassy plains of the Barkly Tableland are

124

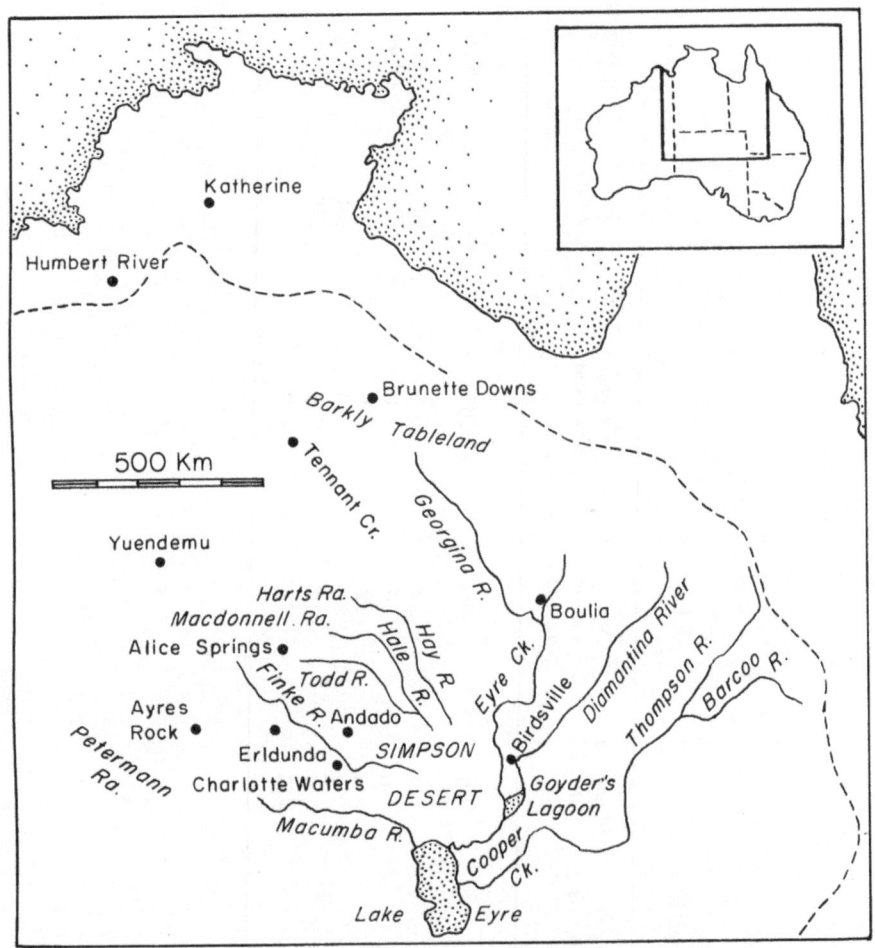

Fig. 5. Localities mentioned in text and the extent of the arid zone.

distinctive. The soils of the plains are mostly deep, black, cracking clays supporting a mixture of grasses, predominantly the perennial Mitchell and Flinders grasses (*Astrebla pectinata* and *Iseilema membranacea*). These permanent grasses grow well after rains, to a height of about 0.5 m, covering the plains. During the dry season (winter), the grass dies back and is often grazed down by cattle. The few trees (*Acacia* and *Eucalyptus*) that grow mostly do so along water-courses or in the occasional patches of red sandy soil. The climate is sub-tropical and sub-monsoonal. Summer is very hot, and often humid, for that is the rainy season, between 35 and 55 mm of rain falling on average. Winters are mild and dry.

Four species of rodent are known. The chief is a large rat, *Rattus*

Table 4. Rodents caught in central Australia*

Habitats	Pseudomys hermansburgensis	Pseudomys forresti	Notomys alexis evardensis	Rattus villosissimus	Mus musculus	Trap-nights
No. 1. Mitchell Grass Plains	0	0	0	207	0	2026
No. 2. Acacia Woodlands	14	2	2	1	59	4921
No. 3. Shrub (Trapped Steppe	12	0	10	0	29	968
(Hand-caught	2	0	72	0	4	—
No. 4. Sand-dune Desert	1	1	0	0	151	724
Totals. Trapped	27	3	12	208	239	8639
Hand-caught	2	0	72	0	4	—

* From March 1966 to December 1971.

villosissimus (placed as a subspecies of *R. sordidus* by Taylor & Horner, 1973), termed locally, the plague rat, for that is its renown. The other rodents, not recorded in this study (see Table 4) are *Notomys alexis, Pseudomys forresti,* and *Ps. hermannsburgensis.*

HABITAT 2: ACACIA WOODLANDS

Red clayey loams about 150 to 300 km south of Habitat 1 support an extensive woodland of low (10–15 m) trees, predominantly mulga *(Acacia aneura)*. Between the trees and on open plains adjacent to the MacDonnell Ranges (included in this Habitat because mulga is common there also) grow several grasses *(Astrebla, Aristida, Chloris, Eragrostis, Enneapogon)*. Summer rains promote growth of grasses and winter rains (about 25% of the total – see Table 1) promote herbaceous plants, mainly Compositae and Chenopodiaceae. The climate is drier than on Habitat 1, but not quite so hot, being further south and elevated at around 600 m above sea-level. The MacDonnell Ranges rise a further 500 to 1000 m above the plain. They shed water onto the mulga woodlands to the north and send floods down rivers to the south onto Habitat 4 (see below).

The rodent fauna of this Habitat is diverse with ten native species recorded. There are two species of medium-sized hopping rodent, *Notomys alexis* and *N. longicaudatus;* the former is not really common in this Habitat (see Table 4) and the latter not recorded since 1901. There are four species of small native rodent from the area. *Pseudomys hermannsburgensis* and *Ps. forresti* were both recorded in this study (Table 4). *Ps. desertor* was collected recently to the north-west of Habitat 2 (Watts, 1972) and is recorded for Alice Springs (see Fig. 5). None was caught in this study. One other small rodent, *Ps. fieldi,* is known from only one specimen taken near Alice Springs in 1894.

The three medium-sized rats of the region are also rare. One *Zyzomys pedunculatus,* the only fat-tailed desert rodent (in contrast to the several species of small fat-tailed marsupials there), is known from one recent specimen taken in 1960 to the west of the Habitat. Another, a stick nest rat *(Leporillus apicalis)*, is now almost certainly extinct. *Leporillus* built a nest of sticks in thickets and caves much as the North American pack-rat *(Neotoma)* does. Reports from early explorers (precised by Parker, 1973) indicate that nests were not all that uncommon in places. Never-the-less, the distribution was patchy, probably being hear the northern limits of its range. Finlayson (1941) considered its disappearance due to over-hunting by Aboriginals, but the introduced European fox *(Vulpes vulpes)* probably accounted for them away from settlements. Then there is *Rattus tunneyi,* which was collected near Alice Springs probably between 1894 and 1897, but not since. *R. tunneyi* is largely a tropical rat (Taylor &

HORNER, 1973), and the central Australian population must have been an extreme southern outlier.

Rattus villosissimus is known to occasionally invade Habitat 2, as it did in 1954 (D. R. STEPHENS, pers. comm.), and during this study (see Table 4). The other invader is a true alien, the house-mouse *(Mus musculus)*. It is now a resident and doing well (see Table 4). *Mus* was the only species to form plagues in this Habitat during this study.

HABITAT 3: SALT-BUSH STEPPE

Further south again, this Habitat is a steppe formation of sandy soil overlying limestone that out-crops here and there. Salt scalds are common in the low-lying areas. Salt-bush shrubs (*Atriplex* spp., Fam. Chenopodiaceae) were once common but now have been mostly eaten out by cattle. The usual grasses (as for Habitat 2, excepting *Astrebla*) grow after rain. There are large areas of desert sand-plain covered with spiny tussock-grasses of the xeric spinifex *(Triodia)* which sets much seed after rain. Being further south again in the study-area, the climate is a little drier than in Habitat 2. The annual rainfall averages between 20 and 25 cm, with slightly more of the rain in winter, hence the dominance of Chenopodiaceae.

The rodent fauna is poorly known but may well be as for Habitat 2, except for *R. villosissimus*. However, only three native species were recorded in this study (Table 4). *N. alexis evardensis* (identified by Mr. J. MAHONEY) was the commonest rodent caught, being in plague simultaneously with *Mus*. *Ps. hermannsburgensis* was also caught.

HABITAT 4: THE SIMPSON DESERT (see Fig. 5)

This is the most arid region of all in the study-area, the annual rainfall being around 18 cm on average, one third of it coming in winter (Table 1). The land-form is a sandy dune-field. To the north-west where most trapping was conducted, the land between the tall dunes is clayey and can produce a good flush of grass and other herbage (species mostly as in Habitats 2 and 3). The scant rainfall recorded for this desert does not really tell the truth about its water regime. A series of rivers, the Finke, Hale, and Hay, pour flood-waters into the desert filling swales between dunes should good rain fall on the MacDonnell Ranges 400–500 km away to the north and north-west in Habitat 2 (see Fig. 5). Of the six native species recorded from this Habitat, only two were caught in this study, *Ps. hermannsburgensis* and *Ps. forresti*.

One other, *Notomys amplus*, is known from only two specimens caught in 1896. Two other *Notomys, cervinus* and *fuscus*, were caught in the Habitat at that time, also at the extreme southern edge. These two species are based mainly in the Lake Eyre Basin further south and east where they form occasional plagues. Their presence in the study-area is probably sporadic and marginal, as visitors. One visitor comes often enough into the Habitat, however, so that it cannot be ignored, *Rattus villosissimus*, the plague rat. Though not caught in this study, it was found in the northern part of the Simpson Desert at the time (A. NEWSOME & D. STEPHENS, unpublished), and a colony has been found in the study-area recently, near Andado (see Fig. 5).

Around Birdsville (see Fig. 5), 200–300 km to the east of Habitat 4, *N. alexis* lives sympatrically with *N. cervinus*. They are probably separated by micro-habitat, *alexis* living on the sand-dunes and *cervinus* on the clayey loams and gibber plains (R. E. MACMILLEN, in litt.). *N. fuscus* is another common species there, but there are few records of the fourth species (not present in the study-area at all), *N. mitchelli*. *Ps. hermannsburgensis* also occurs around Birdsville (R. E. MACMILLEN, in litt.).

Another species inhabiting the region around Lake Eyre is the medium-sized *Ps. australis* (= *Ps. minnie* of FINLAYSON, 1939) it has also been known to form plagues (FINLAYSON, 1939). *Ps. desertor* lives in the same region and has been reported in a vast population there by AITKEN (1972).

WATER PHYSIOLOGY

Of the four native species caught in this study (*Pseudomys hermannsburgensis*, *Ps. forresti*, *Notomys alexis*, and *Rattus villosissimus*), the water physiology of the first and third only has been investigated (MACMILLEN & LEE, 1967, 1970; MACMILLEN *et al.*, 1972). They are well-fitted to live in deserts. Indeed, *Ps. hermannsburgensis* and *N. alexis* have the greatest known ability to concentrate urine among mammals, respective average concentration being 4710 and 6550 m Osmole/l, and maximum values of 8970 and 9370 mOsmole/l (MACMILLEN & LEE, 1967). *N. alexis*, the true desert dweller, is better at concentrating urea than *Ps. hermanns-burgensis*.

Notomys' ability to concentrate urine may compensate for a pulmonary water loss slightly higher than for some other desert rodents (MAC-MILLEN & LEE, 1970). The lowest absolute pulmonary loss was found to occur at temperatures between 29 and 33 °C, probably the maximum temperatures that prevail in underground burrows. The temperature at the base of a vertical burrow 50 cm deep held constant at 24 °C while surface temperatures ranged from 17 °C to 40 °C (STANLEY, 1971). And temperatures in nesting burrows in Habitat 3 fluctuated between 23.3 and 23.7 °C (R. BAUDINETTE, pers. comm.). *Pseudomys*' ability to concentrate urine may also compensate for a high evaporative water loss (MACMILLEN *et al.*, 1972).

The water physiology of *Mus musculus* also fits it for a desert existence (FERTIG & EDMONDS, 1969). The average concentration of urine

129

produced on a diet of dry seeds in the laboratory was 3799 mOsmole/kg*
(FERTIG, 1968). This value is about 80% of the average value for *Ps. hermannsburgensis*. Other house-mice fed a high protein diet but no water produced a maximum concentration of 6620 mOsmole/kg though these animals finally died. The indications are, therefore, that the house-mouse has a concentrating ability within range of *Pseudomys* and *Notomys*. At any rate, they survive along with them in the Australian desert (see Table 4).

Populations

Ecologically, little is known of these inland rodents. Aside from FIN-LAYSON's (1939) early work, there is a short note of DUNNET (1956) and a study by CARSTAIRS (1971) on plagues of *R. villosissimus*, though a long-term intensive study of this species by K. WILLIAMS & T. RED-HEAD (Division of Wildlife Research, CSIRO, Darwin) is current. CARSTAIR's (1971) results are incorporated in the study presented here.

As explained above, an hypothesis for irruptions derived from a study of *Mus* in wheatfields is pivotal for the study. Several factors have influenced me to take this approach. The southern wheatlands lie just outside (100 km) the arid zone and endure annually a long summer drought of six months or more, the climate there being Mediterranean. The droughts are predictably relieved every winter, however, in contrast to droughts further inland which have no predictable end. All the same, the weather in central Australia does not differ so much in kind from that in the southern wheatlands as in degree. I have therefore, assumed that the ways in which house-mice survive droughts and sometimes generate high numbers are similar in both places. That house-mice plagues in central Australia sometimes coincide with plagues of native rodents, e.g. of *Pseudomys australis* (= *minnie* in FINLAYSON, 1939) and of *Notomys alexis* (this study), also encouraged me to use *Mus* as a model.

THE HOUSE MOUSE: AN ECOLOGICAL MODEL

Permanent populations of house-mice studied in southern wheatlands occupy the minority habitats. The majority habitat, the huge wheatfields, are occupied opportunistically. Emigrants from the permanent populations colonise the wheatfields in summer. In hot summers with no rain (the usual state of affairs), population density in the fields is low; mice are wanderers, and rarely breed. They get their chance to breed and multiply when good rain falls in summer, a very unusual event. The

---

* Since FERTIG measured osmolality at low dilutions and then calculated real values by simple proportion, his values of osmolality can be compared directly with those of osmolarity given above for native rodents (D. BRADFORD, pers. comm.).

130

moistened soil becomes soft, mice promptly burrow, become resident and breed up. Since food is usually abundant in wheatfields in summer, the rain essentially converts marginal into excellent habitat. Thus, the following hypothesis for the formation of mouse plagues was erected.

Two resources are of great ecological importance to mice, shelter and food, in that order. They need to be available abundantly and simultaneously for three to five months for mice to generate high numbers. Such coincidence is rare. The supply of resources usually is out of phase, soil usually being suitable for burrowing in winter, and food abundant in summer. Extraordinary summer rains provide the opportunity for plagues, as explained above, and also protect the mice from the summer heat and aridity.

This hypothesis was tested experimentally in the field by providing abundant food (NEWSOME, 1970). The highest numbers, the equivalent of 875/hectare (the level of abundance in plagues), was generated when the soil was moist and friable during wet weather. Numbers were lowest when the soil was hard and dry in summer, the mice being restricted to moist soil. The hypothesis received natural tests when chance good rain fell in the mid-summer of 1964/65 in South Australia, and in 1968–70 when good rains kept falling throughout two consecutive summers over an enormous tract of southeastern Australia. Minor plagues arose three to five months after the rains in the first instance (NEWSOME & CROW-CROFT, 1971), and great plagues lasting off and on for 18 months over several thousand square kilometers were induced by the second series of heavy summer falls (NEWSOME, unpublished).

In essence, then, the hypothesis demands three simultaneous events for a period of three to five months: (i) moist, friable soil; (ii) food abundantly available; and (iii) mice.

For such an hypothesis to have any general validity in central Australia, we must first look, therefore, for extraordinary rains as precursors to them.

HISTORIES OF SOME RODENT POPULATIONS 1966–71

Methods

As an indicator of the abundance of small mammals available for dingoes *(Canis familiaris dingo)*, small mammal traps (SHERMANS & LONGWORTHS) have been set in Habitats 1 to 4 throughout the study-area (See Fig. 4) almost monthly from March 1966 to December 1971 (NEWSOME & CORBETT, unpublished). Twenty traps were set about 17 m apart in two lines. If possible, each was placed near a fallen log, grass tussock, etc. Traps were set at night and checked in the morning for an average of seven nights in any locality trapped. Traps were set on a total of 389 nights in the six years, generating 8639 trap-nights.

The chief localities visited in three of the Habitats (and whose weather

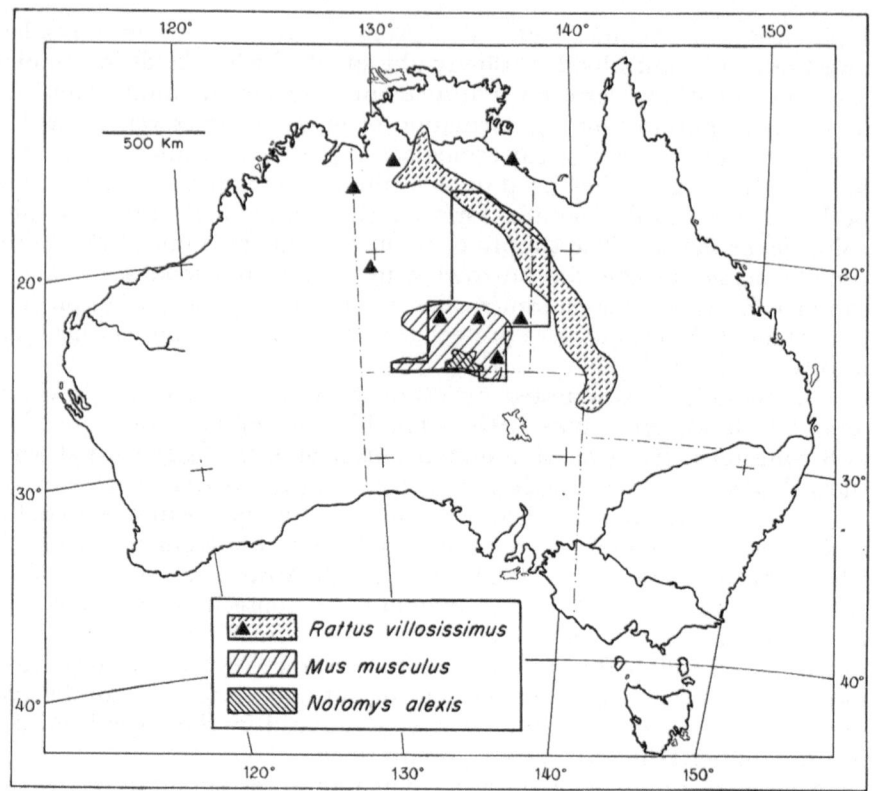

Fig. 6. Geography of rodent plagues 1968–1969 in central Australia and the study-area.

records are used below) were the cattle stations Brunette Downs in Habitat 1 (eleven other stations on the extensive grassy plains of the Barkly Tableland were also visited), Erldunda in Habitat 3, the Shrub Steppe, and Andado in Habitat 4, the Sand-dune Desert (three neighbouring stations were also visited) (see Fig. 5). The Acacia Woodlands of Habitat 2 extend for about 400 km from east to west and about 150 km from north to south. Twenty-six cattle stations were visited in these extensive woodlands, but work was concentrated in the middle of the upper half of the range, within 200 km of Alice Springs (see Fig. 5) whose rainfall records are used here to represent Habitat 2.

Results

Five species of rodent providing 489 specimens were caught (Table 4). Another 78 rodents were caught by hand using spotlights at night. (A

132

few small insectivorous marsupials were also caught: *Sminthopsis crassi-caudata* (seven specimens), *S. froggati* (one specimen), *Antechinomys spenceri* (13 specimens) plus two other unidentified *Sminthopsis*, probably *froggati*).

By comparison with rates of capture in North American deserts (W. Z. LIDICKER, Jr. and J. L. PATTON pers. comm.), these results are meagre. But compared with the usual trapping success in Australian deserts, zero, the results are spectacular. As explained above, we trapped right through plagues of three species of rodent, *Rattus villosissimus*, *Notomys alexis*, and *Mus musculus*. The geography of those plagues is shown in Fig. 6. The data for house-mice are analysed first, to see if the irruption hypothesis for that species can be extended from wheatland to desert.

Mus musculus

House-mice were the commonest rodent caught. There were 239 of them in all; but the majority, 173 mice, were caught in the six consecutive months, from December 1966 to May 1967, during the plague. Note that no house-mice at all were caught in Habitat 1 (see Table 4).

Trapping results are given in Fig. 7 separately for Habitats 2, 3, and 4, the data being standardized as the numbers of mice caught per 100 trap-nights. Also presented and compared in Fig. 7, are periods favouring growth and seeding of native pastures in the three Habitats respectively, calculated from local rainfall (NEWSOME, 1966). Periods of growth began when rainfall exceeded values calculated to be adequate for different localities and months (SLATYER, 1962). (It takes, for example, 5.6 cm of rain in Habitat 1 to keep soil moist and pastures growing in the summer month of January but 6.5 cm in Habitat 4. The respective figures for the winter month of July are 2.8 cm and 1.5 cm). The potential evaporation of water from the soil (0.2 that from a free water surface (SLATYER, 1962) was then subtracted from the actual rainfall week by week to estimate the lengths of growing periods, a technique that proved useful in the study of the red kangaroo *(Megaleia rufa)* in central Australia (NEWSOME, 1965). During calculated periods of growth, it was assumed that the soil was moist and soft enough for burrowing, in accordance with the *Mus* model presented above. And during that time, of course, the desert herbage grew and seeded magnificently.

The widespread heavy rains of January, 1966, broke a long severe drought in central Australia, the worst on record, and one which lasted 6.5 years (Commonwealth of Australia, 1965). Just as rainfall was low in Alice Springs during that drought, amongst the lowest ten percent of recorded annual rainfalls (e.g. 8 cm in 1964 compared with 25 cm on average), so was it high for the subsequent few years: 33 cm in 1966, 24 cm in 1967, and 50 cm in 1968. Such good rains had not been seen in central Australia since the 1947–48's. Native grasses and forbs no sooner seeded than more rain fell inducing fresh bursts of growth and

133

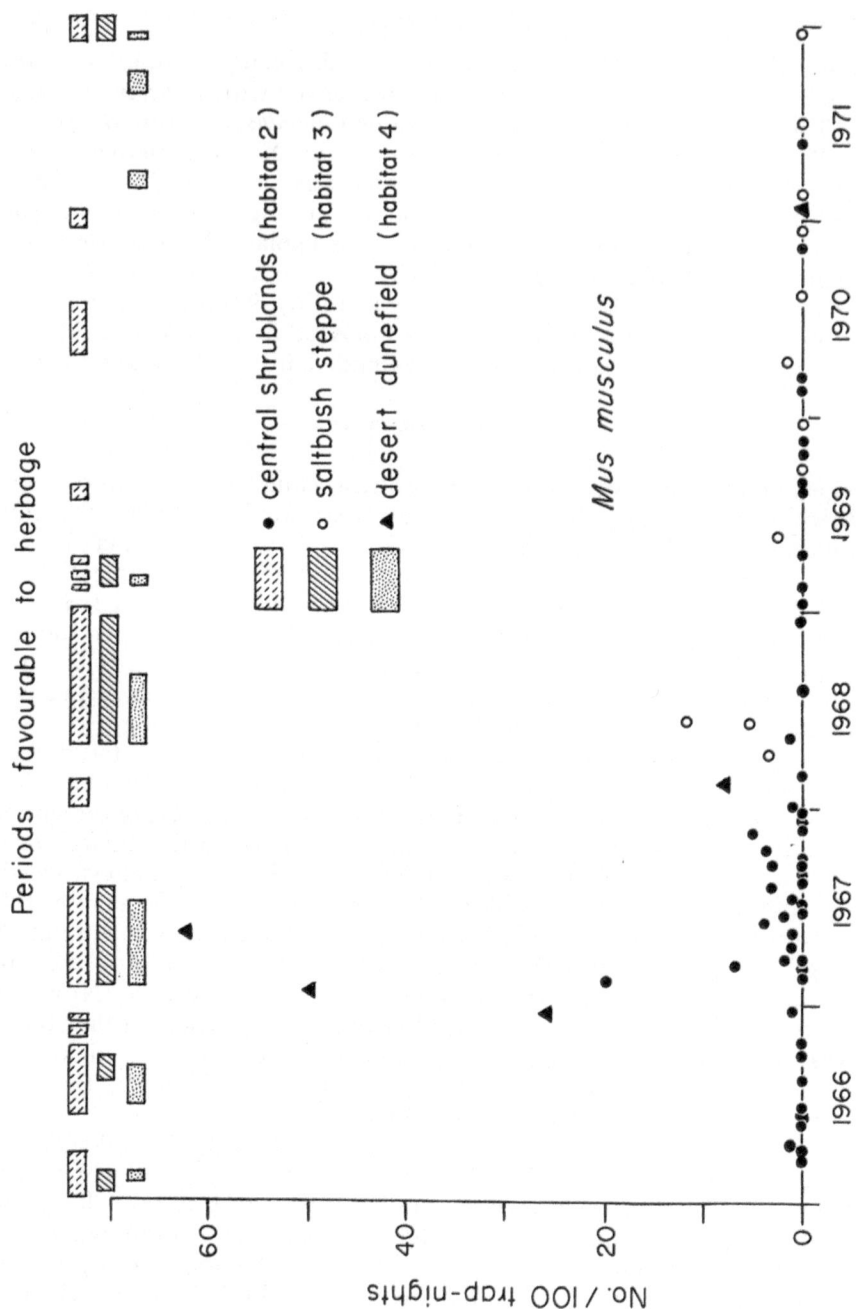

Fig. 7. Population trends of *Mus musculus* on Habitats 2 to 4 in the study-area compared with periods favouring growth and seeding of ground vegetation.

crops of seeds. And this process of rebirth continued for the best part of three years (see Fig. 7).

(i) Habitat 2. Though the mouse plague was first recorded in the Simpson Desert (Habitat 4), results from the Acacia Woodlands (Habitat 2) are discussed first because trapping results from there are most adequate. Note first that few house-mice were caught for about a year after the drought broke, despite flush seasons. By 1967, house-mice had become relatively common throughout the area, though trapping success mostly lay below 5 per 100 trap-nights, rising to 20 per 100 trap-nights in a small valley in the MacDonnell Ranges.

The vegetation continued flush and seeding through 1967 because of further good rains presumably aiding mouse populations. Thereafter, trapping results returned to normal, i.e. mostly to zero, with an almost consistent failure to catch house-mice in the Acacia Woodlands between 1968 and 1971. Given the initial response to rainfall, this decline in abundance was puzzling because it persisted despite further good rains in 1968 (see Fig. 7) that kept vegetation flush right through until early winter in 1969. As explained above, the land had in that time one of the flushest seasons in memory.

It is possible that the relatively dry summers of 1967/68 (see Fig. 7), may have proved inimical to the house-mice but surely no more so than the 6.5 year drought that preceded 1966. If the summers were too hot and dry, perhaps we should look for a delayed numerical response to the second bout of favourable weather similar to the initial response. There is no evidence for it. House-mice became rare in mid-1968 and remained so to the end of 1971. The failure of *Mus* to respond numerically a second time is discussed below.

(ii) Habitat 3. The long drought of central Australia was not broken properly in Habitat 3 (the Saltbush Steppe) until early 1967. There are no trapping results from that Habitat during that year, but residents did not report rodents to be numerous until early 1968. So it seems that populations of house-mice had not built up during the first year after the drought broke in Habitat 3, just as in Habitat 2. Once numerous, however, the mice remained so for about a year, also as in Habitat 2. Then, following the pattern in Habitat 2 again, house-mice became rare with no apparent response to the second good season that came in 1968.

(iii) Habitat 4. Note the results from Habitat 4, the sand-dune desert. The numerical response of the mice did not reflect the local weather as on the other two habitats. Although big rains did not fall there until early in 1967, the peak response of the mice coincided with it. There was apparently no year's delay in response. However, the drought in the Simpson Desert was not broken by the good rains of 1967, but early in 1966, by floodwaters from Habitat 2 to the north. Huge volumes of water flooded into the desert throughout 1966 along usually dry creeks and rivers, notably the Finke, Todd, Hale, and Hay, that originate in the

Table 5. Selected Results for House-Mice

Habitats	Trapping effort			Catch	
	Total trap-nights	Total nights	Total no	Average per 100 trap-nights	Maximum per 100 trap-nights
1. Mitchell Grass Plains	2026	92	0	0	0
2. Acacia Woodlands	4921	219	59	1.2	20
3. Shrub Steppe	968	43	29	2.1	11.2
4. Sand-dune Desert	724	35	151	20.9	62.5
Totals	8639	389	239	2.9	62.5
5. Southern Wheatland*	9314	261	1670	17.9	87.3

* Newsome (unpublished).

MacDonnell and Harts Ranges 250 to 700 km away (see Fig. 5 above). Our trapping on the first two visits to the Simpson Desert was in fact on the flood-plains of two of these streams where the vegetation was lush and soils moist from the extensive floods, and not from local rain. Here, then is an example of the way in which the drainage systems of inland Australia distribute rainfall to benefit distant arid land as FINLAYSON (1939) reported.

Based on average rainfall, vegetation and geomorphology, the Simpson Desert is undoubtedly the most arid of Habitats 1 to 4, yet densities of house-mice there were the highest of any place trapped in central Australia. Table 5 compares selected trapping results for house-mice in the four Habitats studied, and presents data from a plague in the southern wheat-lands as well. The maximum catches are probably the fairest comparison because they remove biases introduced by trapping Habitat 2 for the greatest period of time. Thus, though the reverse was indicated on average catches, the maximum catch of house-mice was greater in Habitat 2 (the Acacia Woodland) than in Habitat 3 (the Shrub Steppe). But the supremacy of Habitat 4 (the Simpson Desert) as a favourable environment for house-mice in good seasons remains. At its best, the desert environment was almost as productive of *Mus* as the fertile wheatlands of Southern Australia at their best.

These results present the paradox that the best habitat for house-mice in central Australia appears to be the least fertile and most arid, the dune-fields of the Simpson Desert, whereas the most fertile, better watered, grassy, black-soil plains of the Barkly Tableland (Habitat 1) yielded no house-mice whatever. This too will be discussed below.

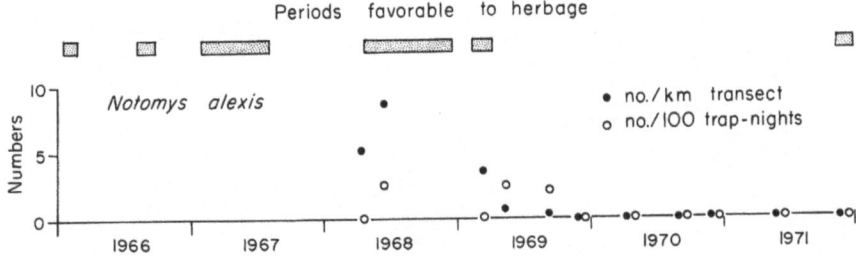

Fig. 8. Population trends of *Notomys alexis* on Habitat 3 (Saltbush Steppe) in the study-area compared with periods favouring growth and seeding of ground vegetation.

Notomys alexis evardensis

No matter where traps are set in central Australia, the chances of catching *Notomys* of any species is remote unless populations are in upsurge. Only one species of the five recorded hopping mice was plentiful during the study in 1968 over a large area of Habitat 3 (the Shrub Steppe), and that was *Notomys alexis*. It was also common about 450 km to the north near Yuendemu (see Fig. 5) in mixed habitat of sparsely treed *(Acacia aneura)* plains and tussock grass *(Triodia)* desert (P. F. AITKEN, in litt.).

Results of trapping *N. alexis evardensis* in Habitat 3 are presented in Fig. 8, along with counts made along an average of 70 km of unpaved roads at night searching with spotlights. Though there are no records for 1966 and 1967, it is known from local inhabitants that the animals were not common then, especially in 1966 because the long drought still persisted there then.

When work first began on Erldunda, *Notomys* was already quite abundant, though none was caught in eight nights trapping (six *Mus*, three *Pseudomys hermannsburgensis* and four *Antechinomys spenceri* (a *Notomys*-sized hopping insectivorous marsupial were caught). We did see 4.3 *Notomys* per km of transect at night. It seems, however, that *Notomys* is a shy trapper. Compare, for example, numbers of four small mammals caught by hand and in traps presented in Tables 4, 6. Note in particular the relative difference between *Notomys* and *Mus*. The former is more readily caught by hand than by trap whilst *Mus* (and *Pseudomys*) are probably the reverse, being less obvious in the abundant ground cover at the time.

That the conventional methods of estimation were too low is sharply illustrated by that hand-caught sample. It was taken in May 1969, when few mice at all were being caught or seen (see Fig. 7). But a heavy down-pour of rain one night flooded the country side for an hour or two, flushing out rodents which were everywhere. In two to three hours, three workers using headlights and spotlights caught 78 rodents, 72 of them *Notomys* (see Table 6)! Also doing well that night was a feral fox *(Vulpes vulpes)*.

137

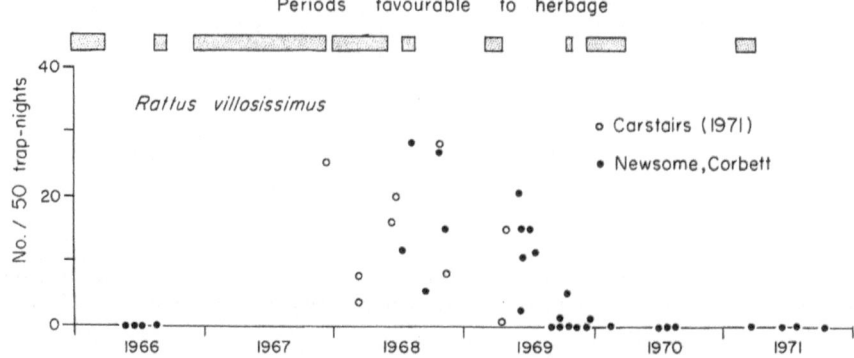

Fig. 9. Population trends of *Rattus villosissimus* on Habitat 1 (Mitchell Grass Plains) in the study-area compared with periods favouring growth and seeding of ground vegetation.

Estimates from both conventional methods are similar, however, and probably demonstrate trends. Populations appear to have been increasing early in 1968, reaching a peak around midyear just as *Mus* did on that Habitat (see Fig. 7). Populations declined throughout 1969 despite a wet year in 1968. Just as described for *Mus* above, there does not seem to have been a response of the same order to this second flush period as for the first (see Fig. 8). Then, during 1970 and 1971, no *Notomys* whatever were caught or seen on Habitat 3 despite 955 trap-nights and about 650 km of night transects. These years were ones of severe drought (see Fig. 8).

Rattus villosissimus

Though known as the plague rat, none may be seen for years. Then, suddenly, populations irrupt into great and extensive plagues (see Fig. 6). The plague discussed here was first reported on the Barkly Tableland (Habitat No. 1) late in 1967, and the rat was quite abundant around Brunette Downs (see Fig. 5) (CARSTAIRS, 1971). That year was so wet that travel on the Tableland, which was mostly sticky mud, was almost impossible, hence the absence of data for that year. The combined results (CARSTAIRS, 1971; NEWSOME & CORBETT, unpublished), indicate, however, the general trend in numbers (Fig. 9). Numbers rose from great rarity in 1966 to great abundance a little over a year later. The plague became extremely widespread (see Fig. 6) and the rats remained abundant for 18 months or more.

After the big rains totalling 92 cm fell in November and December, 1967, the landscape was one enormous series of swamps which, from an altitude of 10,000 m could be seen stretching to the horizon. More heavy

138

rain followed. Monthly totals for January, February, and March, 1967, were 26, 310, and 76 cm respectively. The response of the ground vegetation was amazing and lasted about a year. Then, more good monsoon rain fell the next summer, including a cyclone in March 1968. Grass remained plentiful and seeding from December 1966 well into 1968, drying out gradually thereafter, especially after the poor monsoons of summer 1968/69.

As with *Mus* and *Notomys*, then, great rains induced an excellent and prolonged growing season before the outbreak of *villosissimus* plagues. The rats were a great nuisance. Cattle and horses had hooves nibbled, saddlery was eaten, and people camped away from settlements were bitten in their sleep. To walk about at night with a torch was to see rats every few paces scurrying along worn trails to numerous holes, to see them foraging on the open grassland, and to hear them squeaking and fighting in every quarter. There was no other word than fantastic to describe it all. The scene was reminiscent of a rabbit plague in miniature. A line of 19 traps set over four nights on the western edge of Habitat 1 caught 42 rats in August 1968. Even towards the end of the plague, rodents remained abundant, 84 rats being caught on Brunette Downs in June 1969, with 300 trap-nights.

With the continued drought and poor monsoons of 1969/70, the rats gradually disappeared. Indeed, none has been trapped or reported there since late 1969, though there are locally high populations elsewhere (see below).

The rats appear to have become abundant first in the heart of Habitat 1 centred on Brunette Downs, but soon were numerous throughout the Barkly Tableland. Then, in the middle of 1968, plague rats appeared in large numbers in several unexpected places. They invaded farmland around the township of Katherine 600 km away in the Tropics (see Fig. 5) where they had never been known before. Other reports came in from similar distances to the west, on Humbert River Station (see Fig. 5) where, again, the rats were unknown, not even to old Aboriginals born there (C. SCHULTZ, pers. comm.). Rats became common even further west, at the Western Australia border, a distance of about 450 km from the Barkly Tableland. In late 1968, we found a rat a similar distance to the south in yet another alien habitat, Habitat 2, the Acacia Shrublands of central Australia. Others were found 800 km south of the northern edge of the Simpson Desert (Habitat 4) (A. NEWSOME & D. STEPHENS, unpublished) where their tracks were extremely abundant in some small localised areas. In 1972, a colony was located even further south again in that most unlikely habitat. These and other (PARKER, 1973) outlying localities are indicated by triangles in Fig. 6.

Another very large unidentified rat was seen at night in Habitat 3 after that sharp downpour of rain in May 1969, mentioned above. There is no rat usual to the locality that is so large. It was too elusive to catch, but it would have been no surprise if it had been *R. villosissimus*, which was so widely dispersed at the time. Also, a litter of four baby rats (probably *R. villosissimus*) were caught 600 km further north where *N. alexis* was also in plague (J. EDWARDS, pers. comm.; P. AITKEN, in litt.).

Regardless of the uncertainties, *R. villosissimus*, in a plague rated as one of the greatest on record, suddenly appeared in localities where they had never been seen before. Peripheral records appear to have been the results of a great and rapid dispersal over many hundreds of kilometres radiating out from the Barkly Tableland. Had such large rats been familiar to residents in these areas, they would have been recorded long since, and certainly have been no strangers to Aboriginals. Colonising waves of *villosissimus* have been recorded by FINLAYSON (1939) in the region of Goyder's Lagoon (see Fig. 5) having come down the rivers from higher up, following flood-waters. Waves of invading rats have also been reported on other river-systems of the L. Eyre Basin, e.g. the Macumba River (see Fig. 5) (G. PAGE, pers. comm.).

One other recent invasion of alien habitat by *R. villosissimus* is worth recounting. Following very good rains in summer 1952–53 on the Barkly Tableland (Habitat 1) plague rats passed through the Acacia Woodland (Habitat 2) in 1954, in a front about 60 km wide passing 100 km north of Alice Springs (see Fig. 5) moving from east to west (D. STEPHENS, pers. comm.).

A recent outbreak of *R. villosissimus* arose near Boulia (see Fig. 5) in January 1972 (B. BOLTON, pers. comm.). This plague demands special comment, for no rains preceded it locally. Cyclonic rains outside the arid zone about 800 km to the north had sent great floods down usually dry rivers and creeks, spreading water out over large areas of land that had had no rain. Presumably, residual populations of rats from the earlier plagues there (see Fig. 6) were able to capitalise on this bounty, just as *Mus* appears to have done in the Simpson Desert in 1967.

Summary of the three rodent plagues

Results presented here indicate that the recent plagues of *Mus*, *Notomys* and *Rattus* in central Australia were the result of prolonged bouts of extra-ordinarily favourable environmental conditions that were decidedly non-arid. Great and continuing rains locally, or flood-waters from heavy rain falling hundreds of kilometers away, produced spectacular flushes of herbage. Grasses and forbs were stimulated into a cycle of production, one burst of growth and flowering being followed by another, as more rains fell. The result was a high level of seed production and moist friable soil for periods of a year or more. These are the elements of the irruption hypothesis for *Mus*. High densities of rodents prevailed for long times, culminating in a great dispersal of *R. villosissimus* at least into distant alien habitats.

140

It was no surprise that the rodents disappeared when the normally dry weather returned. But why did population declines begin long before the dry weather returned, indeed, in the presence of continuing good seasons? Why too did populations take so long to respond to the onset of favourable conditions? The answer may lie in predation.

PREDATION

The mammalian bounty provided by the rodent plagues attracted predators. During them, barn owls *(Tyto alba)* became abundant. They were often seen, and their harsh screech, usually so rare, became a common nightly event. Mammalian predators, the dingo *(Canis familiaris dingo)*, the feral European fox *(Vulpes vulpes)* and feral domestic cat *(Felis cattus)* were certainly eating rodents, and their numbers grew.

There is not the array of predators in Australia as in other continents, far from it. Yet it has been possible to detect predator-prey relationships of importance to the prey's life-cycles. These relationships will be analysed in detail elsewhere (NEWSOME, CORBETT & BEST, unpublished), but some aspects are introduced here.

A Simple Predator-Prey Set

The simplest predator-prey set to analyse came from Habitat 1, the Barkly Tableland. There, only one rodent was abundant, *R. villosissimus* (see Table 1), and the dingo was far and away the main mammalian predator, though the European fox and domestic cat are wild there. Avian predators included the letter-wing kite *(Elanus scriptus)* and the barn owl *(Tyto alba)* (PARKER, 1973). Data is presented here for the dingo and the rat.

Dingoes were caught in fourteen samples between 1968 and 1970, during the rats' increase and subsequent decline to rarity. Fig. 10 compares the proportion of dingoes eating rats with their abundance (number per 50 trap-nights).

The relationship is not quite the functional response of SOLOMON (1949) but nearly so; and there are comparisons with HOLLING's (1961) theoretical predator-prey curves which will be discussed elsewhere. The important point is that there were different curves for the two phases of the rats' population growth. During the rats' decline, the curve shifted sharply to the left, co-inciding with the ordinate for much of the length. Thus, even though little or no sign of rats remained, up to half its dingoes were eating them indicating severe predation upon declining rat populations. Such predation pressure should hasten the decline of prey populations and perhaps prevent their early upsurge, as found for voles *(Microtus californicus)* in California (PEARSON, 1966, 1971).

If dingoes could survive on alternate food after rats became rare (and

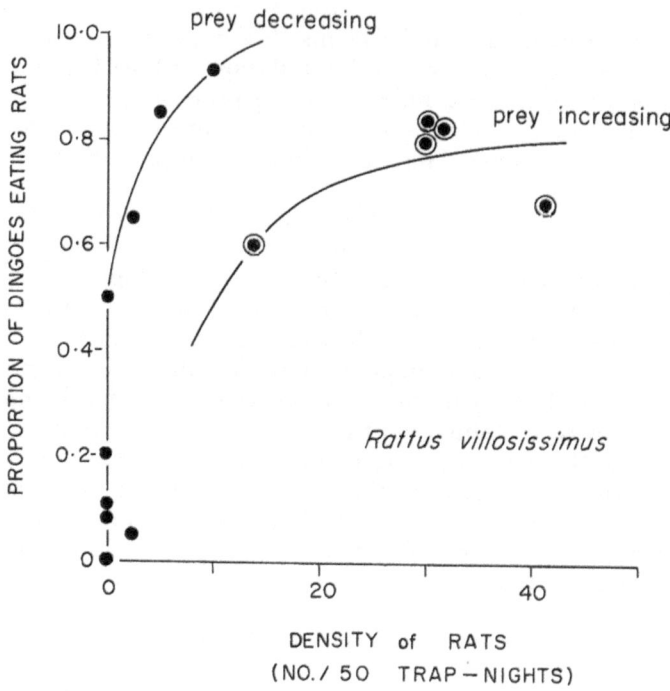

Fig. 10. Predator-prey relationships between *Canis familiaris dingo* and *Rattus villosissimus* as rat populations increased and decreased on Habitat 1 (Mitchell Grass Plains of the Barkly Tableland).

they do) but never-the-less retain their specific search image (TINBERGEN, 1960) for them (two generations of dingo pups would have been reared entirely on rats in 1968 and 1969), such severe predation pressure could prevent the scattered remnants of surviving rat populations from taking advantage of any return of suitable environmental conditions. It also could conceivably delay the initial response of rats to a favourable resource base.

A Complex Predator-Prey Set

Results from the other Habitats (2 and 4) provide predator-prey relationships too complex for analysis here, for there were several species of both predator and prey. Results indicate, however, that, whilst all mammalian predators (dingo, fox, cat) were preying upon *Mus* and *Notomys* (and *Pseudomys*) initially, dingoes eventually turned almost completely to the more slowly increasing but ultimately abundant European rabbit *(Oryctolagus cuniculus)* thus relaxing the pressure on the

142

rodents (NEWSOME & CORBETT, unpublished). Yet the rodents were unable to capitalise on the second bout of excellent seasons. The answer may lie with the other predators.

The European fox and domestic cat and the barn owl became common during these plagues as mentioned above. And so, probably, did snakes though few were seen. Some species would hunt rodents, e.g. the brown snake *(Demansia textilis)*, the mulga snake *(Pseudechis australis)* and several members of the Pythonidae. It is suggested that their combined efforts on the combined populations of rodents may provide a complex example of the predator-prey set described in Fig. 10. Predation pressure may have kept prey populations down despite the second and even third consecutive rounds of excellent environmental conditions.

It is standard folklore in inland Australia that plagues of rodents are followed by plagues of snakes, and then wild domestic cats. During the period under review here, no plagues of snakes were reported, but the cats were numerous. After the abnormal good seasons passed and the usual arid weather returned to central Australia, not only did we see cats often, but also reports of cats in unusually large numbers around isolated settlements were common. For example, in March and April, 1970, around Ayres Rock (see Fig. 5), a National Park to the west of Habitat 4 in desert country where rodents had been abundant, no less than 180 cats were shot around the settlement in a period of five weeks (W. T. HARE, pers. comm.). Stomach contents (analysed by D. HOWE) revealed that they had been scavenging on the rubbish dump. All were in poor physical condition and eight were found dead, apparently of starvation. These cats must have been forced out of the surrounding desert when the rodents, which had been numerous till 1968, disappeared. Few cats have been seen since.

Conclusions and Discussion

CAUSATION AND THE GENERALITY OF THE MUS HYPOTHESIS

The results of the study confirm FINLAYSON's (1939) conclusion that population irruptions of inland rodents are caused by unusually prolonged vegetative flushes. They also confirm that such plagues can be the result of flood-waters from distant heavy rains. FINLAYSON's conclusions were based on irruptions of two native rodents, *Pseudomys australis* and *Rattus villosissimus*, as well as on the alien *Mus musculus*. Results presented here extend the conclusion to include *Notomys alexis* also.

These conclusions permit the extension of the *Mus* hypothesis for plague formation in southern wheatlands to *Mus* in arid and semi-arid lands further north, and to three species of native rodent as well. In all probability, all plagues of inland rodents have the same cause – a prolonged super-abundance of resources. And that super-abundance, as shown here, is in turn caused by unusually favourable weather, or series of heavy rains (or floods rising from them) erasing the usual aridity. Such rains come as chance climatic events in central Australia, once every ten to fifteen years.

The points of departure from the *Mus* model are: (i) the year-long delay in response of the rodents in the arid zone (compared with only three to five months in wheatlands); and, (ii) more importantly, not all bouts of favourable weather result in plagues. The extra factor suggested for these differences is predation.

PREDATION

Predation in central Australia is seen as an intense modifier of population trends for all three rodents studied here much as PEARSON (1966; 1971) found for voles *(Microtus californicus)* in California. The predator's impact was probably not so much on peak populations as on low populations, or those in decline. The shifting predator/prey relationships are seen as follows:

At the end of the long drought that preceded this study, predator populations were likely to be concentrated on small scattered pockets of surviving prey populations. The predator/prey ratio probably favored the predators in most places then, delaying any response of rodent populations to the sudden advent of favourable conditions. With a prolonged and widespread increase in the resource base, a general increase in rodents from breeding and/or the emigration of a few predators to sites where alternative prey populations (birds, lizards and insects) were also increasing, could have shifted the ratio in favour of the prey over a few square kilometers – and that may have been enough to release the prey. Predator populations, however, also increased. Eventually, the rodents ran into trouble, perhaps through the return of drought or floods, and the ratio shifted in favour of the predators once more. No amount of extra rain, food, etc., at such a time could build up the prey populations again.

With continued scarcity of prey, the predators too could be expected to run into trouble. Indeed, feral cats in the desert appear to have been starving (see above). Their influx into isolated settlements came roughly two years after the peak in rodent populations in the area. Presumably, populations of cats had been supported meanwhile on alternative prey, especially lizards and insects.

The predator/prey system in central Australia appears to have, then, some classical elements found in the Northern Hemisphere. The results are not cyclical, however, as in California (PEARSON, 1966, 1971) and further north (PITELKA, 1958), the central Australian weather being far too erratic. In California at least, the weather is favourable every year allowing maintenance of high populations of diverse predators, whereas

144

Table 6. Numbers Caught by Different Methods at Erldunda*

	Notomys	*Pseudomys*	*Mus*	*Antechinomys***
Caught by hand	72	2	4	1
Trapped (333 trap-nights)	8	9	22	2

* Data amalgamated for June 1968, and March and May 1969.
** A small marsupial.

in central Australia, predators will exert a major suppressing effect only when two favourable periods are not too far apart.

The indications from this study, from the failure of rodent populations to respond a second time and from the feral cat populations, are that the time intervals between successive favourable periods in central Australia would need to exceed at least two years for predation not to suppress rodent populations. In the mono-culture of southern wheatlands where predators are few, the main effect on mouse populations is the weather (NEWSOME, 1969a, b; 1970), hence the rapid response (3–5 months) of *Mus* there.

Rather interestingly, plagues of *Rattus villosissimus* more recent than those analysed here have irrupted in two isolated localities despite widespread floods and rains that appear to have been adequate for much more widespread plagues. These localities were near Boulia (B. BOLTON, pers. comm.) and Andado (see Fig. 5), in 1972. Elements of a mono-culture may have been introduced in these two places artificially. Boulia is in an area where dingo populations had been severely reduced in 1971 by poisoning, and Andado is in one of our general study-areas for dingoes where populations are cropped regularly, about every three months (NEWSOME, CORBETT & GREEN, unpublished).

REFUGIA

Though this hypothesis for plague formation (and prevention or modification by predators) in inland Australia requires substantiation with careful population studies, there are a few associated problems that can be discussed here, e.g., where and how do rodents survive long droughts, and are there permanent refugia? Let us look at *Rattus villosissimus* first.

The evidence, though patchy, indicates that the Barkly Tableland and the rivers draining to Lake Eyre (see Fig. 5) must provide refugia for remnant populations of *R. villosissimus* during drought. These remnants must breed up in favourable times, disperse, and colonise the surrounding grassy plains, an extensive, usually marginal habitat improved by ex-

145

Table 7. *Rattus villosissimus* caught in Different Habitats as the Plague Subsided

	Rats	Trap-nights
Open clayey black-soil plains	8	80
Lightly timbered, sandy soils	62	200
Lightly timbered black soil flanking a water-hole	9	11
	79	291

cellent rains. Apparently, also, rats disperse into alien territory as well, presumably enabling them, in a great colonising drive, to locate and occupy refugia for many hundreds of kilometers around the out-break centre.

Analysis of trapping data in different habitats on the Barkly Tableland (Habitat 1) in June and July 1969, during the population decline of *R. villosissimus*, indicate a patchy distribution of rats (see Table 7). Though the rats infested the vast open black-soil plains in large numbers during plagues, such areas supported fewer individuals during declines than the minority habitats, and none during drought (see Fig. 8). Minority habitats all had a light covering of *Acacia cambadgei* woodland on a sandy substrate where rats could burrow deeply, or else had a better moisture regime near a waterhole. The clayey plains develop deep wide cracks in them in dry weather allowing hot dry air to penetrate deeply, whereas the depths of the sandy soil may be better protected. PARKER (1973) reports nuclear colonies inhabiting reedy springs and other wet densely vegetated spots in the drainage systems in the north-east of the Lake Eyre Basin. Refugia for *R. villosissimus* may be, then, the moist minority habitats as with *Mus* in wheatlands (NEWSOME, 1969b). It might be added at this point that *villosissimus* may possibly be a sub-tropical ecological analogue of the introduced house-mouse. Perhaps for this reason, the two species do not appear to co-exist on the Barkly Tableland (Habitat 4), but see below.

It is likely that *Mus* and *Notomys* face similar perhaps more severe problems in the more arid habitats. There are no data for *Notomys*, but some for *Mus*. During the dry years, 1969 and 1970, a few mice were caught here and there; but, given their previous abundance in many habitats and localities, some mice might have been expected to hang on for a short while, and these captures do not necessarily indicate refugia. On the other hand, it came as no great surprise when the exact same sand-dune in Habitat 4 that yielded so many house-mice in May 1967 provided none in January 1971. Indeed, it was hard to see just where mice could have been harboured in that region during the prevailing heat-wave and drought.

146

An occasional mouse has been dug out of holes near stock-waters that are spaced every 5 to 15 km across the land. A mouse was found in a rather bizarre spot near such a water-hole, under a cow carcass. Possibly the mouse was foraging for food from the dead cow's paunch. Such foraging by small mammals may not be unusual because D. LINDNER (unpublished data) caught about 50 specimens of a small insectivorous marsupial *(Planigale ingrami)* under cow carcasses near a bore in Habitat 1 in March 1971, living on dermestid larvae. Mice living around stock-waters may have a ready supply of seeds provided in cattle dung deposited copiously around waters. Cattle ranging up to 10 km away from waters will harvest seed and deposit some undigested near the water. Such un-prepossessing bounty is certainly utilised by two species of pigeon, the flock pigeon *(Histophaps histrionica)* of Habitat 1, and the crested pigeon *(Ocyphaps lophotes)* of all four Habitats.

In 1964, during the long drought that preceded this study, Dr. and Mrs. W. Z. LIDICKER Jr. were trapping on Habitats 2 and 4, and have very kindly made their data available. In open plains in Habitat 2 (Acacia Woodlands), 384 trap-nights yielded no *Mus;* but in minority habitats, rocky gorges and valleys in the MacDonnell Ranges, 363 trap-nights yielded 5 *Mus* (the highest capture-rate in Habitat 2 reported in this study was also in those Ranges: 20 *Mus* for 100 trap-nights). It so happened that there had been one or two cm of rain a month or so before the LIDICKERS trapped there, enough for the winter Compositae to sprout and flower in places. The mice caught were breeding, males being active and one female exhibiting a copulation plug. But, in Habitat 4, the arid dune-field desert, their catch was best just as in this study, with 20 *Mus* in 216 trap-nights, yet there had been no very recent rain there, the grasses all being dry. Perhaps there had been some a few months earlier for the mice showed evidence of recent breeding. Some were sub-adult indicating successful breeding within the previous couple of months, and some females caught had placental scars. If there had not been any early rain, then it may be that some sand-dunes are refugia.

In thinking about possible refugia for any of these three species of rats (and presumably other desert rodents as well) it is impossible not to highlight moisture in so arid an environment and to link plagues and dispersal to survival during subsequent droughts. The bigger the plague and the wider the dispersal, as in *R. villosissimus* for example, the greater the chance of locating suitable refugia and ensuring the survival of a nuclear population somewhere. It is not just that individual rodents have to survive drought, but also that some must breed. Rodents are too short-lived and some droughts too long for it to be otherwise. Their widespread dispersal is probably an evolutionary reflection of the uncertain rainfall, which leads us to suspect that, though there may be fixed habitats most likely to provide refugia, the actual refugia may *shift* from time to time, depending on the geographic variability of the rainfall. Rodents may survive in one place one time, but not the next. Predation cannot be under-estimated either. The data presented here indicates very strong predation pressure on declining populations, and presumably even more so on residual populations, perhaps leading to local extinctions. Widespread dispersal may serve to evade predation to extinction.

In summary, it is suggested that refugia probably exist more often than not in special minority habitats, but not all possible refugia are suitable micro-habitats in all droughts. Whereas population irruptions may depend on one set of stochastic events, a run of unusually heavy rains, survival of nuclear populations of rodents may depend on another, the geographic pattern of the small showers that make up the usual rainfall.

Rarity, spasmodic plagues, low species diversity, and the super-position of mus

Let us now return to the four problems posed in the Introduction. It has been concluded that the general rarity of the native rodents coupled with their spasmodic plagues are due in the main to the widely and erratically fluctuating resource base, and to predation. The other two problems are less tractable. Figure 10 summarises relationships between estimated maximum densities and biomass of *Mus* and the sympatric native rodents caught in the four Habitats in this study. The linear log-log relationships indicate a steady rate of increase in *Mus* with a steady rate of decrease in the native contingent, and *vice-versa*. This relationship implies an increasing competitive displacement by *Mus* with decrease in abundance and biomass (not diversity) of the native species. Least inroads have been made against the large, numerous *R. villosissimus* in the sub-tropical grasslands and most in the Simpson Desert where native rodents are few and uncommon. The biggest faunistic difference in the other two Habitats is that the medium-sized *Notomys alexis* forms plagues in the Saltbush Steppe but not in the Acacia Scrub. *Mus* was less abundant on the former than the latter. So, though there may be physio-ecological problems facing *Mus* as an invader in the various Habitats, particularly towards the north, the major determinant of success may be the biomass of native fauna.

After the Simpson Desert, where faunal complexity is low (three species, *R. villosissimus*, *N. cervinus* and *N. fuscus* are probably interlopers), the next best Habitat for *Mus* was the Acacia Scrub where the faunal complexity is at its highest with nine native species endemic and a tenth an occasional invader. The large and medium-sized local rodents are all rare (one, *Leporillus*, is perhaps extinct), and only a few of the small rodents, two species of *Pseudomys*, are at all common. Table 8 compares the abundance and size of endemic rodents within the study-area.

It is striking that, whereas all size-ranges have species that are rare

148

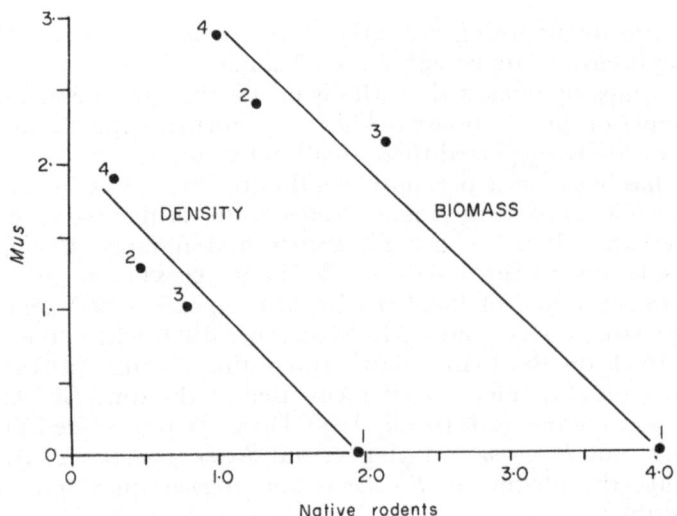

Fig. 11. Log-log plot of maximum densities and biomass of *Mus musculus* and of native rodents caught in Habitats 1 to 4 during the study. To avoid zeros, one was added to all values.

Table 8. The Relative Size and Abundance of Native Rodents in the Study-Area

Size	No. of Species			
	Total	Forming Plagues	Caught Occasionally	Not Caught
Large (50–200 g)	2	1[a]	0	1[b]
Medium (20–40 g)	7	1[c]	0	6[d]
Small (10–15 g)	4	0	2[e]	2[f]

[a] *Rattus villosissimus* (Habitat 1; invading Habitats 2 and 4).
[b] *Leporillus apicalis* (Habitat 2 and ? 3), probably extinct.
[c] *Notomys alexis* (Habitats 1, 2, and 3).
[d] *N. longicaudatus, Zyzomys pedunculatus, Rattus tunneyi* (Habitat 2); *N. cervinus, N. fuscus* (probably interlopers), *N. amplus* (Habitat 4).
[e] *Pseudomys hermannsburgensis* (Habitats 1 to 4), *Ps. forresti* (Habitats 1 and 2).
[f] *Ps. desertor, Ps. fieldi* (Habitat 2).

149

and those forms irrupting are either large or medium-sized, those that are in the intermediate category are all small.

It is perhaps significant that *Mus* is in this size group and that it does form plagues on the study-area. These facts further support the idea that *Mus* has partially displaced these small native mice.

If *Mus* has been a usurper, it follows that the small *Pseudomys* too should have the capacity to form plagues. Indeed, two of them appear to do so. Mr. P. AITKEN (1972) reports *Ps. desertor* in abundance, having found a 'vast population' in the Lake Eyre Basin P. AITKEN (in litt.). He also reports 'an explosion' of *Pseudomys hermannsburgensis* near Yuendemu (see Fig. 5) in late August and early September 1966 when 'they were incredibly thick on the loamy flats'. Interestingly, *Mus* was caught only in and around Yuendemu settlement itself at the time (R. AITKEN, in litt.), i.e., as a commensal. In July 1968, Dr. C. WATTS visited Yuendemu, and found, not *Pseudomys* in plague, but *Notomys alexis* (P. AITKEN, in litt.). Thus, the plagues of *Pseudomys* and *Notomys* may have been displaced in time.

Thus there is evidence that four of the eleven native rodents permanently resident in the study-area are *r*-strategists (MACARTHUR & WILSON, 1967; PIANKA, 1972). They are *Rattus villosissimus*, *Notomys alexis*, *Pseudomys hermannsburgensis* and *Ps. desertor*. Into this mélange, however, stepped a superb *r*-strategist, *Mus musculus*, apparently displacing the latter two species and possibly *Ps. forresti* and *Ps. fieldi* also. The other rare rodents *Rattus tunneyi*, *N. amplus*, *N. longicaudatus*, *Zyzomys pedunculatus* and *Leporillus apicalis* are all of medium size. The cause of their rarity is not known.

Given so capricious an environment as central Australia, it is hard to dismiss the possibility that the only way for a rodent to survive there would be as an *r*-strategist. The converse would be that *K*-strategists (MACARTHUR & WILSON, 1967; PIANKA, 1972) could not survive or could do so poorly. So, alongside such likely causes of rarity as increased predation* and the super-position of *Mus* as a competitor with the small species at least, must be ranged changing climate. If, as seems likely, inland Australia was even more arid in mid-Recent times (estimates vary anywhere from 4000–10,000 years B.P. (see MERRILEES (1968) for a review)) than now, the climate ameliorating a little since then but not beyond the present arid and semi-arid regime, perhaps we could expect *K*-strategists to be rare or even to disappear leaving only the opportunists. Such could be the explanation of low species diversity of rodents in central Australia.

---

* From the alien European fox and cat, and possibly also from dingoes whose numbers may have increased as the cattle industry became established (NEWSOME, CORBETT, BEST & GREEN, unpublished).

Acknowledgements

Many people have generously helped with the preparation of this Chapter.

First and foremost, our grateful thanks must go to other members of the Dingo Project who did most of the sampling in central Australia, often under arduous conditions, in particular, Mr. L. BEST, Mr. P. HANISCH, Mr. H. WAKEFIELD and Mr. R. BURT. We are also most grateful to them for the assemblage of much of the raw data. Mr. P. F. AITKEN, Mr. J. H. CALABY, Dr. W. Z. LIDICKER, and Mr. J. MAHONEY have been most generous and helpful with faunal lists, distribution and identifications.

We are most grateful also to various members of the Museum of Vertebrate Zoology, University of California, Berkeley, California, especially Dr. W. Z. LIDICKER, Dr. O. P. PEARSON, and Dr. F. PITELKA for their considerable help in discussions. Dr. and Mrs. W. Z. LIDICKER most generously provided results of their trapping in central Australia in 1964. Drs. LIDICKER and PEARSON very kindly read the manuscript, and their criticism and ideas have helped it greatly. The splendid figures were drawn by Mr. G. CHRISTMAN, and Mrs. I. O'CONNOR typed the manuscript.

The extensive rodent trapping in central Australia could not have been possible without financial support for the Dingo Project from the Australian Meat Research Committee Projects CS 17, CS 85.

To all these people and others who have helped, we extend our most grateful thanks.

REFERENCES

AITKEN, P. F. 1968. Observations on *Notomys fuscus* (Wood Jones) (Muridae-Pseudomyinae) with notes on a new synonym. *S. Aust. Nat.* 43: 37–46.

AITKEN, P. F. 1972. *Planigale gilesi* (Marsupialia, Dasyuridae); a new species from the interior of south-eastern Australia. *Rec. S. Aust. Mus.* 16: 1–14.

CARSTAIRS, J. 1971. The correlation of anatomical changes with population changes in *Rattus villosissimus* during the 1966–1969 plague. Ph. D. Thesis, Monash University.

CLELAND, J. B. 1918. Previous phenomenal visitations of rats and mice in Australia. *Proc. Roy. Soc.* N.S.W. 52: 123–165.

Commonwealth of Australia. 1965. Meteorological aspects of the recent drought in central Australia. Meteorological summary (September 1965). Commonwealth Bureau of Meteorology, Melbourne).

151

DUNNET, G. M. 1956. Preliminary note on a rat plague in north-west Queensland. *CSIRO Wildl. Res.* 1: 131–132.

FERTIG, D. S. 1968. Water relationships of feral house mice, *Mus musculus* Linnaeus, from a coastal salt-marsh habitat. Ph.D. Thesis, Univ. So. Calif., Los Angeles.

FERTIG, D. S. & EDMONDS, V. W. 1969. The physiology of the house-mouse. *Scient. Amer.* 221: 103–110.

FINLAYSON, H. H. 1939. On mammals of the Lake Eyre Basin Pt. V. General remarks on the increase of murids and their population movements in the Lake Eyre Basin during the years 1930–1936. *Trans. R. Soc. S. Aust.* 63: 348–353.

FINLAYSON, H. H. 1941. On central Australian mammals Pt. II. The Muridae. *Trans. R. Soc. S. Aust.* 65: 215–231.

GIBBS, W. J. 1972. Meteorology and climatology. In. PERRY R. A. (ed.). Studies of the Australian Arid Zone pp. 33–54 (CSIRO Australia).

HALL, E. R. & KELSON, K. R. 1959. The Mammals of North America (New York; 1083 pp.).

HOLLING, C. S. 1961. Principles of insect predation. *Ann. Rev. Ento.* 6: 173–182.

MACARTHUR, R. H. & WILSON, O. 1967. The Theory of Island Biogeography. Monographs in Population Biology 1. (Princeton).

MACMILLEN, R. E. & LEE, A. K. 1967. Australian desert mice: independence of exogenous water. *Science*: 158: 383–385.

MACMILLEN, R. E. & LEE, A. K. 1970. Energy metabolism and pulmocutaneous water loss of Australian hopping mice. *Comp. Biochem. Physiol* 35: 355–369.

MACMILLEN, R. E., BAUDINETTE, R. V. & LEE, A. K. 1972. Water economy and energy metabolism of the sandy inland mouse, *Leggadina hermannsburgensis. J. Mamm.* 53: 529–539.

MEIGS, P. 1953. World distribution of arid and semi-arid homoclimates. In. Arid Zone Programme I. Reviews of Research on Arid Zone Hydrology: 203–210. (UNESCO. Paris).

MERRILEES, D. 1968. Man the Destroyer: Late Quarternary changes in the Australian marsupial fauna. *J. Roy. Soc. West. Aust.* 51: 1–24.

MOORE, R. M. & PERRY, R. A. 1970. Vegetation. In Moore, R. M. (Ed.). Australian Grasslands. pp. 59–73. Sydney.

NEWSOME, A. E. 1965. Reproduction in natural populations of the red kangaroo, *Megaleia rufa* (Desmarest), in central Australia. *Aust. J. Zool.* 13: 735–759.

NEWSOME, A. E. 1966. Estimating the severity of drought. *Nature* 209: 904.

NEWSOME, A. E. 1969a. A population study of house-mice temporarily inhabiting a South Australian wheatfield. *J. Anim. Ecol.* 38: 341–359.

NEWSOME, A. E. 1969b. A population study of house-mice permanently inhabiting a reed-bed in South Australia. *J. Anim. Ecol.* 38: 361–377.

NEWSOME, A. E. 1970. An experimental attempt to produce a mouse plague. *J. Anim. Ecol.* 539: 299–311.

NEWSOME, A. E. & CROWCROFT, W. P. 1971. Outbreaks of housemice in South Australia in 1965. *CSIRO Wildl. Res.* 16: 41–47.

PARKER, S. A. 1973. An annotated checklist of the native land mammals of the Northern Territory. *Rec. S. Aust. Mus.* 16: 1–57.

PEARSON, O. P. 1966. The prey of carnivores during one cycle of mouse abundance. *J. Anim. Ecol.* 35: 217–233.

PEARSON, O. P. 1971. Additional measurements of the impact of carnivores on California voles *(Microtus californicus)*. *J. Mammal* 52: 41–49.

PITELKA, F. A. 1958. Some characteristics of microtine cycles in the Arctic. *Eighteenth Ann. Biol. Coll. Oregon State College* pp. 73–88.

PERRY, R. A. & CHRISTIAN, C. S. 1954. Vegetation of the Barkly Region. *CSIRO Austr. Land. Ser.* No. 3: 78–112.

PERRY, R. A., MABBUTT, J. A., LITCHFIELD, W. A. & QUINLAN, T. 1962. Land Systems of the Alice Springs area. In Lands of the Alice Springs Area, Northern Territory, 1956–1957. *CSIRO Aust. Land Ser.* No. 5: 20–108.

PERRY, R. A. 1970. Arid Shrublands and Grasslands. Sydney. In MOORE, R. M. (Ed.) Australian Grasslands, pp. 246–259.

PIANKA, E. R. 1972. r and K selection or b and d selection. *Amer. Nat.* 106: 581–588.

PLOMLEY, N. J. B. 1972. Some notes on plagues of small mammals in Australia. *J. nat. Hist.* 6: 363–384.

RIDE, W. D. L. 1970. A Guide to the Native Mammals of Australia. Melbourne (249 pp.).

SLATYER, R. O. 1962. Climate of the Alice Springs area. In: General Report on Lands of the Alice Springs Area, Northern Territory, 1956–1957. *CSIRO Aust. Land Res. Series No. 6*: 109–128.

SOLOMON, M. E. 1949. The natural control of animal populations. *J. Anim. Ecol.* 18: 1–35.

STANLEY, M. 1971. An ethogram of the hopping mouse, *Notomys alexis*. *Z. Tierpsychol.* 29: 225–258.

TAYLOR, J. M. & HORNER, B. E. 1973. Systematics of native Australian *Rattus* (Rodentia, Muridae). Results of the Archbold Expeditions No. 98. *Bull. Amer. Mus. Nat. Hist.* 150: 1–130.

TINBERGEN, L. 1960. The natural control of insects in pinewoods. I. Factors influencing the intensity of predation of songbirds. *Arch. neerl. Zool.* 13: 217–228.

TROUGHTON, E. 1947. Furred Mammals of Australia. New York (374 pp.).

WATTS, C. H. S. 1972. Handbook of South Australian Rodents and small Marsupials. Adelaide (29 pp.).

Manuscript written whilst the senior author was a Research Associate of The Museum of Vertebrate Zoology, University of California, Berkeley, California, U.S.A.

VII. OBSERVATIONS OF ARGENTINE DESERT RODENT ECOLOGY, WITH EMPHASIS ON WATER RELATIONS OF *ELIGMODONTIA TYPUS**

by

M. A. MARES

Introduction

The ecologist's predictive powers concerning the adaptedness of desert rodents to the aridity component of their niche are quite limited. Adaptation in this niche parameter can be visualized as occurring along a continuum from few and simple behavioral traits at one extreme (e.g. nocturnality, obtaining water from vegetation), to a combination of behavioral and/or physiological mechanisms (e.g. nocturnality, burrowplugging, burying food in humid microenvironments, urine-concentrating ability, and so forth) at the other extreme. Thus dependence on free or vegetational water in an arid environment is near one endpoint, exemplified perhaps by *Neotoma* or *Peromyscus* of the southwestern United States (MACMILLEN, 1964; SCHMIDT-NIELSEN, 1964), grading to complete independence of free water and the complex adaptive mechanisms that this entails. Examples near this endpoint would include *Dipodomys* and *Perognathus* of the United States and Mexico (see SCHMIDT-NIELSEN, 1964 for review) or *Dipus* of the Egyptian desert (KIRMIZ, 1962 as cited by SCHMIDT-NIELSEN, 1964).

Since evolution is often a complicating process, it is tempting to equate the position a species occupies along this theoretical continuum with its degree of adaptation (STERN, 1970). Yielding to the temptation, I will define the species more adapted to a particular niche parameter as those that would evidently have the greatest probability of transmitting their adaptations to their offspring if the ecological effect of that parameter were magnified. If two species were to suddenly come into a situation in which they were competing with one another for some limited resource, we would expect that one species would eliminate the other or cause it to use resources differently, or to use different resources. The winning species would have a greater amount of competitive ability than the loser as far as that resource axis of the n-dimensional niche is concerned (e.g.

---

* This chapter is dedicated to the memory of ABEL FORNES who was killed by poison gasses while studying methods of controlling vampire bats. His numerous publications, and personal collection of over 5000 mammal specimens, attest to the scientific loss suffered by South America with his death at the age of 33.

it would be more adapted to using that resource in that particular habitat). Similarly, in desert-inhabiting species, we might ponder which ones could continue to inhabit a particular area if it were to become drier over a time period too short for evolutionary changes to occur within the species. If the continuum is real, we should be able to predict which species could exist in a particular desert by knowing the water regimens of the area (other things being about equal) and, more importantly, we might be able to predict the types and frequencies of adaptations that will be exhibited by the mammals inhabiting a particular area. This whole method assumes, of course, that species will respond to similar environments in a finite number of similar ways and that in similar faunally-distinct environments we will not find species-types occurring in a rag-tag manner.

As PIANKA (1967) noted, water is the master limiting factor in desert areas, and the scarcity of water must surely be one of the more severe evolutionary hurdles that a species must overcome if it is to successfully inhabit such environments.

Interest in desert ecology has increased within recent years, and most of the world's deserts have had at least one or more of their rodent species examined ecologically and/or physiologically. It is from such studies that the aforementioned continuum might be constructed. Many dry areas are found in South America, but, with the exception of KOFORD's (1968) paper on *Phyllotis gerbillus* of Peru, the physiological adaptations of rodents from these areas are unknown. Argentina possesses habitats ranging from grasslands and rainforests to montane coniferous *Araucaria* forests, but the bulk of the country is composed of arid or semiarid areas (BURGOS, 1963). The Argentine thorn scrub, or Chaco, the Andean puna, Patagonia and the Monte (HUNZIKER, 1952; CABRERA, 1953; MORELLO, 1958), while differing thermally and vegetationally, share a common denominator of aridity. Although the land occupied by such habitats is immense, a dearth of ecological studies concerning their small mammal inhabitants makes these regions among the world's most poorly known in this regard.

This study presents some preliminary results of current ecological investigations concerning the rodents of the Argentine Monte, and deals with three species found throughout parts of the Monte: the riparian *Akodon varius* Thomas and *Calomys laucha* Olfers, as well as *Eligmodontia typus* Cuvier, a species ubiquitous throughout many of the more arid parts of the Monte.

The Argentine Monte and some of its Rodents

The Monte is an arid region limited largely to a narrow strip of western Argentina, lying between longitudes 66° and 70° and extends more than 2000 kilometers from Salta Province in the north to Chubut

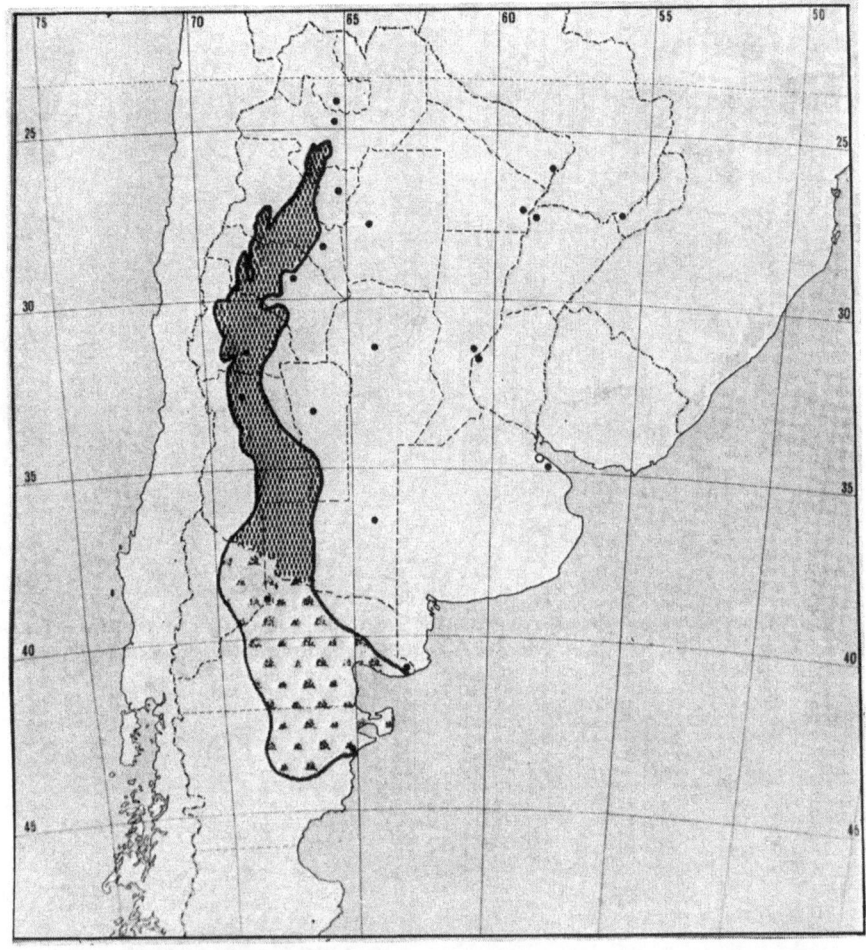

Fig. 1. The Argentine Monte, extending from Salta Province in the north, to Chubut Province in the south, is divisible into northern and southern sections on the basis of rainfall patterns. (After MORELLO, 1958).

Province in the south (Fig. 1). Physiognomically, parts of the Monte bear a striking resemblance to the Sonoran and Chihuahuan deserts of North America (Figs. 2, 3, 4 and 5) with the major floral components consisting of shrubs of the families Zygophyllaceae *(Larrea, Bulnesia, Plectrocarpa)* and Leguminosae *(Acacia, Prosopis, Cassia, Geoffroea)* and various members of the families Bromeliaceae and Cactaceae. Numerous species of the Graminae and Compositae are also present.

MORELLO (much of the following discussion is from MORELLO, 1958) divided the Monte into northern and southern sections at about 37 °S

157

Fig. 2. These cactus-bromeliad covered slopes are fairly representative of the situation found in parts of the northern Monte. A regular inhabitant of these areas is *Graomys*.

Fig. 3. Eligmodontia typus is the only non-fossorial rodent commonly found in the *Larrea cuneifolia* flats such as this one.

Fig. 4. A row of tall *Prosopis* and *Acacia* trees occuring along a dry riverbed; habitat of *Microcavia* and *Graomys*.

Fig. 5. In the valley of Andalgala, the most diverse group of rodents (*Phyllotis*, *Graomys*, *Akodon*, *Microcavia*, *Calomys* and *Oryzomys*) is found along small permanent rivers.

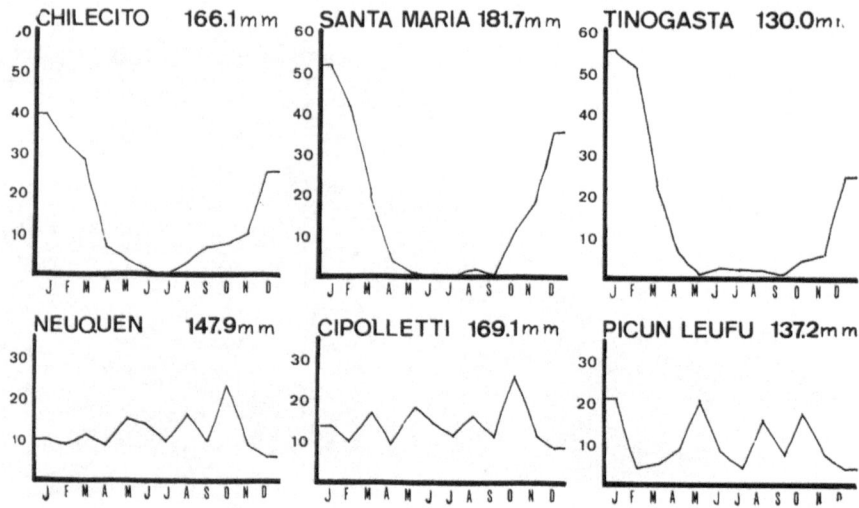

Fig. 6. Rainfall patterns in three locations in the northern Monte (Chilecito, La Rioja Province; Santa Maria and Tinogasta, Catamarca Province) and three from the southern Monte (Neuquen and Picun Leufu, Neuquen Province; Cipoletti, Rio Negro Province). After MORELLO, 1958.

latitude on the basis of rainfall patterns. The northern Monte (the region of interest in this study) is subject to droughts of up to nine-months duration, the rainy season occurring during the summer. The southern Monte exhibits a Mediterranean type of climate with the percentage of winter rainfall increasing with south latitude so that at the southern extreme of the Monte winter rainfall exceeds summer rainfall. Figure 6 illustrates the rainfall patterns in three different localities of the northern and southern Monte. Particularly evident is the greater seasonality of rainfall in the northern area, as well as the low annual totals (generally less than 200 millimeters for the Monte).

The northernmost locality of the hottest section of the Monte is the Andalgalá area in Catamarca Province, which has been the focus of much of my current research. Figure 7 depicts rainfall patterns in three different localities near Andalgalá. The curve for the 'La Toma' site is based on seven years of weather records, the 'estacion' site on 13 years of data, and the 'Andalgalá' site on 21 years of records. All sites could be listed as rainfall data from Andalgalá, but, as in many deserts, rainfall is often quite localized. The La Toma area is about 10 kilometers north of the estacion site and is located at the base of mountains in a tall, complex forest with a fairly closed canopy, whereas the estacion site is situated in the midst of a *Larrea* flat. The Andalgalá weather station is about midway between the other two sites. Rainfall increases as one

160

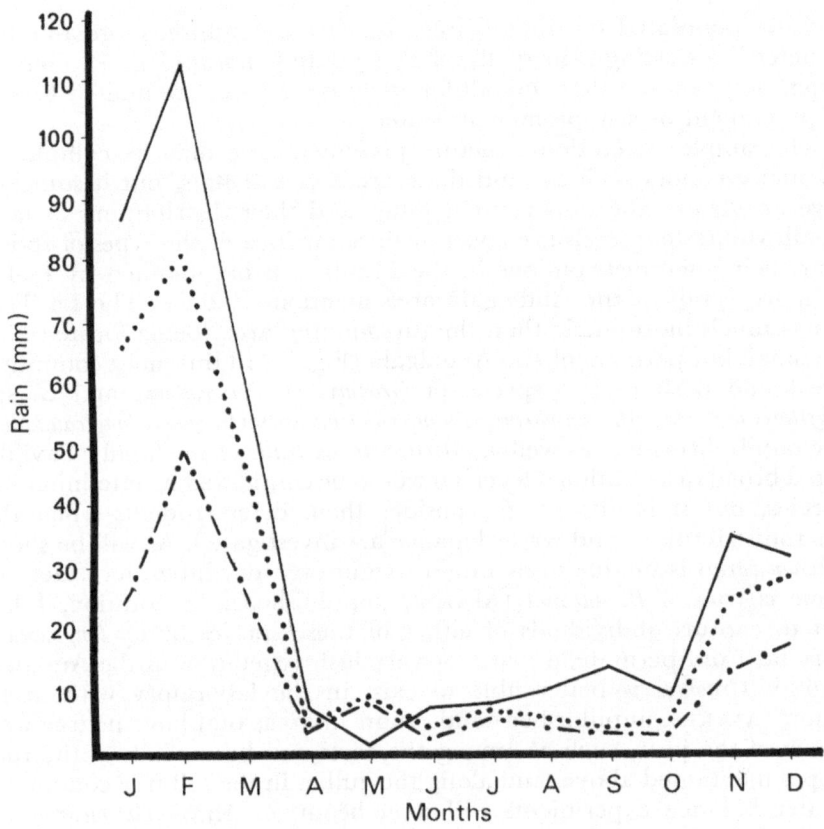

Fig. 7. Rainfall patterns at three locations within six kilometers of the town of Andalgalá. The La Toma site (solid line) is located in a tall forest along the permanent Rio Andalgalá and is six kilometers north of the Andalgalá site (dotted line). The latter is four kilometers north of the Estación site (broken line), which is in a *Larrea* flat.

approaches the mountains and much of the water received in the lower sites is runoff and largely limited to the immediate vicinity of gullies.

Thus, rainfall data can give a picture, albeit a simplified one, of the distribution of plant communities in a northern valley of the Monte and, ultimately, an idea of small mammal distributions as well. Standing on an upper slope (which are almost invariably very steep) of a 'typical' northern Monte valley, one would be amidst rocky hillsides covered by dense mats of bromeliads and tall cacti (*Trichocereus, Cereus*) in abundance (Fig. 2). This grades rapidly at the base to a mixed-shrub community on the upper bajada and finally to the low (less than 2 meters) *Larrea*-dominated community of the lower bajada and flats (Fig. 3). Gullies leading from the mountains cut into the valley in numerous locations

161

and are populated by the tall tree and shrub community (greater than 5 meters) consisting principally of *Prosopis* and *Acacia* (Fig. 4). Generalizing, one can say that the tall forests of the Monte are limited to areas of permanent or semipermanent water.

The simple vegetational picture presented here may be complicated by many factors such as sand dune areas or salt flats, but it suffices to give an idea of the important habitats and their distributions as far as small Monte mammals are concerned. Some idea of the types of rodents and their microdistributions in the Monte can be obtained by looking in more detail at the Andalgalá area mentioned above. The La Toma site is much more mesic than the surrounding area, being located along the small but permanent Rio Andalgalá (Fig. 5). In this moist community are found perhaps two species of *Graomys* (*G. griseoflavus* and *G.* sp.), *Phyllotis darwini, Akodon varius, Calomys laucha* and *Oryzomys longicaudatus* of the family Muridae, as well as *Microcavia australis* of the family Caviidae. On a broad distributional level, all would be considered Monte-inhabiting species, but it is difficult to consider them desert rodents when their microdistributions and water balance are investigated. As will be shown, *Akodon varius* is unable to maintain weight on a dry laboratory diet. The same is true of *P. darwini* (MARES, unpublished). In addition, I have yet to capture individuals of either of these species or *O. longicaudatus* very far from permanent water or very lush vegetation in the Andalgalá region. *Graomys* is better able to exist in the laboratory without free water (MARES, unpublished) and occurs in areas that have no free water much of the year, such as among the cacti and bromeliads of the rocky slopes mentioned above, and along the gullies in the tall tree community. Water balance experiments will soon begin on *Microcavia australis;* its apparent water needs are indicated by its habitat selection, which, apart from permanent water sources, include the tall, forested wash communities. Both *Graomys* and *Microcavia* are very arboreal, a trait which could aid them greatly in avoiding the roaring flood waters that inundate the habitat during the rainy season. The large caviid, *Dolichotis patagonum*, is often found in the *Larrea* areas, but appears to feed mostly along the gullies.

The habitats along the permanent rivers and forested gullies do not seem to impose very severe problems of water procurement or heat stress for mammals that inhabit them. The plants are good water sources and consumed readily by both species occurring there (*Microcavia* and *Graomys*). The humidity of such areas may be only slightly higher than the surrounding desert but ground temperatures are much lower during the day, a factor important to the diurnal *Microcavia* which spends most of its time under or in tall *Prosopis* and *Acacia*.

The Monte is subject to very high and low temperatures. Table 1 lists monthly maxima and minima for 10 years for the town of Andalgalá. Note particularly the winter ranges, which may be as great as 40°C

Table 1. Monthly maxima and minima for 10 years at Andalgalá, Catamarca Province Argentina.

	Year									
	62	63	64	65	66	67	68	69	70	71
JAN	40	46	40	40	38	42	42	38	40	40
	13	18	13	14	15	12	12	21	18	10
FEB	38	42	40	38	38	39	40	36	39	37
	13	18	10	15	10	8	15	14	10	13
MAR	36	38	34	35	38	39	38	37	37	33
	13	8	4	9	9	8	7	10	9	7
APR	34	34	33	36	34	38	34	36	33	32
	6	8	6	4	7	7	6	5	10	2
MAY	28	32	29	28	32	34	29	26	26	28
	0	4	4	—1	3	6	—4	1	1	0
JUNE	34	34	38	40	28	22	23	28	22	30
	—1	—4	—2	—3	—2	—5	0	—4	4	4
JULY	26	30	36	31	35	30	28	32	30	26
	—3	—1	—1	—4	—3	—3	0	—2	—4	—3
AUG	33	33	36	35	31	30	26	32	34	34
	1	3	—1	—2	—4	2	3	—2	—1	—1
SEPT	32	40	36	35	37	36	26	32	34	34
	4	6	2	—2	0	6	7	2	4	6
OCT	43	40	38	38	38	36	37	36	40	35
	5	6	12	8	8	7	10	5	8	5
NOV	43	41	38	40	39	40	37	38	37	35
	20	6	12	7	7	7	7	13	10	10
DEC	43	40	40	40	42	41	44	36	40	44
	20	10	10	10	10	10	18	13	14	12

during a single month. Soil temperatures may reach 70 °C during hot summer days, although these temperatures are avoided by all animals. The tall shrub gully community offers the most dense – and practically only – shade in the desert region. During the heat of a summer day I have travelled down a wide gully, and within a 20-kilometer distance found tinamous *(Eudromia formosa)*, rheas *(Rhea americana)*, cariamas *(Chunga burmeisteri)*, foxes *(Dusicyon griseus)*, guinea pigs or cuis *(Microcavia australis)* and Patagonian 'hares' *(Dolichotis patagonum)*, all seeking shade under the dense vegetation. Activity during the hot months for these animals is of the crepuscular type, with the possible exception of *Dusicyon* which at times appears to be active all night.

The hottest and driest community, and the one with the greatest soil insolation, is the pure *Larrea cuneifolia* flat, which is inhabited regularly by only one non-fossorial small rodent, *Eligmodontia typus*. (Fossorial *Ctenomys* spp. occur throughout most habitats, probably depending primarily on soil depth.) *E. typus* occurs mainly in the open communities of the flats, only very infrequently in the washes, and not on the rocky

163

hillsides. Free water, while infrequent in the washes, is extremely rare in *Larrea* areas. It is safe to suppose that the vegetation of the former has a higher water content than that of the latter. *Eligmodontia*, then, faces the greatest water stress of the small mammals thus far discussed.

Methods and Materials

The 35 *Eligmodontia typus* (25 ♂♂, 10 ♀♀) were captured 8 kilometers south of Andalgalá, Catamarca Province (2), along the northern rim of El Valle de La Luna, San Juan Province (2), and at the Estancia El Carrizalejo, 32 kilometers west (by El Manzano Monument road) of Tunuyán, Mendoza Province (31). The six *Akodon varius* (4 ♂♂, 2 ♀♀) and the four *Calomys laucha* (3 ♂♂, 1 ♀) were trapped along the Rio Andalgalá immediately north of Andalgalá, Catamarca Province.

Animals were housed individually in wire cages (17 × 13 × 10 centimeters) and were maintained on a 12-hour photoperiod. Sand was provided in each cage for bathing. Birdseed was available ad libitum for *E. typus* and *C. laucha* during the entire experiment. *Akodon varius* was fed rolled oats rather than birdseed. Temperatures in the lab were not controlled and averaged about 25 °C with extremes of 18 °C to 31 °C, while relative humidity ranged from 32 percent to 72 percent, although the latter value was recorded on only one day of the experiment. Generally, humidity was less than 50 percent. All animals were initially provided with free water and lettuce until it had been demonstrated that they were either maintaining or gaining weight for at least a five-day period, after which experiments were begun. Water was provided from inverted graduated cylinders fitted with L-shaped glass drinking tubes. All animals rapidly learned to accept water from the tubes. A small wad of cotton was provided for bedding material.

Ad libitum water consumption was measured by noting the amount of tap water drunk every other day during the course of an experiment. Animals maintained or gained weight during studies of ad libitum water consumption. During all water consumption experiments an additional graduated cylinder was placed near the cages to measure evaporation, but this proved to be so slight as to be insignificant. Only rarely did an animal spill water from the drinking tube (noticeable by a wet spot on the sand), but when this occurred the measurement was not counted.

Restricted water consumption was measured by halving an animal's water allotment every six days until weight was just maintained for a six-day period. During studies of water deprivation, animals were allowed no water other than what was available in the air-dry food or from metabolism of food.

Eligmodontia typus was tested for the ability to utilize cactus as a water source. A number of cacti (*Trichocereus candicans*) were collected near El Manzano Monument, Mendoza Province, and brought to the lab.

164

This cactus is common in areas inhabited by *E. typus*. Animals were allowed free water for a time and then this was replaced by cactus pads that had been split open and cut into sections. The succulent cactus sections were replaced every other day by freshly opened cactus pads. After being allowed this water source, the cactus was also removed and the weight response of the animals noted. Oven drying at 95 °C revealed that the cactus contained 775.2 percent water by weight (percent expressed on a dry weight basis). Birdseed used in the experiments contained 8 percent water by weight, while the rolled oats contained 10 percent water by weight.

Results and Discussion

Ad libitum water consumption. Ad libitum consumption was greater in *E. typus* (17.3 ± S.D. 7.2 percent body weight per day) than in *C. laucha* (13.7 ± S.D. 0.27 percent body weight per day), although the reverse might logically have been expected when the animals' habitats are considered (Fig. 8). It is known that laboratory conditions may stimulate animals to drink excessively (RICHTER & MOSIER, 1954; CHEW, 1965; BOICE, 1968) and that animals often segregate into two categories: 'drinkers' and 'non-drinkers' (CARPENTER, 1966; MAC-MILLEN & LEE, 1968). Nevertheless, ADOLPH (1949) felt that ad libitum water consumption was a predictable value, irrespective of habitat, if only the body weight of an animal were known, by use of the formula (modified),

$$\log I = \log 0.24 + 0.88 \, (\log W),$$

where I = cc H_2O/day and W = body weight. HUDSON (1962) showed that the predictive efficiency of the formula was far from perfect when he noted that only eight of 14 small mammal species' calculated ad libitum water consumption values agreed with the predicted values. Other workers (ODUM, 1944; LINDEBORG, 1952; GETZ, 1963, 1968) believed that ad libitum values were correlated with the amount of moisture in an animal's environment. Similarly, CARPENTER (1966) was able to correlate ad libitum consumption with habitat of *Dipodomys agilis* and *D. merriami*, although the former drank water at twice the predicted rate. MACMILLEN (1964), however, determined the ad libitum rate of *D. agilis* to be about 11 percent of body weight per day, or approximately 25 percent less than the predicted value. Also, he found that of seven semidesert rodent species examined, five consumed at about the same rate, while *Neotoma fuscipes* exceeded all other species and the desert-adapted *Perognathus fallax* hardly consumed water at all.

Table 2 lists the ad libitum water consumption for *E. typus* and *O. longicaudatus*, as well as for some other desert and semidesert inhabiting species. Predicted values calculated from the literature may be approxi-

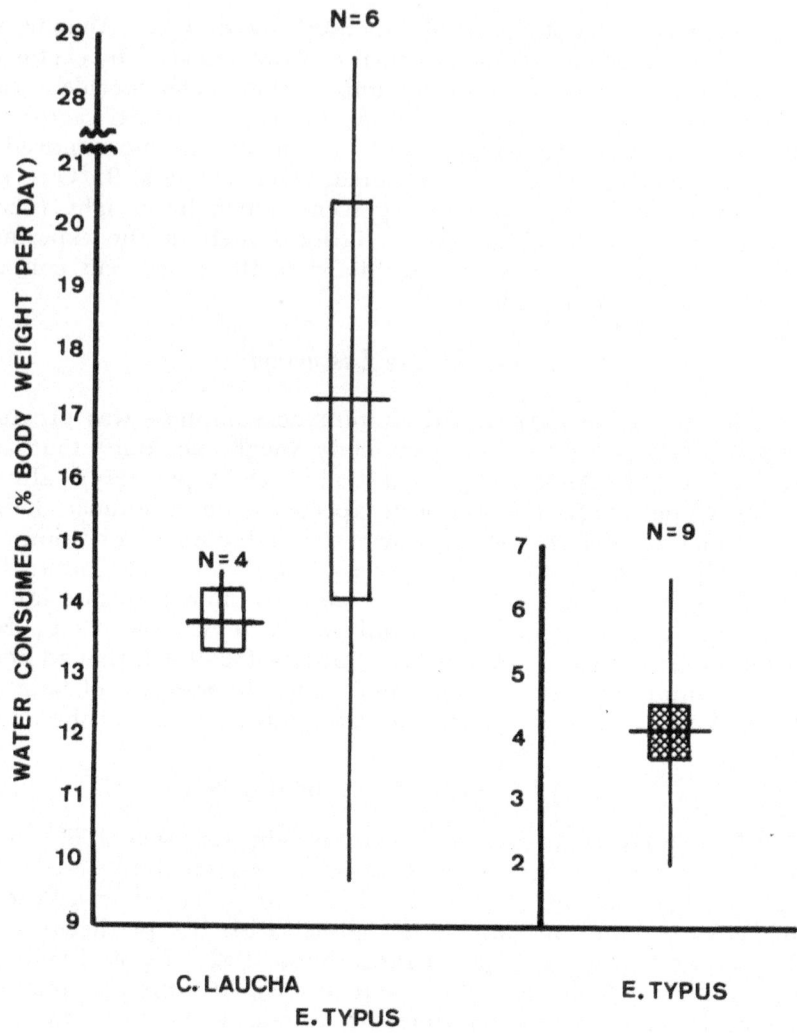

Fig. 8. Ad libitum water consumption of *Calomys laucha* and *Eligmodontia typus* (open rectangles) and minimum water consumption of *E. typus* (shaded rectangle). Horizontal bar is the mean, vertical bar is the range. Rectangles enclose $\pm$ two standard errors.

mations due to the use of means rather than individual data in the calculations and having to infer values from published figures. Of the 12 species listed, five drink water at rates below those predicted, four drink at about predicted levels, and two drink more than predicted by body weight. As mentioned above, *D. agilis* is reported to drink both more and less than predicted.

Table 2. Predicted and calculated ad libitum water consumption values for various desert and semidesert rodents. (Water consumption listed as percent body weight/day).

Species	Mean Body Weight (gm)	N	Water Consumption		Exist Without Free Water	Source
			Predicted	Observed		
Eligmodontia typus	26.5	6	16.2	17.3	Somewhat	This Study
Calomys laucha	24.7	4	16.3	13.7	Unknown	This Study
Liomys irroratus	51.8	10	14.9	7.5	Yes	HUDSON & RUMMEL (1966)
L. salvani*	54.5	5	14.8	5.5	Yes	HUDSON & RUMMEL (1966)
Dipodomys merriami	33.9	10	15.2	15.6	Yes	HUDSON (1962)
D. agilis	—	8	14.2	28.4	Somewhat	MacMILLEN (1964) CARPENTER (1966)
D. agilis	51.9	10	14.9	12.1	—	HUDSON & RUMMEL (1966)
Perognathus fallax	19.1	9	16.7	1.7	Yes	MacMILLEN (1964)
Neotoma fuscipes	188.0	9	12.8	35.5	No	MacMILLEN (1964)
N. micropus	294.4	10	12.7	12.1	Unknown	BIRNEY & TWOMEY (1970)
Meriones unguiculatus	77.2	4	14.3	8.9	Yes	BOICE & WITTER (1970)
M. unguiculatus	85.3	7	14.1	4.7	—	ARRINGTON & AMMERMAN (1969)
Notomys alexis	29.0	10	16.0	27.0	Yes	MacMILLEN & LEE (1969)
N. cervinus	33.7	9	15.7	14.0	Yes	MacMILLEN & LEE (1969)

* Not a dry area species.

167

Table 3. Rainfall patterns over a three-month period during the wet summer of 1969–1970. (Data from town of Andalgalá)

Month		Rain (mm)	Month		Rain (mm)
December	6	12	February	4	6
	8	15		5	20
	10	30		15	30
	13	18		16	18
	19	6		21	35
	22	2		28	30
	28	4			
January	4	6	March	1	25
	8	9		2	20
	19	4		3	20
	24	4		8	15
				10	10
				16	15
				17	5

An interesting pattern emerges if the ability to exist without free water is examined. One of the species reported to drink at rates greater than those predicted can exist without free water. This is *Notomys alexis*, a highly desert-adapted Australian murid. Of the rest of the species consuming ad libitum water at or below expected levels, all are able to exist on very small amounts of free water. *E. typus*, as will be discussed below, fits the pattern as well. *C. laucha*, not as yet investigated, would be expected to be independent of free water, if the pattern holds. The pattern may only apply to arid and semiarid rodents.

Eligmodontia typus inhabits *Larrea* flats subject to long droughts and, during the rainy season, sudden showers spaced some days apart. Considering this fact, there is another possible reason to account for such animals that seem to drink excessively in the laboratory when free water is available. As CHEW & HINEGARDNER (1957) suggested for water-deprived white mice, in nature a small drink of dew or rainwater every few days could be very important in allowing a mouse to eat and masticate its food properly during intervening days of extreme water shortage. Thus, *E. typus* encountering the rainfall regimen of Table 3 during the hot desert summer could reap great selective advantage by drinking water whenever it is available. A species associated with free water in a moist habitat would have no need of drinking each time that free water was encountered, since such field encounters would be numerous. This could account for the high water consumption values exhibited by species which are – in the laboratory at least – able to exist without free water. It is also possible that if the ability to exist without free water at different

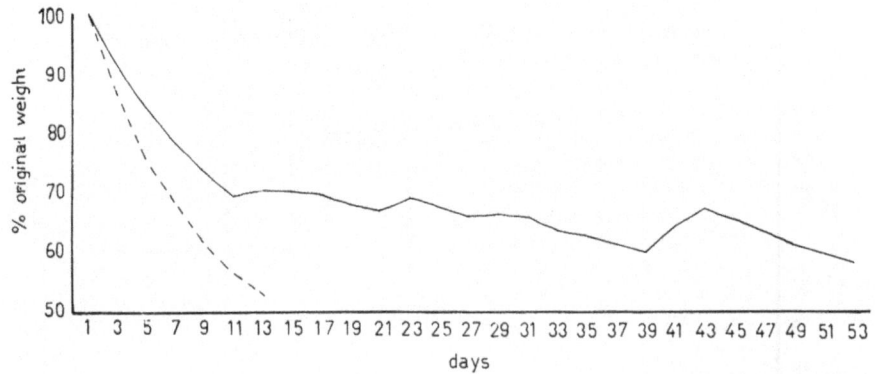

Fig. 9. Body weights of *Eligmodontia typus* (solid line) and *Akodon varius* (dashed line) on an air-dried seed diet with no access to free water. Initial sample size of *A. varius* = 5. E. typus sample size = 17 (day 11), 9 (day 29), 4 (day 53).

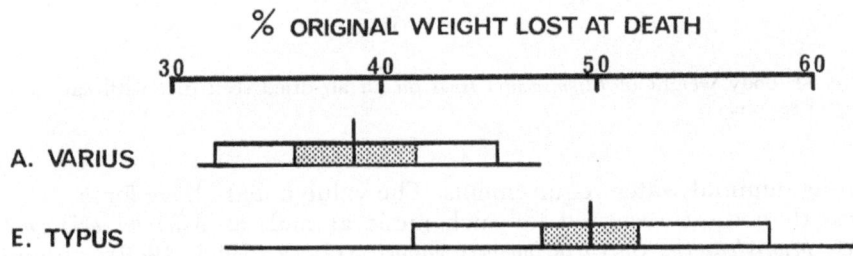

Fig. 10. Percent original weight at death for *A. varius* and *E. typus*. Vertical bar is the mean; horizontal bar is the range. Shaded rectangles enclose ± two standard errors; open rectangles enclose ± one standard deviation.

temperatures and humidities were examined, the animals which exhibit the lowest ad libitum consumption rates could be most able to survive more rigorous experimental conditions.

Minimum water consumption. It is likely that the minimum-water allotment on which an animal can maintain body weight is a better measure than the preceding of its actual water needs in nature. GETZ (1963, 1968) and other workers have demonstrated that, between species, those requiring less water to maintain weight will often inhabit drier microclimates. *E. typus* (Fig. 8) can maintain weight on 25 percent of the ad libitum consumption (0.83 ml/day/animal, or 0.041 ml/g/day). CARPENTER (1966) listed the minimal water requirements of various desert rodents. These ranged from no water needed *(Perognathus, Microdipodops, Dipodomys)* to 10.2 percent body weight needed per day *(Neotoma fuscipes)*. *E. typus* would thus be intermediate between the heteromyids and *Neotoma*

169

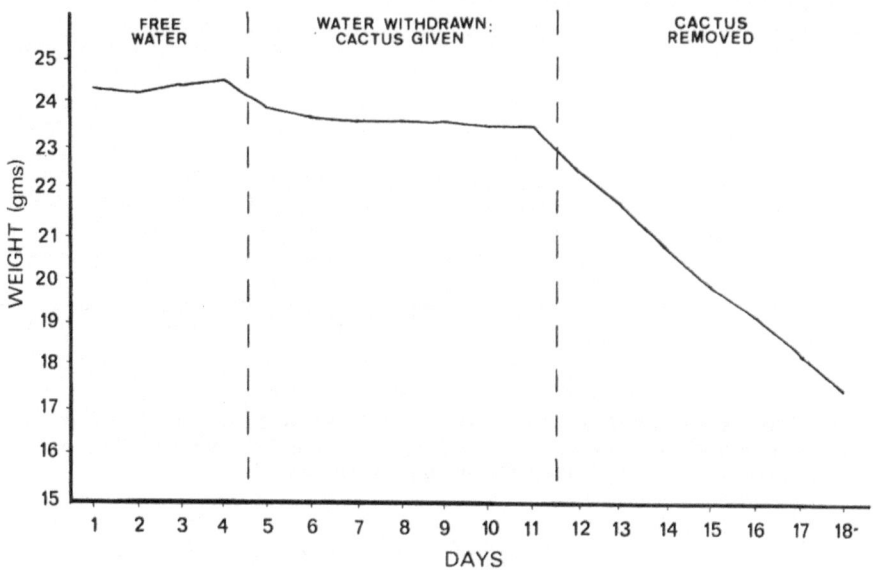

Fig. 11. Body weight of *Eligmodontia typus* on an air-dried seed diet with cactus as a water source.

in its minimal water requirements. The value noted above for *E. typus* is less than those recorded for such mesic animals as *Microtus ochrogaster*, *M. pennsylvanicus* or *Clethrionomys gapperi* (GETZ, 1963, 1968), although it is greater than the amount needed by *Peromyscus leucopus* (GETZ, 1968), a dry woodland-inhabiting murid (cricetine).

Water deprivation. The reactions of *E. typus* and *Akodon varius* to total water deprivation are illustrated in Figures 9 and 10. *A. varius* rapidly lost weight (1.57 g/day) and was able to survive without free water for a maximum of 13 days. *Eligmodontia* also lost weight sharply during the first 11 days of the experiment (0.56 g/day), but thereafter the rate of weight loss was reduced. During the first 13 days of water deprivation, *E. typus* suffered 17.6 percent deaths; 42.8 percent of the remaining animals died over the next 28 days. After 41 days, 52.9 percent of the original sample had died. Mean days to death for *E. typus* was 22.1, while *A. varius* had a mean death day of 8.7. Thus, *E. typus* would appear to have a better ability to resist dessication than *Peromyscus eremicus*, *P. maniculatus* and *P. californicus* (MACMILLEN, 1964). Although its mean death day is somewhat lower than these species, it must be remembered that four *E. typus* were alive after 53 days, whereas all of MACMILLEN's *Peromyscus* were dead after 55 days. In this respect, *E. typus* would seem more closely allied to *Dipodomys agilis* (MACMILLEN, 1964) which had some survivors after 55 days without free water. One

170

of the four *E. typus* alive at the end of the experiment had lost only 28.3 percent of its original weight and probably could have withstood water deprivation for some time longer. The other three animals averaged 44.5 percent of original weight lost at this time and it is doubtful that much more weight loss could have been tolerated. All of these animals were obviously suffering the effects of dessication and lack of food such as having the eyes appear to be crusted shut and being quite listless.

The response of *E. typus* body weight to cactus as a water source is shown in Figure 11. While there was a slight reduction of weight from that observed when free water was available, original weight was practically maintained with the cactus water source. Since Figure 11 shows only average weights, some details are obscured. Three of the six animals gained weight while eating cactus and three lost weight slightly over the same period. One of the latter accepted the cactus only grudgingly, eating minimal amounts. When the cactus was removed as a water source, all animals lost weight rapidly, as in the initial period of the water deprivation experiment.

GENERAL IMPLICATIONS

Of the three Monte species investigated, the very mesic, vole-like *Akodon varius* is the least able to exist in the laboratory on a dry diet. Probably a need for free water or very succulent vegetation limits the distribution of this species in the Andalgalá region. It is not an arboreal species, and the only moist vegetation in many parts of the desert is the trees of the forested arroyos. Without climbing abilities *A. varius* could not obtain moisture from the vegetation in such a habitat, assuming such was a sufficient water source.

Calomys laucha presents a different problem. I have only found the species in very wet vegetation (same habitats as *A. varius*) or a bit more distant from free water, yet still in moist woods. Individuals are apparantly not arboreal and have not been trapped in trees. The fact that this species consumes less ad libitum water than *E. typus* is interesting. If the ad libitum method is valid, some mechanism other than physiological limitations must be sought to account for this species' distribution. Obviously, more physiological and ecological investigation is necessary to clarify *C. laucha* habitat selection.

E. typus has been shown to be able to exist for almost two months on a dry laboratory diet without access to free water, although almost all individuals died or were dying after this period. At least a small percentage of individuals can withstand these stringent experimental conditions, however. Such tolerance to dessication almost surely is important in allowing this species to inhabit some of the more arid parts of the Monte. The willingness of the animals to eat cactus and the resultant weight stabilization with this water source points out what is probably an

171

important factor in the water balance of *E. typus*. Cacti are abundant throughout the northern Monte, and if *E. typus* needs them to successfully inhabit certain dry areas, then it is possible that an association between it and cacti will ultimately be noted. During the driest parts of the year cacti may offer the only dependable water source.

I have observed (unpublished) very stereotyped insect-catching behavior for *E. typus*, and it is quite possible that insects serve as a water source for this species during the summer. Such food habits are well known for other rodents (BAILEY, 1929; HAWBECKER, 1947; EGOSCUE, 1960).

E. typus, then, avoids the desert extremes by being nocturnal and passing the daylight hours in its burrow. Vegetational water is probably its main moisture source throughout the year. This may be supplemented by insects during the summer.

Some discussion on general desert rodent ecology is apropos at this point. A perusal of the literature easily conveys the impression that each desert has its distinctive rodent fauna and that such will consist of highly-adapted desert forms and species that exist only by using available water sources – either vegetational, animal or free water.

In Australia, for example, there are desert marsupials that exist by either avoiding temperature extremes or drinking brackish water and forming concentrated urine (BENTLEY, 1955; EALEY, BENTLEY & MAIN, 1965), while the Australian murid hopping mice have been shown to be animals highly specialized for a desert existence (MACMILLEN & LEE, 1968). Gerbils, whether from Mongolia (WINKELMANN & GETZ, 1962; ARRINGTON & AMMERMAN, 1969) or India (GHOSH, PUROHIT & PRAKASH, 1964) are known to be great water conservers.

Each desert thus far investigated has seemed to fit the pattern of having some species conforming to the heteromyid-dipodid habit, with the bipedal locomotion and kidneys of great concentrating abilities.

The study by KOFORD (1968) on two Peruvian desert *Phyllotis gerbillus* was not extensive enough to determine whether or not the animals could maintain body weight on a dry seed diet. KOFORD felt, however, that the abilities of the mice in this regard were much less marked than the well-known, highly-adapted desert forms. Indeed, KOFORD showed that house mice *(Mus musculus)* were as efficient at surviving dessication as were *P. gerbillus*. *E. typus* has likewise been shown to be less physiologically adapted than many rodents from other deserts. It appears, therefore, that South American deserts lack true desert rodents (i.e., great water-conserving, bipedal forms) as conspicuous faunal elements. *E. typus* possesses some hopping abilities and has rather elongate hind legs; with its silky pelage and large ears it resembles a hypothetical cross between a *Peromyscus* and a *Perognathus*. Its physiological adaptations seem to place it in a similar relationship to these two genera, more adapted than the former and less so than the latter.

At this point it is difficult to explain why the Argentine desert (and South American deserts in general) should lack classic desert rodents, or indeed any widespread, highly-adapted desert species. Of the Monte mammals discussed, *Microcavia australis* and *Graomys* are Monte invaders from the eastern Chaco and exist in the Monte by inhabiting its Chaco-like habitats. *Dolichotis patagonum*, by virtue of its distribution, entered the Monte from the southern areas.

E. typus occurs throughout most of southern and western Argentina, parts of the Pacific Andean slopes of Chile, and as far north as Peru and Bolivia. GREER (1965, 1968) has collected the species in moist and forested habitats and PEARSON (1951, 1957) has trapped *E. typus* in the Peruvian highlands at elevations of up to 15,000 feet. It is definitely not an exclusively Monte species and may have invaded this desert from the south.

C. laucha, *A. varius* and *P. darwini* are sylvan and/or montane forms and enter the Monte's slopes and river valleys in this manner. Probably none of the rodents mentioned in this report (and, excluding *Ctenomys*, few other possibilities exist) have evolved *in situ* as Monte species.

Whether or not the lack of an autochthonous Monte rodent fauna is a result of the series of isolated northern valleys limiting colonization from any direction is unknown. If colonization of the arid parts of the Monte occurred from the south, it would have been difficult for Monte species to form since there is little geographic isolation in this area and Chaco leads to Monte with an intermediate ecotonal area. Given enough time, perhaps *E. typus* of the northern valleys could approach the classic desert-rodent type, but as yet it is not at that stage of evolutionary development.

Acknowledgements

This study was facilitated by many Argentine friends too numerous to mention. Special thanks are due my friend Sr. JUAN CARLOS SCHIAPPA DE AZEVEDO, who not only allowed me to trap regularly on his estancia, but indeed opened his home to me while I was in the field. Dr. VAUGHN SHOEMAKER provided suggestions and helpful criticism regarding the manuscript. Ing. VIRGILIO ROIG suggested various trapping localities for collecting Monte rodents and offered the use of the Instituto de Investigaciones de Las Zonas Aridas y Semiaridas in Mendoza. For help and encouragement throughout the study I thank Dr. W. F. BLAIR. Dr. C. C. OLROG aided in identification of the species and offered comments on their distributions and ecology. Mr. DAVE BRADFORD was of indispensible aid providing needed suggestions and scientific reprints from the United States. Part of the research was conducted at the Universidad Nacional de Cordoba while I was a faculty member there.

All aspects of this investigation would have been much more tedious and much less fun without the unflagging help of my wife, Lynn. This work was supported as a part of the IBP Integrated Research Program on Origin and Structure of Ecosystems through a National Science Foundation grant GB 27152 to The University of Texas at Austin.

REFERENCES

ADOLPH, E. F. 1949. Quantitative relations in the physiological constitutions of mammals. *Science* 109: 579–585.

ARRINGTON, L. R. & AMMERMAN, C. B. 1969. Water requirements of gerbils. *Lab. Animal Care* 19: 503–505.

BAILEY, V. 1929. Life history and habits of grasshopper mice, genus *Onychomys*. *Tech. Bull. 145, US Dept. Agric.*: 1–15.

BENTLEY, P. J. 1955. Some aspects of the water metabolism of an Australian marsupial *Setonyx brachyurus*. *J. Physiol.* 127: 1–10.

BIRNEY, E. C. & TWOMEY, S. L. 1970. Effects of sodium chloride on water consumption, weight, and survival in the woodrats, *Neotoma micropus* and *Neotoma floridana. J. Mamm.* 51: 372–375.

BOICE, R. 1969. Water intake as a function of ease of access in *Neotoma. J. Mamm.* 50: 605–607.

BOICE, R. & J. A. WITTER. 1970. Water deprivation and activity in *Dipodomys ordii* and *Meriones unguiculatus. J. Mamm.* 51: 615–618.

BURGOS, J. J. 1963. El clima de las regiones áridas en la Republica Argentina. Rev. Investig. Agric., vol. 8, no. 4. INTA. Buenos Aires.

CABRERA, A. L. 1953. Esquema fitogeográfico de la Republica Argentina. *Rev. Mus. La Plata (nueva serie), Bot.* 8: 87–168.

CARPENTER, R. E. 1966. A comparison of thermoregulation and water metabolism in the kangaroo rats *Dipodomys agilis* and *Dipodomys merriami. Univ. Calif. Publ. Zool.* 78: 1–36.

CHEW, R. M. 1965. Water metabolism of mammals. *In* Physiological mammalogy (MAYER, W. V. & VAN GELDER, R. G., eds.), Academic Press, N.Y. pp. 43–178.

CHEW, R. M. & HINEGARDNER, R. T. 1957. Effects of chronic insufficiency of drinking water in white mice. *J. Mamm.* 38: 361–374.

EALEY, E. H. M., BENTLEY, P. J. & MAIN, A. R. 1965. Studies on water metabolism of the Hill Kangaroo, *Macropus robustus* (Gould), in northwest Australia. *Ecology* 46: 473–479.

EGOSCUE, H. J. 1960. Laboratory and field studies of the northern grasshopper mouse. *J. Mamm.* 41: 99–110.

GETZ, L. L. 1968. A comparison of the water balance of the prairie and meadow voles. *Ecology* 44: 202–207.

GETZ, L. L. 1965. Influence of water balance and microclimate on the local distribution of the redback vole and white-footed mouse. *Ecology* 49: 276–286.

GHOSH, P. K., PUROHIT, K. G. & PRAKASH, I. 1964. Studies on the effects of prolonged water deprivation on the Indian Desert gerbil, *Meriones hurrianae. In* Environ. Physiol.

174

Psychol. in Arid Conditions. Proc. Lucknow Symp. (1962), UNESCO, Paris, 1964. pp. 301–306.

GREER, J. K. 1965. Mammals of Malleco Province, Chile. *Publ. Mus. Mich. State Univ., Biol. Ser.*, 3(2): 51–151.

GREER, J. K. 1968. Mamíferos de la provincia de Malleco. Publ. del Mus. 'Dillman S. Bullock', Angol, Chile. 114 p.

HAWBECKER, A. C. 1947. Food and moisture requirements of the Nelson antelope ground squirrel. *J. Mamm.* 28: 115–125.

HUDSON, J. W. 1962. The role of water in the biology of the antelope ground squirrel *Citellus leucurus. Univ. Calif. Publ. Zool.*, 64: 57–96.

HUDSON, J. W. & RUMMEL, J. A. 1966. Water metabolism and temperature regulation of the primitive heteromyids, *Liomys salvani* and *Liomys irroratus. Ecology* 47: 345–354.

HUNZIKER, J. H. 1952. Las comunidades vegetales de la cordillera de La Rioja. *Rev. Investig. Agric.* 6(2): 167–196.

KOFORD, C. B. 1968. Peruvian desert mice: water independence, competition, and breeding cycle near the equator. *Science* 160: 552–553.

LINDEBORG, R. G. 1952. Water requirements of certain rodents from xeric and mesic habitats. Contrib. Lab. Vert. Biol. Univ. Mich. No. 58, 32 p.

MACMILLEN, R. E. 1964. Population ecology, water relations and social behavior of a Southern California semi-desert rodent fauna. *Univ. Calif. Publ. Zool.* 71: 1–59.

MACMILLEN, R. E. & LEE, A. K. 1969. Water metabolism of Australian hopping mice. *Comp. Biochem. Physiol.* 28: 493–514.

MORELLO, J. A. 1958. La provincia fitogeográfica del Monte. Univ. Nac. Tucumán, Inst. 'Miguel Lillo', Opera Lilloana II: 155 pp., 58 plates.

ODUM, E. P. 1944. Water consumption of certain mice in relation to habitat selection. *J. Mamm.* 25: 404–405.

PEARSON, O. P. 1951. Mammals in the highlands of southern Peru. *Bull. Mus. Comp. Zool.*, 106: 117–174.

PEARSON, O. P. 1957. Additions to the mammalian fauna of Peru and notes on some other Peruvian mammals. *Breviora* No. 73, 7 p.

PIANKA, E. R. 1967. Lizard species diversity. *Ecology* 48: 333–351.

RICHTER, C. P. & MOSIER, H. D., Jr. 1954. Maximum sodium chloride intake and thirst in domesticated and wild Norway rats. *Amer. J. Physiol.* 176: 213–222.

SCHMIDT-NIELSEN, K. 1964. Desert animals. Oxford Univ. Press, New York.

STERN, J. T., Jr. 1970. The meaning of 'adaptation' and its relation to the phenomenon of natural selection. *In* Evolutionary Biology (DOBZHANSKY, TH., HECT, M. K. & STEEVE, W. C., eds.), Appleton-Century-Crofts, New York. Vol. 4. pp. 39–66.

WINKELMANN, J. R. & GETZ, L. I. 1962. Water balance in the Mongolian gerbil. *J. Mamm.* 43: 150–154.

VIII. LA DIVERSITÉ DES GERBILLIDÉS

par

F. PETTER

Les Gerbillidés sont une famille de rongeurs myomorphes dont toutes les espèces habitent des régions désertiques ou semi-désertiques d'Afrique et d'Asie.

Ils sont caractérisés extérieurement par leurs yeux toujours volumineux, leur pelage doux le plus souvent de couleur adaptée au sol dans lequel ils vivent, et leur queue non écailleuse, très souvent longue, velue, et terminée par un pinceau de poils. Malgré cet ensemble de caractères qui sont le fait de beaucoup de genres et d'espèces, quelques représentants de la famille se distinguent par un pelage grossier ou une queue courte et nue.

Le crâne de tous les Gerbillidés actuels est caractéristique de la famille: il montre toujours notamment une plaque zygomatique étendue vers l'avant et des bulles tympaniques dont le développement, quelquefois extraordinaire, est toujours important.

Cette famille de Rongeurs, qui constitue une entité bien caractérisée pour les zoologistes par l'ensemble de ses caractères anatomiques, est directement issue des Cricétidés, famille qui fut florissante dans l'Ancien Monde jusqu'à ce que les nouvelles familles auxquelles elle a donné naissance ne la remplacent peu à peu.

L'étude des structures dentaires de l'ensemble des Gerbillidés actuels permet de rapporter le plan de leurs trois dents molaires supérieures et inférieures au 'plan cricétin' décrit par Stehlin et Schaub pour les Cricétidés. Il a donc été possible d'envisager, sinon une filiation réciproque des divers représentants actuels de la famille, au moins une série de niveaux d'évolution dentaire que l'on discerne d'un genre à l'autre, d'une espèce à l'autre, ou même au cours de la vie d'un même individu, d'un stade d'usure dentaire à l'autre. Ce sont ces lignes d'évolution que nous nous efforcerons de préciser ici en fonction essentiellement des structures dentaires. Il faut souligner toutefois que la hiérarchie des caractères dentaires n'est pas suffisante et qu'il convient également de tenir compte de certains caractères fondamentaux de la structure du crâne. L'un de ceux-ci, l'aplatissement, ou plutôt l'absence de gonflement de la partie mastoïde de la bulle tympanique, paraît caractériser ensemble les genres *Tatera*, *Taterillus* et quelques genres voisins. Les paléontologistes nous diront probablement dans l'avenir si cet ensemble peut correspondre à un phylum isolé dans la famille des Gerbillidés ou s'il s'agit de convergences.

On sait maintenant, grâce à l'étude concomitante des chromosomes et des dents des rongeurs myomorphes actuels, l'importance qu'il faut attribuer, du point de vue évolutif, à la moindre variation de l'un des caractères de la morphologie dentaire. Tout se passe comme si chaque dent était modelée dès la naissance sous l'influence d'un grand nombre de gènes répartis sur des chromosomes différents. Le maintien d'une formule chromosomique constante est le garant d'une certaine identité morphologique des individus et de leur inter-fertilité dans les limites de la variation individuelle de l'espèce. Au contraire, une modification de la formule chromosomique, corrélative d'une interstérilité, correspond au passage d'une espèce à l'autre et s'accompagne, dans le cas général, de modifications dentaires.

Nous sommes donc bien fondés à suivre ces modifications dentaires à la façon des paléontologistes, comme les témoignages d'une évolution. Le véritable problème auquel se sont attaqués, jusqu'ici sans résultat probant, un certain nombre d'auteurs est de savoir dans quelle mesure les modifications dentaires que l'on constate correspondent pour elles-mêmes à des spécialisations évolutives analysables. Les écologistes montrent en général, au contraire, que les modifications dentaires ne sont que des caractères associés à des caractères anatomiques ou physiologiques, sur lesquels a porté réellement la sélection qui a permis un progrès adaptatif. Celui-ci s'est toujours traduit, dans le cas de la naissance d'une nouvelle espèce, par l'extension de répartition du genre auquel elle appartient. Le morphologiste qui n'examine que les dents ignore tout de l'animal vivant et ne peut donc apprécier que l'une des faces du problème de l'évolution de la famille.

La structure des dents de Gerbillidés est dérivée du schéma primitif que l'on connaît chez les Cricétidés. Ce schéma est particulièrement net sur la seconde molaire supérieure qui comprend typiquement 4 cuspides, correspondant respectivement aux protocone, paracone, métacone et hypocone. Chez les Cricétidés, ces 4 cuspides sont réunies entre elles par des crêtes d'émail qui se poursuivent en avant et en arrière de la dent par un bourrelet ou 'cingulum'. C'est ce cingulum qui subit un développement particulier sur la Ière molaire supérieure qu'il prolonge vers l'avant. Il est transformé en une ou plusieurs 'cuspides' qui constituent la partie antérieure de cette dent et lui donnent, probablement pour des raisons mécaniques, une structure à peu près symétrique. La 3ème dent molaire est habituellement réduite par rapport à la seconde. Elle est plus réduite chez les Gerbillidés qu'elle ne l'est chez les Cricétidés. Les dents inférieures qui s'opposent aux supérieures par leur face occlusale ont des proportions équivalentes et on y reconnaît des schémas comparables.

Chez les Gerbillidés qui présentent le dessin le plus primitif, on retrouve les cuspides caractéristiques des Cricétidés assez bien individualisées, mais la plupart des crêtes qui unissent les cuspides chez ceux-ci ont disparu.

Ce sont les plus petites espèces de Gerbillidés que l'on rencontre en

Mauritanie ou en Somalie, donc sur les marges du désert saharien, qui paraissent avoir conservé les caractéristiques dentaires les plus primitives: cuspides partiellement distinctes, traces de la crête longitudinale, réduction relativement faible de la troisième molaire de sorte qu'elle possède encore parfois les 4 cuspides primitives. Bien que le genre *Monodia* ait été défini par ces caractères, on ne les trouve pas constamment réunis sur les individus de même espèce; il est donc difficile de distinguer les espèces chez lesquelles ces particularités sont fréquentes, des espèces voisines chez lesquelles on les trouve rarement. Il est maintenant admis que le genre *Monodia* n'est pas bien défini puisque ses caractères propres sont inconstants. Les expèces que l'on croyait pouvoir lui rapporter sont référables au genre *Gerbillus*.

Si l'on examine, par rapport aux autres genres de Gerbillidés, l'évolution du dessin dentaire des *Gerbillus* selon les stades d'usure au cours de la vie des individus, on constate qu'on peut y retrouver des dessins caractéristiques de certains autres genres de Gerbillidés à des âges différents de la vie des individus. Ainsi, si le stade le plus jeune de *Gerbillus* présente le dessin le plus primitif que l'on puisse observer dans la famille, un stade 'vieux' de *Gerbillus* correspond au contraire à un stade jeune du genre *Meriones*. Il est d'ailleurs facile de constater, en comparant les dessins d'usure des molaires de tous les Gerbillidés, que tous aboutissent plus ou moins tard, au cours de l'usure de la couronne, à un stade de type *Meriones*.

Gerbillus et *Meriones* peuvent donc être considérés comme les termes extrêmes de l'évolution du dessin de l'émail des molaires dans la famille des Gerbillidés.

Pour évaluer la parenté relative de toutes les espèces de Gerbillidés que l'on peut rapporter à cet ensemble *Gerbillus-Meriones*, et les placer, le cas échéant, à un niveau hiérarchique compatible avec notre systématique, il est nécessaire de discuter l'importance qu'il faut donner à d'autres caractéristiques que le dessin d'usure des molaires: caractères du crâne, de la queue, ou du revêtement pileux. Il faut noter que les auteurs se sont finalement accordés pour ne donner qu'une importance subgénérique à la pilosité des soles plantaires chez les Gerbillidés. Cette évaluation a été faite et les auteurs ont ainsi reconnu parmi les genres les plus proches de *Gerbillus*, les genres *Microdillus*, *Pachyuromys*, *Sekeetamys*, *Ammodillus* et *Dipodillus;* parmi les plus engagés dans le sens *Meriones*: *Brachiones*, *Psammomys* et *Rhombomys*. Une troisième série pourrait comprendre *Gerbillurus*, *Tatera* et *Taterillus* ainsi que *Desmodillus* et *Desmodilliscus*.

Bien entendu, ces trois groupes ne correspondent qu'à des tendances qu'il est pratique de distinguer. Il est encore impossible de décider si un genre comme *Gerbillurus*, par exemple, a plus d'affinité avec *Gerbillus* ou *Taterillus*.

Gerbillus Desmarest, 1804

Les stades les plus jeunes (les moins usés) des plus petites espèces de *Gerbillus* (celles qui avaient été rapportées à *Monodia*, notamment) montrent si clairement des cuspides individualisées qu'on peut les rapporter à un schéma molaire de Cricétidé. Les différences de hauteur relative de chaque cuspide par rapport au collet de la dent, sont responsables de la variation du dessin de l'émail sur la table d'usure et correspondent souvent à des différences spécifiques. La M^3 a, le plus généralement, 3 cuspides, parfois 4. La queue des *Gerbillus* est toujours plus longue que le corps, et terminée par un pinceau de poils.

Gerbillus est réparti depuis le nord de l'Inde, à l'est, jusqu'au sud de la Mauritanie à l'ouest, et jusqu'au Kenya au sud.

Microdillus Thomas, 1910

Si l'on refuse à *Monodia* HEIM DE BALSAC, 1943, la valeur de genre distinct de *Gerbillus* à cause de l'inconstance des caractères qui le définissent par rapport à la variation individuelle de M^3, il devrait en être de même de *Microdillus*. En effet, si *Microdillus*, genre monospécifique, a pu d'abord être distingué de *Gerbillus* d'après des spécimens dont la M^3 présentait deux lames dont la seconde montrait encore la trace de deux cuspides initiales fusionnées, de récentes captures montrent bien là aussi une variabilité telle que la 2ème lame peut n'être indiquée que par une très petite cuspide chez certains spécimens. Il serait donc logique de considérer *Microdillus* comme un représentant de *Gerbillus* si tous les exemplaires connus n'avaient une queue plus courte que le corps, couverte de poils courts et serrés; ceci distingue encore *Microdillus* de toutes les espèces de *Gerbillus*.

La répartition de *Microdillus* est limitée à la Somalie.

Pachyuromys Lataste, 1880

Ce genre également monospécifique, dont les cuspides dentaires sont tôt réunies deux par deux en lames parallèles, est cependant très proche de *Gerbillus* par ses molaires. Toutefois son crâne et surtout sa queue courte et en massue sont caractéristiques. *Pachyuromys* est propre au Sahara.

Sekeetamys Ellerman, 1941

Egalement monospécifique, ce genre a des caractères dentaires comparables à ceux de *Pachyuromys*. Toutefois, la partie libre des lames est plus courte si bien que très vite la dent est de structure prismatique et les dessins des tables d'usure sont très tôt comparables à ceux que l'on observe chez les *Meriones*. La queue des *Sekeetamys* est recouverte de poils comme celle d'un loir, ce qui la distingue de celle de tous les autres Gerbillidés. La répartition de *Sekeetamys* est limitée à la presqu'île du Sinaï.

Ammodillus Thomas, 1904

Le stade dentaire que représente ce genre monospécifique par rapport à *Gerbillus* et *Meriones* est comparable à celui des genres précédents. Quelques particularités du contour des dents, l'absence de processus coronoïde sur la mandibule, la convergence postérieure des rangées dentaires distinguent cependant *Ammodillus* de tous les autres Gerbillidés. La répartition d'*Ammodillus*, comme celle de *Microdillus*, est limitée à la Somalie.

Dipodillus Lataste, 1881

Longtemps considéré comme un sous-genre de *Gerbillus* et attribué par erreur à toutes les espèces à soles plantaires nues, ce genre caractérise en fait un groupe de petites formes nord-africaines dont les molaires sont à peu de détails près celles de *Meriones*, alors que leurs crânes sont caractéristiques de *Gerbillus*. La répartition de *Dipodillus* est limitée à l'Afrique du Nord, du Maroc à l'Egypte.

Gerbillurus Shortridge, 1942

Les espèces que l'on rapporte à *Gerbillurus* étaient préalablement référées à *Gerbillus*. Bien qu'elles aient d'abord comme *Gerbillus* des cuspides individualisées, elles passent très vite par un stade qui, par beaucoup de points, rappelle *Taterillus*. Par ailleurs, ces petits Gerbillidés à soles plantaires peu poilues ressemblent, par leurs proportions et leur pelage, à des *Gerbillus*. Ils se répartissent tous deux dans le sud et le sud-est de l'Afrique.

Desmodillus Thomas et Schwann, 1904
Desmodilliscus Wettstein, 1917

Les dents de *Desmodillus* et de *Desmodilliscus* ont une couronne élevée et restent laminées très longtemps, ce qui permet de les rapprocher de la lignée qui a conduit à *Tatera*. Toutefois, *Desmodillus* se rapporte à une espèce d'Afrique du Sud dont la morphologie générale rappelle celle de *Pachyuromys* et qui est caractérisée par la constitution de son crâne et la pilosité de ses soles plantaires.

Desmodilliscus n'est représentée que par une très petite espèce propre aux limites méridionales du Sahara occidental chez laquelle la M_3 est absente. Ses bulles tympaniques sont hypertrophiées. Ses soles plantaires sont nues et sa queue est plus courte que le corps.

Taterillus Thomas, 1910
Tatera Lataste, 1882

Ces deux genres se distinguent mal l'un de l'autre par la longueur des foramens palatins. Ceux-ci sont relativement courts chez la majorité des espèces de *Tatera*, toujours longs (de la longueur de la rangée molaire) chez toutes les espèces rapportées à *Taterillus*. Dans l'ensemble, *Taterillus* est représenté par des espèces de plus petite taille que *Tatera* et plus sveltes et plus agiles, dont la queue est longue et terminée par un pinceau développé. Quelques espèces de *Tatera* ont des caractéristiques identiques. La majorité, cependant, ont un corps lourd et une queue plus courte. Les molaires des *Taterillus* et *Tatera* sont toutes à couronnes élevées. Elles sont constituées de lames qui restent longtemps séparées.

Meriones Illiger, 1811
Brachiones Thomas, 1925
Psammomys Cretzschmar, 1828
Rhombomys Wagner, 1841

Les molaires de très jeunes *Meriones* ne montrent jamais de cuspides individualisées, mais des lames qui sont, pour la plupart, très rapidement soudées ensemble pour donner la structure prismatique caractéristique. C'est aussi cette structure qu'on observe chez les genres *Brachiones*, *Psammomys* et *Rhombomys*. Ces trois derniers sont monospécifiques. Au contraire *Meriones* s'est bien diversifié en espèces qui se distinguent mal morphologiquement et qui se remplacent écologiquement. Pour cette raison, la répartition de *Meriones* s'étend sur tous les déserts et semi-déserts de l'Afrique du Nord, du Moyen-Orient et de l'Asie paléarctique. *Brachiones* se distingue par la morphologie de son crâne et son pelage sec. Sa répartition est propre au désert de Gobi. *Psammomys*, dont le régime végétarien est hyperspécialisé, habite le Sahara. *Rhombomys*, dont la croissance des molaires est prolongée, vit en colonies dans le nord-est de l'Iran et les régions voisines.

Conclusion

L'examen des structures dentaires des Gerbillidés et de la répartition géographique des genres et des espèces actuels montre que toutes les formes qui présentent des caractères primitifs sont africaines et vivent dans des zones ouvertes non désertiques. La multiplicité des types morphologiques se traduit par un buissonnement de genres monospécifiques voisins de *Gerbillus* et de *Meriones*, et par la multiplicité des espèces à l'intérieur de ces deux genres. Le nombre relativement important d'espèces de *Gerbillus* et de *Meriones* correspond à leur adaptabilité écologique et leur a permis de conquérir tous les milieux du désert et de s'étendre fort loin vers les tropiques, et même jusqu'à des régions asiatiques de climat continental. Il est vraisemblable que les *Tatera* et *Taterillus* ont bénéficié de leur côté de facultés d'adaptation génétiques comparables qui ont amené une pluralité d'espèces extrêmement proches-parentes à conquérir des territoires géographiques de plus en plus lointains. L'exemple le plus extrême de cette faculté d'expansion est celui de *Tatera indica*, seul représentant non africain du genre, dont la répartition s'étend depuis la Mésopotamie jusqu'à Ceylan. Il est toutefois intéressant de constater que la concurrence écologique que se font le groupe *Gerbillus-Meriones* et le groupe *Taterillus-Tatera* n'a pas permis à ce dernier de conquérir les milieux réellement désertiques auxquels se sont adaptés *Meriones*, *Gerbillus* et certains genres voisins. Le dernier mot restera aux paléontologistes lorsqu'ils pourront nous expliquer la véritable histoire des Gerbillidés.

IX. SOME OBSERVATIONS ON ECOLOGICAL ADAPTATIONS OF DESERT RODENTS AND SUGGESTIONS FOR FURTHER RESEARCH WORK

by

A. DE VOS

Introduction

Ecologically, the desert communities and their association into biomes are of great interest because of the extreme degrees of their adaptation to the desert environment.

Burrowing rodents living in deserts have in the course of evolution acquired physiological characteristics which enable them to live and thrive in an environment that is hostile and uninhabitable to other closely related forms.

Compared to studies on the physiological adaptations of desert rodents, work on ecological adaptations of these interesting mammals has remained rather restricted. It is, therefore, rather difficult to write a well-balanced review of the subject. For this reason I have chosen the above-mentioned title and included a section on research needs in this paper.

The Desert Environment

Before delving into the matter of ecological adaptations of rodents to desert environments, it is necessary to outline briefly some of the main environmental characteristics of this environment.

Deserts are low rainfall areas where both plants and animals suffer acute water stress over most of the year. Although this is the dominant feature, other important characteristics include high levels of insolation, mobility and resorting of surface materials, limited shade, large areas with low levels of soil fertility, sparse and often temporary surface water resources, and salinity of soil and underground water (CHRISTIAN, 1964).

Desert environments impose limitations on biological productivity in kind, total amount and stability. Plant species either escape drought by regenerating only when moisture is adequate and, therefore, have short peak periods of production, or resist drought through a number of mechanisms, some of which make them less palatable to animals. Rates of plant production in desert environments may be high for short periods, but the fluctuations in conditions lead to low average levels. Annuals are only prominent after heavy rainfalls; their growth is rapid.

185

As both flora and fauna are poorly to very poorly represented, the structure of the biocenoses will accordingly remain rather simple.

For animals, and particularly for rodents, it is not so much the general climate as the microclimate surrounding them that enables them to live. In a hot climate, daily maxima are more important limiting factors to sustaining animal life than are daily means.

Ecological Adaptations

Especially adapted desert forms of rodents are found in all the major deserts of the world. Although these rodents may belong to the different taxonomic groups, they are all similarly adapted to their environment, both physiologically and ecologically. I will restrict myself to a discussion of the latter.

ADAPTATIONS TO THE CLIMATE

Desert rodents must protect themselves against exposure to extreme heat and desiccation.

Exposure to excessive temperatures may be avoided by adopting nocturnal habits and/or spending the hottest part of the day in burrows. Most desert rodents have, in fact, adopted nocturnal or at least crepuscular habits and live in burrows. They are certainly always able to find a locality where the climatic conditions are tolerable. The need for special adaptation to the adverse climatic conditions is, therefore, reduced.

The great daily range in temperature fluctuations that is found in a desert environment is much reduced in burrows. By burrowing deep an animal can avoid the extreme heat of the day. It can also escape a peak temperature in the burrow by leaving it at night, since at a foot or so beneath the surface there is a lag of about twelve hours in the time of maximum temperature (WILLIAMS, 1954). BOLWIG (1958) recorded that in the Kalahari desert in a burrow of *Parotomys*, 18 inches from the entrance and 18 inches below the surface there was a temperature variation of 2 °C only, as compared with a variation of 30 °C five inches above ground. Another interesting adaptation to reduce temperature fluctuations in burrows is the plugging of their entrances with sand. This is reported in the paper by HAPPOLD in this book for the jerboa, *Gerbillus pyramidum*, in the Sudan.

Another method to reduce exposure to extreme heat is aestivation. This occurs for instance amongst ground squirrels in arid environments in North America. HEIM DE BALSAC (1936) points out that although no case of aestivation has been described for mammals in the Sahara region, it may well occur among jerboas, *Eliomys* and other species.

Weather conditions affect the foraging activities of desert rodents. An example is given of that by HAWBECKER in this book. The speed with which California ground squirrels *(Citellus beecheyi)* forage and whether they eat on the spot may depend somewhat upon the weather. Pleasant days are spent out of the burrow feeding and apparently moving over quite a large area. Extremely hot and cold days are spent in the burrows with only enough time spent above ground to get food, and that during the mildest temperature of the daylight hours.

Desert rodents have a great tolerance to water shortage and it appears that most of the necessary moisture for the sustenance of life is obtained through the ingestion of succulent vegetation. HAWBECKER *(op. cit.)* gives an example of that for *Citellus nelsoni*. Particularly during hot, dry periods moisture-producing materials, green as well as animal, are ingested with greater frequency.

ADAPTATION TO THE SOIL

Soil conditions generally influence the distribution pattern of desert rodents. These may be direct influences, such as suitability of the soil for burrowing or indirect influences on the type and distribution of the vegetation.

HAWBECKER *(op. cit.)* states that the distribution of *Citellus nelsoni* is limited to loams and sandy loams. Nowhere is this squirrel to be found where alkali is abundant. According to HAPPOLD (in this book), the Spiny Mouse *(Acomys cahirinus)* is restricted in the arid areas of the Sudan to jebels (rocky hills) where they hide during the day in large crevices. They do not leave those jebels to feed on the sandy parts of the desert.

OTHER ECOLOGICAL ADAPTATIONS

There is a rather wide range in the adaptability of desert rodents to environmental conditions; some species have very narrow ecological requirements while others do not. HAPPOLD (in this book) gives an example of an adaptable species. In Egypt, *Acomys cahirinus* is mostly associated with human habitations and it is not as common on jebels as in the Sudan. In Libya it is found in areas of loose sand near the bases of palm trees in oases; it does not appear to be common on jebels, nor is it commensal.

Sympatric species may reduce competition by different patterns of using food and space.

HAPPOLD *(op. cit.)* gives an interesting example of some substantial ecological and behavioural differences between the jerboa, *Jaculus jaculus butleri*, and the gerbil, *Gerbillus pyramidum* at Khartoum (in this book). ROSENZWEIG *et al.* (in this book) indicate that the nocturnal desert rodent species of southern Arizona exhibit habitat differences which can be interpreted to be the result of competition. The specializations deal with sculptural features of vegetation.

Need for further Research

Gaps in existing information clearly indicate the need for further research and for this reason I will review some of the subject matter areas in which further work should be undertaken.

1. More data should be collected about the characteristics of desert communities and of the interrelations between populations forming these communities.

2. More careful studies should be made about the microclimate conditions under which desert rodents are living. This will throw more light on their habitat requirements and behavioural adaptations to environmental conditions.

3. Specific studies should be made of the size, shape and use of burrow systems and of variations in temperature and humidity in these throughout the year.

4. More information should be gathered about the influences of soil conditions on the distribution of desert rodents.

5. How do human influences (overgrazing, cutting wood for charcoal) affect desert rodents?

6. Effects of predation on population dynamics and behaviour of desert rodents.

Many more points can be added to this list. It is clear that more and particularly more intensive studies should be made of the ecological adaptations of desert rodents.

REFERENCES

BOLWIG, N. 1958. Aspects of animal ecology in the Kalahari. *Koedoe*, no. 1, p. 115–135.
HEIM DE BALSAC, H. 1936. Biogéographie des mammifères et des oiseaux de l'Afrique du Nord. *Bull. biol.*, supplement 21.
CHRISTIAN, C. S. 1964. Desert environments. IUCN Bull. No. 4.
WILLIAMS, C. B. 1954. Some bioclimatic observations in the Egyptian desert. In: CLOUDSLEY-THOMPSON, J. L. (ed.), Biology of deserts, p. 18–27. London, Inst. of Biology.

X. THE BEHAVIOR PATTERNS
OF DESERT RODENTS

by

J. F. EISENBERG

Introduction

It is difficult to define a desert in a precise fashion. Arid to semi-arid areas may vary in the form of their vegetation; this ultimately reflects the variation of soil conditions, elevation, temperature, and periodicity of rainfall. Such variation in physical and vegetational features will be paralleled by variations in the species and diversity of rodent faunas exploiting them (ZAHAVI & WAHRMAN, 1957; CHEW & CHEW, 1970). Arid grasslands with intermixed *Artemesia* characterize a semi-desert habitat over much of the extreme western edge of the Great Plains of the United States. Semi-desert areas supporting mixed growth of low shrubs (termed chaparral) defines much of the semi-arid country in the southwestern, coastal portion of North America. True North American deserts have been classically divided into low elevation and high elevation life zones, the Lower Sonoran and Upper Sonoran respectively (MERRIAM & BAILEY, 1910).

To generalize, the true desert areas of the world are characterized by low stature, dispersed vegetation forms and extremely low rainfall. Hence, small rodents adapting to these environments are faced with scattered, unpredictable food resources, lack of continuous cover which would aid in predator avoidance, and a persistent shortage or absence of free water. The temperature of the world's deserts is highly variable depending on altitude and/or latitudinal position. The northern deserts may be bitterly cold during the winter when small mammals are confronted with heat conservation problems. In the lower latitudes, deserts may be subjected to extremely high daytime temperatures with concommitant problems of heat dissipation to be faced and solved by the diurnal mammalian inhabitants (SCHMIDT-NIELSEN & SCHMIDT-NIELSEN, 1952).

The world's deserts have their greatest distribution in the Eastern Hemisphere, in particular, North Africa, the Middle East, and Central Asia. These deserts cover a tremendous span of latitude and exhibit contrasting temperature extremes. Most of the Asian and African desert areas have counterparts in North America, for example, the Sahara-Sindian life zone (BODENHEIMER, 1935) is comparable to the Colorado desert of North America. Central and western Australia is dominated by a variety of desert forms which are climatically and geophysically comparable to the North American and Asiatic series. Only South America exhibits a limited true desert area, the Atacama Desert confined to the coast of Peru and Chile. This desert is of relatively recent origin and does not display a series of Cricetine rodent forms exhibiting the diversity and degree of specialization that one finds among Heteromyidae of North America, and the Dipodidae and Gerbillinae of Africa and Asia.

When comparisons are made between desert rodents from two separate geographical areas, one is continually struck with the patterns of convergence. Similarities in color, form, ecology, and behavior demand some functional explanation, but equally intriguing are the differences. Attention will be focused on behavioral convergences but the importance of behavioral discontinuities begs that additional questions be posed.

Rodents are to be found in all of the world's desert areas; however, the most highly adapted desert rodents with the longest continual history of evolution and adaptation to desert life are to be found in North America and Asia. To a lesser extent, rodent evolution in Africa parallels the Asian adaptive radiation and (in part) the African desert rodent fauna is derivable from Asia's. The Australian desert rodent fauna shows some remarkable convergences toward that of the major continental radiations but the diversity of forms is lower.

In this review the major functional classes of rodent behavior patterns will be described for several species. I will employ a functional classification because I wish to emphasize that behavior patterns are mechanisms which serve to maintain the organism and allow selective exploitation of a suitable micro-habitat. Behavior patterns are adaptive strategies which are as much the products of natural selection as are the bones and muscles utilized as reference points by anatomists. The stereotyped, neuromuscular coordination patterns or 'fixed action patterns' which comprise a species' behavioral repertoire may be compared from one species to the other with useful inferences and predictions deriving from such comparisons (EISENBERG, 1967). For the purpose of this chapter, the behavior patterns of the heteromyid rodents will be taken as baselines to which comparisons of other species will be made.

The Evolution of Desert Adaptations

ECOLOGICAL NICHES

If one turns to the semi-desert habitats of southwestern North America, a variety of rodent species exhibiting a variety of adaptations can be discerned. The following ecological niches can be described (see Tables 1 and 2).

190

Table 1. Semi-desert and Desert Rodent Faunas.

	Libya (Mediterranean)[1]	California (Californian)[2]	Libya (Saharan)[1]	California (Sonoran)[2]
Ground Squirrels	—	Sciuridae *Citellus beecheyi*	—	Sciuridae *Citellus mohavensis* *Citellus leucurus*
Voles	Microtinae *Microtus mustersi*	Microtinae *Microtus californicus*	—	—
Fossorial Forms	Spalacidae *Spalax ehrenbergi*	Geomyidae *Thomomys bottae* Cricetinae	—	Geomyidae *Thomomys bottae* Cricetinae
Small Quadrupedal	Murinae		Gerbillinae	
Granivore/Insectivores	*Aconys cahirinus*	*Peromyscus boylei* *Peromyscus maniculatus* *Neotoma fuscipes*	*Gerbillus gerbillus* *Gerbillus campestris* *Meriones crassus* *Meriones caudatus*	*Peromyscus eremicus* *Onychomys torridus* *Neotoma lepida*
	Gerbillinae *Gerbillus eatoni* *Gerbillus henleyi* *Meriones libycus* *Pachyuromys duprassi* *Psammomys obesus*	Heteromyidae *Perognathus fallax* *Perognathus inornatus* *Perognathus californicus*		Heteromyidae *Perognathus formosus* *Perognathus penicillatus* *Perognathus longimembris*
Bipedal Saltators	Dipodidae *Allactaga tetradactyla* *Jaculus orientalis*	Heteromyidae *Dipodomys heermani* *Dipodomys venustus* *Dipodomys agilis*	Dipodidae *Jaculus jaculus* *Jaculus deserti*	Heteromyidae *Dipodomys deserti* *Dipodomys merriami*

[1] Biotic Provinces and faunal elements from RANCK, 1968; BODENHEIMER, 1935.
[2] Biotic Provinces and faunal elements from DICE, 1952; INGLES, 1965.

191

Table 2. Micro-habitat Preferences for Some Desert Rodents

Substrate Type	Gaza, Israel[1]	Walker Lake, Nevada[2]
Rock	*Sekeetamys calurus* *Gerbillus dasyurus*	*Neotoma lepida* *Peromyscus crinitus* *Perognathus formosus*
Sand	*Jaculus jaculus* *Gerbillus gerbillus*	*Dipodomys deserti* *Microdipodops pallidus*
Soil	*Meriones crassus* *Gerbillus nanus*	*Perognathus longimembris* *Citellus leucurus* *Peromyscus eremicus* *Dipodomys microps*

[1] ZAHAVI & WAHRMAN, 1957; BODENHEIMER, 1935.
[2] Data from EISENBERG.

Diurnal, surface foraging granivores

This niche is occupied by various species of the genus *Citellus (Spermophilus)* including *Citellus mohavensis* and *C. leucurus* (see HUDSON, 1962, and BARTHOLOMEW & HUDSON, 1960).

Nocturnal, insectivore-granivore

This is a very specialized niche occupied by *Onychomys* in North America.

Nocturnal, granivore-herbivore*

a) Adapted to chaparral habitats; most species are dependent on free water in succulent plants. Examples: *Neotoma lepida, Peromyscus maniculatus, P. eremicus, Perognathus californicus, Dipodomys agilis* (see MAC-MILLEN, 1964, for a discussion of this rodent community). b) Adapted to flat, pebble deserts in extreme xeritic conditions. Examples: *Perognathus formosus, Dipodomys microps, Microdipodops megacephalus.* c) Adapted to sand deserts with spaced vegetation and a minimum of cover. Examples: *Perognathus penicillatus, Microdipodops pallidus, Dipodomys merriami, Dipodomys deserti.*

Completely fossorial forms: herbivores

Example: *Thomomys bottae.*
Such a series of forms may be found under similar ecological conditions

* Some insects taken as prey.

Table 3. Convergent Evolution of Bipedal Saltating Rodent Genera in Arid Areas

Geographical Area	Family	Genera
Africa	Dipodidae	*Jaculus*
		Allactaga
	Pedetidae	*Pedetes*
Madagascar	Muridae	*Macrotarsomys*
Asia	Dipodidae	*Allactaga*
		Alactagulus
		Euchoreutes
		Salpingotus
		Cardiocranius
		Pygerethmus
		Dipus
		Scirtopoda
		Paradipus
		Eremodipus
		Jaculus
Australia	Muridae	*Notomys*
North America	Heteromyidae	*Microdipodops*
		Dipodomys

in Africa, including *Paraxerus* as a counterpart of *Citellus*, *Acomys* in place of *Peromyscus;* some species of *Meriones* and *Gerbillus* in place of *Neotoma* and *Perognathus;* and *Jaculus* in place of *Dipodomys*. A similar series of forms could be established for Asia, including *Citellus* and some species of *Meriones* as counterparts of *Citellus* in North America; *Acomys* as a counterpart of *Peromyscus;* *Gerbillus* as a counterpart of *Perognathus;* *Salpingotis* for *Microdipodops*, and a variety of the Dipodid genera, *Jaculus*, *Paradipus*, *Dipus*, and *Scirtopoda*, as counterparts of some species of the North American *Dipodomys*. In Australia, *Notomys*** appears to be a reasonable counterpart of the smaller species of *Dipodomys* in North America (see Table 1).

The degree of convergence is a function of a) the different origins of the rodent stocks currently occupying the different continental areas, b) the length of time it has taken for divergence and adaptive radiation to occur, and c) the degree of similarity in the environments in which the adaptive radiations have taken place. Convergence in form and behavior results from not only physical similarities in environments but also from the similarity of associated species or populations which have been in competition with one another throughout the evolutionary history of their adaptations to desert areas. Suffice to say, however, that a remarkable convergence in niche occupancy has occurred as can be seen

** *Macrotarsomys* of the south central desert region of Madagascar is very convergent with *Notomys*.

193

when rodent families are compared for the continental deserts of the world (see Table 3).

The Evolution of Desert Adaptations within the Family Heteromyidae.

Sometime during the Oligocene, the crust of the earth began to buckle initiating a period of mountain-building in North America and Asia. This process of elevation was followed by the development of grassy plains during the early Miocene replacing the wet forests of the Eocene and early Oligocene. From the beginning of this period, a gradual increase in aridity became the dominant trend. During the Pleistocene this drying-out oscillated with the advance and retreat of the glaciers, but the trend toward aridity persisted. In North America, a family of rodents, the Heteromyidae, was evolving in step with these environmental changes. The adaptive radiation culminated in a bipedal, ricochetal form, the genus *Dipodomys*. Today the Heteromyidae consist of 5 genera, grouped into 3 subfamilies. The first subfamily, the Heteromyinae diverged in the early Oligocene and later in the Miocene split into two lines leading to each of the present genera, *Liomys* and *Heteromys*. The other stock diverged in the Miocene forming the lines giving rise to the Perognathinae and the Dipodomyinae. These first divisions corresponded to the more arid conditions of the middle and later Pliocene (WOOD, 1935).

Today the genera are distributed as follows: *Heteromys* is a tropical form confined to the moist areas of Central and South America. *Liomys* ranges in the subtropical, wet to semi-arid areas of coastal Mexico. *Perognathus* is divided into 2 subgenera 1) *Chaetodipus* of the more arid areas of the southwestern United States and Mexico, and 2) *Perognathus* ranging from the Lower Sonoran of the western United States well into the Great Plains. *Microdipodops* is confined entirely to the Great Basin. *Dipodomys* ranges all over the semi-arid and arid areas of western North America. This genus has also invaded the Transition life zone of California but in general is an inhabitant of the Sonoran life zone. Thus we have represented with this family a morphological spectrum from a relatively generalized rodent form in *Heteromys* to a very specialized form in *Dipodomys* (see Fig. 1). The trend toward specialization seems to indicate increased adaptation to an arid habitat culminating in *Dipodomys deserti* as a saltator of the sand dunes in California and Nevada (WOOD, 1935).

Bipedal saltating locomotion shown by *Dipodomys* and *Microdipodops* is one extreme specialization to desert life and has evolved convergently in all rodent populations found in the major deserts of the world (HATT, 1932; HOWELL, 1932) (see Table 3). Bipedal saltation is a means of gaining speed for short durations, but it is wasteful of energy since momentum may be lost when the animal strikes the ground. However, bipedal saltation permits the animal to change direction rapidly and this may be its chief selective advantage. Further the ability to spring upwards may be important in avoiding the strike of a snake. It would

Fig. 1. Bipedal, ricochettal locomotion by *Dipodomys venustus*. Photograph of animal taken at the instant of impact. Note the prominent white tail tip and that the lateral white band on the tail forms an almost continuous line with the white hip stripe and white border of the ventrum. The long vibrissae are in contact with the substrate during slow, bipedal locomotion.

appear to me that predators (cursorial, avian, and reptilian) have been the primary selective forces in producing saltating forms whenever a rodent stock has begun to adapt to patterns of foraging in the open away from cover. Similarly predators have been responsible for the convergence in color patterns which one finds when desert rodents are compared from one continental area to the next.

To round out our discussion of morphological convergence among desert rodents, some mention should be made concerning the pinna and the tympanic bulla. The desert or steppe adapted rodents generally show a relative increase in either the size of the tympanic bullae (*Micro-dipodops* and *Dipodomys*) or the external ear (*Allactaga* and *Jaculus*). Occasionally some species show a proportionate increase in both the bulla and the pinna (OGNEV, 1959; HOWELL, 1932). It would appear that an enlarged pinna enhances the perception of low amplitude sounds by focusing sound energy at the external auditory meatus. The expanded bulla definitely increases the sensitivity of the ossicles in *Dipodomys* to

195

frequencies of 1 to 3 KHz and may aid in detecting predators (WEBSTER, 1960, 1962). Morphological changes in the cochlea correlating with the expanding tympanic bullae have been established for the Gerbillinae by LAY (1972) and for the Heteromyidae by PYE (1965). The exact manner in which the bulla functions to maintain sensitivity to low amplitude sounds is only partially understood but an hypothesis has been advanced by WISNER & LEGOUIX (WISNER et al., 1954; LEGOUIX et al., 1954). LAY (1972) should be consulted for a general review.

A Comparison of Behavior Patterns

The behaviour patterns of desert rodents were first systematically studied and recorded by FERNAND LATASTE (1886–1889). After this pioneer effort, ethological studies of mammals were of sporadic occurrence until a renewed effort was made by I. EIBL-EIBESFELDT (1951). During the last decade rodent behaviour studies have increased greatly, and several summaries of behavioral repertoires are now available (EIBL-EIBESFELDT, 1958; EISENBERG, 1967). The following references will serve as a guide to the desert rodent literature: a) Interspecific relationships (MAC-MILLEN, 1964; WAGNER, 1961); b) the genus *Peromyscus* (EISENBERG, 1968; KING, 1968); c) the family Heteromyidae (EISENBERG, 1963a; EISENBERG & ISAAC, 1963); d) *Citellus* – general behavior (BALPH & BALPH, 1966; BALPH & STOKES, 1963); e) *Citellus leucurus* (HUDSON, 1962); f) *Citellus mohavensis* (BARTHOLOMEW & HUDSON, 1960); g) *Xerus* (EWER, 1965); h) *Onychomys* (RUFFER, 1968); i) *Gerbillus* (KIRCHSHOFER, 1958); j) *Meriones* (EIBL-EIBESFELDT, 1951; RAUCH, 1957; PETTER, 1961; BARAN & GLICKMAN, 1970; THIESSEN & YAHR, 1970; KUEHN & ZUCKER, 1968); k) *Jaculus* (KIRMIZ, 1962; HAPPOLD, 1970); l) *Notomys* (MARLOW, 1969; STANLEY, 1971); m) *Acomys* (DIETERLEN, 1962). Useful comments on the natural history of the Dipodidae are summarized by OGNEV (1959, 1963).

To maximize the utility of this brief review, I will choose only 5 rodent genera for intensive comparisons (i.e. *Gerbillus*, *Meriones*, *Perognathus*, *Dipodomys* and *Jaculus*). This series of forms includes species which are specialized for the exploitation of extremely arid habitats and offers a sample from both bipedal and quadrupedal forms. To avoid repetition in the description of behavior patterns, it should be emphasized that the basic behavioral repertoires for most rodent species are remarkably similar and behavioral inventories or ethograms for different species show a basic uniformity. Terminology and refined descriptions are included in EISENBERG (1967, 1968), EIBL-EIBESFELDT (1951, 1958), and GRANT & MACKINTOSH (1963).

Jaculus and *Dipodomys*

Since the bipedal saltating rodents are an end point in adaptation to

196

desert habitats, I will confine the most detailed descriptions to these forms. In particular, I wish to compare and contrast *Dipodomys* with *Jaculus*. This is an instructive comparison since the animals are similar in external morphology but differ profoundly with respect to some of their behavior patterns.

General Comparison

Dipodomys merriami and *D. deserti* are roughly comparable in size to *Jaculus jaculus* and *J. orientalis*. All 4 species are adapted to extremely arid habitats. These animals can exist on metabolic water and require no succulent plants for maintenance of body weight (KIRMIZ, 1962). They are primarily granivores and forage at night for seeds and small arthropods. *Jaculus* displays torpor during cold periods and accumulates fat reserves which appear to supply it with energy during inclement periods. It caches very little. By contrast, *Dipodomys* is a persistent hoarder of seeds and, in common with all the Heteromyidae, possesses externally opening, fur-lined cheek pouches. These pouches are employed to transport seeds from collection points to caching areas in or near the burrow.

A further difference between *Jaculus* and *Dipodomys* concerns the gestation and development of the young. As noted in a previous publication (EISENBERG, 1963a) the genus *Dipodomys* shows both the longest gestation (29–33 d.) and on the average produces the smallest litters in the family Heteromyidae.* The net result would appear to be the production of rather precocial young which show eye opening times from 12 to 17 days and weaning at 21 to 29 days. These data contrast sharply with the maturation data for the Dipodoidea. The primitive Zapodinae *(Sicista)* have long gestations (28–35 d.) and long developmental times for the young (eyes open at 28 days for *Sicista betulina*) (MOHR, 1954). Those specialized jerboas which have been studied indicate the retention of similar tendencies for both a long gestation and a long developmental period for the young. *Jaculus orientalis* has a gestation period of from 28 to 30 days as does *Dipus sagitta* (OGNEV, 1963). The eyes do not open until 5 weeks of age and coordinated locomotion with bipedal hopping does not occur before 6 weeks of age (see also KIRMIZ, 1962).

By contrast, coordinated bipedal locomotion manifests itself in *Dipodomys* from 3 to 3½ weeks of age. The rapid maturation of *Dipodomys* young results in early attempts at locomotion and the young of *D. nitratoides* begin to crawl in the maternal tunnel system at 10 days of age. The maternal retrieving response is highly developed in *Dipodomys* and errant young can be quickly re-assembled in the maternal nest. On the other hand, the young of *Jaculus orientalis* locomote very little, even at 4 weeks, and it would appear that the maternal retrieving response is very weak (see also PETTER, 1961).

---

* *Notomys* exhibits a similar reproductive trend when compared to more typical Muridae.

Burrowing and nest construction are highly characteristic patterns for both *Dipodomys* and *Jaculus*. Basically the animals employ the forepaws and incisors to loosen soil. Soil which accumulates under the body is either kicked to the rear with the hind feet or the animal turns and pushes the earthen pile to the rear employing its forepaws and chest. In contrast to *Dipodomys*, the pro-odont dentition of many dipodids permits gnawing into a flat surface.

Burrow walls are typically packed by *Dipodomys* employing a pushing and patting motion with the forepaws. *Allactaga* and *Jaculus*, however, utilize the incisors and snout to 'tamp down' the soil by raising and lowering the head in the vertical plane thus bringing the snout and incisors in repeated, forceful contact with the soil. Typically both *Jaculus* and *Dipodomys* plug the entrances to their burrows with earth at the cessation of their nocturnal activity cycle.

The construction of a nest from dried plant material varies from species to species in *Dipodomys* and the dipodids (EISENBERG, 1963a, 1967); however, females with young typically build a nest.

Assembly of foodstuffs differs markedly in *Dipodomys* and *Jaculus*. Whereas *Jaculus* will gather some plant material into its burrow system, all species of *Dipodomys* collect and cache seeds and plant parts. Most species of *Dipodomys* form small, surface caches in the vicinity of the burrow and in addition cache seeds in special chambers within the burrow. The kind and magnitude of the caches vary from species to species (EISENBERG, 1963a). It is worth re-emphasizing that the reduced caching tendency in the dipodids is probably correlated with their capacity for assuming hibernation or torpidity. Thus, the ability to establish fat reserves in the dipodids may be an alternative to caching as a means of passing through periods of food scarcity.

LOCOMOTION, DAILY ACTIVITY, AND EXPLORATION

Moving on a plane surface may involve 3 basic patterns in small rodents 1) diagonal limb coordination when the contralateral limbs are in synchrony, 2) quadrupedal saltation where the forelimbs alternate with the hind limbs in striking the ground simultaneously, and 3) bipedal locomotion which may be expressed as either a) the walk where the hind limbs support the weight alternately, or b) bipedal saltation when the hind limbs strike the ground simultaneously after each successive hop (HATT, 1932; EISENBERG, 1963a).

Dipodomys and *Jaculus* employ quadrupedal saltation when moving slowly or foraging for seeds but bipedal saltation predominates as a mode of rapid forward progression. When foraging for seeds the animals may shuffle in a bipedal walk.

Both *Jaculus* and *Dipodomys* are nocturnal. Upon awakening in the burrow, the animal generally yawns and stretches and shakes itself. It may then proceed to a special chamber in the burrow system and urinate. If some scraps of food have been cached in the burrow system, these may be nibbled or alternatively the animal may move up the tunnel to the burrow entrance where it pauses before leaving and beginning to explore. Exploration of a foreign environment generally involves cautious departures from the nest only to return to the safety of the burrow before proceeding out again. Once the animal has become reasonably confident in a new environment, it will begin exploratory activity. Initially the body may be tensed either in a quadrupedal or in an upright stance. As it becomes more relaxed, the animal will assume more rounded body contours and its movements will become more purposive.

Upon leaving the burrow, the animals generally sandbathe at a specific locus. This serves to dress the pelage and serves also to mark the sandbathing spot with odors from glands situated at various points on the body. Once the animal has settled down in a new environment, it generally initiates foraging behavior. From time to time while moving about the environment, the animal will depress the hind quarters bringing the ano-genital region into contact with the substrate. This 'marking behavior' is very similar in both *Dipodomys* and *Jaculus*.

CARE OF THE BODY SURFACE

Auto-grooming (washing) occupies an important role in the activity of most small rodents (BÜRGER, 1959), since grooming activities involving the teeth, forepaws, and tongue serve to remove ectoparasites from the animal's fur. Bipedal desert rodents, such as *Dipodomys* and *Jaculus*, generally have shortened forelimbs when contrasted with their quadrupedal counterparts. The shortened forelimbs impose certain restrictions upon the animals when they wash their face, since wiping movements behind the ears generally cannot be performed simultaneously by both forepaws. Instead, the head must be turned alternately to one side to allow the short, ipsilateral forelimb to sweep behind the ear and over the face nearest to it.

The sequence of movements employed during washing are very similar for *Jaculus* and *Dipodomys*. The tongue is used to lick the forepaws and fur while the teeth are generally used to nibble the coat. Forelimbs are employed in 2 ways 1) the fur may be brushed by the inside of the wrist and forearm as it is passed over the surface of the body, and 2) the claws and digits are used to manipulate the tail, comb at the fur, and grasp while wiping the nose, ears, and vibrissae. In general the forepaws are first held under the mouth and licked and then the nose and vibrissae are wiped. Employing lateral up-and-down strokes, the animal may wipe its face with the inner sides of its forelimbs; gradually lengthening the

Fig. 2. Extension on the ventrum during the ventral-rubbing phase of sandbathing by *Dipodomys merriami.*

strokes until they reach behind the ears. The combing movements of the forelimbs are then extended to the flank and belly areas and often the combing movements of the forelimbs are combined with licking and nibbling. A complete washing sequence usually ends when the animal grasps the tail and brings it up to its mouth, licking and nibbling the tail from base to tip.

In addition to the typical washing sequence, the animal may intersperse washing bouts with scratching. The hind limb is moved with rapid pendular strokes and directed at various parts of the animal's body. When scratching, the head generally receives the most attention. The nails of the hind foot are usually cleaned with the animal's tongue and incisors between bouts of scratching.

Sandbathing in desert rodents is a method of dressing the pelage and has the derived function of marking. There are 3 distinct components: 1) the animal initiates a sandbathing bout by digging in the substrate with its forepaws and then rubbing either its side or its ventrum in the sand. 2) The side-rub consists of lowering the side of the face to the substrate and gliding forward by extending the body; the body is then

Fig. 3. Flexion on the side during the side-rubbing phase of sandbathing by *Dipodomys merriami.*

alternately flexed and extended while the animal remains on its side. 3) In the ventrum-rub the body is extended and flexed while the animal lies with its ventrum pressed against the substrate. Typically *Dipodomys* shows sandbathing by combining ventrum- and side-rubs into a rather predictable sequence (EISENBERG, 1963b) (Figs. 2 and 3). On the other hand, the dipodid rodents, including *Jaculus* and *Allactaga*, almost invariably show sandbathing involving the side-rub alone and the ventrum-rub is either shown rarely or may be shown independently in combination with other marking behaviors (EISENBERG, 1967).

ANTI-PREDATOR BEHAVIORS

The jerboas and kangaroo rats are relatively large in size when compared to the other nocturnal rodents which share the same microhabitats. As a result they are somewhat more conspicuous. Perhaps the origin of bipedality is in some way a result of predator selection for a species which can exhibit greater speed when under attack by an aerial or terrestrial predator. No doubt the remarkable convergence in the color patterns of desert rodents from the various areas of the world has resulted from predator selection since there is a strong correlation between the reflectance of the pelage and the soil from which the species has been obtained.

The eyes of jerboas and kangaroo rats are reasonably large and visual detection of predators may be important in predator avoidance. Mention has already been made of the tendency for either an increased pinna size or inflated mastoid bullae; it may be that these modifications are related to predator detection. Indeed, the experiments of WEBSTER (1962) indicate that the ability to perceive low-frequency sounds of low amplitude produced by rattlesnakes is related to the presence of an inflated mastoid bulla.

Predator avoidance involves the ability to detect the presence of the predator and then to follow through with several alternative courses of action. The bipedal ricochet permits kangaroo rats and jerboas to leap into the air, thus avoiding the strike of a terrestrial predator. At the same time, bipedality permits the animals to move rapidly and change direction quickly, thus avoiding both aerial or terrestrial pursuit.

Other predators, especially snakes, can be avoided by plugging the burrow. Of course, plugging the tunnels in the burrow is probably also related to the maintenance of a high humidity within the burrow system to reduce water loss. However, burrow-plugging also occurs in terrestrial rodents not adapted to arid environments.

When disturbed by a mild stimulus in their burrow, such as falling sand grains in the tunnel system, kangaroo rats and jerboas frequently approach the source of the disturbance. Upon approaching such a minor tunnel disturbance, the animal will generally begin digging and kicking back and pushing and patting at the walls of the tunnel. This can serve 2 purposes: 1) it can reinforce the tunnel walls during incipient cave-ins, and 2) the digging and pushing activity can serve to plug the burrow. Thus burrow-plugging and the maintenance of burrow walls are interrelated activities elicited by minor disturbances to the soil of the tunnel walls. Such a disturbance could be caused 1) by a predator attempting to enter the burrow, and 2) by slight earth movements occurring from the cracking and drying of the tunnel walls.

When mildly disturbed on the surface, jerboas and kangaroo rats have a tendency to approach a novel object and to pause at a certain distance surveying it. If the stimulus does not induce flight, the animal may exhibit displacement activities including digging. In the kangaroo rat, *Dipodomys deserti*, this behavior pattern is ritualized and has become an anti-predator mechanism. The presence of novel objects will elicit not only digging but the directed kicking back of sand or soil onto the stimulus. If a snake is involved, it may strike (whereupon the kangaroo rat avoids the strike by leaping up) or move away to avoid the sand storm. Hence, kicking sand at a novel stimulus has adaptive value in that it forces the predator to reveal its identity through movement or to move away. Such ritualized sand-kicking has not been noted as yet for *Jaculus*.

Kangaroo rats and jerboas have characteristic responses to both conspecifics and small predators which invade their home range, including the assumption of an alert posture, pilo-erection, and tooth-chattering. *Jaculus* and *Dipodomys* will also thump with the hind foot while facing a variety of alien animals. This tooth-chattering, pilo-erection, and thumping form part of an intimidation display. Of course, any sustained attack by a small predator produces immediate escape flight. Drumming by *Dipodomys deserti* is highly ritualized, including at least 3 variants: slow drumming, fast drumming, and the roll. These form a graded series and probably represent different levels of arousal (EISENBERG, 1963a).

COMMUNICATION MECHANISMS

Before moving on to a discussion of social behavior, we must establish a basic assumption; namely, that posture, sound, and marking behaviors have a significance in communicating information to conspecifics. The inference that information transfer occurs, results from observations of 2 animal encounters where predictable responses occur in the presumptive receiver to known activities of the presumptive sender. The investigation of communication mechanisms requires much more experimentation before the exact significance of many behavior patterns can be ascertained (EISENBERG, 1963a; 1967).

Many of the configurations shown by 2 interacting animals involve the exchange of tactile information but this is very difficult to distinguish from the exchange of chemical information. Thus, the 2 modalities may be combined.

Visual communication mechanisms are probably of minimal importance in nocturnal rodents. However, the eye is quite large in *Dipodomys* and *Jaculus* and certainly on moonlit nights, movement can be perceived. There is a remarkable convergence in the marking patterns on the body of jerboas and kangaroo rats. Regardless of the color of the dorsal pelage, the ventrum is white and the terminal tuft of the tail generally consists of one, proximal, black band demarcating a white tail tip (see Fig. 4). This contrasting white portion on the tip of the tail would appear to serve as an orientation point for a male when engaged in the sexual pursuit of a female.

The white ventrum is displayed during the ritualized upright postures employed in fighting and in courtship. The display of this white ventrum produces a sharply contrasting, reflective surface and may aid orientation during a sequence of mutual uprights and sparring behavior; however, the use of the white ventrum as a signal is not unique for either the kangaroo rat or the jerboa (EISENBERG, 1967).

Aspects of auditory communication involve 3 classes of sounds: a)

Fig. 4. *Jaculus orientalis*. Note the white tail tip bordered proximally with black. When relaxed, the animal sits with the heel touching the substrate. When hopping bipedally, the animal rises on its toe tips as is indicated by the position of the left foot as the animal prepares to move.

sounds produced by the activities of the animals themselves, namely, the sounds of digging and kick-back. These may have incidental communicatory significance. b) Non-vocal communication by ritualized movement patterns including tooth-chattering or drumming with the hind feet. As indicated in the previous section, drumming with the hind feet is generally shown when the animal is suddenly startled or is in a thwarting context. Tooth-chattering generally occurs in a threat situation. c) The production of vocalizations by expelling air through the animal's glottis.

In both the jerboa and the kangaroo rat, one can distinguish 5 classes of vocalizations: 1) a high squeal which may be shown in a defensive situation; 2) a low grunt; 3) a cry emitted in a young animal when alone, cold, and isolated; 4) a growl or buzzing sound which may be given by an aggressive animal; and 5) a 'comfort' peeping produced by the neonate when the female grooms it.

Male dipodids, including *Allactaga* and *Jaculus*, are prone to produce a 'courting sound'. It tends to be of a buzzing quality in *Jaculus;* however, in *Alactaga*, the sound consists of a buzz component followed by

204

Table 4. Some Rodent Vocalization for *Dipodomys* and *Jaculus*.[1]

Species	Sound	Context(s)	Implied Function	Sample (N)	Structure of Sound	Frequencies with Greatest Energy (Hz)	Duration of Pulse (sec.)
Dipodomys panamintinus nitratoides	1. Squeal High intensity	When injured, when defensive	Inhibits attack	N.R.	—	—	—
	Chirp	When defensive	?	10	Harmonic	1500–2500	.18–.21
	2. Growl	Thwarting situations	Warning of attack	10	Blurred harmonics to no harmonics	800–1800	.07–.09
	3. Grunt	Startle	?	N.R.	—	N.R.	N.R.
panamintinus	4. Cry of abandoned young (Day 1)	Young out of nest	Induces retrieval	9	Harmonic repetitive	2000; 3000	.1 –.21
Jaculus orientalis	1. Squeal High intensity	When injured, when defensive	Inhibits attack	N.R.	—	N.R.	N.R.
	Low intensity						
	2. Growl or Buzz (Courting)	Thwarting situations	Warning of attack	N.R. 4	Non-harmonic repetitive	N.R. 1000–1800	N.R. .1 – .2
	3. Grunt	?	?	N.R.	—	N.R.	N.R.
	4. Cry of abandoned young (Day 22)	Young cold	Induce return of female	2	Harmonic repetitive	1500; 3000	.22

? = Function incompletely understood.
N.R. = Not recorded.
1 = All recordings analyzed to 20 Khz only.

a high squeal. This sound appears to have similarities to sounds produced in clearly agonistic situations. Kangaroo rats rarely produce a courting sound, although occasionally during courtship pursuit the female will produce a rather grating, chirp sound. Table 4 compares the sounds of *Dipodomys panamintinus* and *Jaculus orientalis* with respect to the neonate abandonded cry and the 'courtship call'. A remarkable similarity exists between these 2 genera since the fundamental frequency and the frequencies emphasized are roughly the same for both species. The calls have a harmonic structure in the young animal although the harmonics may be blurred in the adult, thus giving rise to the buzzy or noisy quality of the sounds. It is interesting to note that the fundamental frequency and greatest energy concentration in the sound tends to be concentrated at the level to which the ear is most sensitive in *Dipodomys* (see WEBSTER, 1962).

Olfactory communication is undoubtedly of great importance in coordinating the social behavior of rodents (MYKYTOWYCZ, 1970; WHITTEN & BRONSON, 1970). Marking by dragging the ano-genital area over the substrate appears to impart information of use to conspecifics. Furthermore, sandbathing at the same locus establishes an area where odor traces of the sandbather can be left behind for conspecifics to find. The repeated use of the same locus for sandbathing by an individual and the frequent sandbathing of other individuals at the same locus reinforces the conviction that some form of chemical communication is going on (EISENBERG, 1963a; 1967). Marking by depressing the ano-genital region is often accomplished at this same spot. This permits not only leaving traces from the genitalia and/or urine at the marking spot but also implies that the animal may mark its own fur with such traces.

AGONISTIC BEHAVIOR

In bipedal rodents, such as *Jaculus* and *Dipodomys*, the upright posture is frequently employed. An animal in a low crouch generally has a high tendency to attack. An extreme upright posture at almost 90° to the substrate, however, indicates an animal in an ambivalent motivational state which is not likely to attack an opponent immediately (EISENBERG, 1963a). Assumption of an extreme upright posture is often accompanied by hopping back and forth on the hind legs, termed jockeying. A highly aggressive animal may hop immediately toward an opponent inducing flight behavior. If both animals clash together, they may lock together gripping one another while kicking and biting. Fights are extremely brief and generally a subordinate animal will move away. Under confined conditions of captivity, however, severe wounding and death can result from fighting (EISENBERG, 1967).

206

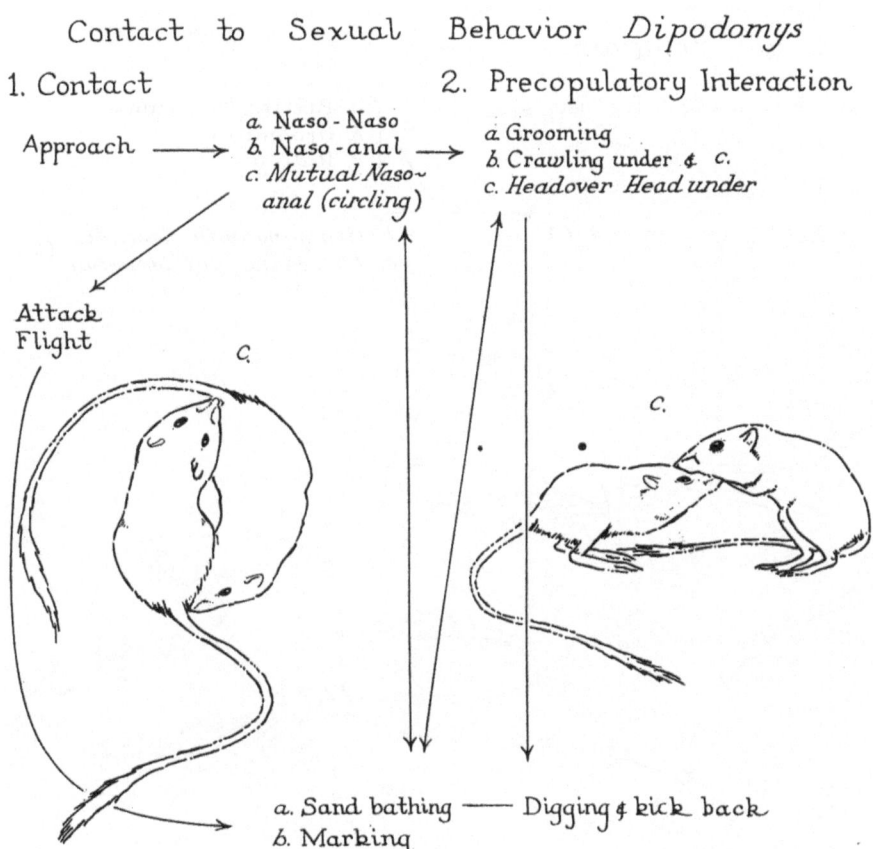

Contact to Sexual Behavior *Dipodomys*

1. Contact

Approach ⟶
a. Naso-Naso
b. Naso-anal ⟶
c. *Mutual Naso-anal (circling)*

Attack
Flight

c.

2. Precopulatory Interaction

a. Grooming
b. Crawling under & c.
c. *Head over Head under*

c.

a. Sand bathing —— Digging & kick back
b. Marking

Fig. 5. Contact and precopulatory interaction by *Dipodomys*. Two common contact-promoting postures are illustrated including mutual naso-anal investigation and head-over head-under (from EISENBERG, 1963a).

COURTSHIP AND MATING

Courtship behavior generally involves interactions which promote contact. Certain elements of agonistic behavior may be shown in the initial phases of courtship but in general contact is established between the 2 animals. Prominent in contact-promoting behavior is the slow approach of one animal to another, placing the head of a subordinate animal beneath the head of a dominant, and then either crawling under the dominant animal or submitting to grooming by the dominant. Initial phases of contact often involve naso-anal investigation by the partners which may be mutual (Fig. 5). Sexual behavior generally involves a driving process where the male will approach the female, touch nose-to-

3. Sexual Behavior

♂ Following & Driving ⟶ { a. Patting the rump
                                                          b. Grooming
                                                           c. Riding

a. Neck grip & Mount (1)   b. Grasping with forelimbs (2)
                                                        c. Thrusting, Intromission

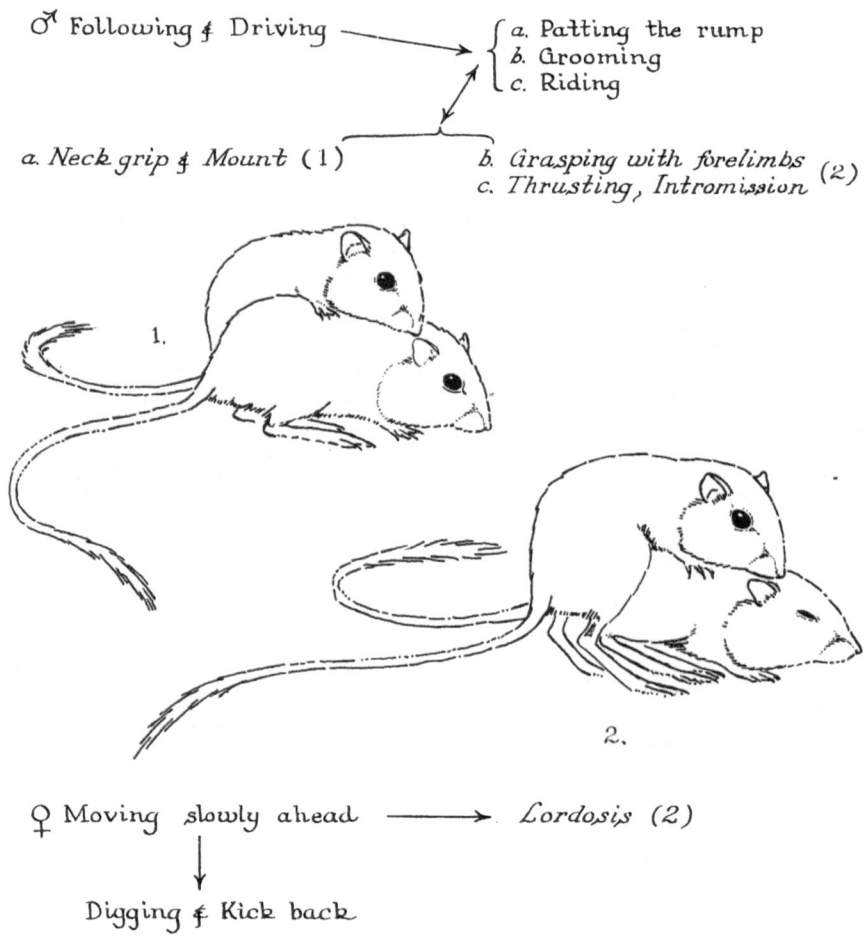

♀ Moving slowly ahead ⟶ Lordosis (2)

Digging & Kick back

Fig. 6. Sexual behavior in *Dipodomys*. Preparation for mount involves a neck grip by the male. During mounting behavior the female assumes a lordosis posture and moves her tail to one side (from EISENBERG, 1963a).

nose, then proceed to the rear of the female to perform a naso-genital investigation. If the female is slightly unreceptive, she may move away, giving rise to the driving pattern whereupon the male follows the female. If the female is receptive, she will stop suddenly and assume a lordosis posture, remaining quadrupedal but raising the hind leg. Upon the assumption of lordosis, the male will generally attempt to mount (**Fig. 6**). A neck grip is employed by the male kangaroo rat to secure the female but such a behavior pattern is not generally shown by the male *Jaculus*. Vast differences in the temporal patterning of copulatory behavior may be noted when various rodent species are compared (EISENBERG, 1963a; 1967), but an exact quantification of the copulatory patterns for *Jaculus* remains to be done.

PARENTAL CARE BEHAVIOR

The genus *Dipodomys* is characterized by a high intraspecific aggressiveness. The parturient female generally drives the male out of the burrow and raises the young alone. Although pair tolerance is higher in *Jaculus jaculus* and *J. orientalis*, the female still has a tendency to nest alone during the early development of the young. Nest-building behavior generally increases after the birth of the young and the female spends a great deal of time crouching over the young keeping her back arched to prevent crushing them. This allows the young to nurse while they are kept warm. The female shows a high level of grooming activity, licking the young and removing the feces and urine from the anal and genital regions. The retrieving response is highly developed in the genus *Dipodomys* but only weakly shown in *Jaculus*. During retrieving a pup that has wandered out of the nest is picked up by a fold of the skin with the incisors and carried back to the nest site.

GENERAL SUMMARY

Several remarkable convergences can be noted when the dipodids are compared with heteromyids. Considering *Jaculus* and *Dipodomys*, one notices a convergence in the locomotor patterns with corresponding morphological modifications including shortening of the neck, shortening of the forelimbs, and lengthening of the hind foot and tail. Color patterns including the white tail tip bordered by black and the white ventrum show remarkable similarity between the 2 genera. The expansion of the tympanic bulla in the 2 genera seems to indicate a similar function in predator detection. Marking by depressing the ano-genital region on the substrate and marking by means of sandbathing appear to be convergently evolved in the 2 genera although the integration of the patterns and the form of the patterns are slightly different.

The construction of burrows and the plugging of burrows is no doubt related to survival in the arid, hot desert areas but burrow plugging would appear to be a possible anti-predator response to snakes. The food-caching behavior of *Dipodomys*, although strongly developed is not paralleled by a similar development in *Jaculus* which appears to rely more on fat reserves for passing through periods of food scarcity.

The tonal structures of adult and neonatal vocalizations in *Jaculus* and *Dipodomys* show similarities. The development of the young in *Jaculus* is conservative reflecting some of the trends manifest in more primitive

209

dipodids such as *Sicista*. Development of the neonate in *Dipodomys* shows a specialized advance over more primitive members of the family Heteromyidae (EISENBERG, 1963a). Differences in food-caching behavior appear to be reflected in the social behavior, since the pair tolerance of *Jaculus orientalis* is higher than that displayed by *Dipodomys deserti*. In *Dipodomys* the social system appears to be predicated on individual defense of a burrow system and defense of associated caches.

A Comparison of Behavior Patterns in:
Meriones, Gerbillus and Perognathus

Gerbillus and *Meriones* are members of the subfamily Gerbillinae which show close affinities with the Cricetinae. Four species were studied by the author, *Gerbillus nanus*, *Gerbillus gerbillus*, *Meriones unguiculatus*, and *Meriones hurrianae*. *Meriones unguiculatus* is the best studied of the gerbiline rodents and a rather extensive literature has accumulated within the past 10 years (GULOTTA, 1971). Unfortunately very little is known concerning its ecology in the wild. The genus *Perognathus* has been the subject of a variety of field studies and behavior investigations. In the main my remarks will be confined to *Perognathus parvus* and *Perognathus californicus* (see Figs. 7, 8 and 9).

GENERAL COMPARISONS

Perognathus californicus and *P. parvus* are comparable in size to *Gerbillus gerbillus* and *G. nanus*. The species of *Meriones* are somewhat larger animals and some species (e.g., *M. hurrianae*) resemble the smaller ground squirrels, such as *Citellus leucurus* of the Sonoran Life Zone in North America. In behavior and ecology, *Gerbillus nanus* and *Perognathus parvus* resemble one another more closely than is the case when either *Perognathus* or *Gerbillus* species are compared with the genus *Meriones*.

Certain species of the 3 genera exhibit the capacity to exist on metabolic water alone. *Meriones unguiculatus* can survive without free water (WINKELMAN & GETZ, 1962; KUTSCHER, 1968). A similar ability has been demonstrated for *Gerbillus gerbillus* (BURNS, 1956). Comparable phenomena are reviewed for *Perognathus* by SCHMIDT-NEILSEN & SCHMIDT-NIELSEN (1952). *Gerbillus* and *Perognathus* very strongly exhibit the trait of caching seeds. The 2 species of *Meriones* which were studied, although caching seeds, exhibited more variability and less persistence in the formation of caches in their burrow systems.

Reproduction is quite similar for the 3 genera. *Meriones* typically shows a gestation of approximately 25 days; eye-opening time in the young may be from 14–19 days of age. *Gerbillus nanus* shows a gestation of 23 days with eye-opening from 14–15 days of age. *Perognathus* shows gestation periods of 24–25 days with eye-opening times from 14 to 16 days of age.

210

No profound difference appears to exist in the development of the young as was the case when *Jaculus* was compared with *Dipodomys*.

MAINTENANCE ACTIVITIES

Gerbils, jirds, and pocket mice construct extensive tunnel systems for caching food and constructing nests. A given individual may build more than one burrow system, some of the systems being quite simple and used only occasionally. Parturient females apparently construct more extensive burrow systems, as has been reported for *Gerbillus nanus* (KIRCHSHOFER, 1958) and *Perognathus parvus* (SCHEFFER, 1938). Digging in the substrate involves the teeth, forepaws, and hind limbs. Once tunnels have been excavated, the tunnel walls are packed by pushing and patting with the forepaws. *Meriones* differs somewhat from the smaller species of *Gerbillus* and *Perognathus* in that the walls of the tunnels are packed by employing both the forepaws and the nose. The whole body of *Meriones* is jerked back and forth while holding the forepaws rigid on either side of the nose; thus, the nose pad and forepaws strike the soil and serve to tamp it firmly into place. *Gerbillus* and *Perognathus* push and pat with the fore-paws as the principal method for reinforcing the tunnel walls. Nests are generally constructed within the tunnel system by assembling dried plant material. The tendency for nest building increases at the time of parturition.

As noted in the previous section, all species of the 3 genera assemble foodstuffs and cache it within the tunnel system. The species of *Meriones* are prone to cut vegetation into small pieces for caching. This tendency to bite vegetable matter, stalks, roots, and pods into small pieces has been termed Häckseln by EIBL-EIBESFELDT (1951). This pattern has been noted by the author in *Meriones unguiculatus*, *M. hurrianae*, and *Tatera indica*.

LOCOMOTION AND EXPLORATION

In contrast to the bipedally adapted dipodids and kangaroo rats, pocket mice, gerbils, and jirds do not habitually locomote bipedally. Instead, they perform either the diagonal limb coordination pattern or saltate quadrupedally where the fore and hind limbs alternately strike the ground. Although these species can stand erect bipedally, bracing the body with the tail and occasionally shuffle forward in a bipedal walk, sustained bipedal saltation is not exhibited.

Gerbillus and *Perognathus* are typically nocturnal whereas both *Meriones hurrianae* and *M. unguiculatus* show tendencies to be active during the early morning and early evening. Although the species of *Meriones* avoid strong light, they may be said to be semi-diurnal or at least crepuscular in their activity (NAUMAN, 1968). This is in keeping with their apparent convergence toward a ground-squirrel-like ecological niche.

211

Sandbathing may be shown by all 4 species either at the entrance to the burrow or upon leaving the burrow. As with the jerboas and kangaroo rats, this behavior pattern serves to dress the pelage and, additionally serves to mark the sandbathing spot with odors from glands situated at various points on the body. Marking by dragging the ano-genital area over selected points on the substrate (perineal drag) while depositing urine traces or exudates from glands associated with the anus or genitalia is characteristic for these rodent species. *Meriones*, however, has a very specialized marking movement involving extending and flexing while depressing the ventrum to the substrate. This serves to bring the large gland located on the ventrum of this genus into contact with specific points in the environment. This ventral rubbing is performed as a distinct movement from sandbathing, which in *Meriones* consists almost solely of side-rubbing. The functional aspects of this behavior will be discussed in the section on communication.

CARE OF THE BODY SURFACE

Auto-grooming (washing) is performed in the typical Myomorph pattern (BÜRGER, 1959). The smaller heads of gerbils and pocket mice permit simultaneous wiping of the face with the forepaws without the necessity of turning the head to one side and wiping alternately as is the case in the jerboas and kangaroo rats. Aside from this difference, the temporal patterning of the activity is remarkably uniform (EISENBERG, 1963a, 1967).

Sandbathing by *Perognathus* differs somewhat from the pattern shown by the gerbils. All species of *Perognathus* combine ventral rubs with side rubs into an integrated sandbathing movement which apparently serves both to mark the substrate as well as to dress the pelage (EISENBERG, 1963b). *Meriones unguiculatus* and *Gerbillus nanus* also perform sandbathing which consists almost entirely of extending and flexing while lying on the side. The ventral rubbing component is rarely expressed as such by *Gerbillus nanus*, but is expressed as an independent marking movement by *Meriones unguiculatus* and *M. hurrianae*. Sandbathing is generally shown at specific loci on consecutive days; thus implying that sandbathing has the potential for chemical communication in addition to marking behavior. Further discussion of differences in the form of sandbathing behavior are included in the publications by EISENBERG (1963b, 1967).

ANTI-PREDATOR BEHAVIORS

Gerbils and pocket mice are relatively small, nocturnal rodents which appear to escape predators by relying on speed and inconspicuousness (Fig. 7). No special anti-predator behaviors have been noted for *Gerbillus* and *Perognathus*. Species of *Meriones* are larger and, to an extent, semi-

Fig. 7. Gerbillus gerbillus. In size, coat color and texture, this species resembles *Perognathus penicillatus* of North America.

diurnal. Foraging during daylight hours certainly exposes them to certain forms of predation shared in common with ground squirrels. Since some species of *Meriones* are often colonial, one would expect convergences in anti-predator behaviors similar to those shown by ground squirrels. The habit of sitting at the burrow entrance in an upright posture and scanning the terrain has been remarked on as convergent toward ground squirrel habits (see ALLEN, 1940, for remarks on *Meriones unguiculatus*).

COMMUNICATION MECHANISMS

The introductory remarks for the discussion of kangaroo rats and jerboas are applicable to these species as well. Chemical and tactile modes of communication are often combined and visual communication mechanisms seem to be of reduced importance (see also EISENBERG, 1963a, 1967).

Auditory communication involves the same 3 classes of sounds discussed previously for the jerboas and kangaroo rats. Tooth-chattering is shown by all species and drumming with the hind foot is found in many species included in this study; however, it should be noted that not all

213

Table 5. Some Vocalization Patterns of *Meriones* and *Tatera*.[1]

Species	Sound	Context(s)	Implied Function	Sample (N)	Structure of Sound	Frequencies with Greatest Energy (Hz)	Duration of Pulse (sec.)
Meriones hurrianae	1. Squeal	When defensive	Inhibits attack	3	Harmonic	1700–4600	.53
M. hurrianae	2. Cry of abandoned young (Day 3)	When cold	Attracts mother	4	Harmonic repetitive	2000–3000	.15–.20
M. hurrianae	3. Comfort peeping (Day 9)	When female grooms young	?	5	Harmonic repetitive	2250; 5000	.10–18
Tatera indica	4. Comfort peeping (Day 9)	When female grooms young	?	5	Harmonic repetitive	4000–6000	.08–11

[1] All recordings analyzed to 20 KHz only.
? = Function incompletely understood.

214

species of *Perognathus* drum with the hind foot (EISENBERG, 1963a). Drumming with the hind foot in *Meriones* appears to occur when the animals have been startled and also is shown by the male during bouts of sexual activity (KUEHN & ZUCKER, 1968).

Vocalizations for gerbils and pocket mice fall into 4 functional classes of sounds: 1) defensive squeal, 2) low grunts, 3) cry of the abandoned young, and 4) a comfort sound produced by the young when the female grooms them. As was noted for the kangaroo rat, the optimum auditory sensitivity for *Meriones* seems to correspond to the frequencies carrying the greatest energy in the calls of *Meriones* (see FINK & GOEHL, 1968; FINK & SOFOUGLU, 1966) (See Table 5).

Varying degrees of tympanic bulla expansion can be noted in the species of *Gerbillus*, *Meriones*, and *Perognathus*. The significance of bulla expansion in *Meriones* has been discussed by LEGOUIX, PETTER & WISNER (1954). It would appear that the bullar expansion aids in maintaining the sensitivity of the ossicles for low amplitude sounds at the resonant frequencies of the ossicles themselves.

Olfactory communication has been investigated to a limited extent for these species. As noted previously, the genus *Meriones* is characterized by a large gland field on the ventrum, first noted by LATASTE. Ventral marking by the Mongolian gerbil is displayed most strongly by the male and maintenance of the glandular tissue is under the control of testosterone. Castration of males reduces the propensity to mark and the gland atrophies (THIESSEN *et al.*, 1968). The function of such marking behavior by *Meriones* has been investigated by BARAN & GLICKMAN (1970). It was concluded that presence of marked objects in the environment, even when marked by aliens, increases the propensity to mark by an individual placed in an alien environment. It was further suggested that the glandular secretions are not necessarily employed in territorial defense but rather have some other function in communicating and advertizing the presence of an individual.

Sandbathing appears to be a form of indirect communication for *Gerbillus nanus*, *Perognathus parvus*, and *P. californicus* (EISENBERG, 1967). Considerable overlap is shown in the use of the same sandbathing spot by different individuals which provides the possibility for indirect communication or exchange of chemical information without the animals coming into physical contact.

AGONISTIC BEHAVIOR

Quadrupedal genera, such as *Meriones*, *Gerbillus*, and *Perognathus*, may adopt bipedal upright postures as part of their threat behavior (EISENBERG, 1963a). Two animals in a bipedal upright can then spar and ward with the forepaws, however, it should be noted that such quadrupedal genera are more prone to show locked fighting behavior than are bipedal species such as kangaroo rats and jerboas which fight extensively on their hind legs and seldom roll and tumble together during fighting bouts. Furthermore, both kangaroo rats and jerboas fight by leaping into the air and kicking down at the partner, a maneuver which is seldom executed by the quadrupedal genera.

The propensity to exhibit agonistic behavior is highly variable when species are compared. *Meriones unguiculatus* appears to be extremely tolerant and semi-communal. *M. hurrianae* shows a high male-female tolerance but a lowered threshold for the exhibition of agonistic behavior toward strange conspecifics. *Gerbillus gerbillus* is similar to *M. hurrianae*, whereas *Gerbillus nanus* is highly antagonistic toward conspecifics of either sex, and pair tolerance is low. Contact between a male and female is generally shown only when the female is in estrus (KIRCHSHOFER, 1958; EISENBERG, 1967). Most species of the genus *Perognathus* are similar to *Gerbillus nanus* is being non-contact animals. In general, adults tend to nest alone in burrows and, although they share foraging areas in common, are disinclined for contact except during the female's estrous period (EISENBERG, 1963a).

COURTSHIP AND MATING

Mating behavior involves a preliminary period of mutual investigation which may often grade into agonistic behavior (EISENBERG, 1967). If the female is in estrus, the male will begin to follow her and attempt to mount. Mounting and thrusting occur in bouts, often including repeated bouts of mounting with intromission preceding a final mount with an ejaculation. *Gerbillus nanus, Perognathus californicus*, and *P. parvus* show a rather similar temporal patterning of mounting leading to a mount with an ejaculation. Although *Meriones unguiculatus* exhibits, in common with *Perognathus* and *Gerbillus*, a multiple mount series prior to ejaculation, the number of mounts with intromission necessary before ejaculation is 3 to 4 times as great (KUEHN & ZUCKER, 1968; EISENBERG, 1967).

PARENTAL CARE BEHAVIOR

The basic patterns of maternal care for these rodent species are similar to those described for kangaroo rats and jerboas. The maternal retrieving response is highly developed in all of these species in contradistinction to *Jaculus*.

GENERAL SUMMARY

Gerbillus nanus resembles the silky pocket mice of the genus *Perognathus* in many respects. It shows strong convergences in morphology, reproduction, water metabolism, and behavior. Their social behavior also

216

shows similarities since they are sensitive to crowding and the male-female relationship is marked by intolerance except when the female is in estrus. This suggests that a very strong convergence at all behavioral levels exists between the silky pocket mice and *Gerbillus nanus*. It further implies that such similarities in behavior, ecology, and reproduction delineate a special adaptation syndrome exhibited by these 2 forms as they have evolved independently to occupy a similar desert niche.

Spacing and Communication

Before considering the use of space by mammals, it is necessary to clarify the use of several terms. Home range applies to the area traversed by an animal in the normal course of its maintenance activities. Territory refers to either all or part of the home range which is habitually defended by the owner against intruding conspecifics (BURT, 1943). In extremely arid areas the carrying capacity of the environment is vastly reduced; thus, some rodent species may exist at exceedingly low densities (PETTER, 1961). Evidence indicates that spacing mechanisms are operative. Such spacing mechanisms theoretically insure an adequate utilization of the habitat and reduction of competition at the intraspecific level. The possibility that interspecific competition can be important at certain seasons of the year must not be overlooked. Research by MACMILLEN (1964) indicates that interspecific spacing may become critical during the period of minimum rainfall in the south coastal, arid zone of California. In his study, the desert wood rat *(Neotoma lepida)* actively defended prickly pear *(Opuntia)* patches in which it nests. Intrusions by conspecifics and other members of the rodent community are reduced by the defensive behavior shown by the wood rat. The prickly pear is an important reservoir of water for the wood rat and is actively competed for by other rodent species which require succulents for survival.

Most studies of desert rodent populations show the existence of a broad overlap among home ranges when the ranges of males are compared with those of females or other males. The home range of an individual female appears to overlap to a much lesser extent with the range of an adjacent female, especially when the females in question are pregnant. Hence, some form of spacing behavior among females is certainly in effect. REYNOLDS (1960), employing a live-trapping technique, discovered that *Dipodomys merriami* females occupy smaller home ranges than the males and that the female's home range was during the breeding season free from the activities of other females. Although the home ranges of male kangaroo rats overlapped, the center of a male's range seemed to be somewhat isolated. DIXON (1959) studied *Perognathus nelsoni*, employing a trapping, marking, and release method. By setting traps in a permanent grid, he was able to elucidate the spacial distribution of individual animals. He noted that, although the males had larger home ranges than

217

the females and the males' home ranges overlapped, there was almost no overlap among the home ranges of individual females. A detailed trap, mark, and release study by IVERSON (1967) indicated that the burrow system of *Perognathus parvus* appeared to be randomly distributed. The home ranges of males were larger than those of females but broad overlap was shown among home ranges of both males and females, although pregnant females appeared to have less overlapping ranges. IVERSON concluded that each adult animal had a permanent burrow system which was occupied singly and defended. The literature for the heteromyid rodents has been reviewed by IVERSON (1967) and EISENBERG (1963a). By and large, it would appear that *Perognathus* and *Dipodomys* are intolerant of conspecifics in the same burrow system. True territorial defense, then, seems to be restricted to a limited area in the vicinity of and including the burrow but foraging areas are shared. Overlap in home ranges thus makes it possible for animals to communicate with one another at specific points but at the same time defense is shown with respect to the burrow system thus insuring that seed caches are utilized by the owner of the cache itself.

Although active burrow defense and defense of a limited area around the burrow appears to be characteristic of most genera of the family Heteromyidae, many desert rodents appear to occupy burrow systems in the form of mother families. Thus an extended family group can be built up prior to the dispersal of the newly recruited young. Communal burrow occupancy by several families, however, appears to be rare. Pair tolerance within the same burrow system is moderately high in *Allactaga elater* and *Jaculus orientalis* of Asia, as well as *Gerbillus gerbillus* of North Africa, and *Notomys mitchelli* of Australia. Nevertheless, even if pair tolerance is shown, as in the preceding species, some form of communication mechanisms promoting spacing or promoting the exchange of information among members foraging over a common area is necessary. It would appear that chemical communication and auditory communication have become rather important in most desert rodents.

When species of rodents either exist at rather low densities or maintain some form of exclusive use with respect to burrow systems, some form of distance communication becomes mandatory. Visual communication or tactile communication, so useful at close range, may be supplemented by forms of long-range communication. This may involve the deposition of scents which persist for a period of time and thus may be encountered and scented by a wandering conspecific. The production of sounds which have a reasonable range also allows perception by a conspecific at a considerable distance. Let us reconsider these two points. Chemical communication by depositing glandular substances, urine, or feces at specific points in the environment is widely employed by desert rodents as a form of communication (EISENBERG, 1963a, 1967). It has already been pointed out that sandbathing as a mode of dressing the pelage is also a means of marking and that discrete sandbathing loci may be used repeatedly by the same individual and shared in common by neighbors (EISENBERG, 1963b). Auditory communication has been little studied in desert rodents but the presence of the inflated mastoid bulla coupled with an extreme sensitivity to low amplitude sounds suggests that such a mechanism may function in the perception of intraspecific signals as well as in the perception of predators. The audible calling produced by the grasshopper mouse, *Onychomys* (HILDEBRAND, 1961), may serve both as a spacing mechanism and as a means of indicating the location of a given individual. This desert mouse tends to live in a dispersed pattern of social organization and, although a series of burrows may be discretely localized in a specific area, such as a small arroyo, 2 adult individuals are seldom trapped at a given burrow site. It would seem that the evening cries of the grasshopper mouse might well be functional in indicating position of such spaced individuals. A similar function has been proposed for the low intensity squeal of the kangaroo rat, *Dipodomys spectabilis* (GIBBS, 1955), and it has been suggested that cries produced by *Jaculus* are employed in intraspecific communication over distance (PETTER, 1961).

Species such as *Perognathus parvus*, which typically exhibit solitary occupancy of a burrow system as adults, very often exhibit marked aggressive behavior when placed together in the artificial confinement of a small cage. Adults of such species will generally fight, often wounding each other severely. Aggressive behavior may be high between a male and a female when the female is not in estrus. Compatible groups may be formed through captive breeding, if littermates are allowed to remain together; however, if littermates of *Perognathus parvus* are separated at approximately 2 months of age for several days and then allowed to re-encounter, they will generally behave aggressively toward one another (EISENBERG, 1967). Such low tolerances for extended proximity in captivity are generally shown by rodents which typically exhibit individual burrow defense as adults in the wild. The testing of social tolerance in captivity will often correlate with distribution patterns exhibited in the field (EISENBERG, 1967). For example, the arid-adapted subfamily Gerbillinae, based on captive studies, would appear to show a range of social types. *Gerbillus nanus* is relatively solitary whereas *Tatera indica* may exhibit a communal structure based on an extended family group. Within the family Dipodidae, *Jaculus orientalis* and *Allactaga elater* show a pair tolerance, however, the female may exhibit a tendency to nest alone at the time of parturition. The desert-adapted family Heteromyidae appears to exhibit an enduring trend toward pair intolerance and individual defense of their burrow systems. By contrast the desert-adapted species of the North American, cricetine genus *Peromyscus* show a range in their social tendencies. *Peromyscus crinitus* is intolerant, whereas the equally xeric-adapted *P. eremicus* exhibits a high pair tolerance (EISENBERG, 1968).

Fig. 8. Perognathus formosus. This species of pocket mouse is adapted to pebble deserts and is often found on soils of low reflectance. In size and habits it resembles several smaller species of the Old World muroid genus *Gerbillus*.

Fig. 9. Microdipodops pallidus. This species is the smallest bipedally saltating heteromyid. The relatively large head reflects the enlarged tympanic bullae. In external morphology *Microdipodops* strongly resembles the Asiatic dipodid, *Salpingotus*. The gerbilline genus *Pachyuromys* shows a similar bullar expansion and in common with *Microdipodops* accumulates fat reservoirs in its tail.

The ultimate conclusion, then, is that adaptation to desert habitats with dispersed food supplies is not necessarily in and of itself conducive to selection for a solitary existence as an adult. Equally important selective forces include other aspects of the species' ecology including its mode of assembly of foodstuffs, its shelter construction, its mode of reproductive behavior, and parental care. Although food caching within the burrow system or in its vicinity is strongly correlated with burrow defense and a solitary way of life within the Heteromyidae, it is not the case that individual burrow defense is always a concomitant of caching. It would appear that the Heteromyidae have retained a phylogenetically ancient trait because of some fundamental adaptive advantage accruing from its individual defense of cached food (EISENBERG, 1963a). Nevertheless, those species of desert rodents which do show individual patterns of burrow defense (such as *Perognathus* and *Dipodomys*) also show forms of chemical and auditory communication which tie the members of a given 'community' into an information exchange system (LEYHAUSEN, 1965). The study of such communication systems offers much promise for the future.

Acknowledgements

This review is based on the author's studies of *Heteromys, Liomys, Perognathus, Microdipodops, Dipodomys, Meriones, Gerbillus, Tatera, Jaculus, Allactaga*, and *Notomys* in captivity. Captive studies were supplemented by the author with some field observations of all Heteromyid genera and *Tatera*. Field data summarized by PETTER (1961) and OGNEV (1963) for the Gerbillinae and Dipodidae are respectfully acknowledged. Research was supported in part by National Science Foundation Grant No. GB-3545.

REFERENCES

ALLEN, G. M. 1940. The mammals of China and Mongolia, in Natural history of Central Asia, vol. 11, pt. 2 (GRANGER, W., ed.), Amer. Mus. Nat. Hist., New York.
BALPH, D. F. & STOKES, A. W. 1963. On the ethology of a population of Uinta ground squirrels. *Amer. Midland Nat.*, 69: 106–126.
BALPH, D. M. & BALPH, D. F. 1966. Sound communication of Uinta ground squirrels. *J. Mamm.*, 47: 440–450.
BARAN, D. & GLICKMAN, S. 1970. Territorial marking in the Mongolian gerbil: A study of sensory control and function. *J. Comp. Physiol. Psychol.*, 71: 237–245.

BARTHOLOMEW, G. A. & HUDSON, J. W. 1960. Aestivation in the Mohave ground squirrel, *Citellus mohavensis*. *Bull. Mus. Comp. Zool.*, 124: 193–208.

BODENHEIMER, F. S. 1935. Animal life in Palestine. Jerusalem, viii + 506 pp.

BÜRGER, M. 1959. Eine vergleichende Untersuchung über Putzbewegungen bei Lagomorpha und Rodentia. *Zool. Gart. Lpz.*, 23: 434–506.

BURNS, T. W. 1956. Endocrine factors in the water metabolism of the desert mammal *Gerbillus gerbillus*. *Endocrinology*, 58: 243–254.

BURT, W. H. 1943. Territoriality and home range concepts as applied to mammals. *J. Mamm.*, 24: 346–352.

CHEW, R. M. & CHEW, A. E. 1970. Energy relationships of the mammals of a desert shrub *(Larrea tridentata)* community. *Ecol. Monographs*, 40: 1–21.

DICE, L. R. 1952. Natural communities. Univ. Michigan Press, Ann Arbor. x + 549 pp.

DIETERLEN, F. 1962. Geburt und Geburtshilfe bei Stachelmaus. *Zeit. für Tierpsychol.*, 19: 191–222.

DIXON, K. L. 1959. Spatial organization in a population of Nelson's pocket mouse. *Southwest Nat.*, 3: 107–113.

EIBL-EIBESFELDT, I. 1951. Gefangenschaftsbeobachtungen an der persischen Wüstenmaus (*Meriones persicus persicus*, Blanford). *Zeit. für Tierpsychol.*, 8: 400–423.

EIBL-EIBESFELDT, I. 1958. Das Verhalten der Nagetiere. Handb. Zool., Band 8, Lieferung 12. Gruyter, Berlin.

EISENBERG, J. F. 1963a. The behavior of heteromyid rodents. *Univ. Calif. Publ. Zool.*, vol. 69, iv + 100 pp.

EISENBERG, J. F. 1963b. A comparative study of sandbathing in heteromyid rodents. *Behaviour*, 22: 16–23.

EISENBERG, J. F. 1967. A comparative study in rodent ethology with emphasis on evolution of social behavior I. *Proc. U.S. Nat. Mus.*, vol. 122, No. 3597, 51 pp.

EISENBERG, J. F. 1968. Behavior patterns, Chapter 12, pp. 451–495, in Biology of *Peromyscus* (KING, J. A., ed.) Serial Publ. No. 2. Amer. Soc. Mamm.

EISENBERG, J. F. & ISAAC, D. E. 1963. The reproduction of heteromyid rodents in captivity. *J. Mamm.*, 44: 61–67.

EWER, R. F. 1965. Food burying in the African ground squirrel, *Xerus erythropus* (E. Geoff.). *Zeit. für Tierpsychol.*, 22(3): 321–327.

FINK, A. & GOEHL, H. 1968. Vocal spectrum and cochlear sensitivity in the Mongolian gerbil. *J. Audit. Res.*, 8: 63–69.

FINK, A. & SOFOUGLU, M. 1966. Auditory sensitivity of the Mongolian gerbil. *J. Audit. Res.*, 6: 313–319.

GIBBS, R. H., Jr. 1955. Vocal sound produced by the kangaroo rat, *Dipodomys spectabilis*. *J. Mamm.*, 36(3): 463.

GRANT, E. C. & MACKINTOSH, J. H. 1963. A comparison of social postures of some common laboratory rodents. *Behaviour*, 21: 246–259.

GULOTTA, E. F. 1971. *Meriones unguiculatus*. Mammalian Species No. 3: 1–5 pp. Amer. Soc. Mamm.

HAPPOLD, D. C. D. 1970. Reproduction and development of the Sudanese jerboa, *Jaculus jaculus butleri* (Rodentia, Dipodidae). *J. Zool. Lond.*, 162: 505–515.

HATT, R. T. 1932. The vertebral columns of ricochetal rodents. *Bull. Amer. Nat. Hist.*, 63: VI.

HILDEBRAND, M. 1961. Voice of the grasshopper mouse. *J. Mamm.*, 42: 263.

HOWELL, A. B. 1932. The saltatorial rodent *Dipodomys*: the functional and comparative anatomy of its muscular and osseus systems. *Proc. Amer. Acad. Arts. Sci.*, 67: 377–536.

HUDSON, J. W. 1962. The role of water in the biology of the antelope ground squirrel, *Citellus leucurus*. *Univ. Calif. Publ. Zool.*, 64(1): 1–56.

INGLES, L. G. 1965. Mammals of the Pacific states. Stanford Univ. Press. xii + 506 pp.

IVERSON, S. L. 1967. Adaptations to arid environments in *Perognathus parvus* (Peale). Ph.D. Dissertation, Univ. Brit. Columbia. 130 pp.

KING, J. A. (editor). 1968. Biology of *Peromyscus*. Serial Publ. # 2. Amer. Soc. Mamm. xiii + 593 pp.

KIRSCHSHOFER, R. 1958. Freiland- und Gefangenschaftsbeobachtungen an der nordafrikanischen Rennmaus. *Zeit. für Säugetierk.*, 23: 33–49.

KIRMIZ, J. P. 1962. Adaptation to desert environment: a study on the jerboa, rat, and man. Butterworth, London. xi + 168 pp.

KUEHN, R. E. & ZUCKER, I. 1968. Reproductive behavior of the Mongolian gerbil *(Meriones unguiculatus)*. *J. Comp. Physiol. Psychol.*, 66: 747–752.

KUTSCHER, C. L. 1968. Plasma volume change during water deprivation in gerbils, hamsters, guinea pigs, and rats. *Comp. Biochem. Physiol.*, 25: 929–936.

LATASTE, F. 1887. Documents pour l'ethologie des mammiferes. Bordeaux, 659 pp. (Reprinted from: Actes Soc. Linn. de Bordeaux, vols. 40, 41, and 43.)

LAY, D. M. 1972. The anatomy, physiology, functional significance, and evolution of the specialized hearing organs of gerbilline rodents. *J. Morphol.* 138: 41–120.

LEGOUIX, J. P., PETTER, F. & WISNER, A. 1954. Etude de L'Auditien chez des Mammifères à Bulles Tympaniques Hypertrophis. *Mammalia*, 18(3): 262–271.

LEYHAUSEN, P. 1965. The communal organization of solitary mammals, pp. 249–264, in Symp. Zool. Soc. Lond., No. 14. Academic Press, New York.

MacMILLEN, R. E. 1964. Population ecology, water relations, and social behavior of a Southern California semidesert rodent fauna. *Univ. Calif. Publ. Zool.*, 71: 1–59.

MARLOW, B. J. 1969. A comparison of the locomotion of two desert-living Australian mammals, *Antechinomys spenceri* and *Notomys cervinus*. *J. Zool. Lond.*, 157: 159–167.

MERRIAM, C. H., BAILEY, V., NELSON, E. W. & PREBLE, E. A. 1910. Zone map of North America. Biol. Survey, U.S. Dept. Agric., Washington, D.C.

MOHR, E. 1954. Die freilebenden Nagetiere Deutschlands und der Nachbarländer. Jena Veb Gustav Fischer. 212 pp.

MYKYTOWYCZ, R. 1970. The role of skin glands in mammalian communication, pp. 327–360, in Advances in chemoreception, vol. I., Communication by Chemical Signals (JOHNSTON, J. *et al.*, eds.). Appleton-Century Croft, New York.

NAUMAN, D. J. 1968. Open field behavior of the Mongolian gerbil. *Psychonom. Sci.*, 10(5): 163–164.

OGNEV, S. J. 1959. Säugetiere und ihre Welt. Berlin: Acad. Verlag. viii + 362 pp.

OGNEV, S. J. 1963. Mammals of the U.S.S.R. and adjacent countries. Vol. 6, Rodents, 508 pp., published by Israel Program for Scientific Translations.

PETTER, F. 1961. Repartition geographique et ecologie des rongeurs desertique. *Mammalia*, 24: 1–219.

PYE, A. 1965. The auditory apparatus of the Heteromyidae (Rodentia, Sciuromorpha). *J. Anat. Lond.*, 99: 161–174.

RANCK, G. L. 1968. The rodents of Libya: Taxonomy, ecology, and zoogeographical relationships. *U.S. Nat. Mus. Bull.* 275, vii + 264 pp.

RAUCH, H. G. 1957. Zum Verhalten von *Meriones tamariscinus* Pallas. *Zeit. für Säugetierk.*, 22: 218–240.

REYNOLDS, H. G. 1960. Life history notes on Merriam's kangaroo rat in southern Arizona. *J. Mamm.*, 41: 48–58.

RUFFER, D. G. 1968. Agonistic behavior of the northern grasshopper mouse *Onychomys leucogaster breviauritus*. *J. Mamm.*, 49(3): 481–487.

SCHMIDT-NIELSEN, K. & SCHMIDT-NIELSEN, B. 1952. The water metabolism of desert animals. *Physiol. Rev.*, 32: 135–166.

SCHEFFER, T. H. 1938. The pocket mice of Washington and Oregon in relation to agriculture. U.S. Dept. Agric., Tech. Bull. No. 608, 15 pp.

STANLEY, M. 1971. An ethogram of the hopping mouse, *Notomys alexis*. *Z. Tierpsychol.*, 29: 225–258.

THEISSEN, D. D., LINDZEY, G. & FRIEND, H. C. 1968. Androgen control of territorial marking in the Mongolian gerbil. *Science*, 160: 432–434.

223

THEISSEN, D. D. & YAHR, P. 1970. Central control of territorial marking in the Mongolian gerbil. *Physiol. and Behavior,* 5: 275–278.

WAGNER, H. O. 1961. Die Nagetiere einer Gebirgsabdachung in Sudmexiko und ihre Bezichungen zur Umwelt. *Zool. Jb. Syst.,* 89(8): 177–242.

WEBSTER, D. B. 1960. Auditory significance of the hypertrophied mastoid bullae in *Dipodomys. Anat. Rec.,* 136: 299.

WEBSTER, D. B. 1962. A function of the enlarged middle ear cavities of the kangaroo rat, *Dipodomys. Physiol. Zool.,* 35: 248–255.

WHITTEN, W. K. & BRONSON, F. H. 1970. The role of pheromones in mammalian reproduction, pp. 309–326 in Advances in chemoreception, vol. I, Communication by chemical signals (JOHNSTON, J., *et al.,* eds.). Appleton-Century Croft, New York.

WINKELMAN, J. R. & GETZ, L. L. 1962. Water balance in the Mongolian gerbil. *J. Mamm.,* 43: 150–154.

WISNER, A., LEGOUIX, J. P. & PETTER, F. 1954. Etude Histologique de L'Oreille d'un Rongeur a Bulles Tympaniques Hypertrophies, *Meriones crassus. Mammalia,* 18: 371–374.

WOOD, A. E. 1935. Evolution and relationship of the heteromyid rodents with new forms from the tertiary of western North America. *Ann. Carnegie Mus.,* 24: 73–262.

ZAHAVI, A. & WAHRMAN, J. 1957. The cytotaxonomy, ecology and evolution of the gerbils and jirds of Israel (Rodentia: Gerbillinae). *Mammalia,* 21: 341–380.

XI. ACTIVITY PATTERNS
OF A DESERT RODENT

by

NORMAN R. FRENCH

Introduction

Small mammal activity has been investigated in the past primarily by evaluation of trapping success. With such results, the influence of weather conditions on the degree and intensity of rodent activity has been evaluated. The procedure gives valid results over short time periods, where the same traps are used in the same locations on a series of nights. In long-term studies, however, intrinsic factors such as available food supply or reproductive activity may influence the pattern of activity (MAZA, FRENCH & ASCHWANDEN, 1973). Response to available traps may also be altered.

Results reported here are part of a study of effects of chronic radiation exposure on the ecology of desert organisms. In these studies the determination of radiation dose rate to individual small mammals in a desert rodent population provide a unique set of data which is a direct expression of the time spent on the surface of the ground by individual members of the population. Accurate integrated monthly activity records are available for animals which carried radiation dosimeters. Since, in burrow-dwelling mammals, a measure of time spent on the surface of the ground is a direct measure of activity, these results lend themselves well to an analysis of the factors which influence activity patterns.

The objective of this report is to utilize the radiation exposure data as an indicator of activity for the principal species of desert rodent (in terms of numbers and biomass) of this study area, the pocket mouse *Perognathus formosus*. By means of statistical analysis, the variation in the index of activity is correlated with environmental variables. Other factors such as the behavior and physiology of these animals is considered, and the interaction of these with the environmental variables is evaluated.

These results were made possible by the assistance of a number of colleagues and associated at the University of California, Los Angeles. Mr. B. G. MAZA and Mr. A. P. ASCHWANDEN participated in all aspects of this study. The work was supported through a contract #AT(04-1) GEN-12 between the University of California and the U.S. Atomic Energy Commission. Logistic support that made this study possible was provided by the Civil Effects Test Organization of the U.S. Atomic Energy Commission, and particularly by representatives of that organization at the Nevada Test Site. Dr. T. BOARDMAN of the Colorado State University Statistical Laboratory assisted and advised on statistical analyses.

Methods

The study area is located in the Mojave Desert of southern Nevada, 12 miles west of Mercury, Nye County, Nevada. The study site was on a gravelly outwash of the Specter Range. It is characterized by widely spaced desert shrubs, such as *Franseria dumosa, Larrea divaricata*, and *Lycium andersonii*. The small mammal fauna is dominated by members of the family Heteromyidae, including two species of pocket mouse (*Perognathus formosus* and *P. longimembris*) and two species of kangaroo rat (*Dipodomys merriami* and *D. microps*). Two species of lesser importance but of regular occurrence in the study areas were the antelope ground squirrel, *Ammospermophilus leucurus*, and the grasshopper mouse, *Onychomys torridus*. All of these are nocturnal, with the exception of the ground squirrel. Rodent populations were under investigation for several years in four separate study areas. Each of the four areas was circular and was 9 ha in extent; three were surrounded by fence which served as a barrier to rodent movements. The fence was a woven wire that the rodents could not pass through, and it was buried beneath the surface of the ground to prevent burrowing. It was topped with sheet metal to prevent the animals climbing over. The only small mammal for which this barrier was incomplete was the ground squirrel. These wide-ranging animals occasionally managed to climb over the top of the barrier. They are not considered in these investigations.

The rodent population of each study area was censused at monthly intervals. The census was conducted by saturating the study area with live-traps and activating these for three successively nights each month. For these rodents three succesive nights of trapping accounted for virtually all of the animals present in the study area. This conclusion was verified by intensive trapping in the enclosed study areas, to determine that unmarked animals appearing in the traps were always additions to the population by recruitment. The live-traps were placed on a grid pattern with 15 m spacing. There were 400 trap stations in each circular 9-ha study area.

In the center of one of the study areas, a large radiation source (cesium 137) was supported on a tower 15 m above the surface of the ground (LUCAS & FRENCH, 1967). Below the source was a circular radiation shield of varying thickness, which served to reduce the intensity of the radiation exposure at ground level near the center of the study area. In this way all animals living in the area, regardless of where their activity was centered, received a low-level chronic exposure to radiation. This level was far below that known to cause any clinical effects on individual

226

animals. The radiation field, however, was not completely uniform. At the periphery of the study area, the dose rate at ground level was approximately two roentgens per day, while near the center of the study area, the dose rate was approximately 12 roentgens per day.

The radiation from cesium 137 is a low-energy gamma radiation. It, therefore, travels through air with only slight attenuation, but is very readily absorbed by any dense material such as the shield below the source or by the ground surface. Beneath a few centimeters of soil, the dose rate was reduced to less than 10% of the dose rate at the surface of the ground. Therefore, rodents in their burrows would be effectively shielded from the radiation. Since the radiation exposure of the rodent population was dependent upon the time spent on the surface of the ground, a study of this activity was conducted in order to determine the range and variability of individual dose rates to members of the population.

Small (1 mm × 6 mm) dosimeters were placed in plastic capsules which could be attached to the skin of individual rodents (FRENCH, MAZA & ASCHWANDEN, 1966). Each plastic capsule was attached by means of a single suture to the skin of the back of the neck of an animal. A dosimeter could be placed inside the capsule when the animal was trapped during the routine census. At the next census the dosimeter could be removed and replaced by a new one. Dosimeters were collected in this fashion and returned to the laboratory for evaluation and analysis. Thus, the radiation exposure over periods of one month were determined for a large number of individuals of the population over a period of two and one-half years.

Records were also maintained of the trap locations where individual animals appeared. This was essential information for determining activity because of the nonuniformity of the radiation field throughout the study area. By relating the total exposure as indicated on the dosimeter to the exposure rate at ground level in the location where the animal was repeatedly trapped, the total exposure on the dosimeter could be converted to hours of exposure to the radiation source at the ground surface.

THE MICRODOSIMETER AND THE INDEX OF ACTIVITY

Microdosimeters were secured to selected individual pocket mice for a period of three years. They were attached to and retrieved from animals at monthly intervals from November 1964 to June 1965. This effort resumed at monthly intervals beginning in October 1965 and continued through November 1966. Due to the low amount of activity expected during the winter season, as indicated by previous results, only one month of exposure was measured during the next winter season, this from January to February 1967. Monthly placement of dosimeters resumed in April 1967 and continued through October 1967.

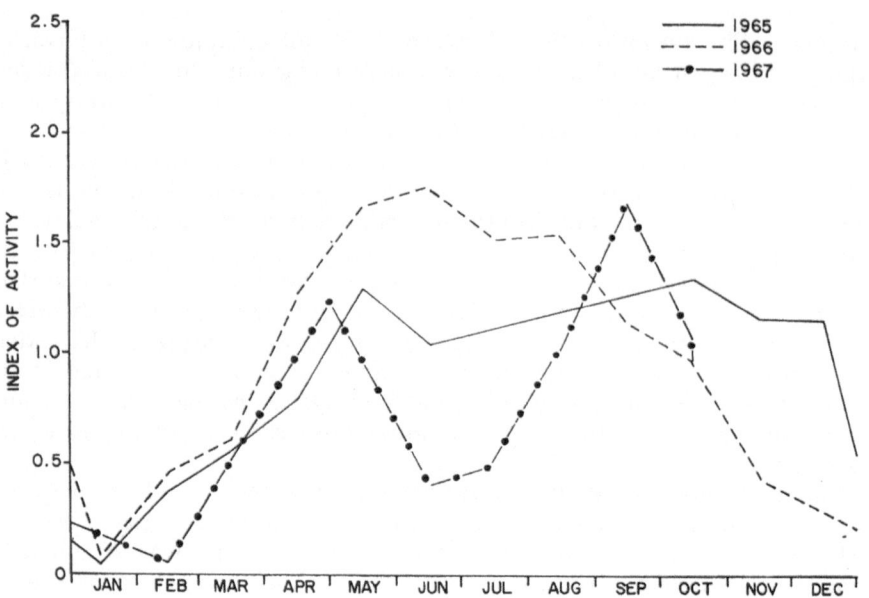

Fig. 1. Activity above ground during three years of the pocket mouse, *Perognathus formosus*, in the Mojave desert.

The results presented are based on 258 microdosimeters that were carried by animals for periods ranging from three weeks to four months. For the most part, records considered in this analysis are based on month-long exposures. The total number of animals used for this evaluation is somewhat less than the total number of dosimeters, since some animals carried dosimeters in two or more time periods. One pocket-mouse provided a total of six monthly exposure measurements.

The mean monthly exposure of animals in the irradiated plot ranged from $1.76 \pm .82$ r/day (N = 16) in June 1966, to $.41 \pm .03$ r/day (N = 6) in June 1967, and to $.05 \pm 0$ r/day (N = 2) in January 1965. This extreme seasonal variation was typical, indicating that these animals are almost wholly inactive during the cold season of the year. Activity regularly increased for three months starting in February when the summer peak of activity was attained. In the fall the declining activity began in September or October. In some years there was a marked decrease in activity in the midsummer period. This was evident in the 1965 data, but was especially pronounced in 1967. The summer season of 1966 was notable in that there was no evidence of a decline in activity during the summer months. It was these pronounced shifts in animal activity that suggested needed analyses for evaluating the role of environmental variables on the modification of activity patterns in this species.

228

Animal activity varied within and between years (Fig. 1). Within years, activity began at a very low level (for all practical purposes, nearly 0) and increased regularly beginning in March through the early summer months. A decline in activity occurred after the months of September or October, approaching zero in typical years in the month of December. The pattern followed a smooth curve within one year of observation, the year 1966. In that year a maximum was obtained in the month of June, with regularly increasing activity prior to that time and regularly decreasing activity after that time. By contrast, the year 1967 showed a distinct decrease in animal activity during the midsummer months of June, July, and, in part, August. In that year maximum activity was actually attained in the month of September. The year 1965 apparently showed a similar midsummer decrease in activity, but missing data at this point makes this conclusion tentative. However, the 1967 pattern, showing a decrease in activity in the midsummer months, may be typical of these desert regions.

CLIMATOLOGICAL VARIABLES AND ANIMAL ACTIVITY

Meteorological measurements that were available for comparisons with animal activity patterns included temperature and precipitation (Table 1). Because these animals are nocturnal, the minimum temperatures or average temperatures may be expected to be more closely related to their rate of activity than would the diurnal maximum temperatures. A plot of the monthly average minimum temperature (Fig. 2) shows a pattern that strongly resembles the pattern of activity in these animals. From a low in December and January, there is a steady increase to middle or late summer and then a regular decline to winter. The highest average minimum temperature, however, occurred somewhat later than the maximum animal activity. In 1966 the highest average minimum temperature occurred in the month of August, while the maximum animal activity occurred in the month of June. In 1967 the greatest average minimum temperatures occurred in the months of July and August. These high temperatures were greater than for the corresponding months in 1966. The difference is even more noticeable in terms of maximum temperature for July and August in these two years. It was during this period of high temperature that the animal activity showed a marked decline.

There is generally an inverse relationship between animal activity and precipitation during what would be considered a typical year. The year 1966 was such a year. Most precipitation occurs in the winter months, although there is some increase in midsummer. Most precipitation, therefore, occurs when animal activity is least. In 1967, however, the reverse situation occurred, with relatively great amounts of precipitation

229

Table 1. Climatological variables and index of activity for each time period.

Time Period	Index of Activity	Temperature (°F) Maximum	Minimum	Average	Precipitation	Days of Precipitation	Days ppt. > 25 mm	Nights min. temp. >10°C	Nights min. temp. >15°C
Nov. 64	.69	57.83	36.80	46.63	0.13	3	0	0	0
Dec. 64	.21	56.36	36.19	45.83	0.0	0	0	0	0
Jan. 65	.05	59.52	37.48	47.67	0.42	4	3	1	0
Feb. 65	.37	61.64	37.21	48.56	0.0	0	0	0	0
Mar. 65	0.55	63.52	39.00	50.73	0.49	6	2	0	0
Apr. 65	0.79	66.50	47.27	57.10	2.37	10	7	11	0
May 65	1.30	76.81	51.84	65.49	0.09	2	0	18	0
June 65	1.04	85.60	59.17	73.79	0.02	2	0	29	2
Nov. 65	1.17	61.86	44.31	52.01	1.94	9	6	5	0
Dec. 65	1.16	52.86	38.27	45.15	2.46	7	2	0	0
Jan. 66	0.09	50.10	30.94	39.30	0.35	4	1	0	0
Feb. 66	0.46	52.86	32.21	41.99	0.64	4	1	0	0
Mar. 66	0.60	66.00	41.26	53.63	0.05	1	0	3	0
Apr. 66	1.26	75.80	46.80	62.02	0.03	1	1	10	0
May 66	1.65	84.90	57.68	72.78	0.34	1	1	26	0
June 66	1.76	91.23	63.93	79.10	0.07	1	0	30	4
July 66	1.52	96.03	68.48	84.07	0.27	2	1	30	19
Aug. 66	1.54	98.59	71.04	86.07	0.15	1	1	31	27
Sep. 66	1.14	88.34	61.96	76.17	0.04	1	0	30	6
Oct. 66	0.98	77.00	52.45	63.85	0.01	1	0	22	0
Nov. 66	0.44	63.70	43.26	52.94	0.10	2	1	5	0
Dec. 66	0.03	53.45	35.29	43.68	0.42	2	1	30	0
Jan. 67	0.02	56.61	35.54	45.01	1.29	3	2	0	0
Feb. 67	0.06	62.29	36.64	48.50	0.00	0	2	0	0
Mar. 67	0.50	64.47	42.70	54.05	0.00	0	0	3	0
Apr. 67	1.00	59.30	35.43	48.05	1.05	3	1	1	0
May 67	1.00	81.71	53.29	68.35	0.21	2	1	18	2
June 67	0.41	85.00	56.65	71.69	0.45	2	1	22	5
July 67	0.50	102.07	73.10	88.49	0.28	2	1	31	27
Aug. 67	1.01	102.25	73.61	88.17	1.98	4	4	31	29

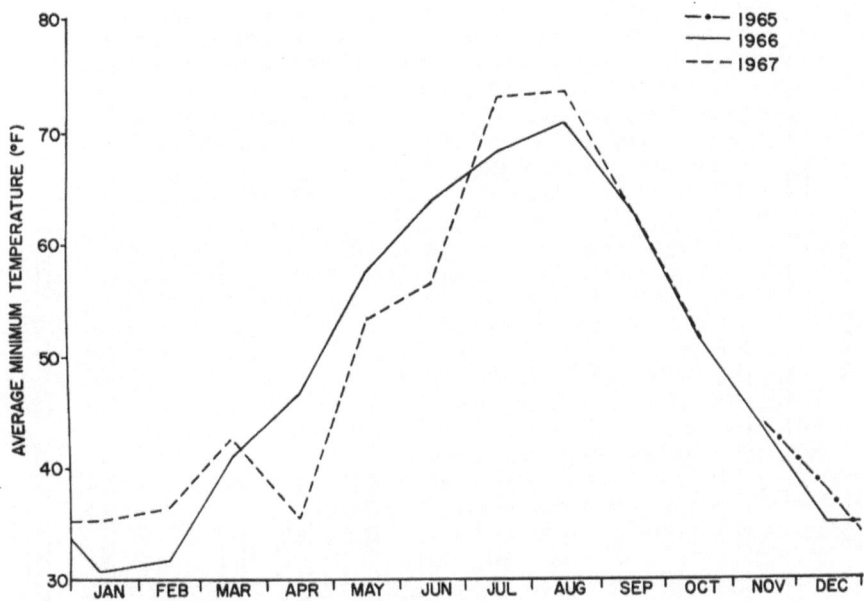

Fig. 2. Monthly average minimum temperatures (°F).

in April and again in August. In that particular year the periods of maximum rainfall more nearly coincided with periods of small mammal activity. Precipitation data for 1965 are incomplete, as are the records for animal movement. However, there was again a very high period of rainfall in April of that year and a high period of rainfall late in the season during November and December.

Since the index of activity is a monthly average determined both by the number of times visits were made aboveground and the duration of time spent aboveground, it is of interest to consider the number of nights which may have been inhospitable for emergence from burrows. To examine this, the total number of days on which rain occurred during each month was tabulated and graphed (Fig. 3). This is related to, but not necessarily exactly the same as, the total precipitation. For instance, in 1965 the total amount of rainfall was greater in December, but the number of days of rain was greater in November. The data for 1966 again suggests an inverse relationship between animal activity and the number of days of precipitation. Again, the reverse situation occurred in 1967, when number of days of precipitation was greater during the period of greater animal activity.

These factors, precipitation and temperature, are the ones considered to be the key factors determining activity of these small rodents. Some manifestation of one or the other, or perhaps some combination with

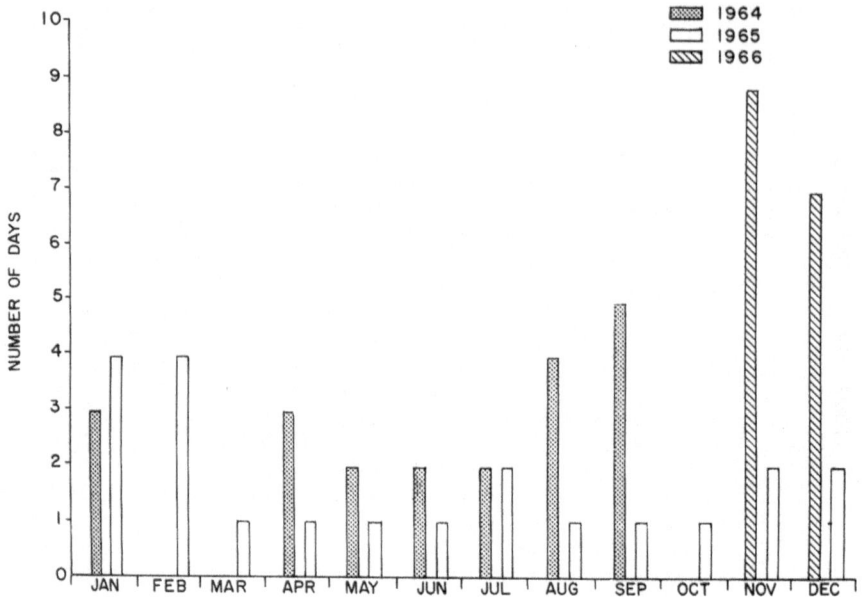

Fig. 3. Number of days each month on which rainfall occurred.

interaction, will determine the total amount of time and the number of occurrences of these animals out of their burrow systems on the surface of the ground. In an attempt to explore these interrelationships, certain statistical analyses of the data were performed.

Analysis

As a first comparison, a test of the significance of the correlation coefficients between activity and indices of environmental variables was performed (Table 2). This approach has limitations because some of the variables under consideration are inter-correlated; therefore, some of the correlation of activities with certain variables may be spurious. Thus, in the correlation coefficients for the year 1966, there is a highly significant correlation between animal activity and three variables which are manifestations of temperature (average minimum temperature, mean temperature, and average maximum temperature). However, these three variables themselves are highly inter-correlated. Therefore, the variable with the highest correlation coefficient with activity, average minimum temperature, is selected as the most important.

There is also inter-correlation between the variables representing rainfall. Here it is evident that the number of days on which precipitation

232

Table 2. Correlation coefficients (r) between index of activity and environmental variables.

	1. Average Min. Temp.	2. Mean Temp.	3. Average Max. Temp.	4. Ppt.	5. No. Days Ppt.	6. Days Ppt. >0.1 inch
			1966			
2.	.993**					
3.	.983**	.997**				
4.	—.416	—.496	.554			
5.	—.539	—.617*	—.663*	.920**		
6.	—.227	—.302	—.342	.766**	.878**	
Activity	.821**	.814**	.789**	—.034	—.251	.017
			1967			
2.	.995**					
3.	.988**	.997**				
4.	.217	.202	.191			
5.	.375	.336	.308	.659*		
6.	.461	.442	.429	.932**	.670*	
Activity	.523	.530	.534	.102	.472	.152

* Significant at the 5% level.
** Significant at the 1% level.

occurred, rather than total amount of precipitation, is the most important factor.

There is a striking difference between the correlation coefficients of activity with environmental variables between the 1966 and 1967 data sets. As activity continued to increase in the summer months of 1966, there was a distinct decline of activity in the summer months of 1967. As a result, there is no significant correlation coefficient of activity for the year of 1967.

Multiple regression analysis of these data by separate years accounts for a high degree of the variability in 1966 ($R^2 = .935$), with the majority of the variation (67.4%) accounted for by average minimum temperature. Other variables included in the analysis are those used for examination of correlation coefficients. Multiple regression analysis of the 1967 data gave a poorer fit ($R^2 = .599$). Both temperature and precipitation are indicated to be important in this year, with average minimum temperature accounting for 27.3% of the variation and number of days of precipitation accounting for 25.4%. None of the correlation coefficients of animal activity with environmental variables was significant in the 1967 data.

The contribution of individual variables to the variance was tested by running separate analyses with the addition or deletion of certain variables.

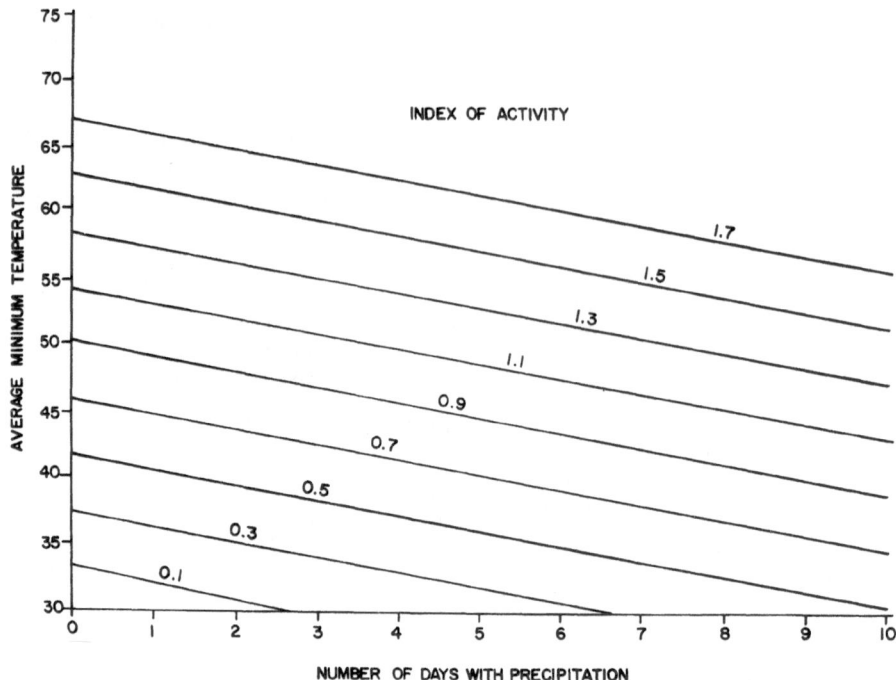

Fig. 4. Relationship between index of activity of the pocket mouse, *Perognathus formosus*, and average minimum temperature and number of days on which precipitation occurred.

Removal of the variables for mean daily temperature and for number of days of precipitation greater than 0.1 inch makes essentially no difference in the R^2 value for regression in either year under consideration. Addition of two variables, the reciprocal of precipitation and the reciprocal of the number of days of precipitation, improved the multiple regression for 1967 data (R^2 went from .60 to .79), but made little difference in the 1966 data.

With the addition of two different variables regression analysis was considerably enhanced. These two variables, number of days on which the minimum temperature exceeded 15 °C (68 °F) and number of days on which the minimum temperature exceeded 10 °C (50 °F), are selected because these temperatures are critical in the metabolic activity of the pocket mice (TUCKER, 1962). In these analyses an additional year, 1965, was also considered. The total regression for all years was improved (R^2 went from .555 to .712). Regression analysis was also improved for the separate years considered previously. The coefficient of determination was somewhat improved for the 1966 data (R^2 went from .935 to .975) and was markedly improved for the 1967 data (R^2 went from .599 to .744). In both analyses the computed t value for the partial correlation

234

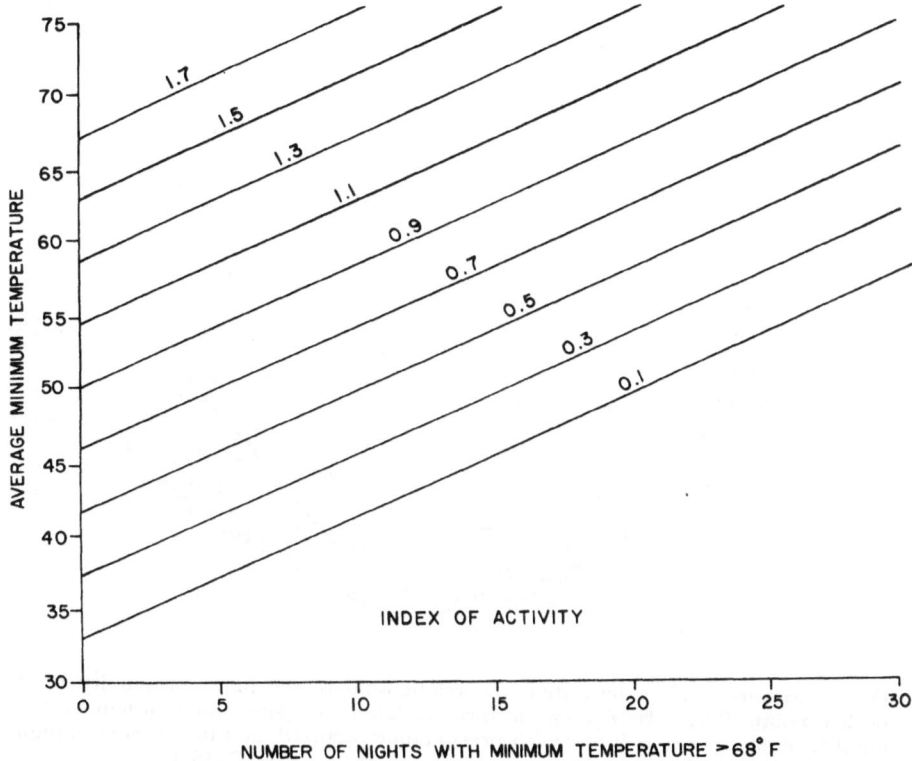

Fig. 5. Relationship between index of activity of the pocket mouse, *Perognathus formosus*, and average minimum temperature and number of nights per month when the minimum temperature exceeded 68°F (15°C).

coefficient was particularly significant for the variables relating to precipitation and to temperature in those years. In 1966 the total precipitation was indicated to be the most significant variable, while in 1967 the number of days on which the minimum temperature exceeded 15°C was most important. This latter variable was highly significant in both years.

Stepwise multiple regression analysis of the data for all three years provides a screening procedure which quantifies their reduction between the actual and predicted values of the dependent variables, removing the variables in order of their importance to the contribution to variation. Only three variables entered produced a significant reduction in the residual variation associated with the value of the dependent variables. These were in order of their importance 1) average minimum temperature, 2) number of days when minimum temperature exceeded 15°C, and 3) number of days of precipitation. These three variables collectively

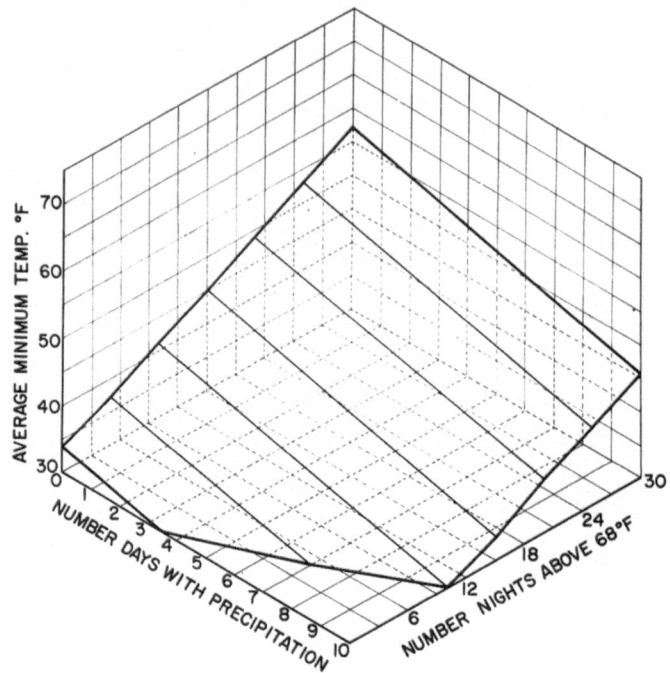

Fig. 6. Response surface indicating the relationship of the index of activity of the pocket mouse, *Perognathus formosus*, to three variables: average minimum temperature, number of days per month on which precipitation occurred, and the number of nights per month on which the minimum temperature was above 68°F (15°C).

accounted for 62% of the variability between the predicted and actual Y values. Therefore, the following linear regression equation can be given as a predictor of animal activity.

$$Y = -1.466 + .047x_1 + .055x_5 - .038x_8$$

where Y = activity index; x_1 = mean daily minimum temperature; x_5 = number of days with precipitation; x_8 = number of days minimum temperature > 68°F.

By the use of stepwise regression, it is evident that of the variables considered in analysis of these data sets, only three are of importance. With this information and the use of a response surface plotting routine available at the Colorado State University Statistical Laboratory, it is possible to display graphically the response of rodent activity through combinations of these variables (Fig. 4, 5, and 6). With straight lines representing varying degrees of animal activity, it is evident that the index of activity increases as average minimum temperature increases, and the activity decreases when the number of days of precipitation

236

increases (Fig. 4). When average minimum temperature is plotted against the number of nights with minimum temperature greater than 15 °C, it is apparent that the activity increases sharply with the increasing number of warmer nights (Fig. 5). These three variables may be combined to show the response surface plotted on three axes (Fig. 6). This shows a plane surface which is tilted to represent increasing animal activity as average minimum temperature increases, number of nights below the critical temperature decrease, and the number of days of precipitation decrease.

Discussion

It is only necessary to have the appropriate measurements in order to reveal the relationship between activity of the desert pocket mouse and environmental variables. In the case of desert rodents, which live in an open environment and are consequently directly exposed to ambient climatic conditions when they are active on the surface of the ground, the standard temperature and rainfall measurements are adequate to demonstrate the existence of such correlation. In the desert the microclimate near the surface of the ground is more closely related to climatic conditions a metre or more above ground than would be the case in a closed or forested environment. It is also evident that there is need to express the climatological variables in appropriate terms, that is, terms that are meaningful to the physiology and behavior of the animal. In this case the critical minimum temperature is evidently in the vicinity of 15 °C. This proved to be the single most important variable entered in the regression analyses. It is also quite possible, of course, that other important variables have not been considered here. These should be selected on the basis of knowledge of the animal's physiological and metabolic requirements and should be tested by similar comparisons.

Through extensive laboratory investigations of related species it has been determined that the metabolic rate, and consequently activity, of these animals is controlled by ambient temperature and by food supply (TUCKER, 1965a, 1965b, 1966). When food supply is reduced these animals may undergo a daily period of torpor. The proportion of time in a torpid condition depends upon temperature and availability of food.

Knowledge of these relationships indicates the importance of an additional variable that should be entered in analysis of the activity pattern of these desert rodents. That variable is a measure or index of food availability. In our studies we have estimates of the production of rodent food in the study areas for three separate years (FRENCH et al., 1974). These results, when compared to estimates of the population energy requirements for the season, indicate that food supply may well be limiting in these populations of desert rodents. There are also profound demographic effects, as well as metabolic or activity effects, resulting from changes in primary productivity in the desert environment. In years of poor plant growth and little or no seed production, reproduction in the rodent population is inhibited. In extreme cases, there is no production of young during certain seasons.

A quantitative measure of available food supply would undoubtedly improve the fit of our regression analysis. The only data available indicate food production in the spring of the year, which tells little about the food availability at other times of the year. Consequently, it seemed inappropriate to enter this variable in quantitative fashion for these analyses. These animals are known to store quantities of food supplies in their burrows below the surface of the ground. How much they can depend on these stores during the fall and winter and how much they must glean from actively searching on the surface of the ground is unknown. Estimation of food availability would require quantitative evaluation of both sources.

The labile response of these animals to environmental conditions makes it difficult to fit a simple linear regression equation to activity data. The extreme midsummer reduction in activity during the year 1967 may in part be a result of high ambient temperatures during that period or may be a result of reduced food supply. The year 1967 was one in which rodent food production was very poor. Animal activity increases with increasing temperature to a point, then there may be a reversal as in the case of 1967 (and perhaps 1965). A satisfactory predictive equation for activity in these animals should, therefore, include this reverse response at critical temperature conditions as well as the important variable representing available food. Hence, there remains considerable work to be done before a satisfactory predictive equation for activity for these desert rodents can be established.

Summary

Estimates of time spent out of their burrows on the surface of the ground are available for a population of pocket mice, *Perognathus formosus*, over a three year period. Activity was measured in the Mojave Desert of southern Nevada by attaching small dosimeters to individual animals. Total exposure of the dosimeters to an elevated radiation source was measured and converted to an index of activity, representing the amount of time the animal spent on the surface of the ground.

Correlation analysis, as well as multiple and stepwise regression analysis of precipitation and temperature data, indicated the importance of minimum temperatures and of rainfall frequency in governing aboveground activity of the rodents. Frequency of days with minimum temperature above 15 °C is closely correlated with activity. A response surface is developed based on linear regression. Additional precision of the empirical model would be attained by inclusion of threshold effects of temperature, response reversal (e.g., torpor), and quantitative information on food availability. With complete data a predictive model of rodent activity could be developed.

REFERENCES

FRENCH, N. R., MAZA, B. G. & ASCHWANDEN, A. P. 1966. Periodicity of desert rodent activity. *Science* 154: 1194–1195.

FRENCH, N. R., MAZA, B. G., HILL, H. O. & ASCHWANDEN, A. P. 1974. A population study of irradiated desert rodents. *Ecological Monographs* 44: 45–72.

LUCAS, A. C. & FRENCH, N. R. 1967. A miniature thermoluminescent dosimeter and its application to radioecology, p. 402–411. In ATTIX, F. H. (ed.). Luminescence dosimetry. U.S. Atomic Energy Comm. Symp. Ser. No. 8.

MAZA, B. G., FRENCH, N. R. & ASCHWANDEN, A. P. 1973. Home range dynamics in a population of heteromyid rodents. *J. Mammal.* 54: 405–425.

TUCKER, V. A. 1962. Diurnal torpidity in the California pocket mouse. *Science* 136: 380–381.

TUCKER, V. A. 1965a. Oxygen consumption, thermal conductance, and torpor in the California pocket mouse, *Perognathus californicus. J. Cellular Comp. Physiol.* 65: 393–403.

TUCKER, V. A. 1965b. The relation between the torpor cycle and heat exchange in the California pocket mouse, *Perognathus californicus. J. Cellular Comp. Physiol.* 65: 405–414.

TUCKER, V. A. 1966. Diurnal torpor and its relation to food consumption and weight changes in the California pocket mouse, *Perognathus californicus. Ecology* 47: 245–252.

XII. PATTERNS OF FOOD,
SPACE AND DIVERSITY*

by

M. L. ROSENZWEIG, BARBARA SMIGEL & A. KRAFT

Introduction

In a band of semi-arid grasslands and shrublands of the high basins of the American Southwest, one can find impressive numbers of nocturnal rodents. Species are both diverse and abundant in the traps of the rodent ecologist. One family in particular, the Heteromyidae, a family endemic to North America, has particularly rich associations: its members are usually far more abundant than those of all other families combined and it quite often displays three of its species on the same patch of ground (less than 1/5 hectare) at the same time. In fact, associations of four or even five species are not infrequently encountered and there is a report in the literature of six species taken in about 1/2 hectare (HOFFMEISTER & GOODPASTER, 1954). Often, associations of Heteromyidae are joined by various cricetines including members of the genera *Peromyscus*, *Onychomys*, *Reithrodontomys*, *Baiomys*, *Sigmodon* and *Neotoma*.

Since many of the species in such associations are heavily dependent on seeds as their food, since they are mostly nocturnal and since many construct burrows for homesites, it is not unreasonable to assume that interspecific competition plays a role in regulating their abundances. This is particularly true of the heteromyids – all nocturnal burrowers and all granivorous. Naturally one immediately wonders whether such apparently close competitors exhibit specializations appropriate to their avoiding competitive exclusion.

The specializations of ecorole which can promote competitive coexistence have been classified by MACARTHUR & LEVINS (1964). We adopt their classification system and expand it to try to make it comprehensive, as follows:

1) Habitat selection. If a species utilizes the various habitats and microhabitats of an environment in some proportion other than that in which they occur, then the species may be said to 'habitat select'. In the extreme form of habitat selection, a species restricts itself to one habitat patch and ignores all others. Space can be selected on a horizontal or a vertical basis. A species can also select habitats in time via migratory movements or metabolic retreat. Conscious habitat choices may occur, but they are

---

* Dedicated to ROBERT H. MACARTHUR for many, many reasons.

not essential to habitat selection; instead the selection may be passive and imposed.

2) Resource allocation. If a species does not utilize the resources it discovers in the proportion in which it discovers them, then that species is a resource allocator. It and its competitors may coexist owing to the allocation. In addition, its living resources may coexist because of it. A living resource might be either a victim in an exploitative (predatory) interaction or a mutualist. A known set of resource allocators is mammalian carnivores (Rosenzweig, 1966); Carnivora allocate resources in that many take victims of restricted body sizes.

Based on our work and on the literature, we believe it is likely that all methods of habitat selection and resource allocation are involved in maintaining the coexistence of desert rodents. However we have not studied all of them. Instead we have concentrated on the exploration of seed allocation and of both vertical and horizontal spatial habitat selection. We have also attempted to integrate what we have discovered about spatial patterns into a model predicting spatial variations in rodent species diversity.

Resource allocation by seed selection

Theories of resource allocation (MacArthur & Pianka, 1966; Emlen, 1966) stress the need of each individual to optimize its use of time and energy. In other words, a species does not allocate resources to do another species a competitive favor. It allocates because if it utilized everything it discovers, it would waste some of its time on relatively unprofitable resources. For another species these same resources may be most profitable.

One way in which this situation might occur is through specialized body sizes. This alternative appeared especially attractive to us in view of the fact that heteromyids can be grouped into size classes and that coexisting sets of species tend to have representatives of several classes, but tend not to have more than one representative of a given class. In the high San Pedro Valley of southeastern Arizona for example, there are six species. Of these, *Perognathus flavus* weighing about 8 to 9 g is the small pocket mouse, *P. penicillatus* at 15 to 20 g is the medium-sized one; and *Dipodomys spectabilis* at 100 to 125 g is the large kangaroo rat. *Perognathus hispidus* at 30 to 45 g is the largest pocket mouse known. But *Dipodomys* has two similar sized species represented: *D. merriami* and *D. ordii* (both about 45 to 50 g).

These two dipos by the way exhibit a body size pattern which is also suggestive of the importance of body size in heteromyid competition. In the San Pedro Valley, they are virtually identical in size (Hoffmeister & Goodpaster, 1954). But the *D. merriami* in the valley appear to have evolved to be about 20% larger than their relatives in warmer deserts (Table 1). This does not appear to be a general phenomenon

242

Table 1. Mean sizes of two sympatric species in different parts of their range.

	D. merriami			P. penicillatus		
	Hindfoot (mm)	Weight (g)	Sample (N)	Hindfoot (mm)	Weight (g)	Sample (N)
S.E. Arizona	37.8	45.0	57	22.5	16.5	16
Central Arizona	36.9	39.6	22	24.8	17.7	9
Student's t	3.64	2.72		7.03	1.28	
Probability	< 0.0005	< 0.005		< 0.0005	< 0.15	

explicable on a climatic basis, because *P. penicillatus* in the valley are smaller than their conspecifics in warmer deserts (Table 1). One way to explain this phenomenon can be derived from the theory of character convergence (MacArthur & Levins, 1967). If this theory applies here, one may draw two implications about these dipos: their convergent morphological characteristics are somehow related to their competitive specialty and the two species are extremely close competitors. Thus the body size patterns suggest the possibility that *D. merriami* and *D. ordii* are intense competitors with which natural selection is trying to fill the same niche and that the ability to fill that niche successfully is determined by body size.

How might body size be so important in determining a species specialty? There are many ways, but two are suggested by theories of resource allocation published by MacArthur & Pianka (1966) and by Emlen (1966). Because it has a certain body size, a species must expend calories at a certain rate to maintain itself. It therefore is not able to handle a food resource efficiently if it is found in too small a 'package', because the energy required to harvest it is better spent looking for larger packages or, in the extreme, because more energy is spent in harvesting than is returned in the food. Clearly it is also possible that a package is so large that a small body stumbles or fails altogether to handle it. Thus the body size of a granivore might determine a special range of seed sizes which it must harvest or it might also determine a special predator or set of predators which it tends to escape or be rejected by.

It seemed worthwhile therefore to investigate the seed size preferences of heteromyids and to examine their relative ability at husking seeds since the husking process reveals their capacity for handling seeds after they discover them. It is also worthwhile to examine differences in their susceptibility to predators but this has been deferred.

Laboratory studies. By timing starved heteromyids of 7 species, Sterner (1968) measured their husking times. He selected seeds that had proved

243

Table 2. Mean seed husking times (sec.)*

Species	Spinach	Sunflower	Squash	Pumpkin
P. flavus	14.2	17.2	70.1	32.0
P. amplus	10.3	9.9	42.5	28.5
P. penicillatus	8.2	7.7	32.4	37.8
P. baileyi	8.4	6.3	49.5	33.1
P. hispidus	6.9	3.8	38.0	22.2
D. merriami	6.3	4.3	37.8	20.5
D. spectabilis	4.6	3.1	25.7	19.7

* See ROSENZWEIG & STERNER, 1970 for more extensive tables.

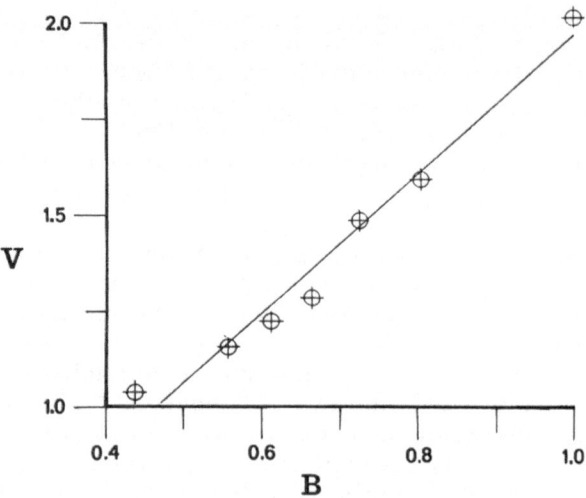

Fig. 1. Larger heteromyids husk and handle seeds faster than smaller ones. The ordinate, V, is a variable which measures relative speed (see text); the abscissa, B, is relative logarithmic body size. From ROSENZWEIG & STERNER, 1970.

palatable to all 7 species and were readily available in bulk and un-poisoned.

ROSENZWEIG & STERNER (1970) analyzed these measurements (Table 2) according to the theory of resource allocation. As one might have guessed, the larger the species, the faster it husks seeds. This is seen from Fig. 1 which is obtained as follows. The mean time it takes the largest species to husk a spinach seed is divided by the mean time it takes the slowest species to do the same. Then the process is repeated for sunflower seeds (where there may be a different slowest species). And it is repeated again for squash seeds and for pumpkin seeds. The four quotients are summed and the sum is divided by 4. This final quotient is called the mean relative husking speed, p, of the largest species. The process is repeated for the next to largest species which in turn yields its p value, and so on for each heteromyid.

The abscissa of the graph, B, is simply the logarithm of the mean body weight of a species divided by a standard, the logarithm of the mean weight of a *Dipodomys spectabilis*. Since *D. spectabilis* is the largest heteromyid in the study, the upper limit of B is 1.00.

The relationship between B and p is quite significantly positive. The value of the Pearson product-moment correlation coefficient, r, is $+0.975$. Student's t for this coefficient is 9.826 which tells us that Fig. 1 has a probability of less than 0.001 of representing a random relationship.

This is all well and good, but one feels it is somehow irrelevant to the processes of natural selection and competitive exclusion. In order for a species to exclude another, it must reproduce faster under all conditions. For a phenotype to displace another, it too must reproduce faster. Speedy husking of seeds is not necessarily speedy reproduction; it is just speedy energy collection.

To convert p into a measure of reproductive success, ROSENZWEIG & STERNER related the husking speed of each species. They showed that a species' energetic profit per second divided by its metabolism per second is linearly related to a variable they called the energy ratio, ER, where

$$\text{ER} = \frac{(\text{kcal husked/sec})_{1,j}}{(70/8.64 \times 10^4)K_j^{.75}} \tag{1}$$

In Eq. 1 i identifies the type of seed, j is the heteromyid and K_j is its mean weight in Kg. Eq. 1 contains the standard metabolic formula of KLEIBER (1961). They felt justified in using it because McNAB (1963) showed that the variation in home range sizes of a variety of mammals can be explained on the basis of that formula. Also HEMMINGSEN (1960) had concluded that the 0.75 power is reasonable for a large variety of taxa. We can also add that although heteromyids are known to have depressed metabolisms, BARTHOLOMEW & MACMILLEN, 1961, have suggested that this depression is about 1/4 for all heteromyids. Thus the exponent still applies and since we studied only heteromyids, the relationship between these species should not be altered by the change in linear constants.

In order to evaluate the relative performances of the seven species on the four seeds, ER was transformed into a relative variable. Each species ER on say sunflower seeds, was divided by the ER of the species most profitably husking sunflower seeds. Thus the highest value of this variable is 1.00 and the lowest is zero. The process of obtaining relative ER was repeated for all four seeds and finally each species four values were averaged.

Two conclusions emerged. First the average relative ER was strongly

245

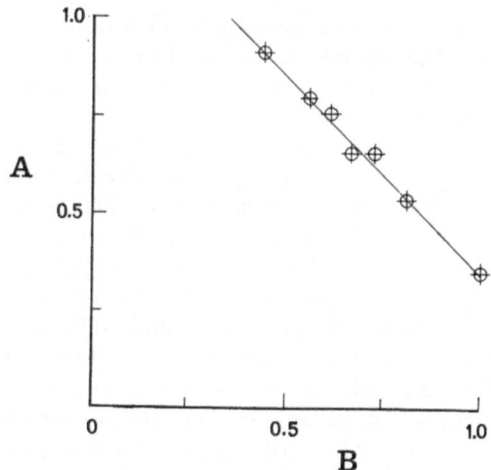

Fig. 2. Smaller heteromyids husk and handle seeds more profitably than larger ones. B as in Fig. 1. The ordinate, A, is a measure of profitability which should be more relevant to natural selection than V. This relationship is quite significant (see ROSENZWEIG & STERNER, 1970).

Table 3. Relative profitability of each seed for seven Heteromyidae. A '1' indicates the most profitable seed.

Species	Sunflower	Pumpkin	Spinach	Squash
P. flavus	2	1	3	4
P. amplus	1	2	3	4
P. penicillatus	1	2	4	3
P. baileyi	1	2	4	3
P. hispidus	1	2	4	3
D. merriami	1	2	3	4
D. spectabilis	1	2	3	4

negatively correlated with body weight (Fig. 2). Second, there was little difference in the rank of a species ER regardless of the seed it was calculated for (Table 3). What do these conclusions mean?

Compared to most seeds we saw in the natural environments of these species, our experimental seeds were very large. Yet the large heteromyids did not handle even these as profitably as the smaller ones. It seems unlikely they could best the small rodents at using untested smaller seeds.

Furthermore, it is clear (on the basis of existing allocation theory) that a species should specialize according to its ER's. If it takes one seed, that seed should have the highest ER. If two, they should be the two highest, etc. So, since the species tended to rank seeds in about the same

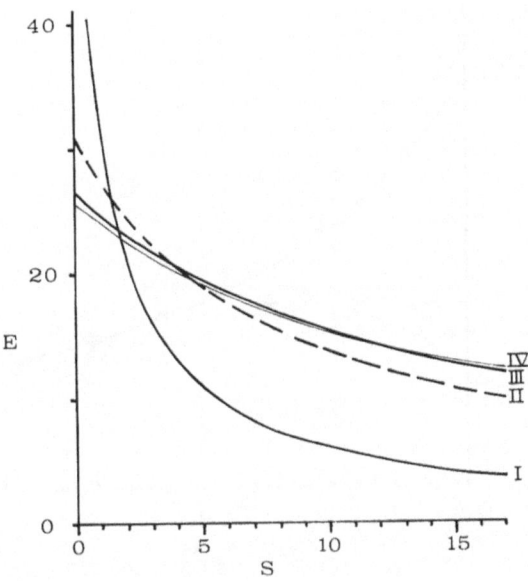

Fig. 3. The rate, E, in calories per second, at which *P. penicillatus* prepares energy while searching for, handling and husking the four seeds investigated. S is the time in seconds for one seed to be discovered. See text for formula. The line 'I' denotes E should the rodent ignore all but its most profitable seed (sunflower). The line 'II' assumes it is using both sunflower and pumpkin seeds. All seeds are used in 'IV'. Note that specialization is more profitable when seeds are abundant and S small.

order, they are unlikely to specialize on different seeds. However there is one plausible way in which coexistence might be explained by already published theory: some might specialize while others are generalists. We now examine this alternative.

For purposes of illustration, let us assume an environment in which the four test seeds and only the four test seeds occur as resources. The seeds are assumed to be equally abundant and randomly dispersed. On the basis of these assumptions, the rate at which an individual of one species, j, prepares energy for consumption is

$$E_j = \left[\sum_{i=1}^{n} e_i - (\text{Metabolism/sec.}) \left(4S + \sum_{i=1}^{n} P_i \right) \right] \bigg/ \left(4S + \sum_{i=1}^{n} P_i \right)$$

where n is the number of kinds of seeds actually harvested; S is average search time per seed; P_i is the time it takes the species to hull and eat a seed of type i and e_i is the energy in a seed of type i. The actual eating time is assumed to be proportional to the seed's size and independent of the species that eats it.

We now allow S to vary (i.e. total seed production goes from infinity

247

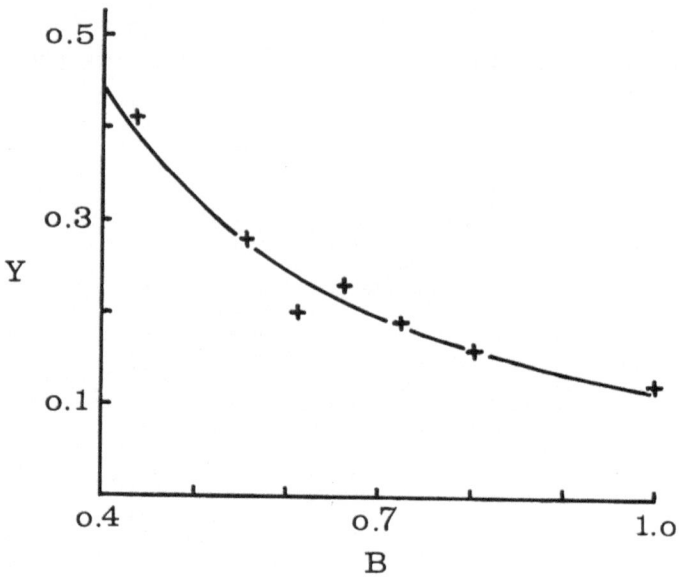

Fig. 4. Smaller heteromyids need to specialize at larger values of S than do larger heteromyids. B as in Fig. 1. Y is directly proportional to a species need to specialize. To obtain Y, one finds the minimum value of S at which a species needs to adopt strategy IV (total generalist, see Fig. 3). Then one multiplies this S by a constant to make the maximum possible value of Y exactly 1.00 (in this case by 0.01675). The regression line depicted is $Y = 0.116 B^{-1.47}$. The relationship has a significant correlation coefficient: $r = -0.981$, $t = 11.30$, df $= 5$ and $P < 0.0005$. This relationship occurs because larger species tend to have less variable husking times from seed to seed. Since that difference seems reasonable in general, this pattern itself may well be general.

to zero) and we calculate E_1 for four strategies: ignoring all but the most profitable seed (I), to ignoring no seed (IV). This is depicted for *P. penicillatus* in Fig. 3. Notice therein how increasing S requires decreasing specialization in order to produce maximal E_1.

Despite the artificiality of Fig. 3, it and others for the other species, are instructive. It demonstrates that even with the data generated in these studies, it is possible to generate seed allocation at certain inter-mediate values of S. The variable previously omitted was switch time: the S at which an organism must increase its generality. Different species have slightly different switch times and so it is possible for one species to specialize where another generalizes. If the specialist is superior on its speciality, which seems reasonable, the two species could coexist.

For this data, the need to generalize is associated with realistic values of S, values which change little if one assumes the seeds are unequally abundant (for example we have tried assuming they fit a MacArthur non-overlapping broken stick distribution). Also, in the present data, the need to generalize is positively correlated with body size (Fig. 4). So, in environments of intermediate values of seed production, allocation based on nested niches may occur. This would be similar to COHEN's balls-and-boxes model (1968) and has previously been suggested (ROSEN-ZWEIG, 1966; SCHOENER & GORMAN, 1968; PULLIAM & ENDERS, 1971 and RECHER, p.c.) for other mammals, for lizards and for birds.

In summary we may say that 1) although larger species do husk large seeds faster than smaller species, 2) they do not do so fast enough to make up for the increased costs of being large, but 3) in a few environments of intermediate production, larger species may be generalists compared to smaller ones and so coexist with them. Otherwise it seems, based on existing resource allocation theory, that smaller species should usually maintain a competitive advantage in collecting and preparing energy.

Field studies. We have learned from other workers that despite our laboratory work, different species of rodents sometimes do take different size seeds. For instance, SMITH (1942) found that two coexisting hetero-myids took seeds of different size. *Dipodomys heermanni*, the larger, ate larger seeds (mostly manzanita, *Arctostaphylos* sp.), whereas *Perognathus parvus* subsisted instead on the smaller seeds of *Ceanothus vetulinus*.

Also, DUNHAM (1968) showed that seeds taken from the pockets of snap-trapped kangaroo rats tended to be bigger if *D. ordii* had collected them as compared to *D. merriami*. The greatest linear dimension of 86% of the seeds collected by *D. merriami* fell between 0.06 and 0.10 cm; 65% of seeds taken from *D. ordii* range from 0.11 to 0.15 cm. This work was done in central New Mexico where the two species are often sym-patric.

In the most systematic collection of seeds from snap-trapped animals, BROWN & LIEBERMAN (1973) confirmed this result with rodents living on sand dunes at a variety of latitudes in the general longitude of the Colorado River delta. There *D. merriami* had seeds whose mean size was somewhere between 0.09 and 0.13 cm, whereas *D. ordii*'s averaged be-tween 0.16 and 0.19 cm.

Moreover, they collected seeds from other heteromyid species and showed that seed size was closely correlated to body size with two striking exceptions. *D. ordii* and *D. merriami* are about the same size, but take mostly different sized seeds. Also *D. merriami* is about twice as heavy as *P. parvus*, but these two take about the same size food.

Using a different technique and a similar pair of species, *D. merriami* and *P. penicillatus*, SMIGEL (1973) has confirmed the latter anomaly. Her technique and results are worth describing, because the technique is likely to prove useful to the field biologist in a wide variety of situations.

An experimental, stable-isotope method for use in the field

Seeds of various sizes (or with other interesting differences) are impreg-nated with salts of various rare elements by soaking them in the appropri-

Table 4. Percentage composition of diet as indicated by three tagging experiments.

Seed	Tracer	D. merriami	P. penicillatus
Oats	Dy	37	37
Millet	In	61	63
Ryegrass	Sm	2	0
Oats	In	14	16
Millet	Sm	84	85
Ryegrass	Dy	0	0
Oats	Sm	12	2
Millet	Dy	86	98
Ryegrass	In	2	0

ate salt solution. Such tagged seeds can then be safely distributed in small quantities in the natural environment, where wild rodents can eat the seeds of their choice.

After scattering such seeds for a period of two to five days, one traps the rodents alive and obtains fecal samples. When such fecal pellets are bombarded by neutrons in a nuclear reactor, they become radioactive and emit gamma rays. Each element emits a characteristic spectrum of gamma rays. Thus, measuring the spectra emanating from the feces allows one to quantify the rodent's diet with respect to the tagged items (SMIGEL et al., 1974).

Using this method, SMIGEL performed a set of experiments involving grass seeds of three sizes. It was carried out in the mesquite association that lines shallow drainage depressions near the San Pedro River, Cochise Co., Arizona (SMIGEL & ROSENZWEIG, 1974).

The results (Table 4) show that great apparent variation in diet comes from varying the method of tagging. We believe this to be a calibration problem, but even in its presence the method show the agreement of the data between species. Within any one experiment, both species took seeds in the same quantitative way.

This work is being repeated and extended both to other seed families and to other heteromyids. But it is already clear (compare also REYNOLDS & HASKELL, 1949 with REYNOLDS, 1950), that two species of strikingly different weight and linear dimensions are selecting quite similar food sizes.

Required: a new theory of resource allocation

The two mentioned anomalies might be an indication that resources are not being partitioned according to published theory. The lab data also suggests that the theory is inadequate.

But clearly allocation occurs. BROWN (p.c.) has suggested that the theory which explains allocation may be one that takes mobility differences into account. He feels the assumption that different sizes of seeds are similarly distributed is crucially wrong. Sensory differences and aggressive differences may also play a role. But something that is known to occur ought to have an explanation.

HABITAT SELECTION IN SPACE

Many studies of terrestrial vertebrates have shown that such species tend to be sensitive to the sculptural characteristics of foliage. MAC-ARTHUR was first to show this with his work on birds (MACARTHUR & MACARTHUR, 1961); PIANKA (1967) followed with lizards, and somewhat to our surprise, we followed with desert rodents (ROSENZWEIG, et al., 1969). As this section will show, there is no doubt that desert rodents belong on the list.

It is possible to take various measurements of the foliage and associate them with measured abundances of various mammals (ROSENZWEIG & WINAKUR, 1969). Repeated experience with *D. spectabilis* for example, confirms its need for open environments in our study areas. Repeated quantitative measurements of *D. merriami* confirms its association with low foliage densities in the foliage layer from 0.25 ft. to 1.5 ft. (Fig. 5). *Perognathus penicillatus* is associated with habitats in which vegetation is relatively dense in the layer over 2 ft. high; in this it resembles other species including *P. baileyi*, *Reithrodontomys fulvescens* and several *Peromyscus* spp. (Fig. 6).

One characteristic of these figures is noteworthy: sometimes a species appears to be much less dense than we predict from foliage measurement. This was especially striking for silky pocket mice whose population irrupted after the summer of 1967 and continued to decline for three years. Many of the places in which they had been common were empty of them by 1970. (By 1971 they had begun to increase and repopulate these areas.) Similarly a plot studied in 1968 had been known to harbor both *D. spectabilis* and *D. merriami* in 1967. In late summer 1967, a badger *(Taxidea taxus)* moved in and was observed digging up the *Dipodomys* burrows. In early 1968 neither species of *Dipodomys* was present, but by mid-summer a *D. spectabilis* had moved in. Thus the proper foliage qualities are only necessary features of the environment; other factors can and do depress populations and diversities temporarily in at least local places.

Vertical habitat studies

Since several desert rodents have consistently appeared to require the presence of some shrubby or bushy vegetation, we decided to attempt to

251

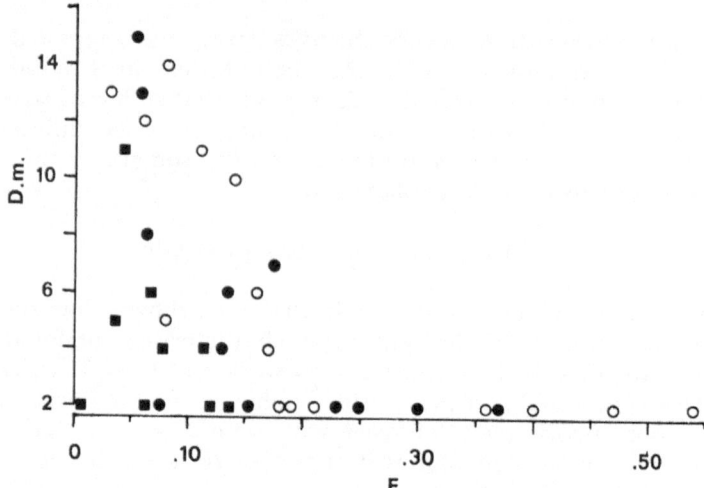

Fig. 5. D. merriami requires sparse foliage from 75 to about 300 mm above ground. Dm is the estimated number of Merriam's kangaroo rats in a plot; F is the relative amount of foliage in this stratum. Legend: open circles, 1967; closed circles, 1968; squares, 1971. Notice that the 1971 densities appear depressed. From this and other indications we believe there was a major *D. merriami* irruption in mid-summer 1971.

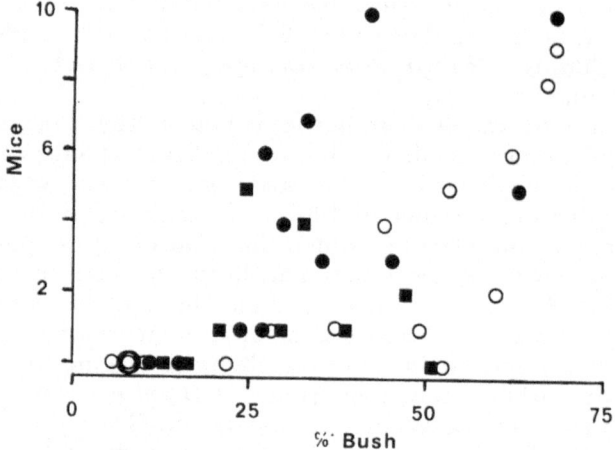

Fig. 6. A number of species appear to require that the bushy segment of the foliage be important. The ordinate is the estimated number of *P. penicillatus* + *Peromyscus* spp. + *Reithrodontomys* spp. The abscissa is the proportion of foliage over 450 mm tall. Legend: as in Fig. 5. The large amount of noise in this graph is due to several things including: a high rate of ear tag loss in 1968 leading to possible overestimates of mouse density in that year; and a misrepresentation of some kinds of bushes as non-bushes if they grow below 450 mm. This noise can be ignored because both vertical trapping and habitat tailoring confirm the truth of the relationship (see below).

252

see if there were parallel differences in the trappability of the various species when traps were set in bushes off the ground. If there were, they might well be part of the ecological differentiation necessary to promote competitive coexistence.

During the 1968 rainy season in the San Pedro Valley, we set two types of vertical trap plots: 'choice' plots and 'take-it-or-leave-it' plots. The 'choice' plots offered the rodents three levels of traps simultaneously open. The other plots had only two levels (one at ground level) and only one level was open on any given night.

In both cases, live-traps were fastened to 4″ × 6″ plywood platforms with rubber bands. The trap's front door was located on the platform itself so that some of the wood of the platform was always exposed in front of this door. Platforms were nailed to ordinary 1″ × 2″ scrap-wooden surveyor's stakes of various lengths and driven into the ground next to or within the foliage of a bush, so that the platform touched or was very near a branch. Mixed bird seed was placed in the traps and on the platforms as bait. Traps were opened near sunset and examined in the morning. Captured animals were returned to the field station for ear-tagging. Animals were released that evening at the exact site of their capture.

Two plots of five stations each were surveyed to be choice-plots. One plot was in a whitethorn acacia *(Acacia constricta)* habitat; the second was in a mesquite *(Prosopis juliflora)* and grass habitat. Both habitats had extensive patches of bare soil, though cover could be said to have been more complete in the acacia because of the relative continuity of its canopy. Trap stations were placed in a bush close to the vertices of a regular pentagon whose sides measured 50 ft. (15 m). At each station three traps were set: one on the ground; one about 25 cm high and one about 60 cm high. Exact measurements taken after the stakes had been inserted are: middle level, 21.4 to 30.0 cm (average, 24.9 cm); upper level, 56.5 to 64.5 cm (average, 60.7 cm). Plots were trapped regularly from July to August, a total of 19 nights × 15 traps/plot = 285 trap nights per plot.

This investigation occurred in the midst of another one in which 12 other plots were being censused using capture-recapture techniques (see below). Of these, 9 had some bushes. These nine were used for our second plot type. On each plot we installed ten stakes with platforms about 45 cm high (exact measurements are: 30.3 to 51.5 cm; average, 42.3 cm). Trap stations were near bushes and clustered within a variable area of less than 0.05 hectare. The first night, all traps were set on the ground under the platforms. Then they were set on the platforms for enough nights (two or three) to enable the number caught on the platforms to equal approximately the number caught previously on the ground. This procedure was continued for two weeks (414 up trap nights and 153

Table 5. Rodent captures at three levels. All levels are set simultaneously.

| | Level (cm) | | |
	0	25	60
P. penicillatus	9	10	0
D. merriami	35	10	2
Sigmodon hispidus	4	16	2
Peromyscus maniculatus	0	10	12
Onychomys spp.	4	6	1

ground trap nights) until time forced us to return the plots to their original use.

The two choice plots indicated definite differences between the species (Table 5). Chi-square tests were run to test the various null hypotheses. Where applicable, YATES's correction was applied, but occasionally even this is inadequate to deal with the small numbers. FISHER's exact test could be used in these cases, but this seems to us to attach too much significance to this data which we present as merely suggestive (especially when compared to the other kinds of data from our work).

Null hypothesis: Rodent captures are equally distributed throughout the three levels. $\chi^2 = 20.01$, d.f. $= 2$; $p < 0.001$. So, rodents as a whole are treating the levels differently.

Null hypothesis: No difference exists between species as to the proportions in which they visit each level. $\chi^2 = 58.9$, d.f. $= 8$; $p < .001$. Hence, species are treating the levels differently.

Moreover the differences between species are reasonably interpreted in most cases as predictable from the data which associated density with foliage. *D. merriami* is caught more on the ground than other species. *Peromyscus* was a bush mouse and *Perognathus penicillatus*, although it never was caught at the 60 cm level, did not appear to discriminate between the ground and the 25 cm level. *Sigmodon hispidus* apparently didn't view 25 cm as a challenge; clearly this result could be an artifact of the experiment: *Sigmodon hispidus* is normally found in grassy stands often in the absence of any shrubbery.

Because of the apparent ambiguity of the 25 cm level, it was abandoned during the 'take-it-or-leave-it' series. 25 cm may require climbing by *Perognathus* or *Peromyscus*, but perhaps not by *Dipodomys* or *Sigmodon*. The 45 cm level apparently removed this ambiguity, although of course the ambiguity itself might be ecologically relevant.

Results obtained from the second series were more taxonomically extensive. They added support to the hypothesis that species whose densities are proportional to bushes, are actually climbing about in them for one purpose or another (Table 6).

Table 6. Rodent captures at two levels. Only one level is open at one time.

Species	Captures		Captures per 100 trap nights	
	0	45 cm	0	45 cm
P. penicillatus	11	20	7.2	4.8
P. hispidus	3	2	2.0	0.5
D. merriami	26	0	17.0	0
D. spectabilis	6	0	3.9	0
Peromyscus maniculatus	5	20	3.3	4.8
Peromyscus leucopus	3	8	2.0	1.9
Sigmodon hispidus	9	4	5.9	1.0
Reithrodontomys fulvescens	2	7	1.3	1.7
R. megalotis	0	1	0.2	0
Onychomys spp.	9	16	5.9	3.9

Null hypothesis: Species all had the same proportion of their captures at ground level. $\chi^2 = 44.04$, d.f. $= 9$, p $< .001$. Thus again, species proved to utilize the two height opportunities in different manners.

That desert rodents as a whole tend to use the brush less than the ground is quite clear from the fact that it took almost three times as long to achieve 78 captures at the 45 cm level as it had to make 74 on the ground. Yet there were several cases of species which went unrecorded on some plots during weeks of intensive ground level work, but were taken when traps were opened at 45 cm. As one might have expected, these cases involved *Peromyscus maniculatus* and *Reithrodontomys fulvescens*.

Because there is no way to set a trap for one and only one species, one cannot pluck out the data of one species and ask whether its distribution conforms to any particular null hypothesis. But the data can be reduced to captures per trap night for each species. The impressions one gains from doing this are again consonant with those one gets from the habitat associations. The 'bush mice', *Peromyscus* spp., *Reithrodontomys* spp. and *Perognathus penicillatus* are about as active at 45 cm as on the ground. The others are caught predominantly or entirely on the ground.

One surprise for us was *Onychomys*. There are two species of this arid-zone, carnivorous cricetine genus present in the valley and they often reach unusual densities (especially in whitethorn acacia within two miles of the river). Although we made no attempt to discriminate them, it is clear the genus climbs. This shows up better in the second set of data than the first, but I doubt the first were really different. Just why they should be climbing is strictly speculative as is any possible difference in this behavior between the two species.

In general, the vertical studies support the hypothesis that some but not all species of desert rodents use the vegetation by climbing in it readily. Those that do are those found in densities proportional to the importance of bushy foliage. There are several ways this might promote coexistence between ground specialists and bush specialists, but as yet these hypotheses are unsorted and untested.

Habitat tailoring experiments: horizontal habitat selection

Having shown that there is probably some direct utilization of shrubby foliage, we wanted to test the impression we had formed that some particular configuration of foliage was necessary in the habitat of each species. We began with two of the most common species: 1) *D. merriami* a ground dweller, which appeared to be as common as foliage from 75 to 450 mm is rare (Fig. 5) and 2) *P. penicillatus*, a species which was much more readily taken in bushes than *D. merriami* and whose population densities seemed to depend on the presence of an important shrub layer (Fig. 6).

We took advantage of the rather sparse nature of arid land vegetation and restructured it so that its new pattern could be predicted to exclude one or the other of these two species. We call this experimental method, habitat tailoring.

First we sampled 36 plots, each 16 m in radius and each at least 200 m from its nearest neighbor. We trapped each plot on two circles of trap stations (one at radius 8 m, the other at 16 m). We continued until we located a set of nine plots such that the following was true: within any period of time, the catch of either *D. merriami* or *P. penicillatus* was about 10 per plot and was equally divided between the inner and outer circles of trap stations. The successful plots were covered with moderately dense stands of the leguminous shrub, *Prosopis juliflora*. We then split the plots carefully into three matched sets of three plots each such that within each set the catch was still uniformly distributed between species and trap circles as specified above. Very few individuals of any other species were taken on these plots.

Next, the three sets were randomly chosen to be 1) a control set, 2) a cleared set, and 3) an augmented set.

Vegetation was removed from the cleared set within an area of radius 12 m and hand-carried to the augmented set. It was scattered within the same 12 m areas of the augmented set and within these areas, the natural vegetation of the augmented set was also cut down. No cut branch was permitted to rise more than 450 mm above the ground (see Fig. 7).

The effect of this on the *D. merriami* captures was just as expected; *D. merriami* remained on all trap circles except the inner augmented ones where vegetation had actually been added (the outer augmented plot's traps were 4 m beyond the augmentation). But *P. penicillatus* continued to use the inner cleared trap circles, contrary to our hypothesis (Table 7).

Fig. 7a. A control plot which typifies the appearance of the nine plots before tailoring.

Fig. 7b. A plot after clearing.

257

Fig. 7c. A plot after augmentation. All views taken looking outward from near the center of the plot.

Table 7. Captures during the first habitat tailoring experiment, 1970.

	Control	Plot type Cleared	Augmented
D. merriami			
Pretailoring			
In	12	14	15
Out	15	14	14
Post tailoring			
In	28	25	4*
Out	18	34	30
P. penicillatus			
Pretailoring			
In	13	14	13
Out	11	15	14
Post tailoring			
In	22	27	41
Out	39	45	28

* Significantly depressed.

258

Table 8. Captures during the second habitat tailoring experiment, 1971.

| | Plot type | |
	Control	Extensively Cleared
D. merriami		
Pretailoring		
In	11	5
Out	19	11
Post tailoring		
In	17	23
Out	24	16
P. penicillatus		
Pretailoring		
In	11	10
Out	9	16
Post tailoring		
In	34	18*
Out	39	73

* Significantly depressed.

After a year, the *P. penicillatus* were still found in abundance on the cleared inner stations and we concluded that we had not yet affected their habitat significantly. We then cleared additional vegetation such that each of the six inner trap stations was surrounded by 8 m of cleared land. The outer trap stations were rotated on their circle so that they remained within untouched vegetation.

This time we had done something to the *P. penicillatus*. In the sampling period after the second clearing, about 80% of the *P. penicillatus* captures on cleared plots were made on the outer circles (Table 8). Apparently a distance of 4 m in the open is not too much for a *P. penicillatus*, but a distance of 8 m begins to be avoided.

A more detailed account of these experiments has appeared recently (ROSENZWEIG, 1973). In it, various statistical tests are executed which confirm the significance of these data. The data are also significant if we record the number of individuals which use a given spot, instead of the number of captures (many individuals were caught more than once). Thus we conclude that these two species actively and consciously recognize the foliage characteristics that are so well correlated with their densities.

Coexistence of *D. merriami* and *P. penicillatus*

The previous data are not sufficient to determine precisely how these two species coexist. But the data are most suggestive. It does appear that these two rodents probably do not allocate seed sizes between them. Instead they exhibit different habitat or microhabitat preferences. And the one that prefers bushes can also be caught in them. How can this difference lead to coexistence?

Of many possible ways, two are perhaps most impressive. 1) *P. penicillatus* might be climbing to collect seeds which are unavailable to *D. merriami*; it might thus preserve itself. But, for this mechanism to exist, *D. merriami* must be the superior competitor on the ground. The seed hulling work suggests the opposite is true. Of course seed hulling may give an inadequate picture of competitive advantage. *D. merriami* might possess some other countervailing advantage. It might be able to exclude *P. penicillatus* by aggression. It might have anti-predator advantages which outweigh its energetic disadvantages. J. BROWN (p.c.) has suggested there might be an important mobility advantage to a dipo's mode of locomotion.

2) *P. penicillatus* may have the resource collecting advantage on or off the ground, but may require shrubs for cover. Its ability to climb would only maintain it in its struggle with congeners like *Perognathus hispidus* and/or *P. flavus*. On the other hand *D. merriami* would be able to coexist with it by virtue of the dipo's superior adaptations for escape in open patches of environment (BARTHOLOMEW & CASWELL, 1951); the dipo has after all, better ears (WEBSTER, 1962), more unpredictable jumping escape behavior and more advantageously positioned eyes.

THE PATTERN OF LOCAL SPECIES DIVERSITY

If habitat variables can predict densities of species, then some arrangement of them into a habitat complexity variable should be associated with rodent species diversity. ROSENZWEIG & WINAKUR (1969) proposed such a complexity variable. They also showed that rodent species diversity was not correlated with plant species diversity.

The proposed variable had three components: one for vertical foliage complexity; another for horizontal variability in density of cover; and a third for variability of the sheer strength of the soil surface. The variability or complexity was in each case developed by using the formula $1/\sum p_i^2$ where p_i is the proportion of any given set of measurements that fall into class i (MACARTHUR & WILSON, 1967, HURLBERT, 1971). Species diversity was also calculated this way.

The vertical component is built on the premise that mammals in a desert will use the ground whether it is covered with or bare of plants. Foliage in strata above the ground should add species. Further, the foliage above ground level is divided into two strata: one from 75 mm high to 450 mm high, another above 450 mm. Maximum complexity (i.e. 2.00) would be reached if a community had all vegetation growing above 75 mm *and* half its vegetation over 450 mm. Minimum complexity (of zero) is recorded if all vegetation is under 75 mm tall. The formula for calculating vertical complexity is $(p_2 + p_3)^3/(p_2{}^2 + p_3{}^2)$, where p_2 is the proportion of foliage in the middle layer and p_3 is the proportion above 450 mm (Rosenzweig *et al.*, 1969).

The horizontal variability is calculated by dividing the measurements of foliage density that are taken near the ground (i.e. at 75 mm) into a group of sparse and another of dense values (Rosenzweig *et al.*, 1969). The formula is $(1/\sum q_1{}^2) - 1.00$, where q_1 is either the proportion of measurements yielding sparse densities or the proportion yielding dense values.

The variable for soil surface complexity may be examined in Rosenzweig *et al.*, 1969. It is not reviewed here since we will show below that it does not appear to be important.

In order to test the validity of these components, we set up 12 study plots in the San Pedro Valley during the summer of 1968. The plots were trapped simultaneously instead of consecutively in the hope this would eliminate any possible differences due to the rapid temporal changes occurring during the rainy season. Another change from previous methods was made: instead of removing the animals as they were captured, they were returned to their capture-sites after being eartagged. Thus we hoped to minimize the effect of transients entering our plots. And we hoped to allow animals to be caught in proportion to their use of our small study plots. Hopefully such data on activity-density would contain less of the error introduced by equating the capture of a transient with that of a resident or the capture of a species living on the periphery with that of a species living squarely in a plot.

Unfortunately this method turned out in practice to be somewhat less refined than we hoped. In the first place, because our plots were scattered about the whole valley and because we were using three times as many plots at once, we had to settle for fewer traps and trap-stations per plot. Instead of the 25 stations used previously, we had only 7: one in the center of a circle of 15 feet in radius and 6 spaced evenly along its circumference. This made for more trap competition than desirable: often every station caught a rodent, and in one plot every trap (of 14) once did. After a while, in this plot rodents actually began to emerge earlier than usual and we saw them literally battling over the 'right' to be caught. This plot also was one of the ones in which our intensive study missed one species entirely (in this case *Peromyscus maniculatus*); the species showed up when we trapped vertically (see above) and was caught regularly while traps were being placed on stakes in bushes. In the second place, the extensive trap-happiness, revealed in our ever-mounting daily hauls, was probably drawing in peripherals on a regular basis, exactly what we wanted to avoid.

But the method is different and its problems were sometimes illu-

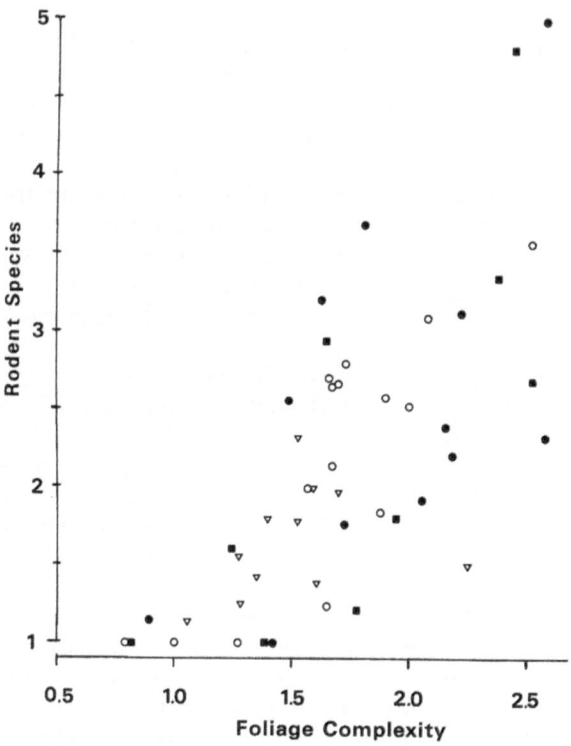

Fig. 8. Nocturnal rodent species diversity as a function of vertical and horizontal foliage complexity. Diversity is expressed in number of equally common species (ordinate). *Onychomys* is omitted. Legend: as in Fig. 5 except, triangles, 1966. See text for formulae. The 1968 data are obtained from counts of individuals (see Fig. 9). The relationship is not altered if the Shannon-Weiner function is used to calculate the ordinate.

minating. In addition, there were plots which appeared to be empty for a week or two and then began producing regular catches. Two of these (which were adjacent) produced large numbers of the supposedly rare *Sigmodon fulviventer* (formerly *S. minimus*), a species never taken in the 4 day removal censuses that we have done before and since. (A third 1968 plot also produced one individual of this species).

So this method had advantages and deficiencies somewhat different from the other. Hence it is doubly interesting that its diversity data are in agreement with the two foliage complexity components.

The results of the 1968 censuses are plotted in two ways in Figs. 8, 9. In one, numbers of individuals are used; in the other, each trap record counts separately even if the individual had previously been taken. There is little to choose between the two methods. Even the ranking of the plots is hardly affected by the difference.

262

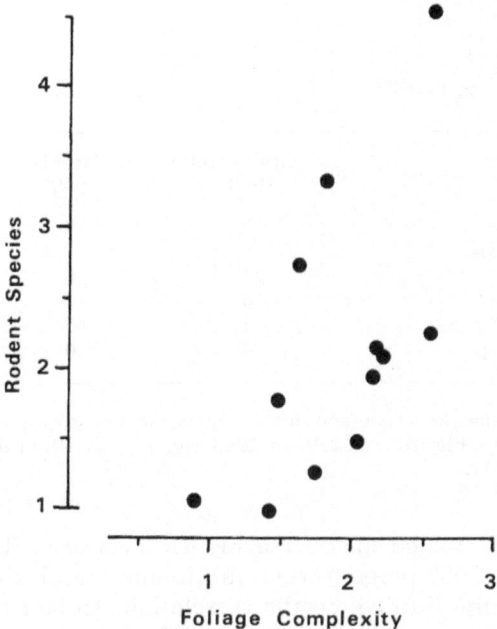

Fig. 9. Nocturnal rodent species diversity as a function of foliage complexity, 1968 data. All calculations as in Fig. 8 except the diversities are calculated from the number of captures of each species.

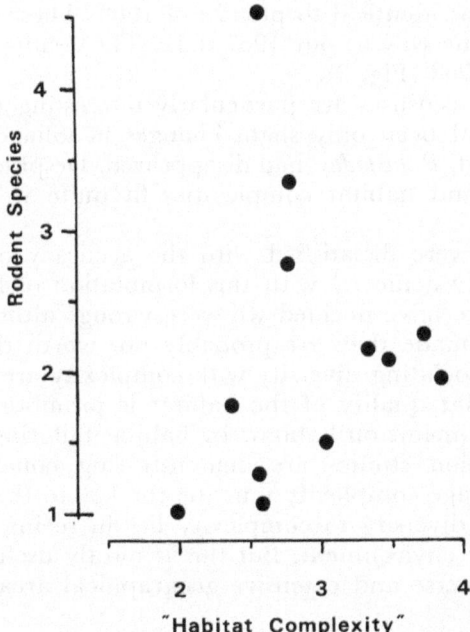

Fig. 10. Nocturnal rodent species diversity as a function of habitat complexity. The 1968 data presented as in Fig. 9 except that the soil strength diversity has been added to the foliage complexity of Figs. 8 and 9 (see ROSENZWEIG & WINAKUR, 1969). The addition of this added information fails to strengthen the correlation; in fact it weakens it.

Table 9.   Censuses on Plot 22.

Species	Activity-Densities 1968	Numbers of Individuals	
		1968*	1971
P. flavus	0	0	3
P. penicillatus	1	1	1
P. hispidus	24	6	0
D. merriami	6	2	2
Sigmodon hispidus	4	1	4
Onychomys sp.	0	0	1

* *Reithrodontomys fulvescens* was taken during the vertical censusing recorded in Table 6. Habitat complexity (Fig. 8) was 2.16 in 1968 and 2.37 in 1971: diversities were 2.38 and 3.33 respectively.

In addition to foliage measurements, we measured the sheer strength of the soil in the 1968 plots. Adding this to our complexity measurements however, certainly didn't help the correlation. In fact as Fig. 10 shows, it weakened it. Alternate formulations of the complexity of soil strengths were tried, but none have proved useful. So we have abandoned this component.

In 1971, we amassed removal census data on nine more plots in the valley. One was identical to plot 22 of 1968. These data were taken in exactly the same way as our 1967 data. The results are similar both to them and to 1968 (Fig. 8).

The plot 22 censuses are particularly interesting (Table 9). In three years there had been only slight changes in foliage and yet the most common rodent, *P. hispidus*, had disappeared. Despite the turnover, these two censuses and habitat complexities fit quite well into the pattern (Fig. 8).

At first we were dissatisfied with the accuracy of the prediction of species diversity achieved with this formulation of habitat complexity. But recently we have decided we were wrong; although improvements can surely be made they are probably not worth the effort. After all, attempts at associating diversity with complexity are undertaken mostly to discover what quality of the habitat is promoting specialization of ecoroles; experiments on habitats by habitat tailoring and on allocation by seed selection studies are demonstrating conclusively that some measure of foliage complexity must be the key to the pattern. A second goal of fitting diversity to complexity lies in seeing how finely species subdivide their environment. But this is mostly useful in studies over a much more diverse and extensive geographical area than southeastern Arizona.

Finally, there are inherent errors in measuring the species diversity itself and these would take Herculean efforts to overcome. Transients come and go and must be sorted out. Peripherals would need to be minimized by studying much larger plots. Temporary depressions which might be caused by all sorts of interference and from which mammals do not appear to recover as quickly say as birds (STEWART & ALDRICH, 1951; HENSLEY & COPE, 1951) or insects (SIMBERLOFF & WILSON, 1969), must eliminated by long term censuses. Also in an area as horizontally patchy as that valley, each habitat is an island of varying size; the local extinctions and rapid turnover that this should cause are strongly reflected in our data. Thus a perfect prediction of diversity would include a correction factor for island size. But islands will be different sizes for different species (the habitat of one often includes only parts of the habitat of another). Moreover when we know enough to find their 'shorelines' we will, with more efficient methods, have already satisfied our curiousity as to the specializations of the species they harbor.

In the course of our work, we have often wondered what effect man has had on the diversity of this marvelous and perhaps unique alluvial valley. Originally we thought he increased it. By overgrazing he added to the area covered with desert shrubs and perhaps raised the horizontal patchiness of the valley.

But much of this patchiness may be natural. The valley is covered with a network of underground and intermittent water courses which likely have 'always' supported shrubby growth and its quota of 'bush mice'. In the grassland, prairie dogs, *Cynomys ludovicianus*, once cleared patches on which *Dipodomys* could have thrived. And a very few kilometers from our study sites, there are natural patches of prairie surrounded by desert thorn bushes. These patches have deep soil and are apparently caused by their being accidentally located in a local drainage basin; once a bit of extra water sinks in, the land is leached away a bit faster than its surroundings causing yet more extra water to enter the even lower depression and setting up a positive feedback system which results in a local patch of grassland. Such a sharp change in vegetation caused by so slight a change in water balance is indicative of the ecotonal nature of the valley.

On the other hand, there is no doubt that man has depressed diversity in this valley in the past half-century. When he plows up fields of the thorn scrub he has produced, he plants them in exotic grasses and they become nearly devoid of rodents. He has exterminated *Cynomys* and their possible diversifying effect with them. There is also strong circumstantial evidence that most of his kind have no truck with *Dipodomys spectabilis* and many of our plots which should have harbored this most magnificent of all desert rodents, didn't. And, finally, he has reduced many of the populations of predators in the valley; we have seen how their presence in force may be necessary for the coexistence of *Dipodomys* with cover-loving *Perognathus*.

But whether man is responsible or not, we have never observed quite the diversity that had previously been reported (HOFFMEISTER *et al.*, 1954). And we have worked very hard in precisely the same spots. All the species they reported seem to remain, but some, especially *D. ordii*, are hanging on by the thinnest of margins. They will probably survive elsewhere, but their continued participation in what may have been the richest local association of rodents in the world is precarious. And so the association itself seems destined for history.

Acknowledgements

Many people have helped with the field work; of them, only RICHARD GARDNER has not previously been thanked. Mr. and Mrs. CARROLL PEABODY deserve our repeated thanks: it is difficult to believe we could have completed any of this work without them and their facilities; we wish them Godspeed in their efforts to preserve as much of the natural wonder of their corner of Arizona as is humanly possible. Essential for the neutron activation analysis was the unstinting cooperation of Dr. WM. JESTER and Dr. J. C. BLOMGREN of the Breazele Nuclear Reactor, Pennsylvania State Univ. We thank Dr. HARRY RECHER for allowing us to see a copy of his important ms. on heron feeding ecology. Discussions with Drs. J. H. BROWN, J. FINDLEY and V. HAYNES have been most enjoyable and helpful. We also thank Dr. FINDLEY for examining the manuscript. The following organizations, of which the National Science Foundation (USA) has been far and away the most generous, have supported the research: NSF; Theodore Roosevelt Memorial Fund (AMNH); Research Committee of Bucknell University; Research Foundation of the State University of New York.

REFERENCES

BARTHOLOMEW, G. A. & CASWELL, H. H. 1951. Locomotion in Kangaroo rats and its adaptive significance. *J. Mamm.*, 32: 155–159.

BARTHOLOMEW, G. A. & MACMILLEN, R. E. 1961. Oxygen consumption, estivation, and hibernation in the kangaroo mouse, *Microdipodops pallidus*. *Physiol. Zool.*, 34: 177–183.

BROWN, J. H. & LIEBERMAN, G. A. 1973. Resource utilization and coexistence of seed-eating desert rodents in sand-dune habitats. *Ecology*, 54: 788–797.

COHEN, J. E. 1968. Alternate derivations of a species-abundance relation. *Amer. Natur.*, 102: 165–172.

DUNHAM, MARILYN. 1968. A comparative food habit study of two species of kangaroo rats-*Dipodomys ordii* and *D. merriami*. MS thesis, Dept. Biology, UNM, Albuquerque.

EMLEN, J. M. 1966. The role of time and energy in food preference. *Amer. Natur.*, 100: 611–617.

HEMMINGSEN, A. M. 1960. Energy metabolism as related to body size and respiratory surfaces, and its evolution. Repts. Steno Memorial Hospital and Nordisk Insulinlab., Gentofte, Denmark, 9 (Part 2), 110 p.

HENSLEY, M. M. & COPE, J. B. 1951. Further data on removal and repopulation of breeding birds in a spruce-fir forest community. *Auk*, 68: 483–493.

HOFFMEISTER, D. & GOODPASTER, W. 1954. The mammals of the Huachuca Mountains, southeastern Arizona. *Ill. Biol. Monogr.*, XXIV(1): 1–152.

HURLBERT, S. H. 1971. The nonconcept of species diversity: a critique and alternative parameters. *Ecology*, 52: 577–586.

KLEIBER, M. 1961. The fire of life. Wiley, N.Y., 454 p.

MACARTHUR, R. H. & LEVINS, R. 1964. Competition, habitat selection and character displacement in a patchy environment. *Pr. Nat. Acad. Sci.*, USA, 51: 1207–1210.

MACARTHUR, R. H. & LEVINS, R. 1967. The limiting similarity, convergence and divergence of coexisting species. *Amer. Natur.*, 101: 377–385.

MACARTHUR, R. H. & MACARTHUR, J. 1961. On bird species diversity. *Ecology*, 42: 594–598.

MACARTHUR, R. H. & PIANKA, E. R. 1966. On optimal use of a patchy environment. *Amer. Natur.*, 100: 603–609.

MACARTHUR, R. H. & WILSON, E. O. 1967. The theory of island biogeography. Princeton U. Press, Princeton, 203 p.

MCNAB, B. K. 1963. Bioenergetics and the determination of home range size. *Amer. Natur.*, 67: 133–140.

PIANKA, E. R. 1967. On lizard species diversity: North American flatland deserts. *Ecology*, 48: 333–351.

PULLIAM, H. R. & ENDERS, F. 1971. The feeding ecology of five sympatric finch species. *Ecology*, 52: 557–566.

REYNOLDS, H. G. 1950. Relation of Merriam kangaroo rats to range vegetation in southern Arizona. *Ecology*, 31: 456–463.

REYNOLDS, H. G. & HASKELL, H. S. 1949. Life history notes on Price and Bailey pocket mice of southern Arizona. *J. Mamm.*, 30: 150–156.

ROSENZWEIG, M. L. 1966. Community structure in sympatric carnivora. *J. Mamm.*, 47: 602–612.

ROSENZWEIG, M. L. 1973. Habitat selection experiments with a pair of coexisting heteromyid rodent species. *Ecology*, 54: 111–117.

ROSENZWEIG, M. L. & STERNER, P. W. 1970. Population ecology of desert rodent communities: body size and seed husking as bases for heteromyid coexistence. *Ecology*, 51: 217–224.

ROSENZWEIG, M. L. & WINAKUR, J. 1969. Population ecology of desert rodent communities: habitats and environmental complexity. *Ecology*, 50: 558–572.

SCHOENER, T. W. & GORMAN, G. C. 1968. Some niche differences in three Lesser Antillean lizards of the genus *Anolis*. *Ecology*, 49: 819–830.

SIMBERLOFF, D. S. & WILSON, E. O. 1969. Experimental zoogeography of islands. The colonization of empty islands. *Ecology*, 50: 278–296.

SMIGEL, B. W. 1973. Ph.D. Thesis. Dept. Biol. Sci., SUNY at Albany, New York.

SMIGEL, B. W., JESTER, W., BLOMGREN, J., PRASAD, K. N. & ROSENZWEIG, M. L. 1974. Dietary analysis in granivores through the use of neutron activation analysis. *Ecology*, 55: 340–349.

SMIGEL, B. W. & ROSENZWEIG, M. L. 1974. Seed selection in *Dipodomys merriami* and *Perognathus penicillatus*. *Ecology*, 55: 329–339.

SMITH, C. F. 1942. The fall food of brushfield pocket mice. *J. Mamm.*, 23: 337–339.

STERNER, P. W. 1968. Heteromyid husking performance and resource selection strategies. MS thesis, Dept. Biol. Bucknell Univ., Lewisburg, Pa.

267

STEWART, R. E. & ALDRICH, J. W. 1951. Removal and repopulation of breeding birds in a spruce-fir forest community. *Auk*, 68: 471–482.

WEBSTER, D. B. 1962. A function of the enlarged middle-ear cavities of the kangaroo rat, *Dipodomys*. *Physiol. Zool.*, 35: 248–255.

XIII. DESERT COLORATION IN RODENTS

by

DAVID L. HARRISON

Introduction

The phenomenon of desert coloration has been a subject of profound interest to naturalists for many years. It is a familiar and indisputable fact that the majority of desert rodents are pallid in colour, their generally sandy or greyish pelage blending harmoniously with their surroundings. (Figs. 1, 2). It is undeniable also that this coloration is genetically determined, since desert rodents bred in captivity far removed from their normal environment continue to reproduce the same coloration for many generations.

Desert Coloration

It has been widely held that desert coloration is primarily protective, aiding concealment from the sharp eyes of predators in the relatively open terrain of the desert, where cover is so sparse. This view has not, however, always received the unqualified support of those field naturalists familiar with desert life, who have observed numerous inconsistencies and exceptions. Thus BUXTON (1923) doubted the value of protective coloration to nocturnal desert animals, since even by bright moonlight different shades of colour are not appreciated. The experiments of DICE (1947) were of crucial importance, however, in refuting this argument, since he was able to show that owls select victims in darkness, which do not resemble their environment in preference to those that do, this protection continuing to have effect in a degree of illumination imperceptible to the human eye.

It is amongst desert rodents in different parts of the globe that a steadily enlarging body of evidence has been found in support of the primarily protective role of desert coloration. Wherever detailed studies of desert fauna have been made it has been found that rodents inhabiting arid terrain are not in fact uniformly pallid, but that local populations may differ widely from one another in coloration, tending to resemble the predominant colour of the soil or rock on which they dwell. Examples of these 'substrate races' as they have come to be known, are numerous in the literature. Thus DICE & BLOSSOM (1937) noted parallel series of pale, intermediate and dark races of certain New World Rodents (*Perognathus intermedius, Peromyscus eremicus, Neotoma albigula* and *N. lepida*) occurring on pale, intermediate and dark coloured rocks. These authors rightly

269

Fig. 1. a) Habitat of *Gerbillus nanus setonbrownei* at Uhi Batinah Coast, Oman; b) Two specimens of *Gerbillus nanus setonbrownei* (HARR 16.4556.18.4558) on soil sample from their habitat at Uhi; c) Two specimens of *Gerbillus nanus arabium* (HARR 1.511 and 2.1512) on sand sample from their habitat at Buraimi, Trucial Oman.

pointed out that this cannot be due to coincidence, or to climatic variation.

A number of other studies have concerned the rodents dwelling in lava deserts in the New World (SUMNER, 1921; HOOPER, 1941; BAKER, 1960;

270

HOFFMEISTER, 1956 and BENSON, 1933). The results of these studies appear at first to be conflicting. Thus SUMNER (1921) found that *Peromyscus crinitus stephensi* dwelling on a lava field in the Mojave Desert did not resemble the lava in coloration, and did not differ significantly from populations obtained elsewhere on pale soils, and consequently rejected the arguments in favour of protective coloration. In a more comprehensive study of the mammals inhabiting the Durango lava field, however, BAKER (1960) found that fourteen out of twenty eight mammal species obtained had more or less darker coat colours on the lava and significantly these included all but one of the small ground rodent species with limited home ranges, of which individuals could live their whole lives within the lava fields. The dominance of dark coloured populations showed protective value both for diurnal species such as *Spermophilus spinosoma, S. variegatus* and possibly *Thomomys umbrinus* as well as for the remaining eleven nocturnal species. Amongst the rodents *Baiomys taylori* alone failed to show significant colour change. In contrast HOOPER (1941) studied sixty species and subspecies from lava fields and adjacent areas in New Mexico and found only three species tending towards the development of dark races. These contrasting results are attributed to the varying degrees of isolation of the populations on the lava fields concerned. BENSON (1933) also stressed the importance of isolation and noted the occurrence of dark lava races and pale sand desert races in the Tularosa Basin.

RANCK (1968) has recently found much evidence of substrate adaptation in Libyan rodents, showing that the subspecies of *Gerbillus campestris* and *Meriones caudatus* demonstrate this adaptation particularly well. A dark chestnut colored race *G.c.brunnescens* occurs on the darker soils of the Cyrenaican Plateau, while the pallid *G.c.dodsoni* occurs on the paler sand and desert soils to the south and west. The subspecies of *M. caudatus* exhibit a corresponding resemblance to the substrate. HOESCH (1956) found a similar correlation in gerbils of the Namib Desert in South west Africa.

Recent investigation of the Arabian desert rodents has revealed further evidence of the importance of substrate races. Thus HARRISON & SETON-BROWNE (1969) found much evidence of this phenomenon in Oman, where there is a striking difference in the predominant soil coloration on the two sides of the Oman mountains. In the interior of Oman the light reddish or light yellowish sands of the Empty Quarter extend to the foothills of the range, while on the Batinah Coast on the other side of the range the gravel plains and fertile littoral zone are strikingly darker, with a predominantly greyish coloration. A species of Gerbil, *Gerbillus nanus*, occurs on both sides of the range, but the race occurring on the Batinah, *G.n.setonbrownei*, is strikingly darker than the sandy colored *G.n.arabium* occurring in the interior. It is interesting to note that specimens of these two races match perfectly in colour soil samples from their

271

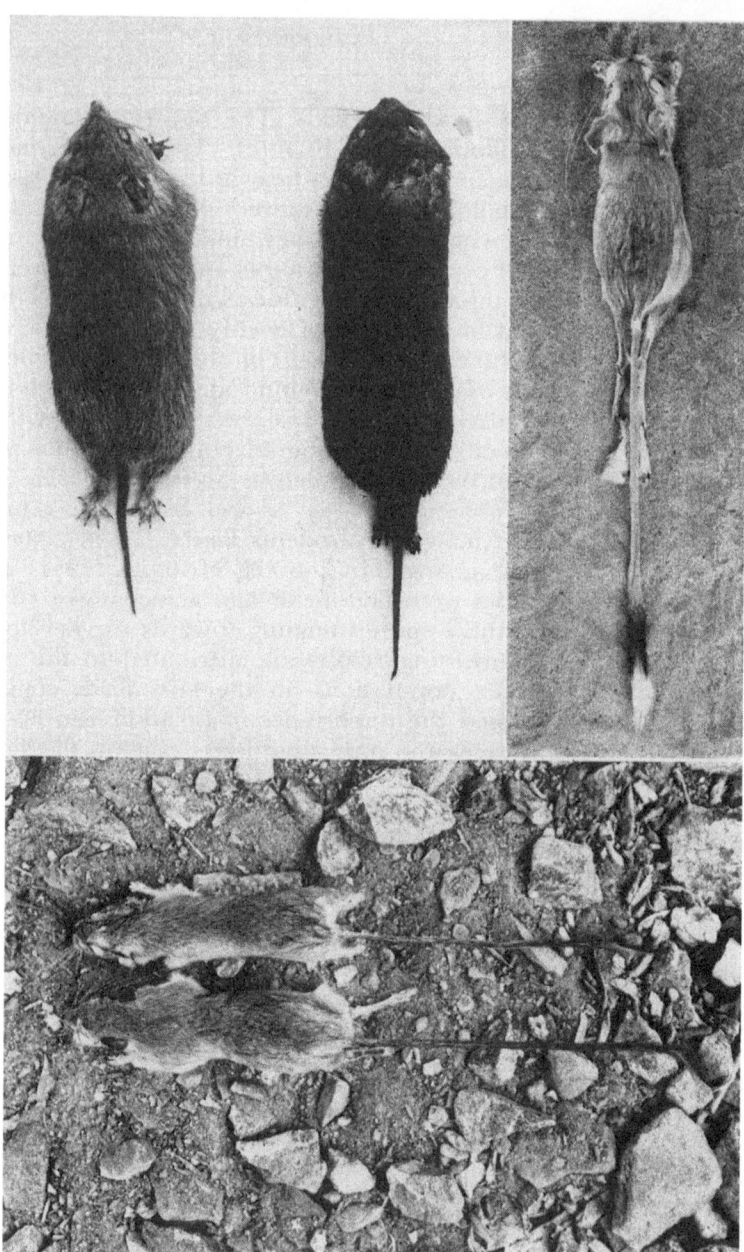

Fig. 2. a) Specimens of the Golden Spiny Mouse *(Acomys russatus)* to illustrate the evolution of a blackish lava-dwelling substrate race. Left: *A.r. russatus* HARR 1.2753 Wadi Raman, Israel; Right: *A.r. lewisi* HARR 6.4585 N.W. of Azraq-Shishan, Jordan; b) Lesser Three-toed Jerboa *(Jaculus jaculus vocator)* HARR 53.6376 on sample of sand from its habitat near Ahmadi, Kuwait. c) Two specimens of *Gerbillus dasyurus gallagheri* (HARR 28.5771, 29.5865) on samples of soil and rock from their habitat at Masafi, Oman.

own terrain, but become conspicuous when transposed on to the native soil of the other. These mice are nocturnal; their habitat is sparsely vegetated and they are often to be seen at night wandering about over the desert at some distance from the security of their holes, when they would be exposed to the attacks of nocturnal predators, such as owls, foxes and wild cats. Recently another local gerbil has been described from Oman, *G. dasyurus gallagheri*. This is a rock-dwelling form (HARRISON, 1971) which differs quite strikingly in colour from the sandy coloured *G.d.dasyurus* of Northern Arabia from which it is isolated by the great sand desert of Rub al Khali. The darker coloration of this gerbil and its blackish tail are clearly an evolutionary response to the reddish and blackish rocks and soil of the Oman range. (Fig. 2c).

Perhaps the most striking substrate form yet found in Arabian desert rodents was recently discovered in Jordan (ATALLAH, 1967). The Golden Spiny Mouse, *Acomys russatus*, is a rock-dwelling, diurnal species, widespread in the rocky mountain ranges of western Arabia south to Southern Yemen (HARRISON, 1972). It is seen darting about amongst rocks and boulders even in the full heat of the day. It's normal coloration is russet brown, but a highly distinctive form has been found living on the black lava desert of eastern Jordan, *(A.r.lewisi)*, which has evolved an almost entirely black pelage, clearly related to the substrate (Fig. 2b). It is interesting that the predominantly nocturnal *Acomys dimidiatus*, also occurring in Arabia, is similarly subject to much colour variation in the peninsula apparently dependent on the substrate.

It is certainly no coincidence that the most beautiful sandy-isaballine dorsal coloration is developed in those Arabian gerbils which are strict psammophiles, such as *Gerbillus cheesmani*, *Gerbillus gerbillus* and *Gerbillus andersoni*. These species are always found in the true sand desert, with dune formation. It is noteworthy that where the terrain of these strict psammophiles impinges on that of other species, the sand-dwelling gerbil is seldom found outside the sand dune area. Thus in the Wadi Araba of Israel ZAHAVI & WAHRMAN (1957) found *Gerbillus gerbillus* inhabiting areas of sand dune, while the adjoining saline flats were densely populated by *G.nanus*. Only on rare occasions was *G.nanus* trapped on the margins of the dunes, and the psammophilic *G.gerbillus* on the other hand, was never obtained on the saline flats. It is not known whether these animals instinctively limit their zone of activity to areas of suitably coloured substrate. This is certainly a field for further fruitful investigation. It is noteworthy that in Arabia the sandy coloured race of the Hare, *Lepus capensis cheesmani* is found in the sand deserts of Rub al Khali, extending to the margins of the sands, while darker coloured races, *(L.c.omanensis, L.c.arabicus)* are found in the surrounding steppe and mountain zones with predominantly darker soils. Again further study of the activity patterns of these races and the selective pressures imposed upon them by predators on the various substrates would be most informative.

It is clear that, as pointed out by MAYR (1963), the more open the country and the more contrastingly colored the substrate, the more striking is the development of these substrate races. It is additionally clear that the degree of isolation of the population involved is also of paramount importance, since this affects the amount of 'gene flow' occurring with surrounding populations.

The length of time required to produce such substrate races is not known for certain. The experiments of DICE *(loc. cit.)* indicated very high possible rates of selection, from which it could be deduced that natural selection can theoretically produce very rapid evolution whenever a genetically variable population, sufficiently isolated from surrounding populations, is exposed to its action. In this connection it is interesting to note that the Durango lava field studied by BAKER *(loc. cit.)* is considered to be of recent origin so that the marked selective adaptation found amongst the rodents there has occurred at most within the last 10,000 years.

The rapidly accumulating evidence of substrate races has clearly disproved attempts to explain the cryptic coloration of 'desert races' on the basis of such climatic factors as heat, dryness or solar radiation, or alternatively as some primarily physiological phenomenon, as suggested by MEINERTZHAGEN (1954). The amazing agreement with the colours of the substrate and the occurrence of blackish lava desert races and pallid sand desert races in close proximity can only be explained by the need for protective coloration. As stated by MAYR *(loc. cit.)* the majority view now is that substrate races are the result of predator selection, the most discordant individuals being captured and eliminated from the gene pool first. The argument advanced by some naturalists (HEIM DE BALSAC, 1936; KACHKAROV & KOROVINE, 1942; MEINERTZHAGEN, 1954, HOESCH, 1956) that predators are too rare in the deserts to be of selective significance appears most unconvincing, since examination of owl pellets from Arabia has revealed numerous remains of such desert rodents as *Meriones, Gerbillus* and *Jaculus*. Certainly there are still inconsistencies and difficulties to be explained, but it seems possible now to assert with some confidence that desert coloration is indeed primarily protective in rodents. NIETHAMMER (1959) has come to the same conclusion regarding the desert birds of the Sahara and Namib.

It is interesting to note that many otherwise cryptically coloured species of desert rodents, as well as birds, possess some striking and predominant marking. Familiar examples are the black and white tail tufts of Jerboas (*Jaculus* and *Allactaga*) and the black tufts of *Meriones* and some gerbils. The effect of these markings is often to become clearly visible when the animal moves and they may thus function as a signal to warn other individuals of impending danger or perhaps to assist in location of widely dispersed individuals. This signal coloration (HARRISON, 1964) seems sometimes to be enhanced by the animal's behaviour, thus

274

the diurnal Sand Rat, *Meriones libycus*, when alarmed and running for the burrow holds the tail vertically up, displaying the black tuft, while the black and white tuft of the jumping Jerboa is readily visible in the desert dusk. The existence of such signal markings underlines the fundamental fact that protective coloration is only effective when an animal is motionless.

It is clear that coloration is very important in the lives of desert rodents, representing one of the many fascinating ways in which they have become adapted to survive in this harsh environment.

Acknowledgements

The author is much indebted to Dr. P. F. HARRISON for her kind assistance in preparing the figures of specimens and soil samples. The Editor of the Zoological Journal of the Linnean Society of London has kindly consented to the reproduction here of Figs. I(b) and (c). S. I. ATALLAH, Major M. D. GALLAGHER, Lt. Col. C. J. SETON-BROWNE and Mr. S. HOWE and his colleagues of the Ahmadi Natural History and Field Studies Group, Kuwait, have all contributed invaluable specimens and soil samples from Arabia, without which this study would not have been possible.

REFERENCES

ATALLAH, S. I. 1967. A new species of Spiny Mouse *(Acomys)* from Jordan. *J. Mammal.* 48(2): 255–261.
BAKER, R. H. 1960. Mammals of the Guadiana Lava Field, Durango, Mexico. *Mich. State Univ. Biol. Ser.* 1: 305–327.
BENSON, S. B. 1933. Concealing coloration among some desert rodents of south-western United States. *Univ. Calif. Publ. Zool.* 40: 1–70.
BUXTON, P. A. 1923. Animal Life in Deserts. E. Arnold Pub. London. pps. 176.
DICE, L. R. 1947. Effectiveness of selection by Owls of Deer Mice *(Peromyscus maniculatus)* which contrast in color with their background. *Contr. Lab. Vert. Biol. Univ. Michigan Pub.* No. 34: 1–20.
DICE, L. R. & BLOSSOM, P. M. 1937. Studies of mammalian ecology in south-western North America with special reference to the colours of desert animals. *Carnegie Inst. Washington.* Pub. No. 485: 1–129.
HARRISON, D. L. 1964. The Mammals of Arabia, Vol. I. Introduction, Insectivora, Chiroptera, Primates. Benn Pub. London. pps 192.
HARRISON, D. L. 1971. Observations on some notable Arabian mammals, with the description of a new Gerbil *(Gerbillus, Rodentia: Cricetidae)*. *Mamm.* (Paris) 35: 1; 111–125.

HARRISON, D. L. 1972. The Mammals of Arabia. Vol. III. Lagomorpha, Rodentia. Benn Pub. London.

HARRISON, D. L. & SETON-BROWNE, C. J. 1969. The influence of soil colour on sub-speciation of mammals in eastern Arabia. *Zool. J. Linn. Soc.* 48: 467–470.

HEIM DE BALSAC, H. 1936. Biogeographie des mammifères et des Oiseaux de l'Afrique du Nord. *Bull. Biol. France, Suppl.* 21: 1–447.

HOESCH, W. 1956. Das Problem der Farbübereinstimmung von Körperfarbe und Untergrund. *Bonn Zool. Beitr.* 7: 59–83.

HOFFMEISTER, D. F. 1956. Mammals of the Graham (Pinaleno) Mountains, Arizona. *Amer. Midl. Nat.* 55: 257–288.

HOOPER, E. T. 1941. Mammals of the lavafields and adjoining areas in Valencia County, New Mexico. *Misc. Publ. Mus. Zool. Univ. Mich.* No. 51: 1–47.

KACHKAROV, D. N. & KOROVINE, E. P. 1942. La vie dans les désert Payot Pub. Paris. pps. 360.

MAYR, E. 1963. Animal Species and Evolution. Harvard Univ. Press Pub. pps. 797.

MEINERTZHAGEN, R. 1954. The Birds of Arabia. Oliver and Boyd Pub. Edinburgh and London. pps. 624.

NIETHAMMER, G. 1959. Die Rolle der Auslese durch Feinde bei Wüstenvögeln. *Bonn. Zool. Beitr.* 3/4: 179–197.

RANCK, G. L. 1968. The Rodents of Libya. Taxonomy, Ecology, and Zoogeographical Relationships. *Smith. Inst. Un. St. Nat. Mus. Bull.* 275: 1–264.

SUMNER, F. B. 1921. Desert and lava dwelling mice and the problem of protective coloration in mammals. *J. Mammal.* 11: 75–86.

ZAHAVI, A. & WAHRMAN, J. 1957. The cytotaxonomy, ecology and evolution of the gerbils and jirds of Israel (Rodentia: Gerbillinae) *Mammalia* (Paris) 21: 341–380.

XIV. THE BIOLOGY OF SOME DESERT-DWELLING GROUND SQUIRRELS

by

A. C. HAWBECKER

Introduction

The approximately 30 species of North American ground squirrels range from the Arctic to Mexico and from the Pacific Ocean to the eastern half of the continent and the Gulf of Mexico, but they do not naturally reach the Atlantic Ocean. Most of the species have representatives that are found in the Sonoran Desert, and a few species are found only there. This account will deal with *Citellus nelsoni*, the Sonoran form studied in detail by the writer, and other Lower Sonoran forms will be compared with that ground squirrel. Other species, though not necessarily desert species, will be referred to where pertinent. HOWELL (1938), MILLER & KELLOGG (1955) and HALL & KELSON (1959) should be consulted for an evaluation of all species.

The taxonomy of this group has been approached in different ways, but MILLER & KELLOGG (1955) will be used as the authority here. In the genus *Citellus* there are two sub-genera that are confined to, or nearly confined to, the Lower Sonoran deserts, and these will receive the most consideration. They are the following:

Sub-genus Ammospermophilus
 species – *harrisii*, Harris' Antelope Squirrel: SE Calif.; S. Ariz. to New Mex.; NW Sonora.
 species – *leucurus*, White-tailed Antelope Squirrel: SE Ore.; SW Idaho; SE & E Calif.; Nevada; Utah; W Colorado; NW New Mex.; N Ariz.; Baja Calif.
 species – *interpres*, Texas Antelope Squirrel: S Central New Mex.: W Texas; Coahuila.
 species – *insularis*, Espiritu Santo Island Antelope Squirrel: Espiritu Santo Island, Baja California only.
 species – *nelsoni*, Nelson's Antelope Squirrel: SW San Joaquin Valley, Calif. only.

Sub-genus Xerospermophilus
 species – *mohavensis*, Mohave Ground Squirrel: N end of Mohave Desert, Calif.
 species – *tereticaudus*, Round-tailed Ground Squirrel: S Nevada; SE Calif.; W Sonora; NE Baja California.

The sub-genus Ictidomys contains the species *mexicanus* of E Mexico, S Texas, and SE New Mexico, and the species *spilosoma* of central U.S. and central Mexico. The sub-genus Otospermophilus contains the species *variegatus* of SW U.S., and much of Mexico. These species are largely confined to the Sonoran desert also, but will be considered only generally here.

Fig. 1. Citellus nelsoni. From a mounted specimen by W. T. Shaw.

Fig. 2. Citellus leucurus. From a mounted specimen by W. T. Shaw.

278

Only those factors pertaining to the life cycle and the habitat relations will be considered. It is assumed, from the other titles, that hibernation, aestivation, water requirements, and response to temperature will be covered elsewhere.

Reproduction

C. nelsoni is an inhabitant of open rolling land and the sparsely shrubby, gentle slopes of southern and western San Joaquin Valley, California. The vegetational and climatic aspects of the habitat can be described as Sonoran Desert, with saltbush *(Atriplex)* and joint-fir *(Ephedra)* as the principal shrub associates. Other desert ground squirrels are found in similar habitats and associated with similar plant forms, such as greasewood *(Sarcobatus)*, mesquite *(Prosopis)*, and creosote bush *(Larrea)*. All species seem to prefer open or sparsely shrubby terrain which may be quite rocky or dissected by usually dry waterways. *C. tereticaudus* occupies probably the most arid habitat of all. In most of the habitats green, herbaceous material is available only during the January–April period, as there are few late spring, summer or fall rains, so that is when breeding occurs. Those species found in higher, cooler habitats may breed later as the annual green vegetation becomes available.

In general study of *C. nelsoni* 291 individuals were caught, examined, marked and released, and many of these have been retaken. Two hundred seventy eight have been sacrificed for internal examination. Very few of the other species have been examined.

BREEDING CYCLE OF THE MALE

TOMICH (1962), in his discussion of the male California Ground Squirrel *(C. beecheyi)*, says there is often an "elaborate preparatory period" occupying several months in rodents having a short annual breeding season, and that is the case in *C. nelsoni*. The testes of adult males reach their breeding condition early in the fall much before the female's ovaries are beginning to develop. On 11 November 1951, forty days before the start of new plant growth fully developed sperm were found in sectioned testes and epididymes. Stages in the male cycle can be seen by referring to Figure 3, and are also demonstrated by the following marked specimens. One taken 3 August had a dark, short-haired scrotum, but no testicular development. Two taken 22 August each had a small scrotum and 4 mm testes. Four taken 11 September had scrotal 6–10 mm testes and pigmentation, while two taken 27 November had 21 and 15 mm scrotal testes with the scrotum black. All do not follow this cycle as a 15 October specimen had 24 mm testes, but most males collected any November were found to have testes of 21 to 25 mm.

The annual period of fertility is a long one, but confined to the cooler

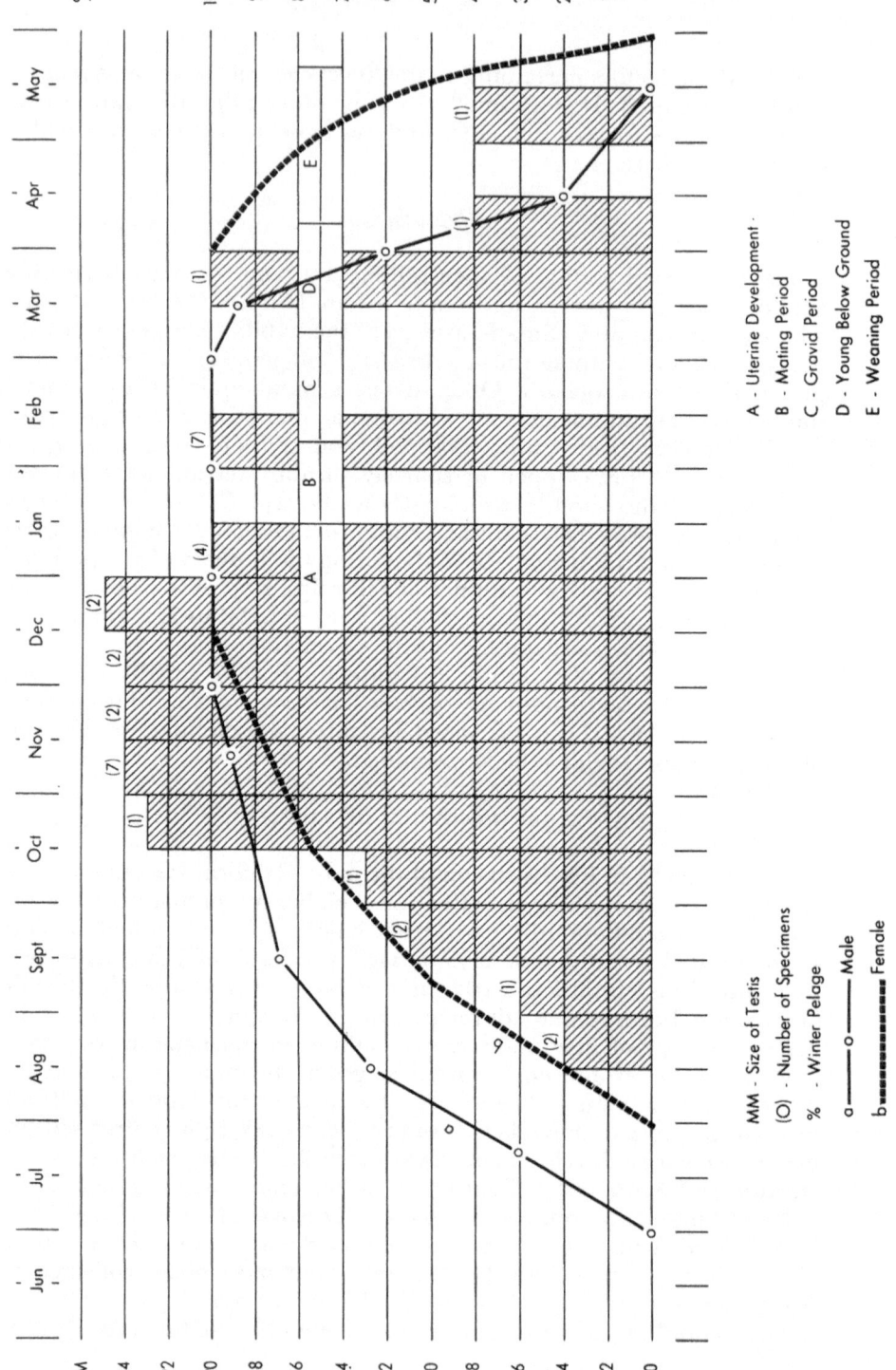

MM - Size of Testis

(O) - Number of Specimens

% - Winter Pelage

o———o Male

b▪▪▪▪▪▪▪ Female

A - Uterine Development

B - Mating Period

C - Gravid Period

D - Young Below Ground

E - Weaning Period

portions of the year. Testes 18–23 mm were found in three squirrels on 11 November, 1942, and 6 March 1943 nearly four months later, five were taken with 16–22 mm testes. Sectioning showed sperm present in both groups. Regression is obvious by the end of March and here decrease in testis size accompanies molt to summer pelgae (Figure 3). By mid-April the testes are 4–8 mm, in most cases, and this size is maintained through the summer period. Both young and marked adults taken during the summer have small, unpigmented scrota and small testes that can be moved manually from scrotum to peritoneal cavity. TAYLOR (1916) collected two males 7 and 28 May 1911 that had enlarged testes, so the breeding season may be even longer in some cases. NEAL (1965) found live spermatozoa in *C. tereticaudus* from 7 January to 8 April, and in *C. harrisii* between 4 November and 16 June.

It is impossible to correlate size and testis length. Two marked adults taken on the same day weighed 207 and 150 gms, and had 20 and 23 mm testes, respectively. All males do not become sexually active as 6 out of 7 males taken 12 November 1948 were well developed, but the seventh was entirely undeveloped. Captive males showed little development, but data of any sort from captives are unreliable. Marking shows that some of these undeveloped individuals are young of the year, but not all. Young of the year have longer, lighter hair over the scrotum than adult males. It appears that young of the year do not breed, but this needs further study.

MOLT

Before discussing the breeding cycle of the female, it is necessary to consider the apparent tie-in between molt and the sexual cycle, as the winter molt coincides with sexual development and the summer molt with regression (Figures 3 & 4). According to HOWELL (1938) all species of *Citellus* do not have 2 molts, but some achieve color change by wear.

After maturity and during the winter or breeding pelage phase the male's scrotum is gray to black and covered with very short, black hair. In the immature and summer or non-breeding pelage phase the scrotum is smaller, much less pigmented and may be covered with long, tawny hair. LINSDALE (1946), studying the California Ground Squirrel *(C. beecheyi)*, suggested that this pigmentation shields the testes from excessive heat. The critical breeding period of this squirrel, as well as other desert forms, is in the winter, so heat is not a factor. The testes are inguinal or abdominal during the heat of the summer. White-tailed ground squirrels *(C. leucurus)* may breed much later in the year, as 6 mm embryos are found as late as 4 April, and an early heat wave could be a factor here. *C. nelsoni* appears to be derived from *leucurus* so the sharing of such a color characteristic can be understood.

Winter pelage persists until the breeding season is completed for each

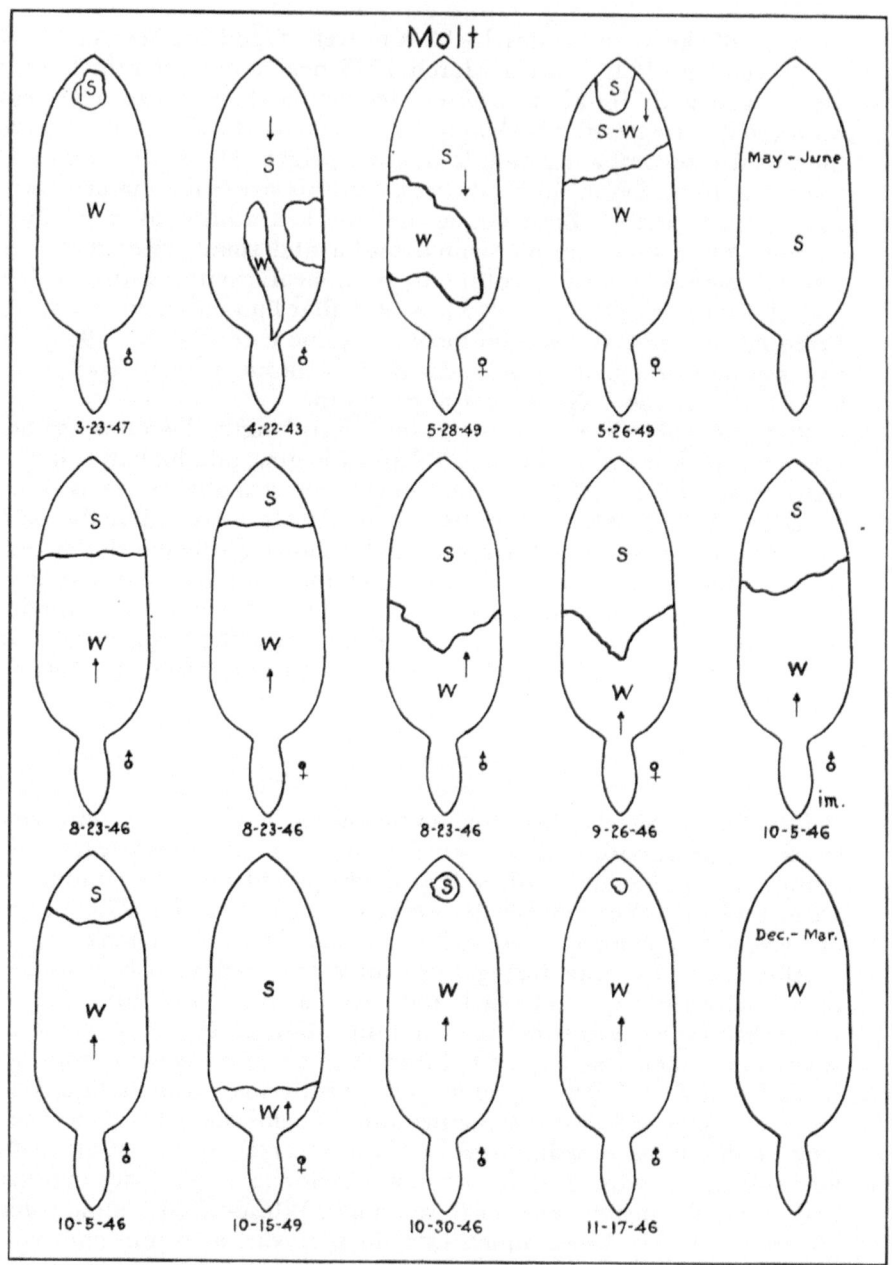

Fig. 4. Molt in *Citellus nelsoni*. This chart can be correlated with 'a' and 'b' on figure 3.

sex, and completion is not simultaneous in the sexes (Figure 4). The female's coat does not start its change until all duties pertaining to the young are completed, thus she molts about one month later than the male. Young follow the course of the female. Both sexes examined in mid February showed changes in the head region. Both sexes examined in late April showed the male almost completely changed, but the female and young either just starting or not yet started. By mid-May the young are completely changed, and the females have most of their summer coat. The summer coat can be felt coming in beneath the winter pelage.

Molting starts again in the summer as 22 August males had molt lines near the rump and enlarging testes, but the females were still in their summer pelage. By 26 September captive females had the posterior half of the body changed. An adult 5 October male had summer pelage on the head, while a young male had such pelage across the shoulders. By mid-October females are about 60 percent molted and males 90 percent (Figures 3 & 4). Summer pelage starts at the head and works back to the tail, while winter pelage starts at the rump and works to the head. Summer pelage is sparser and lighter than winter. This may allow better air circulation and light reflection, while the darker, heavier winter coat absorbs and retains heat.

Mating readiness in the female

The breeding cycle in the female is characterized by a much shorter period of maximum development of reproductive organs than in the male. The cycle, however, includes care of the young, and that extends the full period to the same length as that of the male.

Stages in the sexual change are demonstrated by the following examples (Figure 3): Marked females taken from 15 August to 15 December showed no external or internal development. A 22 December specimen showed a slightly enlarged uterus, with no external change, but another on the same date showed the exact opposite. Marked females showed no further external sign and sacrificed females showed no change internally until the last 10 days of January. Four 21 January females gave 2 with enlarged or open vulvas, but 2, apparently young, with no development. Another sacrificed on that date showed maximum uterine development, and 1 taken 2 days later had an open vulva. All examinations made in other years indicated no earlier open vulva. All examinations made in other years indicated no earlier development, so breeding readiness is postulated to begin near 20 January. All, however, are not ready this early; those that are ready do not mate at the same time; some adults do not develop in a given year; some ready to mate do not mate. Four females taken 22 February had fully developed, but not lactating, mammary glands, vulva and uterus, but gave no signs of having mated or borne young. It would appear likely that such deviations would be

283

correlated with some disturbance or food failure, but unfortunately this has not been determined. This readiness lasts until 5 March; after this date no female has been found without signs of regression. Mating activity in the female seems to be confined to the mid-January early-March interval with little activity before or after, which would give a 45 day mating period. GRINNELL & DIXON (1918) found embryos in *C. leucurus* as early as 20 February and as late as 24 April, which would indicate a longer period of mating readiness in that form. ALCORN (1940) showed that breeding might start as early as mid-February and continue until approximately mid-March in the *C. townsendii* he studied.

NEAL (1965) worked on the breeding habits of *C. tereticaudus* and *C. harrisii* and gives the following data relative to readiness. *C. tereticaudus* had mated by the last week in February, which is his earliest record for this squirrel. Data here would indicate that possibly mid-March would be the last breeding date, unless a previous brood was lost, but he cites other workers that would possibly extend this last date to the end of June. *C. harrisii* also appears to mate from the last week of February but some data indicate a January or possibly a December date. Mid-April seems to be near the last breeding date, but there are data extending this a month or possibly longer.

Gestation period

The period of gestation is unknown. The earliest mating in *nelsoni* is indicated by a 16 February female with 54 mm embryos, which is near their length at birth. No female in that area has ever been found ready to mate before 21 January, so if the 10 days of January are added to the 16 of February, it is possible to deduce a 26 day gestation period. This would approximate the 23–24 day period of the Columbian Ground Squirrel *(C. columbianus)* as determined by SHAW (1925); and the estimated 30 day period of the California Ground Squirrel *(C. beecheyi)* as determined by TOMICH (1962). NEAL (1965) deduces at least 25 days and probably between 28–35 days for *C. tereticaudus*; and around 30 days for *C. harrisii* as gestation periods for these species. ALCORN (1940) suggests a 24 day gestation period in the *C. townsendii* he studied.

As shown, breeding usually takes place in *C. nelsoni* around the latter part of January or early February. Embryos range in number from 6 to 11 with an average of 8.9. One female was examined that had 12 foetal scars. As live-trapping shows (HAWBECKER, 1958), there is a high mortality rate for each year, so the high biotic potential is needed to sustain the species. NEAL (1965) shows that *C. tereticaudus* has from 4 to 9 young with a 6.5 average per litter. He found *C. harrisii* had from 6 to 10 young with an average embryo count of 7.3. The mean number of young based on 34 records was 6.5 in the latter species, which includes all work done to his date. HOWELL (1938) reports 6–12 embryos in *C. tereticaudus*.

Fig. 5. Citellus nelsoni. From western San Joaquin Valley, California.

Fig. 6. Citellus nelsoni. From western San Joaquin Valley, California.

285

GRINNELL & DIXON (1918) give the average number of embryos for *C. leucurus* as 9 with 5 and 14 as extremes. ALCORN (1940) found *C. townsendii* averaged 8 to 10 embryos with extremes of 8 to 16. He found 1 nest that contained 17 young.

The embryos are not equally divided between the horns of the uterus, as the left horn has the greater number. This is discussed by LINSDALE (1946) in relation to the California Ground Squirrel *(C. beecheyi)*, where he found the opposite condition. He explains it on the basis that the stomach lies on the left side of the body cavity and in many cases may be filled enough to upset the ratio of embryos. This apparently is not the case here as even when resorption takes place it may do so on the right side.

Resorption is common in rodent reproduction (TOMICH, 1962), but here it seem correlated with some sort of psychological or environmental upset. Five females taken at an early date in the breeding season and kept during the period of embryonic development produced no young, though almost all specimens examined in the field were pregnant at this time. This was repeated 3 times other years without success, but females taken just before time for birth would produce young. Free females examined early in the breeding season would often have a number of embryos of uniform size, then one or more much smaller. Specimens examined just before birth of young show only full sized embryos. There is little evidence available but it is possible that resorption is due to reduction in food. The growing seasons of 1958–1961 were deficient in rainfall and during this period the annual plant growth and the numbers of antelope ground squirrels decreased greatly. Females sacrificed during the later part of this dry period showed more embryos being resorbed than in better food years. It would appear that in the wild a healthy, undisturbed female has as high as 12 embryos but a decline in the habitat may result in resorption to a point where the female can maintain herself and the remaining embryos. This problem would seem particularly serious in this desert habitat where food serves as both food and water.

Embryos must be present in *C. nelsoni* in late January but all examinations show most development is concentrated in February and early March, as embryos have been found only in that period. Captives have given birth, in most cases, from 4 to 12 March, and foetal scars are found around these dates in sacrificed females. Earlier and later dates are indicated by foetal size, but the first lactating wild females were found 4 March in 3 different years; that date is also indicated as near the birth date of most young. *C. tereticaudus* (NEAL, 1965) gave birth 9 April as the earliest known date for this species while most of the young are born between mid-March and mid-April. He records the earliest date in *C. harrisii* as 4 March with both species having but one litter. HOWELL (1938) reports young of *C. tereticaudus* are born in March and April, with possibly a second litter, but VORHIES (1945) maintains there is only one litter in this species. *C. leucurus* have been collected 4 April with 6, 9, and 12 mm embryos in different individuals. The 9 young of the captive female, *nelsoni* born on 4 March had a total weight of 38 gms, or an average of 4.88 gms at birth; two were measured at 58 mm. They were hairless and if clawed only minutely so, but were quite active and able to make mewing sounds. By 10 March they totaled 53 gms, or a gain of about 1.7 per young. Claws were present, slight pinnae discernible, and hair was darkening the skin of the head; there was less mewing but more activity. By 14 March they weighed 61.3 gms, with two weighing 6.5 and 7.3 gms. One measured 70 mm with the head much darker, and a crease where the eyes were to open. All were very active. On 16 March the tail was up over the back in characteristic fashion. Two days later the female had killed all, and partially eaten some of her young. Other females ate their young almost at birth. HAROLD LEROY WILSON (communication) of Coalinga, California, had a female that bore 8 or 9 young 7 March. By 6 April their eyes were still closed but they were well furred and bit when teased. Females he worked with also allowed their young to die.

Young *nelsoni* are found above ground about 30 days after their birth date, or from 4 April on. GRINNELL & DIXON (1918) give 23 April as the earliest date *C. leucurus* young are seen above ground, but 10 May is nearer the usual birthdate. Embryos up to 12 are found, but never more than 7 young *nelsoni* have been found following a single female. Mortality must begin even before the young come above ground. Two young males live-trapped on 12 April, approximately 8 days after emergence, weighed 40 and 42 gms, the pelage was fine and dark but not thick or adult. Three live-trapped 16 April weighed 70.6, 70.7, and 43 gms, with the last having a bad eye. The healthy ones measured 183 mm. Two young females collected 20 April weighed 86 and 103 gms, and another collected 7 May weighed 120 gms and measured 224 mm. A male examined on the same date weighed 132 gms and measured 230 mm. Both May animals had summer pelage, and were near adult length but not weight. Seven young caught 10 May were in various stages of molt around the head. Change from juvenile to adult pelage begins about the time they become independent, and usually by mid-May the adult summer coat is gained.

LACTATION AND WEANING

As mentioned earlier, lactation was first noted on 4 March. In all the years of study of *nelsoni* no young were found in March, but most of the females were lactating. Groups of females were examined 14, 20, 21, 30, and 31 March, and all were nursing young. The nipples found on each lactating female appear to take up all the available space. Seventeen females, chosen out of many, had the following nipple distribution: Eight had 7 pairs in use; 2 thoracic, 3 abdominal, and 2 inguinal. Eight had

6 pairs in use; 2 pairs in each region, and one had 7 on one side and 6 on the other – 3 nipples in the thoracic region. The large number of young produced, as shown earlier, makes this necessary. GRINNELL & DIXON (1918) note that in *C. leucurus* there are generally 5 pairs, occasionally 6, with rarely 5 on one side and 6 on the other.

The mentioned 4 April brood of 7 were nursing as they came above ground, but nursing had apparently ceased by 20 April. Females trapped 12 April may or may not be nursing. Two taken 25 April appeared to have nursed recently, but these would be the latest as well as the only ones. One female with young may be nursing, but another taken the same date with young may not. Weaning may be started or even completed before coming above ground, but is in no way consistent after emergence, and even those young seen above at the earliest date were busily gathering food. A female *C. tereticaudus* collected 6 May was lactating.

During this weaning period the behavior of the female toward her young changes, as was traced in a family group first seen 4 April. The female preferred to feed alone and would leave any young that tried to approach her. She kept contact by occasional visits or by using the calls by which family groups communicated. She would not allow them to nurse even though they nuzzled her from time to time. Another female observed on 16 April paid little attention pushing her young away when they tried to approach. She was seen to go down a different burrow from theirs when they went in for the night. On 23 April a female was seen to work slowly away from her young, then come back from time to time, but then to show little interest in them. A group watched 26 May showed this family still together, but the female did not return when feeding, or communicate; she did, however, spend the night in the same burrow. GRINNELL & DIXON (1918) state that young *C. leucurus* are on their own after they are half grown. Generally, the young are ignored by the end of April, but in actuality they appear to be independent most of the time after emergence, as the female spends most her time feeding. After the drain of caring for the young she appears to be trying to build herself up with fat before the lean, hot summer and fall months when it is too hot and dry to be too active. She is also molting to the summer pelage at this time, as when the young emerge she is quite ragged. It appears then that in this desert environment breeding duties are completed while it is still cool and while the annual growth is still partially green or there is an abundance of seed and insects.

Food Habits

Two studies, BRADLEY (1968) and HAWBECKER (1947), are available relative to the food of low desert ground squirrels. Apart from these there is a good deal of scattered information that appears to substantiate the

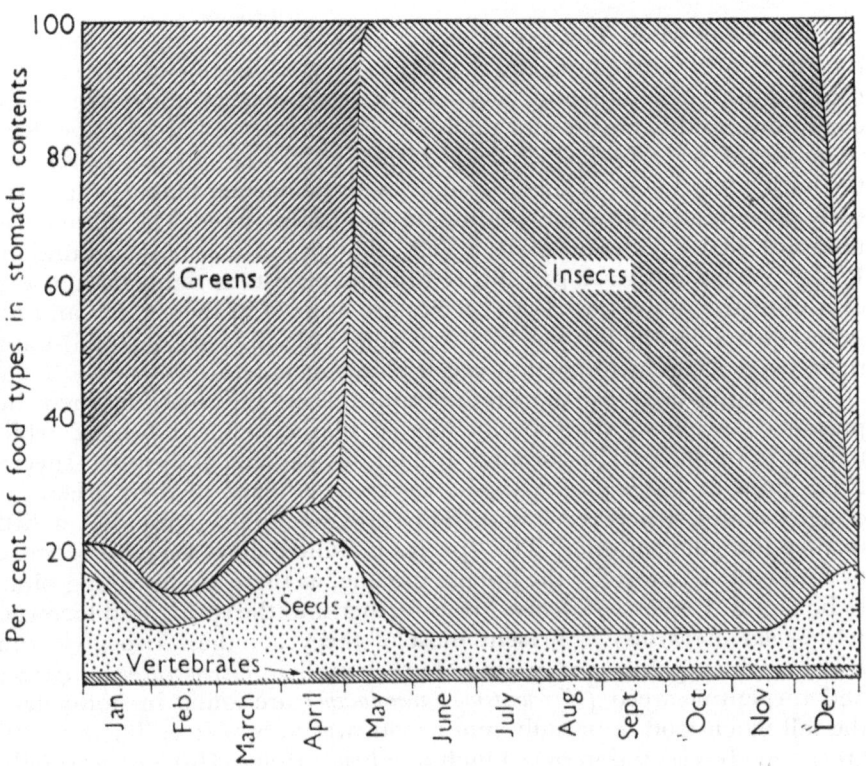

Fig. 7. Food of *Citellus nelsoni*. During years of insect failure the incidence of 'seeds' and 'insects' on the chart would be reversed. In the latter situation a chart for *Citellus leucurus* (BRADLEY, 1968) would not be unlike the above chart.

conclusions of these two sources. Figure 7 shows the occurrence of the various food types, and the approximate percentages of these types, that might be found in any one month in the stomach of *C. nelsoni*. Two hundred and twenty-five ground squirrel stomachs have been examined to date. No attempt has been made to keep exact amounts, as early studies showed that the varying amounts and types of food may depend upon the time of day, what a squirrel might happen upon, and the availability of the food type for that year or that time of year. Figure 7 shows the amounts of food types during the various times of the year. Often, however, there is a great decrease in the number of insects and seed becomes the dominant summer food with insects in a subordinate role.

GREEN MATERIAL

The green growth begins in December, or even January, and lasts until

April or early May. The extent of this growing season depends on when the first winter rains come, and when the spring rains stop and the summer heat begins. Growth is noticeably slow to start, except under the best of conditions, and begins to dry up at the first breath of hot weather. Red-stemmed filaree *(Erodium cicutarium)* and red brome *(Bromus rubens)* are the most conspicuous of these annuals, with the former making the better early growth and the latter coming on and lasting later. The filaree dries up until little evidence of it is left, though seed remains abundant. Except where cattle are present, red brome persists upright, though very dry, until the rains start. Both species are utilized in their green and dry states, especially the former, by sheep and cattle. BRADLEY (1968) found these two species as important green plants in his study.

The plants used by the ground squirrels in the order of their importance are: first, the aforementioned filaree and red brome, which appear to be of equal importance and far outstrip any of the other species. Next, joint-fir *(Ephedra californica)* seeds are utilized in their green and doughy states, wherever found. This provides some moist diet a little further into the summer. This species is limited however and seems to be limited even more by the intense grazing, as there are no young plants to be found. Clover *(Trifolium)* lasts further into the spring and locoweed *(Astrogalus oxyphysus)* is present most of the year; these two give some moisture during the warm period. Turkey mullein *(Eremocarpus setigerous)* and turpentine weed *(Trichostema lanceolatum)* are found in stomachs in the fall when food, especially moist food, was very scarce. There is little green woody vegetation over much of *nelsoni*'s range. It seems reasonable that any herbaceous or succulent green material will be utilized as long as it is available. BRADLEY (1968) found the same utilization of green vegetation during the spring, and the further utilization during the summer of what was available. It is during this period, as recorded earlier, that reproduction takes place.

INSECTS

As far as can be determined this squirrel has no preference as far as insects are concerned, but occurrence in stomachs depends on availability. Decapitated, partly eaten, unidentified Coleoptera of several species are found about squirrel frequented systems almost any time of year. The Jerusalem cricket *(Stenopelmatus longispina)* occurred twice in January taken squirrels in Ortigalita Canyon, and other remains that looked like this insect were found in stomachs in the Panoche Creek area during most of the spring months. Camel crickets *(Ceuthophilus californicus)* are often found when systems are dug out and was found in the stomachs during the spring months. These two insects seem to be the winter and early spring mainstays of the insect diet. The California June beetle *(Phyllophaga errans)* was found out as early as 18 February and in abundance on 21 March to 15 April. This was before grasshoppers had appeared and while green material was still available. This species disappears when hot weather comes, but during the stay they are found in the stomachs, the elytra are found commonly about the mouth of squirrel frequented burrows, and they are frequently seen being pursued by the squirrels. A squirrel taken on 12 April contained, among other things, a species of Carabidae. The same warm weather that killed the herbaceous annuals brought out the nymphs of the grasshopper *(Oedaleonotus enigma)* as early as 8 April with vast numbers present by 21 May. This species was present until the advent of cold weather. Live ones were seen until 18 December and a male squirrel was shot on 29 December carrying a blackened, dead grasshopper. Filarees and grass were just beginning to come on that date.

Grasshoppers or other insects last until the same rainy cold weather begins that brings on the green vegetation. After the green vegetation is gone insects are found in almost all stomachs in large numbers as soon as they are available, and are the dominant item until they are no longer to be found. Harvester ants *(Pogonomyrmex)* however, were not found in the diet until the grasshopper failure of 1946, though they may have been used. It seems that this species may be used only when other insect food, and other moist food, is very scarce. A species of Tenebrionidae, probably *Eleodes*, was present in numbers but was never found to be used even in time of extremity. Squirrels have been noticed turning over dried cow dung apparently in search of insects. BRADLEY (1968) found arthropods present anytime but grasshoppers were utilized most heavily in late spring and summer with arthropods in general important in the late summer and fall. Along with what moist vegetation might be present in the summer, insects would supply moisture to the diet (SCHMIDT-NEILSEN, 1964).

SEEDS

Seed material is found in a chewed up, pasty mass in almost all *nelsoni* stomachs every season of the year, though in some cases it forms only a small percentage of the total contents. As far as can be determined this material is mostly filaree, red brome, *Ephedra*, or any combination of these. This is evidenced by the finding of at least some of the husks of these seeds in the stomachs, the actual observation of the squirrels eating them, or the finding of them in the cheek pouches. GRINNELL & DIXON (1918) record filaree seeds in pouches on 11 May. A male collected on 16 December had red brome seeds readily identifiable in its stomach. In the Panoche area *Ephedra* seeds are taken, both green and dry. They are picked off of the shrub in the spring and picked up dry in the fall, at which time husks are commonly found around squirrel frequented systems. Several young of the year were seen gathering and gleaning among husks on 5 October when other food was scarce. Sprouted seeds from kangaroo rat *(Dipodomys)* surface caches are dug, apparently for the seed rather than for the shoot. It seems reasonable that any seeds, palatable to the squirrel, will be taken but that the most easily obtainable, usually filaree and red brome, are the most commonly used.

Seeds are apparently not the favorite food of *nelsoni*, but serve as a bridge or stop gap. The usual time that seed material reaches its highest percentage is when the more easily gathered green material or insects are absent. This happens in the spring between green material and insects and in the fall between insects and green material. *C. nelsoni* appeared to prefer both green material and insects to seeds, even when seeds were abundant. Part of this preference may be due to the water content of the favored items. BRADLEY (1968) indicates a preference by *C. leucurus* for seeds of shrubs, but *nelsoni* may have few shrubs in much of its habitat. Like *leucurus*, however, *nelsoni* is influenced by the availability in seed utilization. In both cases seed of *Ephedra* and *Bromus rubens* were extremely important in the diet. Seeds, of course, are everywhere abundant over the areas, and seem to be a necessary adjunct to the diet, but generally seem to be considered too hard to gather if something else suitable is present. Squirrels seen out during the summer of 1946 were far more active than those seen out during previous summers. This was probably due to the necessity of gathering seeds rather than the easier-to-gather grasshoppers. *C. tereticaudus* does not utilize dry seeds, according to VORHIES (1945) and does not store food. They feed on green material and reproduce when such material is most abundant but are quite inactive, during the hot months.

VERTEBRATES

In most species vertebrates are found in a small number of stomachs throughout the year. BRADLEY (1968) found the diet about equally divided between rodents and lizards, with some rodents possibly taken as carrion and lizards taken during hibernation. He found them robbing traplines and feeding on road kills. The same can be said for *C. nelsoni*. Both species will feed on their own kind.

NEED OF VARIOUS FOOD TYPES

In considering the types of food, it appears that certain foods are necessary to the well-being of the squirrel and are sought even though another food may be abundant. It will be noted that insects, flesh, and seeds were sought even when green material was very abundant. During the summers that grasshoppers were absent and the squirrels were forced to subsist mostly upon seeds, almost the only insects present in numbers were harvester ants *(Pogonomyrmex)*, which were then found in the diet.

This seems to indicate a need for such material, possibly for the moisture present. It will also be noted that seed material was taken at all times of the year, even when grasshoppers or green material were very abuns dant. This was most noticeable during the hot weather when grasshopper- were in the shade of every blade of grass, and the cooler foraging time was very limited.

The following has not been substantiated by actual weighing, but by observation alone: Adult squirrels kept in captivity lost weight noticeably when kept on a seed and greens diet, though it is possible there was some other reason for the loss. Young squirrels fed a well rounded diet appeared sleek and well fed, but others did not develop as well on a seed-only diet. Captive young squirrels taken in the spring and fed on a seed diet, did not compare favorably in size with those of the same year that were left to grow in the field and collected in the fall. Captives never looked quite as good, or weighed as much, as squirrels of approximately the same age that lived in the wild, however. It seems reasonable that certain, as yet unknown, food components are furnished by each food type, making each type necessary to the diet over and above more abundant food.

It appears that these squirrels are omnivorous and will eat whatever is available, and particularly what is available in large quantities. Some items appear to be necessary to the diet, and moisture-producing materials seem particularly necessary in the summer. Data gathered by HOWELL (1938) would indicate the food habits as described above are quite common in most species of ground squirrels.

FORAGING HABITS

The antelope ground squirrel does not use the same method of obtaining food as does the California ground squirrel *(Citellus beecheyi)*. The latter has a particular hole that serves as a base of operations and, though it may use others for temporary refuge, it always returns to the home burrow, and always returns to it at once if possible. Rather than use one burrow or system as a base, the antelope ground squirrel appears to have a route that it covers, and along this route are a number of holes that it feeds around, and that it runs into if danger threatens. So far all these appear to be of kangaroo rat origin as described later.

These squirrels do not stay long in one place, but move rapidly from refuge to refuge peering about and sampling. Everything is put to the mouth, chipmunk fashion, where it appears it is smelled or tasted, and accepted or discarded. Sometimes actual grazing takes place, where the food is sampled and then when found to be all right, the material is bitten off directly. When the cheek pouches are filled the squirrels retire to holes along the route and eat what they have gathered. A non-breeding female and a non-breeding male, collected on 16 April, had well-filled cheek pouches of 1.0 and 4.1 grams respectively, and empty or nearly empty stomachs which would indicate that they did not eat while foraging. This seed, filaree, had the outer husk removed indicating that this was done en route. Other squirrels have been collected with cheek pouches well-filled with filaree seed. No sign of storage has been found and none of the above had young, so it appears that they gathered the food and then ate it in a safe place. On 5 October bunches of turpentine weed branchlets were found partly dragged into a burrow. These may have been brought to a safer place to eat. Insect, green material, and red brome seeds have seldom been found in pouches, making it appear that they may be eaten at once or carried to the retreat. The cheek pouches are extremely thin, thin enough that the contents can readily be told without taking them out, and so thin that if empty they are readily missed in skinning. Thus it is easy to see that insects or long, sharp seeds might not be readily carried in such pouches.

On 26 May a group of three young squirrels, well grown by that time, were seen using grasshoppers for food, but were discarding the large legs in a heap about the mouth of their system. On 16 April a squirrel was seen to catch a June beetle and run into a burrow with it. These instances indicate such things are eaten at a burrow. They must be eaten as no sign of storage of animal material has been found.

After gleaning for a while, the length of time depending upon success or type of food, the squirrel will go down a hole and stay some time before starting out again.

The speed with which they forage and whether they eat on the spot may depend somewhat upon the weather. Pleasant days are spent out of the burrow feeding and apparently moving over quite a large area. Extremely hot and cold days are spent in the burrows with only enough time spent above ground to get food, and that during the mildest temperature of the daylight hours.

When green *Ephedra* seeds are gathered, the squirrels readily climb into the shrubs; indeed it was this action that first called attention to their using its seed. Caged specimens show great agility in climbing about their cage, even walking across the tops upside down clinging with their claws to the hardware cloth. They have been seen to catch grasshoppers by leaping with the insect until both came down at the same place at the same time. It seems probable that the squirrels must also stalk grasshoppers but it has never been observed. A June beetle was caught by a squirrel which ran along on its hind legs and pawed the air until it hit the low-flying insect. Another was caught by a sudden pounce. Caged young squirrels adeptly pounce upon grasshoppers put into their cage.

Population Characteristics

SURVIVAL

Survival in *C. nelsoni* is limited largely to the year in which the individual was taken. Two hundred forty squirrels were trapped and marked between November 1947 and December 1954, and of these 193 were not taken again later than 9 months after their first capture. This indicates that 80 percent do not survive from one year to the next. In a specific case of 49, mostly young, marked 16–18 June 1953, only 15 or slightly more than 30 percent were retaken another year. Many other available samples could confirm this, and it is apparent that the critical period is the hot summer, i.e., they were trapped in the late spring or early summer, but by fall or winter had disappeared. Survival through the summer is clearly critical with squirrels of all ages, not only the young. Adults that had survived one summer and marked during the winter appeared to run the same chance of survival as the young. Of 55 such marked during the winters of 1947–1951, only 10 were picked up after the following summer, as several repeats may be made in the winter, but all track is lost during or after the following summer. Twenty-nine were caught after one complete year, but only 12 of these after 2 complete years, which indicates a mortality of 60 percent in adult squirrels with established territories.

There is evidence that a superabundance of food may aid in survival. This, of course, reduces the amount of time necessary to get food whether winter or summer, and the time out and vulnerable to predators. The winter season of 1951–1952 was wet and warm, a combination that produces a heavy annual growth the following spring. Trapping in the fall of 1952 and early 1953 demonstrated that where only 8 squirrels per day were caught previously, this year 18 were caught. In one month 70 were trapped, far more than any comparable effort had produced. The spring of 1953 showed a much larger number of young than any previous spring. The following seasons did not produce such growth and in 1954 and 1955 conditions and survival of young were back to the pre-1952 level.

AGE AND SEX RATIO

Fifty-five squirrels have been trapped and marked that might have been recovered up to at least five years later, but only three have lasted that long. Two kept in captivity were taken as young in September, 1946, and were found unaccountably dead during the fall of 1952. The five-year-olds had not been taken for at least a year interval. This shows that although five to six years may be the maximum age, very few live that long. There is no difference in survival between the sexes.

If the system employed in this study gives a true picture, mortality is very high. Two hundred and forty animals were marked up to December, 1954, and of these 193 were not taken after the same season in which they were marked. Twenty-nine were taken one year after marking, none three years after marking, three after four years and three after five

295

years or more. Reasons for this mortality have already been discussed.

Of a total of 272 squirrels marked, 116 (approximately 43 percent) were females. This figure is questionable, however, as trapping records show that if both sexes were taken at one site, almost invariably the male was taken first. This has also been the experience of HAROLD LeRoy WILSON, of the California Department of Agriculture, who has trapped many of these squirrels. This may result in fewer females being caught or in their actually being driven away from the trap. It appears that the sex ratio is even.

HOME RANGE

This is interpreted in many different ways by different workers (HAYNE, 1949); here it is considered as that area enclosed by points of capture. After trapping and merely observing these animals for the last 30 years, this seems to be as good a method as any. More important, however, are the areas in that range where trapping seems to indicate that activity is centered.

In an earlier study (HAWBECKER, 1947) a single squirrel, sex unknown, was watched for $3\frac{1}{2}$ hours. During this time it made a circuit of 1,250 feet. From one end of this plot to the other was actually 400 feet. The spot to which the squirrel finally returned suggested that this was just one radius of a range that could be assumed to be 800 feet in diameter, and about 11 acres in all. There is evidence that the squirrel covers but one half of its range per day, or less. Squirrels are seldom trapped in the same place the following day, but usually are the day after that. BRADLEY (1968) found a range of 14.9 in C. leucurus with a 4 acre daily range.

By the end of 1954, 240 squirrels were trapped. Of these, 117 were taken but once, 58 twice, 29 three times, 17 four times, 4 five times, 4 six times, 3 seven times, 1 each, eight, nine, and ten times, 2 eleven times, and 1 each twelve, thirteen and sixteen times. Some of the retakes were spread over several years while in other cases only from a few days to one year were covered. Length of time, however, seemed to be of little significance.

Captures of one male were spread over one complete year but were contained in an area of 600 by 800 feet, or roughly 11 acres. The first two captures, on the same day, were at about the same place. The third capture, one week later, was 600 feet from the first, and then the fourth, on the same day, was back at the original site. It appears that this spot was a preferred area, or burrow, and when disturbed the squirrel returned. A week later, the squirrel was trapped about 400 feet north of this spot and all captures for the next few months were centered here. Six months later the center of interest was moved back near the first site and captures for the next four months were made there. The last capture, 3 months later, was back in the northern area. It is seen that the squirrel was always in the same general area, but shifted around in and ranged over it, somewhat. No unusual movement was noted during the breeding period.

There have been variations on this behavior. A male first taken in January, 1948, was not retaken for $5\frac{1}{2}$ years, when it was caught three times 2,200' NE which is unusual.

Twenty-nine adult males were caught a varying number of times over one complete year. Twenty-two of these showed no movement outside the 11-acre range. Seven ranged much further. Eight male adults were recaught up to five years later and, of these, five were always recaught in the same range, but three had moved. Fourteen immature males showed no movement in one year, while four did; and three immature males picked up after the first year had not moved.

Females show much the same behavior as males. An adult was taken six times from November, 1951, to March, 1954, and was never found more than 300 feet away. Another female taken the same number of times, over a period of four years, was never taken outside a 400-foot radius.

Habitat Factors

HABITAT FACTORS INFLUENCING *C. nelsoni*

The study of food habits indicates that red-stemmed filaree *(Erodium cicutarium)* and red brome *(Bromus rubens)* are the most important items in the diet of the squirrel. There is much evidence that the squirrel is very omnivorous, as vegetable and animal food of many types are taken, which might indicate that lack of a particular foot item, or even group of items, would not prevent its existence or spread. Certainly, its main food items extend much further than does the squirrel. Thus there is little evidence that the squirrel is prevented from extending its range by the lack of suitable food.

A second factor possibly involved is the apparent dependence of the squirrel on widely spaced shrubby vegetation. A study of the squirrel throughout its range during the breeding season shows that the burrows most used, whether original breeding burrows or not, by adult females and their newly-emergent young, are burrows under some shrub, such as the *Ephedra* or desert saltbush. Much brushless land was investigated during the breeding season, but squirrels were found in only a few places, and then seldom in a breeding condition. Squirrels are occasionally found in such places, but they may be the more 'venturesome' ones found along the margins of any population. South of Panoche, where the hills are more commonly covered with desert saltbush, squirrels with young were relatively easy to find. This shrub is found to extend north and west almost as far as the present records for the squirrel. Local patches of

Fig. 8. Habitat of *Citellus nelsoni*. From western San Joaquin Valley, California.

Fig. 9. Habitat of *Citellus nelsoni*. From western San Joaquin Valley, California.

298

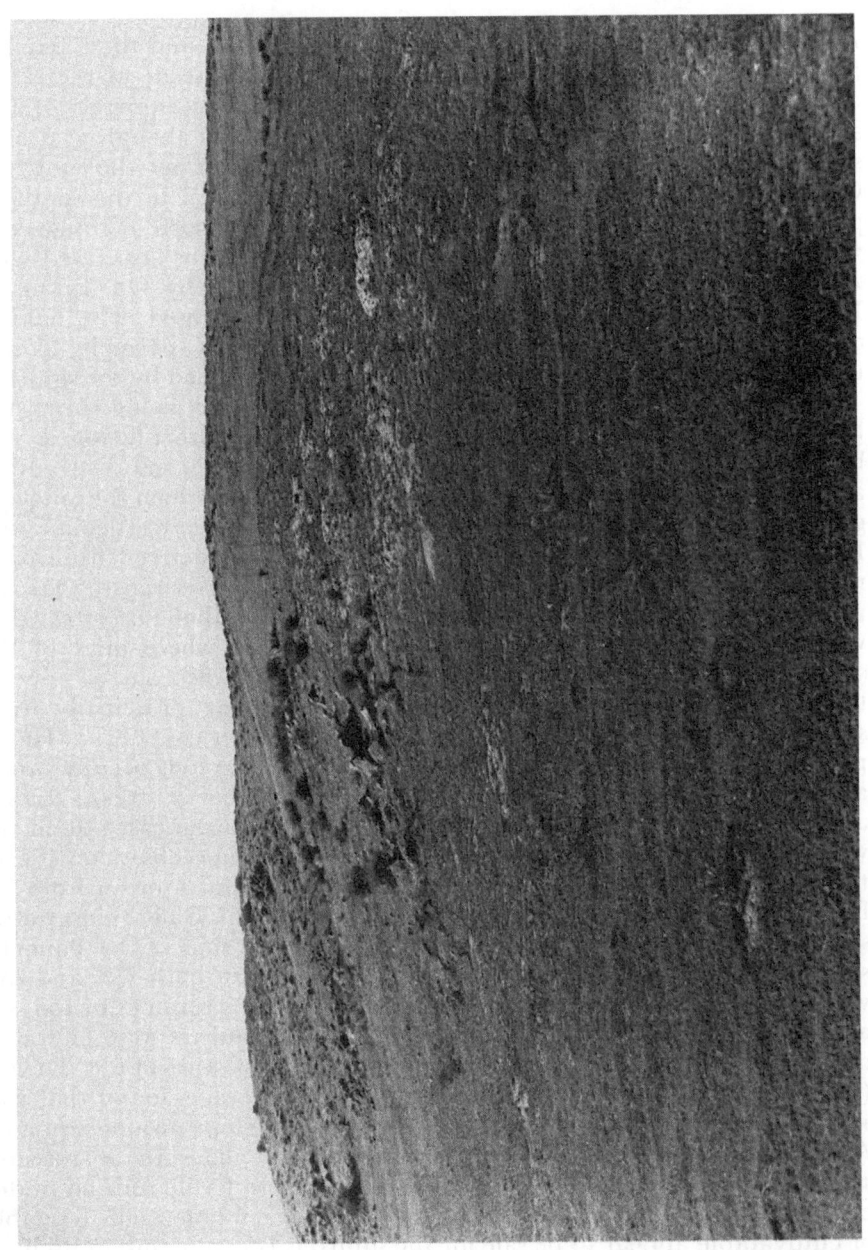

Fig. 10. Habitat of *Citellus nelsoni*. From western San Joaquin Valley, California.

Ephedra harbor squirrels where the saltbush is absent, and these patches account for many of those records for the squirrel outside of the range of the saltbush. Other factors must be emphasized. Squirrels of this species are not normally found on steep slopes even though scattered shrubs are present; also, close spacing of shrubs keeps out the squirrels. GRINNELL & DIXON (1918), who studied the squirrel in the southern part of its range, always associated it with desert saltbush or rough, dry washes. In the more northerly part of its range *nelsoni* is associated with *Ephedra* in addition to the two habitats mentioned by GRINNELL & DIXON. The few rock outcrops present may also be used. The habitat of the spiny saltbush *(Atriplex spinifera)* is invaded but sparingly, as it is usually found in alkali and such soils appear to be avoided by the squirrel.

A third major factor that must be considered in discussing the distribution of this squirrel is soil type. This may be an indirect factor, as will be pointed out later, but, even so, the relation between soil distribution and range of the squirrel is apparent. The soil types on which the antelope ground squirrel is found almost exclusively, are Panoche, Kettleman, and Fresno loams and sandy loams, in that order. Soils derived from sedimentary rocks are favored, and of these the Panoche series is by far the most acceptable. Nowhere is the squirrel found, other than just a straggler, where alkali is abundant. Panoche loam, the soil where most of the squirrels are found, is 'derived from recent alluvial deposits of gray or light-gray to light brownish gray color and coming principally from sedimentary rocks. The soils are low in organic-matter content . . . Hardpan or dense subsoils are not typically present. These soils occupy broad alluvial fans and alluvial valley slopes and small areas of stream terrace and stream bottom' (HOLMES *et al.*, 1919). Soils on the east side of the valley are of organic content. Whether there is some characteristic of these types of soils that prevents the antelope ground squirrel from inhabiting them is not known. It does not prevent the California ground squirrel *(Citellus beecheyi)* from living in them. The soils of the Panoche series are usually well drained and are 'diggable' in both wet and dry weather – that is, they do not become too hard in the summer or too wet and sticky in the winter. The California ground squirrel may be found on the same series, but is in no way limited by it. GRINNELL & DIXON (1918), in their study in the southern part of the range, found that this soil type seemed essential to the habitat of the antelope ground squirrel. As these squirrels are found in the better soils – alkali-free loams and sandy loams – they are being crowded out as these fertile soils go under cultivation. Only those parts of the range that are not potentially irrigable or cultivatable appear to be safe for the squirrel. It can maintain itself on pasture land, but not, as can the California ground squirrel, on irrigated or cultivated land.

This affinity for alluvial and loamy soils may not be due to the digging habits of the squirrel itself. Repeated excavation of burrows in which

300

squirrels were noted revealed that seldom, if ever, did the squirrel occupy burrows dug by itself. All evidence of construction and appearance indicated that the antelope ground squirrel utilized burrows dug by kangaroo rats. Thus, one of the factors involved in the spread of the squirrel might be the spread of other animals that the squirrel depends on for burrows. A study of the range of the Tulare kangaroo rat *(Dipodomys heermanni tularensis)* by GRINNELL (1922) shows that the range of this animal not only includes, but also extends beyond the range of the squirrel. The range of the giant kangaroo rat *(Dipodomys ingens)* as delineated by GRINNELL *(op. cit.)*, is included within the range of the squirrel but is not as extensive. It appears, however, that even though the burrows of these two species are used, the squirrel is not entirely limited by them, but by some other factor or factors.

A fourth factor, obviously the one upon which the others depend, that may affect the ground squirrel directly is climate. A study of two climatic constituents, heat and rainfall, using data from the Weather Bureau publications, shows that there is a gradual decrease in mean annual temperature and increase in mean annual rainfall as the north-western end of the range is approached. Special note should be taken of the abrupt change between Los Banos, Merced County, and Newman, Stanislaus County. Here, within 20 miles, there is a change of 2.13 inches in rainfall. This is taken from cumulative records, and the variation may be much more, or much less, from year to year. There is here, then, a change from desert to non-desert conditions. PIEMEISEL & LAWSON (1937) attribute this sudden change to the influence of the break in the Coast Range at the level of San Francisco Bay. They suggest that this change has inhibited the northern spread of desert and spiny saltbush. This increase in rainfall, and possible increase in humidity, may also be responsible for the limitation of the range of the ground squirrel as nowhere is the squirrel found where the rainfall reaches nine inches. Significantly, the ranges of the desert species of plants and many of the associated animals, such as the giant kangaroo rat and *Dipodomys nitratoides*, also lie within this boundary. Thus, this ground squirrel may be limited to low desert conditions. There appears to be less of a tie-in with temperature, as adjacent areas of like temperatures have not been invaded. The nearest area of like precipitation is across the Sierra Nevada-Tehachapi chain, where the white-tailed antelope ground squirrel *(C. leucurus)* is found.

The California ground squirrel *(C. beecheyi)* appears to be a competitor of the antelope ground squirrel under marginal conditions. GRINNELL & DIXON (1918) show that they appear almost side by side and yet occupy slightly different habitats. Observations have suggested both possibilities. Instances of close contact between the two species are quite common. In the Little Panoche and Ortigalita study area, both species could be collected from the same station by shooting. There were locations within

301

these areas, however, where only a single species could be found. Work here later showed that in one location, formerly occupied by both, only antelope ground squirrels were found, and in another only California ground squirrels. These locations are considered, on the basis of comparison with locations where the most successful colonies are found, as marginal, and this may account for the finding of a species one time and not another. The most successful colony studied, to date, is the one in the Panoche area. Here no California ground squirrels were seen until 1948. Two of this species established themselves near the western end of the colony and were seen until 1950, when another was seen near the middle of the colony, but in 1951 no California ground squirrels could be found. There have been few times when it has been possible to observe the two species together, yet it seems that some factor other than size must be at work. If size were all, the larger, seemingly more adaptive California ground squirrel would have 'flooded out' the smaller antelope ground squirrel.

REFERENCES

ALCORN, J. A. 1940. Life history notes on the Piute ground squirrel. *J. Mamm.*, 21: 160–170.

BRADLEY, W. G. 1968. Food habits of the antelope ground squirrel in southern Nevada. *J. Mamm.* 49: 14–21.

GRINNELL, J. 1922. A geographical study of the kangaroo rats of California. *Univ. Calif. Publ. Zool.*, 24: 1–124.

GRINNELL, J. & DIXON, J. 1918. Natural History of the ground squirrels of California. *Monthly Bull. State Com. Horticulture (California)*, 7: 597–708.

HALL, E. R. & KELSON, K. R. 1959. The mammals of North America. Ronald Press, New York, 1: xxx + 546 + 79.

HAWBECKER, A. C. 1947. Food and moisture requirements of the Nelson antelope ground squirrel. *J. Mamm.*, 28: 115–125.

HAWBECKER, A. C. 1959. Survival and home range in the Nelson antelope ground squirrel. *J. Mamm.*, 39: 207–215.

HAYNE, D. W. 1949. Calculation of size of home range. *J. Mamm.*, 30: 1–18.

HOLMES, L. C., ECKMANN, E. C., NELSON, J. W. & GUERNSEY, J. E. 1919. Reconnaissance soil survey of the middle San Joaquin Valley. Gov't Printing Office, Washington, D. C.

HOWELL, A. H. 1938. Revision of the North American ground squirrels, with a classification of the North American Sciuridae. *N. Amer. Fauna*, 56: 1–256.

LINSDALE, J. M. 1946. The California ground squirrel. Univ. Calif. Press, Berkeley, Calif. xi + 475.

MILLER, G. S. & KELLOGG, R. 1955. List of North American Recent mammals. *U.S. Nat. Mus. Pub.* 205, xii + 954.

NEAL, B. J. 1965. Reproductive habits of round-tailed and Harris antelope ground squirrels. *J. Mamm.*, 46: 200–206.

PIEMEISEL, R. L. & LAWSON, F. R. 1937. Types of vegetation in the San Juaquin Valley and their relation to the beet leaf-hopper. U.S. Dept. Agric., Tech. Bul. No. 557. Washington, D.C.

SCHMIDT-NIELSEN, K. 1964. Desert animals. Oxford Univ. Press, New York and Oxford, 1–277.

SHAW, W. T. 1925. Breeding and development of the Columbian ground squirrel. *J. Mamm.*, 6: 106–113.

TAYLOR, W. P. 1916. A new spermophile from the San Joaquin Valley, California with notes on *Ammospermophilus nelsoni nelsoni* Merriam. *Univ. Calif. Publ. Zool.*, 17: 15–20.

TOMICH, P. Q. 1962. The annual cycle of the California ground squirrel *Citellus beecheyi*. *Univ. Calif. Publ. Zool.*, 65: 213–282.

VORHIES, C. T. 1945. Water requirements of desert animals in the Southwest. Ariz. Agr. Exp. Sta. Tech. Bull. 107.

XV. REPRODUCTIVE BIOLOGY OF
NORTH AMERICAN DESERT RODENTS*

by

H. DUANE SMITH & CLIVE D. JORGENSEN

Introduction

The desert environments, as pointed out in previous discussions, have many unique characteristics, but they are also extremely diverse on seasonal, biological, and daily bases. Desert inhabitants, potential inhabitants, and/or invaders have had to confront these unique characteristics and this diversity, and solve their many problems with either morphological, physiological or behavioral adaptation to survive. Since survival is dependent upon this adaptation and the species concomitant successful reproduction, knowledge of reproductive biology is important for a complete understanding of the rodents that occupy North American desert habitats.

Although many writers have discussed various aspects of reproductive biology for certain species of North American Rodents, their reports have not been comprehensive; thus, our knowledge of this diverse group of rodents is generally incomplete and what is available is poorly known. Information on reproduction for most species is rather superficial and even in the few detailed studies, much is yet to be reported. Also, most of the studies were restricted to the laboratories where inferences were then made concerning the natural environment. This fragmentation and general superficiality coupled with the seeming variability in the reported reproductive cycles and behavior negates making many broad generalizations and necessitates a somewhat encyclopedic approach to its review. In this regard, each species group (genus) or species where information is sufficiently well known will be discussed separately.

The literature most pertinent to the reproductive biology of the North American Rodents, as far as we could determine, is summarized in this report. Also, an effort has been made to relate the reproductive phenomena (Table 1) to possible regulatory factors in the environment. Where appropriate, certain species will be summarized in more detail; whereas, others will be excluded or not discussed as completely – particularly if the species are from marginal desert environments or the reproductive biology has not been reported. Thus, this review may be considered as substantially complete and should reflect the emphasis of past research

* The literature search for this review was finished Dec. 21, 1971.

Table 1. Summary of the information available concerning the reproductive biology of North American desert rodents.

Species	Age at Puberty (Days)	Breeding Season	Sexual Cycle[a] (Type)	Gestation Period (Days)	Litter	
					Size	No./Year
Ammospermophilus interpres	—	Feb	PE	—	5–14	1–2
Ammospermophilus leucurus	—	Feb–July	PE	—	5–11	1–2
Baiomys taylori	40	All Year	PE	20	1–5	10
Cynomys spp.	365	Feb–June	—	28–32	2–10	—
Dipodomys deserti	—	Feb–Apr	—	29–32	3–5	—
Dipodomys merriami	—	Jan–July	PE	17–23	1–5	1–2
Dipodomys microps	143	Feb–May	PE	—	1–4	1–2
Dipodomys ordii	90	Jan–June	PE	29	1–6	1–2
Dipodomys spectabilis	—	Jan–Aug	PE	27	1–3	—
Lagurus curtatus	—	Apr–Aug	PE	—	3–6	1–up
Liomys pictus	—	Apr–July	—	21–28	2–5	—
Microdipodops pallidus	—	Mar–Sept	PE	—	1–6	1–2
Microdipodops megacephalus	—	Mar–Nov	PE	—	2–7	1–2
Neotoma albigula	180	Jan–Sept	PE	38	1–3	2–up
Neotoma lepida	60–90	Apr–July	PE	30–36	1–5	1–2
Onychomys leucogaster	89–91	All Year	PE	32–47	2–6	2–up
Onychomys torridus	63	Jan–Sept	PE	26–35	1–7	2–3
Perognathus amplus	—	Mar–June	PE	—	1–5	1
Perognathus baileyi	—	Jan–May	—	—	2–4	1
Perognathus flavus	—	Mar–Aug	PE	—	2–6	2–up

Table 1. (Continued)

Species	Age at Puberty (Days)	Breeding Season	Sexual Cycle[a] (Type)	Gestation Period (Days)	Litter	
					Size	No./Year
Perognathus formosus	—	Mar–July	—	—	2– 6	1
Perognathus intermedius	—	Feb–June	—	—	3– 6	1
Perognathus nelsoni	—	Feb–June	—	28–32	2– 4	1
Perognathus longimembris	—	Mar–June	PE	—	2– 8	1
Perognathus merriami	—	Apr–Nov	PE	—	3– 6	2–up
Perognathus parvus	—	Apr–Aug	PE	—	3– 8	1–2
Perognathus penicillatus	—	Feb–Aug	PE	—	2– 6	—
Peromyscus crinitus	—	Apr–Aug	—	30	3– 5	—
Peromyscus eremicus	35–55	Jan–Oct	PE	21	1– 5	2–up
Peromyscus maniculatus	35–55	All Year	PE	22–35	1– 9	2–11
Peromyscus truei	35	Apr–Aug	PE	25–40	3– 5	2–7
Reithrodontomys megalotis	100	All Year	ME	21–24	1– 7	1
Spermophilus harrisii	—	Dec–June		30	5– 9	—
Spermophilus mexicanus	—	Mar		—	—	2
Spermophilus spilosoma	—	Feb–July	ME	—	2– 7	1
Spermophilus tereticaudus	133	Jan–Apr		25–35	6–12	—
Spermophilus townsendii	—	Mar	PE	20–27	5–15	2–5
Thomomys bottae	—	All Year		19	2–11	—
Thomomys talpoides	—	—		—	4– 6	—

[a] PE refers to polyestrous, and ME refers to monestrous.

on reproductive biology of North American Rodents. The species which are included in this report are:

Ammospermophilus interpres	Perognathus baileyi
Ammospermophilus leucurus	Perognathus flavus
Baiomys taylori	Perognathus formosus
Cynomys gunnisoni	Perognathus intermedius
Cynomys leucurus	Perognathus nelsoni
Cynomys ludovicianus	Perognathus longimembris
Dipodomys deserti	Perognathus merriami
Dipodomys merriami	Perognathus parvus
Dipodomys microps	Perognathus penicillatus
Dipodomys ordii	Peromyscus crinitus
Dipodomys spectabilis	Peromyscus eremicus
Lagurus curtatus	Peromyscus maniculatus
Liomys pictus	Peromyscus truei
Microdipodops pallidus	Reithrodontomys megalotis
Microdipodops megacephalus	Spermophilus harrisii
Neotoma albigula	Spermophilus mexicanus
Neotoma lepida	Spermophilus spilosoma
Onychomys leucogaster	Spermophilus townsendii
Onychomys torridus	Spermophilus tereticaudus
Perognathus amplus	Thomomys bottae
	Thomomys talpoides

REPRODUCTIVE BIOLOGY: SPECIES SUMMARIES

Ammospermophilus interpres and A. leucurus

The Antelope Ground Squirrels, *Ammospermophilus* spp., are abundant in both hot and cold deserts of North America, but little detail has been reported concerning their reproductive biology. The Texas species, *A. interpres*, begin breeding in Feb., and after an unspecified gestation period gives birth to litters of from 5–14 young (DAVIS, 1966). The young are born in an underground nest approximately 5 in. wide, 7 in. long, and 4 in. high; lined with rabbit fur, shredded bark, feathers, dried grass, and bits of cotton (when available). The nest is usually located midway between entrances to the burrow system. Most females have one litter per year but occasionally a female may have two (DAVIS, 1966).

Ammonspermophilus leucurus begins breeding in Feb. (HALL, 1946 and BRADLEY, 1965) and continues through June, with peak activity during Feb. and March (BRADLEY, 1965). BRADLEY (1965) found females lactating in early March, in Las Vegas Valley, Nevada, with a peak occurring in May; and although the gestation period has not been reported, it very likely approximates 30–35 days. The earliest date that young were captured by BRADLEY (1965) was May 2, and this continued through Aug. 20. Since all squirrels are classified as adults by Sept., the maturation rate must be rather rapid, particularly if females can have two litters with the second being born in June. HALL (1946) reported litter size ranges from 5–11 with a mean of 7.8 and also that the reproductive peak activity may lag at higher elevations.

Longevity in captivity exceeds 48 months, but no such longevity has been reported in nature (BRADLEY, 1965). More than 80% of the ground squirrels that are captured first during the summer months do not survive until the following spring, indicating either a high natural mortality or a high rate of emigration. Eighty-five percent of the marked animals are not recovered after 200 days. BRADLEY (1965) noted that highest population densities were reached during July, Aug., and Sept. when 90 animals per square mile were estimated, mainly due to influx of the young. Populations were lowest during the prime reproductive period where the density was calculated to be 16.5 individuals per square mile.

Baiomys taylori

The pygmy mouse, Baiomys taylori, is a marginal desert dweller found only in grass-covered, hot desert areas. It breeds all year long in favorable weather conditions, and in captivity one female had nine litters in 195 days (DAVIS, 1966). The gestation period is approximately 20 days and the litter size varies from 1–5 with a mean of three. Young are born in surface nests which are usually depressions made in the ground and filled with dry grass – usually under a log or a rock overhang. At birth the young are naked, blind, virtually helpless and weigh about one gram. Their eyes open in 12–15 days, and they are weaned or at least seek their own food at 18–22 days of age. Sexual maturity reportedly occurs in females at 40 days (DAVIS, 1966) but 70 days are required for males (WALKER, 1964). Both parents aid in caring for the young.

Cynomys gunnisoni, C. leucurus, and C. ludovicianus

Prairie dogs, Cynomys, as inferred by their common name are not primarily associated with the desert, but C. ludovicianus and C. gunnisoni occupy hot, grass-covered areas much the same as does Baiomys; and C. leucurus is found in the eastern part of the northern desert (HALL & KELSON, 1959). They are colonial but each sub-group maintains its own nesting burrow within the system. The breeding season begins in Feb. and continues into July, but with a gestation period of 28–32 days most of the young are born in March and April, although some are born in Aug. (COCKRUM, 1962). The litters, usually one per year, range in size from 2–10. The young are born blind, naked, and virtually helpless as are most species, but maturation seems to be somewhat slower in this genus than in many of the others. The eyes do not open until 33–37 days of age, and weaning does not occur until 40–43 days of age. Sexual maturity occurs at approximately one year of age, but they do not reach full growth until about 15 months. At sexual maturity, the males usually leave the colonial sub-group but the females remain.

309

Dipodomys deserti, D. merriami, D. microps, D. ordii, and *D. spectabilis*

Kangaroo rats, *Dipodomys*, are one of the most common rodents of the North American deserts, and as such one would think they would be the most studied. Although considerable work has been done, beginning with VORHIES & TAYLOR (1922), there is still much lacking from our knowledge of their reproductive biology, particularly in nature.

Although the various species of *Dipodomys* often occupy rather different habitats and geographic distributions, they do have some similarities in their reproductive biologies. Breeding activity sometimes occurs all year round, but with two peaks – one in the spring (around May) and the other in the later summer (July–Sept.) (REYNOLDS, 1958).

Dipodomys deserti breeds almost all year round, but most of the young are born in April or May after a 29–32 day gestation period (BUTTERWORTH, 1961a). Litter sizes range from 2–5 with a mean of four (BUTTERWORTH, 1961a and HALL, 1946). The young are born blind, naked, unpigmented, and completely dependent on their parents. Pigment appears by day five (BUTTERWORTH, 1961b), but no record is available for visible hair. The eyes open between days 15 and 17 (CHEW & BUTTERWORTH, 1959) and the young begin pouching seeds at day 21 (CHEW & BUTTERWORTH, 1964). BRATTSTROM (1960) reported a longevity of five years, five months for a *D. deserti* captured as a subadult and retained in the laboratory, but there are no field records reported.

Dipodomys merriami breeds actively all year round, but is most active from Feb. to May (DAVIS, 1966). Gestation periods of 17–23 days are the shortest reported for any *Dipodomys* (CHEW, 1958), and the majority of the young are born from early March to early July (HALL, 1946). ALCORN (1941) suggested there are two litters per year, ranging in size from 1–5 with a mean of three. Young are born naked, blind, unpigmented, and completely dependent on their parents. REYNOLDS (1960) reported that pigmentation occurred in two days and CHEW & BUTTERWORTH (1959) in three days, and hair is visible at 5–6 days (CHEW & BUTTERWORTH, 1959 and REYNOLDS, 1960).

According to CHEW & BUTTERWORTH (1964), the young begin packing seeds as early as 15 days, whereas earlier reports suggested 18–23 days (CHEW & BUTTERWORTH, 1959 and REYNOLDS, 1960). Field estimates of longevity based on mark and recapture methods include a maximum of about two years (CHEW & BUTTERWORTH, 1964), but recapture percentages after one year were a low 12–19%.

Dipodomys microps breeds from late Jan. to May, but since most of the young are born before May (HALL, 1946) the peak likely corresponds to the Feb.–March peak of other *Dipodomys* species. The young are born blind, naked, unpigmented, and completely dependent on their parents, but maturation data are not available. Litter sizes range from 1–4 with a mean of 2.3 (HALL, 1946).

Dipodomys ordii are reproductively active all year long, but their peak occurs from late Jan. to May (HALL, 1946). Since the gestation period is 29 days (DAY, EGOSCUE & WOODBURY, 1956), most of the young are born in April, May, and June, but a peak often occurs in late May and/or June (DAVIS, 1966), due to the likelihood of two litters per year (HALL, 1946). Litter sizes range from 1–6 (HALL & KELSON, 1959) with a mean of 3.5. The young are born blind, naked, unpigmented, and completely dependent on their parents, but further maturation data are lacking.

Dipodomys spectabilis, Bannertailed kangaroo rat, begins breeding in Jan. and continues into Aug. (DAVIS, 1966). The gestation period is in excess of 27 days which accounts for the appearance of young in late Feb. and early March (HOLDENRIED, 1957). Litter sizes range from 1–3 (VORHIES & TAYLOR, 1922). The young, which are blind, naked, unpigmented, and completely dependent on their parents at birth, are born in an underground nest chamber lined with vegetation and chaff refuse from food (DAVIS, 1966). Pigmentation appears on day three and hair is visible by day seven. The eyes open on day 14, and pouching of seeds occurs on day 25 (BAILEY, 1931). *Dipodomys spectabilis*, like *D. merriami*, has a very short life span (maximum of two years) – only 14–25% recaptures were observed after one year (HOLDENRIED, 1957).

Lagurus curtatus

Lagurus curtatus, Sagebrush voles, evidently breed between March and late July since most young appear in late May through Aug. Litter sizes range from 3–6 with a mean of 4.9. Females are polyestrous with multiple litters per year (HALL, 1946).

Liomys pictus

Liomys pictus begins breeding in Feb. and March and most of the births occur later in spring and early summer. The gestation period is 21–28 days, and litter sizes vary from 2–5; and there may be more than one litter per year (WALKER, 1964).

Microdipodops megacephalus and *M. pallidus*

The kangaroo mice, *Microdipodops megacephalus* and *Microdipodops pallidus*, begin breeding in early Feb. but most of the young and pregnant females do not appear until June. The gestation period is unknown but two litters per year are likely which range in size from 1–7 with a mean of 3.9 (HALL, 1946).

The white-throated woodrat, *Neotoma albigula*, is found throughout the hot desert areas of North America where their abundance and indirectly reproduction is restricted primarily by the number of suitable shrubs available for shelter (Vorhies & Taylor, 1940). Any shrub species affording good protection may be used for dens to provide nesting areas for the young. Several *N. albigula* dens may occupy the area around a shrub or in clumps of cactus, but this social tolerance is not maintained by *N. lepida*, where only one female and her young occupy each den.

The breeding season of the desert woodrat, *N. lepida*, is similar throughout its range. Breeding occurs throughout the year in Arizona, but activity is concentrated from late Dec., when five or six females were observed to have enlarged uteri, to late Aug. (Vorhies & Taylor, 1940). Also, only one pregnant female was observed between Sept. and Dec. Davis (1966) reported comparable results from Texas. Successive litters may be produced with very short intervals between them throughout the range of *N. lepida*. The gestation period is approximately 38 days, but lack of information on the time required for the young to become independent of their parent makes it difficult to accurately estimate the number of litters per year. Vorhies & Taylor (1940) found a female in a den with two newly born young hanging to the teats, and three half-grown young in the same den. This supports the claim that at least two litters are born each year and possibly three or more (Davis, 1966).

Litter sizes for *N. lepida* vary from 1–3 (Davis, 1966) but the mean litter size varies as a function of the month in which the sample was taken (Vorhies & Taylor, 1940). In Jan. the mean litter size was 1.75, 2.28 in March, 1.66 in April, 2.05 in July, and 1.40 in Aug. This variability likely corresponds to the amount of available food and to the age structure of the reproducing females. Young and old rodent females tend to have small litters while median aged females have large litters. Many females likely have their first litters in Jan., but not until March and again in July are they likely to be in peak reproductive condition. The low Aug. mean likely reflects the first litters of some of the females born in Jan.

The young (11 g at birth) are born blind, virtually naked, slightly pigmented, and relatively helpless. A unique feature at birth is a special adaptive groove on the teeth for holding teats (Davis, 1966). The securely attached young are often dragged from the nest on their backs, when the mother leaves. Growth and development is rapid with the ears opening between days 13–15, the eyes opening between days 15–19, and weaning occurring at 62–72 days of age. At six months it is difficult to distinguish the young from their parents (Davis, 1966).

Reproduction of *N. lepida* in the field is not as well documented as it is for *N. albigula*, even though *N. lepida* is equally as common. *Neotoma lepida* selects its den site much like *N. albigula* except the den sometimes consists of sticks plus a burrow, or if in open country, only a burrow (HALL, 1946). Several nests may exist in any den (STONES & HAYWARD, 1968) but only one is used for reproduction. In nature the breeding season begins in late Jan. or early March as evidenced by new born animals in March, but birth continues only through May (STONES & HAYWARD, 1968). This is not typical of the species, however, since HALL (1946) reported many pregnant females in June, and FINLEY (1959) suggested that there is likely more than one litter per year. Laboratory data fail to define a definite breeding season since females bred every month of the year, except certain females failed to breed during Dec. and Jan. (EGOSCUE, 1957). In the laboratory the average number of litters per year was four with as many as 14 total young.

Litter sizes range from 1–5 both in nature and the laboratory with a field mean of three (HALL, 1946) and a laboratory mean of 2.3 (EGOSCUE, 1957). The gestation period ranges from 30–36 days depending on the lactating conditions of the female.

EGOSCUE (1957) found that in comparison with other rodents, young *N. lepida* are born in a rather advanced stage of development. The dorsum is heavily pigmented and the tips of the hairs can be seen. The eyes open in 11–13 days. Nursing continues for approximately four weeks, but the laboratory reared young began eating solid food about the end of the third week. Laboratory reared females became sexually mature in 2–3 months with several records of young born to animals only 90–113 days of age, but the earliest age when males reach sexual maturity is unknown (EGOSCUE, 1957).

Onychomys leucogaster and *O. torridus*

The northern grasshopper mouse, *Onychomys leucogaster*, is generally found in semiarid scrub deserts throughout much of the southwestern United States, but it is relatively scarce when contrasted with population numbers of other desert rodent species (EGOSCUE, 1960). Consequently, most detailed reproductive biology data has been worked out in the laboratory (EGOSCUE, 1960, RUFFER, 1965, and PINTER, 1970), but pregnancy records from field captured rodents allow a reasonable reconstruction of the reproductive cycles in Texas (DAVIS, 1966) and Nevada (HALL, 1946). Breeding in the field extends from May to Oct., however, the capture of half grown young in dark juvenal pelage from Feb. to Sept. suggests some year-round breeding activity (DAVIS, 1966 and HALL, 1946). In laboratory studies *O. leucogaster* produced litters during all months of the year, but there was a seasonal periodicity, since the largest number of litters was produced between the end of April and the beginning of Oct. (PINTER, 1970). These results support EGOSCUE's (1960) findings of a peak-breeding period between Feb. and Aug.

Litter size was similar in both laboratory and field animals with a range of 1–6 and a mean of approximately four (DAVIS, 1966 and PINTER, 1970). Most young in the field were born in May and June following a gestation period ranging from 32–47 days (DAVIS, 1966 and HALL, 1946). Longer gestation periods were associated with lactating females; whereas, the shortest periods were for young, first litter or non-lactating females. Laboratory gestation periods for non-lactating females ranged from 26–37 days (PINTER, 1970) and from 32–47 days in lactating females (EGOSCUE, 1970 and SVIHLA, 1936).

The young, 2.2 g at birth (HALL, 1946) are born blind, unpigmented, hairless except for prominent vibrissae, and completely dependent on their parents (DAVIS, 1966). They develop slowly with their eyes opening between days 18–20 and then eating solid food by 23 days of age (HORNER, 1968). Sexual maturity is reached at an early age. EGOSCUE (1960) reported that the youngest laboratory bred female to produce a litter was six months old, and SVIHLA (1936) recorded pregnancy in a 95-day-old female, but failed to indicate when the litter was conceived. PINTER (1970) recorded reproductive activity in laboratory bred females between 4–19 months of age, but even though some females were four months old when their first litters were born most of them were 5–6 months of age. The oldest laboratory bred female to successfully rear a litter was only 19 months of age; whereas, a female, live-trapped as an adult, produced 11 litters (42 young), during 27 months following her capture with no signs of a reproductive lag (PINTER, 1970). This might indicate that field reared females begin reproducing at a later date in life, and consequently, do not reach reproductive inactivity as quickly as do laboratory reared females. The advantages of such a breeding pattern for perpetuation of the species are obvious.

PINTER (1970) found that the youngest laboratory reared males to sire litters were four months of age and they generally remained reproductively active up to 24 months of age. An exception to the maximum active age was a male that sired a litter six weeks prior to his death at the age of 4.5 years which is the longest longevity record for *O. leucogaster*.

The southern grasshopper mouse, *Onychomys torridus*, is a small cricetid rodent that generally lives in low, arid desert valleys of southwestern North America where there is little or no intraspecific competition with *O. leucogaster*. HORNER & TAYLOR (1968) reported that in their field studies conducted intermittently over a five year period to selectively trap *O. torridus*, only 6% of the small mammals trapped were *Onychomys*. This paucity might be explained since *O. torridus* is considered to be a secondary consumer which would be expected to have a lower reproductive rate than primary consumers. Reproductive control mechanisms, other than energy, required to maintain low population numbers and reproductive rates have been investigated by several (TAYLOR, 1963, RUFFER, 1965, HORNER & TAYLOR, 1968, and TAYLOR, 1968).

Onychomys torridus is polyestrous and has produced litters every month of the year in the laboratory, but there was a seasonal variation since the greatest activity occurred from Jan. to July (PINTER, 1970). DAVIS (1966) reported that the breeding season in Texas begins in late Jan. or early Feb., since he collected one pregnant female on Feb. 27, and continues until Sept. HORNER & TAYLOR (1968) and TAYLOR (1968) reported that the breeding season, under natural conditions, appeared to be from the last of March through Sept. with a peak in July. The later field breeding observations correspond closely with that reported in Nevada by HALL (1946), where most pregnancies occurred from May through July. Geographic locations and sub-speciation likely account for some differences.

Determining the stage of estrous for *O. torridus* females is more difficult than in some other genera, since the vaginal lips are not subject to cyclically detectable external changes and, additionally, the vaginal orifice can fluctuate readily between the perforate and the imperforate condition. When the female is in estrous the vagina is typically open, but between periods of estrous and throughout most of the pregnancy the vagina frequently closes, only to open again one to several days before parturition. Hence, vaginal closure in the adult does not necessarily denote a quiescent reproductive state; nor does vaginal perforation necessarily indicate that the animal is either in estrous or preparurient. However, in juveniles and maturing young females, vaginal perforation is more informative, since the entire sequence of juvenile development is characterized by an imperforate vagina. The earliest age at which perforation is known to occur is 44 days and it is usually concomitant with estrous, although more commonly both occur later (TAYLOR, 1968).

Estrous is spontaneous whether the female is paired with a male or in isolation, but it is not always regular. The only estrous which was regular was the postpartum estrous experienced within three days following parturition (TAYLOR, 1968).

Litter size varied from 2–7 with a mean of 4.2 in the field (DAVIS, 1966) and from 1–5 with a mean of 2.4 in the laboratory (PINTER, 1970). Since half-grown *O. torridus* have been captured in April, June, July, and Aug., it is likely that there is more than one litter born per year in the field (DAVIS, 1966), but not nearly so many as are born in the laboratory where a single female has produced as many as 15 litters in 17 months. The excessive number of laboratory litters per year is likely the reason for the smaller laboratory mean litter size. In addition, all evidence seems to indicate that females one year old or less are more effective breeders than their older counterparts. This would also influence the low laboratory average since life span in the field is certain to be shorter than in the laboratory. Laboratory breeding records for *O. torridus* indicate somewhat less variability in the gestation period than is evidenced in *O. leucogaster*. TAYLOR (1968) reported that in non-lactating *O. torridus*

315

females, the gestation period may be from 27–29 days, which is comparable to the range of 27–30 days recorded for lactating females.

According to HORNER & TAYLOR (1968) young are born in a nest, after which the male leaves. They observed one delivery where the female sat on her haunches, with hind legs widely spread, vigorously licking her vulva and the emerging young, and using her fore paws in facilitating its passage through the vaginal orifice. The amnion and placenta were eaten as she continued her energetic washing of the young, and within 10 min. of birth, she nursed the litter. The minimum time noted by HORNER & TAYLOR (1968) for the birth of a litter was 30 min.

At birth the young are blind, naked except for the vibrissae, with dorsal pigmentation of the head and body, and relatively helpless. By day three the body fur emerges, and the eyes open in most animals between days 14–16. By day 19 most of the young are capable of eating solid food and they can be weaned by day 20 (HORNER & TAYLOR, 1968).

The earliest laboratory breeding for field-caught *O. torridus* was 7 weeks of age for females (HORNER & TAYLOR, 1968), and 12 weeks for laboratory bred and field-caught males (TAYLOR, 1963). PINTER (1970) found that females often gave birth to their first litter at four months of age; whereas males sired litters as early as three months, but most frequently at four months. PINTER (1970) observed one record case of littermates giving birth at three months, which means both had to be sexually mature at nearly two months of age. Captive *O. torridus* seldom remain reproductively active beyond the age of two years since the oldest captive male known to have sired a litter was 31 months old, while the oldest female to successfully bear a litter was 24 months. All litters produced by older females died prior to weaning (PINTER, 1970).

Perognathus amplus, P. baileyi, P. flavus, P. formosus, P. intermedius, P. nelsoni, P. longimembris, P. merriami, P. parvus, and *P. penicillatus*

Reproductive biology for *Perognathus amplus* is relatively unknown and what little data there are has come from museum labels at the University of Arizona in Tucson. Their breeding season evidently begins in late Feb. and early March when scrotal males have been collected in various parts of Arizona. Pregnant females began appearing in early April, and by April 25, the number of pregnant females was at its peak with 73% of all females caught being pregnant. This percentage dropped to 60% during May and 20% during June, which suggests a single litter per year. Litter size as determined by embryo counts ranged from 1–5 with a mean of 3.25 per litter. There seemed to be no significant difference in litter size during any production month.

Perognathus baileyi was briefly discussed by REYNOLDS & HASKELL (1949), but little detailed information concerning reproductive biology is available. They stated that breeding was confined to spring and summer months, which may in large be true, but museum labels indicate pregnant females as early as Feb. 22. This indicates that breeding activity begins in Jan., but it does not peak until late spring as pointed out by REYNOLDS & HASKELL (1949). The periods of greatest reproductive activity were reported to coincide with the seasons of new vegetative growth. After the late spring breeding peak, reproductive activity dropped off during June and early July, but picked up again to reach a secondary peak in Aug. The lull reported in reproductive activity corresponded with a drought period during June and early July (REYNOLDS & HASKELL, 1949). According to embryo counts and museum labels, litter size ranges from 2–5 with a mean of four. Sex ratios were nearly equal.

The silky pocket mouse, *Perognathus flavus*, inhabits and tolerates a wider range of habitat conditions than other *Perognathus* species since they live among the rocks, clay or rocky soils, and in sand (DAVIS, 1966). Their breeding season begins in early spring (FORBES, 1964 and DAVIS, 1966) and continues until fall. FORBES (1964) reported that a sample of 20 individuals taken during winter contained 4 young adults (14%), 22 adults (76%), and 3 old adults (10%), but no juveniles or pregnant females; whereas, a spring sample of 51 animals contained 4 juveniles (8%), 4 young adults (8%), 30 adults (59%), and 13 old adults (25%). Five of the adults (9%) in the spring sample and six old adults (11%) were pregnant. A summer sample of 322 mice contained 82 juveniles (25%), 64 young adults (70%), 165 adults (51%), and 11 old adults (3%). Of 149 females, only four adults (1%) were pregnant. A fall sample of 44 animals contained one juvenile (2%), and there were no pregnant females. In all cases the sex ratio was approximately 1:1.

The absence of pregnant females in fall and winter samples, and the absence of juveniles in winter samples, suggests that mice born in late summer do not become sexually active until the following spring. This theory disagrees with BAILEY (1931) who suggested that *P. flavus* might be sexually active in the fall and winter. DAVIS (1966) later reported one lactating female in Dec., thus, females seem to be capable of two or more litters of 2–6 young per litter.

Perognathus formosus inhabits much of the desert area in Nevada and Utah, but in spite of its abundance very little reproductive data have been published. Their breeding season begins in the spring depending upon the growth of the spring vegetation and continues into Aug. The litter size ranges from 2–6 with an overall mean of 5.6 (FRENCH, pers. comm.). FRENCH also said that females less than one year old average 5.1 young per litter, year old females 5.8 per litter, two year old females 5.3 per litter, and three year old females 6.0 per litter. The latter average is based on a sample size of three, consequently, it is questionable.

The intermediate pocket mouse, *Perognathus intermedius*, occasionally inhabits shrubby desert slopes on pebbly soils (DAVIS, 1966), or occasionally sandy soils among the rocks (BAILEY, 1931). DAVIS (1966) reported from meager data that breeding began in Feb. or March and continued for several months. Pregnant females have been caught in May, June, and July with litter sizes varying from 3–6; however, young are likely born prior to May. Nearly half-grown young in juvenal pelage have been taken as early as April, which, although the gestation period is unknown, indicates birth in late March or early April.

The Nelson pocket mouse, *Perognathus nelsoni,* seems to prefer rocky habitats with sparse grass (DAVIS, 1966). DAVIS (1966) reported that the breeding season began in Feb. with the pregnancy peak in March, and juveniles entered traps during April in the Big Bend region of Texas. He inferred a gestation period of about one month with the young leaving the nest approximately four weeks later. Pregnant females were captured in each month from March through July with the number of embryos per litter averaging 3.2 and a range of 2–4. The annual population turnover rate reported by DAVIS (1966) was 86% with only 14 individuals surviving from one year to the next in the Big Bend region of Texas.

The little pocket mouse, *Perognathus longimembris,* has had considerable laboratory work done on its reproductive cycle (HAYDEN & GAMBINO, 1966). Estrous cycles start in mid-Jan. and continue through Dec. in captive animals, lasting about 10 days. The incidence of first estrous periods increases from late Jan. to a maximum in March and then gradually declines into June. The sexually active females were observed to have an average of·3.8 estrous cycles per year with a range of 1–11. At least one full estrous cycle was observed in 58% of the animals studied with 42% remaining inactive during five months of daily observations.

HAYDEN & GAMBINO (1966) reported that the vaginal orifice was completely regressed and sealed with epithelium during the anestrous portion of the year. The orifice was alternately opened and closed, with vaginal walls oppressed or sealed with epithelium during the polyestrous portion. The vulva became swollen during the 1–2 proestrous days, but the vaginal orifice remained sealed with only a characteristic transverse line evident. During estrous the vaginal orifice was open, with the edges much enlarged and evaginated, a condition that lasted from a few hours to one day. External genitalia regressed and the vaginal lining sloughed off to form a mucous plug, which was retained within the vagina for 1–5 days during metestrous. These plugs were different from copulation plugs. Estrous occurred as early as 2–3 days postpartum.

Although males with conspicuously descended testes have been observed in the field, none were observed by HAYDEN & GAMBINO in their laboratory work. Several males were reported to have testes that could be forced into the scrotum, but none were retained there. The abdominal position of the testes did not, however, prevent production of viable sperm.

Pairings of females in estrous with males was started on March 1, which corresponds with the peak of reproduction in nature (HALL, 1946

318

and HAYDEN & GAMBINO, 1966). Even though HAYDEN & GAMBINO (1966) began field trapping on March 5, pregnant females were first caught in mid-April. Twenty percent of the females trapped on April 19 were pregnant, 50% on May 6, 28% on May 17, 25% on June 2, and 4% on June 14. Of 355 estrous cycles that occurred in the laboratory, 217 matings were made (61% of potential receptive periods), resulting in 57 litters (26% success). This was considered a high level of success since sexual receptivity occurred only during the first half of estrous, and even when compatible, the females were aggressive.

Females were examined immediately after copulation and found that their external genitalia were no longer swollen and evaginated as before copulation and the vaginal orifice was sealed with a crust (HAYDEN & GAMBINO, 1966). EGOSCUE (pers. comm. 1962) also noted this immediate disappearance of vulval swelling after breeding in *Dipodomys* spp, indicating that superfecundation may not be possible in heteromyid rodents.

The young are born naked, blind, and helpless. In 74% of the pregnancies the gestation period is from 22–23 days, but the gestation period range is 21–31 days. Litters in the laboratory range in size from 1–6 individuals with a mean of four and sex ratios of one female to 1.13 males. Abortion and/or cannibalism occurs (HAYDEN & GAMBINO, 1966).

Development was relatively slow when compared to other small desert species since the litters were not weaned until 30 days of age, but sexual maturity developed rapidly. One female born in the laboratory was observed to have swollen genitalia, but a sealed vaginal orifice at 41 days of age. At 42 days a vaginal plug was formed which indicated that sexual maturity had occurred. This female was in her second estrous 30 days later, which indicates that the early young of the year are capable of breeding the same year they are born (HAYDEN & GAMBINO, 1966); but since one litter per year is most likely (HALL, 1946), it is questionable whether estrous is brought to fruition in these young females.

Merriam's pocket mouse, *Perognathus merriami*, is a common mouse on sandy soils with sparse or short vegetation in the southwestern United States (DAVIS, 1966), but occasionally they occupy gravelly, desert pavement, with sparse vegetation. Breeding begins in early April and continues to Nov. with young in juvenal pelage being caught in June, July, and late Nov. Although the gestation and weaning periods are unknown, the two periods combined take from 6–8 weeks and there are possibly two or more litters born per year (DAVIS, 1966). Three to six young per litter are born naked, blind, unpigmented, and helpless. DAVIS (1966) reported that annual population turnovers in the Black Gap area of Texas were 84% in one study and 75% in another with a maximum life span of 33 months and 22 months, respectively.

Perognathus parvus which inhabits the more northern areas of the North American deserts in Nevada, Utah, and Idaho has a slightly later sexual cycle than the southern species. Breeding begins sometime in April but

319

since the gestation period is unknown, a definite time is difficult to establish. HALL (1946) reported that in Nevada, young are born in May, although SPETH, PRITCHETT & JORGENSEN (1968) reported pregnant females on May 26, but did not report lactation until June 27 with a peak 12 of 13 females lactating by July 2 in Idaho. Young, 3–8 with a mean of 5.5 per litter, are born in May, June, and July in Nevada (HALL, 1946) with a speculated single litter per year; but SPETH, PRITCHETT & JORGENSEN (1968) did not find this to be the case in Idaho. They reported five females carrying young on Aug. 1 while still lactating with first litters. They also reported young by July 15 while still in juvenal pelage. Thirteen juvenal females were lactating although the males failed to mature as rapidly. Only two of 28 juvenal males had scrotal testes by the end of the breeding season (Aug. 23). SPETH, PRITCHETT & JORGENSEN (1968) reported that 94% of the captures were juveniles; therefore, the population turnover rate is likely rather high.

SPETH, PRITCHETT & JORGENSEN (1968) reported that the reproductive cycle is both exogenously and endogenously controlled. They observed that the 1966 trapping season, which was relatively unproductive compared to 1967 produced very few juveniles. They concluded that breeding preparation is possibly endogenously controlled but initiation and maintenance of the breeding cycle is dependent on an appropriate food supply and favorable environmental conditions. BEATLEY (1969) reported similar observations for southern Nevada where HAYDEN & GAMBINO (1966) projected the same conclusion for *P. longimembris*.

The desert pocket mouse, *Perognathus penicillatus*, usually occurs on sandy or soft alluvial soils (DAVIS, 1966), occasionally on gravelly soils (ARNOLD, 1942), and in areas of *Atriplex, Baccharis*, and *Prosopis*.

The breeding season is slightly varied throughout its range. It begins in late Feb. in Big Bend, Texas, with the maximal peak in pregnancies occurring in April (DAVIS, 1966), but REYNOLDS & HASKELL (1949) and ARNOLD (1942) found the cycle to lag one month behind in Arizona and parts of extreme southern California. This lag is likely a seasonal phenomenon associated with precipitation since in all cases the high activity periods correlated with the spring. REYNOLDS & HASKELL (1949) found that breeding dropped off during the summer drought months of June and July, but picked up again in Aug. when new vegetative growth appeared. This agrees with DAVIS (1966) who reported an Aug. reproductive peak. Many young females were pregnant while still in juvenal pelage (DAVIS, 1966) which was during the time that the older females were giving birth to their second litters. The majority of the population in late Sept. was comprised of juveniles, and DAVIS (1966) reported an annual population turnover of 95%.

Litter sizes varied in all studies with ARNOLD (1942) reporting a range of 3–6 with a mean of 4.38, REYNOLDS & HASKELL (1949) reporting a range of 2–4 with a mean of 3.4, and DAVIS (1966) reporting a range of 2–6 with a mean of 3.6. ARNOLD (1942) reported a heavily female dominated sex ratio but nearly equal ratios were reported elsewhere. The young are born relatively naked, blind, and helpless, but evidently mature rapidly, at least sexually, as evidence by the early breeding of the young females.

Peromyscus crinitus, P. eremicus, P. maniculatus, and *P. truei*

A great deal of work has been done on all aspects of the life of the genus *Peromyscus* and has been well summarized in the 'Biology of *Peromyscus*' by KING (1968), but unfortunately most of the data on reproduction have been worked out on eastern or mid-western species as pointed out by LAYNE (1968).

The canyon mouse, *Peromyscus crinitus*, is a widely distributed but little-known rodent of marginal desert areas of western North America. They are largely restricted to ledges, cliffs, and canyons bordering the desert valleys and seem to prefer sagebrush, *Artemesia tridentata*.

Breeding which EGOSCUE (1964) says is likely seasonally polyestrous in nature, begins in early spring since pregnant females were observed in April, but does not reach full activity until May or June (HALL, 1946 and EGOSCUE, 1964). EGOSCUE (1964) reported birth in the laboratory colony every month of the year, but most (109 of 135 litters) were born between Jan. and Aug. Some females in the laboratory gave birth to as many as eight litters per year, but the average, 2.05 litters, more nearly approximates what one would expect in nature. HALL (1946) reported a range in litter size of 3–5 with a mean and mode of four for field caught animals; whereas EGOSCUE (1964) reported litters of 1–5, but only one litter of five.

Gestation records are not available for field caught animals and precise laboratory records were difficult to obtain. EGOSCUE (1964) found that the mice seldom bred when first paired and often required several weeks before breeding occurred. Some pairs never produced young. Most gestation periods reported by EGOSCUE (1964) are for intervals between consecutive litters when females evidently bred shortly after parturition. The shortest interval between two litters, with the first litter reared to weaning age, was 27 days, but most gestation periods during lactation were 29–30 days. EGOSCUE (1964) reported one female that killed or abandoned her litters at birth or shortly thereafter, and her next litter was born 24–25 days later, which may approximate the gestation time for non-lactating pregnancies. Lactation reportedly extends the gestation period in several species of *Peromyscus* (SVIHLA, 1932). Gestation periods for *P. crinitus* are apparently longer by some 3–5 days than in *P. maniculatus*, but young canyon mice are born in a state of development which approximates that of deer mice 2.5–3.0 days of age (EGOSCUE, 1964).

Canyon mice were naked except for vibrissae, blind, lightly pigmented on the dorsal part of the head and body, and the ears were folded at birth. The young were weaned at four weeks of age when they weighed 13.2–15.0 g. Sexual maturity occurred as early as 70 days of age but most young did not reproduce until they were 4–6 months old.

The desert white-footed mouse, *Peromyscus eremicus*, is restricted almost entirely to desert habitats, preferring rocky outcrops which offer retreats and nest sites. They replace *P. maniculatus* as the most common mouse of this genus over the arid parts of the southwestern United States, but along riparian areas both occur together.

The breeding season begins in early Jan. and extends to Oct. with two or more litters being born each year. Litter sizes in Texas range from 1–4 with a mean of 3.0 (DAVIS, 1966) but HALL (1946) reports a litter size ranging from 3–5 with a mean of 3.75 for Nevada. Both of these field determined litter sizes exceed the laboratory mean of 2.42 young per litter reported by (DAVIS & DAVIS, 1947), 2.60 reported by SVIHLA (1932), and 2.22 reported by BRAND & RYCKMAN (1968), who reported a maximum of three. DAVIS & DAVIS (1947) found that mean litter sizes increased gradually from the first litter to the fifth or sixth, after which they declined gradually to the tenth and abruptly thereafter, but it is doubtful that many animals in nature would ever have 10 litters.

The gestation period of one non-lactating female was found to be 21 days (SVIHLA, 1932), but in the study of DAVIS & DAVIS (1947) individual females had litters 28–30 days apart. This indicates that there is postpartum estrous and that lactation likely prolongs gestation. The young are born more advanced than *P. maniculatus* and at birth they weigh about 2.5 g, are naked except for vibrissae, dorsally pigmented, blind, ears folded, and virtually helpless. Development is rapid and the young remain attached to their mother most of the time until 20–22 days of age, when weaning begins. The young continue nursing part of the time up to day 25, when they seem to be fully weaned – at least they eat solid food (BRAND & RYCKMAN, 1968). DAVIS (1966) reported that the young nurse 30–40 days in nature when the litters are produced slowly. Sexual maturity was reached in the study by DAVIS & DAVIS (1947) at 3–4 months of age as evidenced by breeding, but most did not breed until 10 months which is likely what occurs in nature.

The deer mouse, *Peromyscus maniculatus*, is likely one of the most adaptable of all mammals in that it seems to be ubiquitous, occupying a wide variety of habitats in all parts of the United States. Considerable work has been done on the reproductive biology of *P. maniculatus* but most of it deals with sub-species which occupy mixed forest or grassland habitats.

Breeding in nature occurs throughout the entire years, but there are seasonal peaks which seem to correlate with vegetative growth in the region where they occur. DAVIS (1966) reported two peaks of breeding activity in Texas, one from Jan. through April and the other from June through Nov. This differs from HALL (1946) who reports that pregnancy was most common in May and June in Nevada, when one-half of the individuals were pregnant. HALL (1946) did not report a second reproductive peak for Nevada animals.

In general *P. maniculatus* have relatively large litters when compared with the litters of *P. crinitus*, *P. truei*, and *P. eremicus*. Litter sizes range from 1–9 with a mean of 4.0 in Texas (DAVIS, 1966) and from 2–8 with a mean of 5.3 in Nevada (HALL, 1946). The gestation period varies depending upon postpartum estrous and lactation (LAYNE, 1968 and CLARK, 1938). Normally the gestation period is 22–27 days but may be as much as 35 days (CLARK, 1938, DAVIS, 1966, and HALL, 1946).

At birth the young weigh 1.1–2.3 g, are naked except for vibrissae, blind, unpigmented, with ear pinna folded, and relatively helpless. Pigmentation appears dorsally within 24 hr and the ear pinna unfold at three days of age. The eyes open between days 12–17 with weaning occurring at approximately four weeks (CLARK, 1938, DAVIS, 1966, and HALL, 1946). DAVIS (1966), however, reported a young still suckling at 37 days of age. In these cases the mother would not have a second litter, the young of the first litter are cut off and forced to fend for themselves when the second are born. Sexual maturity occurs later than in *P. eremicus*, but the gray juvenal pelage is always lost at 46–51 days of age (CLARK, 1938). Males mature slightly earlier. Females of litters born early in the year may have a litter by late summer which can lead to the high populations of *P. maniculatus* observed in favorable habitats (DAVIS, 1966).

The reproductive biology of *Peromyscus truei* is essentially unknown except for brief reports of CLARK (1938) and HALL (1946). They reported a breeding season from Feb. to June with the peak of pregnancies in May and June, followed by reduced activity and a secondary peak in Aug. and Sept. The young are born following a gestation period of 25–40 days, depending on the occurrence of lactation. Litters range from 3–5 young with a mean size of 4.3 (HALL, 1946). Growth and development as well as sexual maturity are apparently both rather similar to those reported for *P. maniculatus* (CLARK, 1938).

In all *Peromyscus* species, both young and old females tend to have smaller litters than median aged females and there are several similarities in the birth processes at parturition. During actual delivery, female *Peromyscus* typically assume a quadrapedal or bipepal, hunched over, and crouching position (SVIHLA, 1932 and CLARK, 1937). Abdominal contractions often preceed the appearance of the young and some *P. maniculatus* have been observed to pull the skin around the vulva with the forepaws prior to the beginning of delivery. In several deliveries observed, the female used her teeth, fore feet, or both to aid the passage of the young from the birth canal (LAYNE, 1968).

Under some circumstances, birth may proceed rapidly without being accompanied by the typical posture responses or maternal aid described above. SVIHLA (1932) disturbed a *P. maniculatus blandus* in her nest during parturition and two young with their associated placentas were expelled almost simultaneously in the time taken for the female to move from the nest to a corner of the cage five inches away. Although both head and breech deliveries are common in the same litter, it appears that breech deliveries are most prevalent.

The placenta and fetal membranes may be discharged with aid from the parent almost immediately after delivery or a few minutes afterwards, but all published descriptions of normal births in the genus indicate that the placenta is consumed soon after it has been discharged (LAYNE, 1968). The umbilical cord is either broken by stretching when the placenta is being eaten or may be bitten off.

Reithrodontomys megalotis

The western harvest mouse, *Reithrodontomys megalotis*, is a marginal desert dweller that lives in grass covered areas or often along weedy disturbed roadsides. They breed throughout the year, particularly in the warmer desert areas, but show annual periods of increased reproductive activity. Most of the young are born in May or July with fewer in June and Aug., but periodic cold or prolonged drought conditions can affect this annual cycle (DAVIS, 1966 and HALL, 1946). The young which weigh about 1–1.5 g at birth are born blind, unpigmented, naked except for vibrissae, and dependent. Maturation proceeds rapidly since they are completely weaned at 19 days and leave the nest at 21 days of age. Adult size and weight are reached by five weeks but the females usually do not breed until 13 to 17 weeks of age (BANCROFT, 1966). An exception to this is noted by BANCROFT (1966) wherein a female gave birth at 72 days of age, which means she became pregnant when only seven weeks old.

Litter sizes range from 1–7 with a mean of four (SVIHLA, 1931) and vary with the age of the female, how many litters she has had, and the time of the year. BANCROFT (1966) found that older and larger females tended to have larger litters, and that the mean litter size which was 4.3 for the first litters, increased to a maximum of 6.0 for fifth litters, and decreased to 3.3 for eleventh litters. Senility apparently is an important factor in influencing the number of young per litter, although the sex ratios remained relatively constant at 51% males to 49% females (BANCROFT, 1966).

SVIHLA (1932) reported a captive female that gave birth to seven litters (17 young) in a single year, but later BANCROFT (1966) had females give birth to eight, nine, and ten litters. The latter (10 litters) were born in only eight months by a field caught female. The number of litters born in the wild is undoubtedly less since field breeding records show autumn and spring breeding peaks; however, if a female were to have a litter in each month of the two seasons, seven litters would be possible (BANCROFT, 1966) because the gestation period is from 21–24 days with a mean of 23.

Spermophilus harrisii, S. mexicanus, S. spilosoma, S. townsendii, and *S. tereticaudus*

The Harris ground squirrel, *Spermophilus harrisii*, is found in rocky hills or rocky soils in much of the southwestern United States and is a common sight during the daytime throughout the year. Even though *S. harrisii* does not hibernate they seem to have seasonal breeding cycles. NEAL (1965b) found live spermatozoa in the testes between Nov. and June but female reproductive activity was not evident until Feb. when two females with swollen labia, enlarged uteri, but imperforate vulvae were observed. Female activity, however, likely occurs earlier because a pregnant female with six embryos was taken on Feb. 23 (NEAL, 1965b), and several University of Arizona museum labels indicate that breeding may occur in late Dec. and Jan.

The young, 5–9 per litter with a mean of 6.3, are born following a gestation period of about 30 days (NEAL, 1965b). They are naked at birth except for vibrissae, blind, unpigmented, dependent, and the ears are closed. Development proceeds somewhat slowly with pigmentation at one week of age, beginning hair at two weeks, ears opening between 3–4 weeks, and eyes opening in 29–35 days. Weaning does not occur until approximately seven weeks of age. Males have scrotal testes by the four-teenth week of age and are assumed to be sexually mature, but since no females become sexually active until Dec. or later the young males do not breed early in the year in which they are born (NEAL, 1965b). There is only one litter per year.

The Mexican ground squirrel, *Spermophilus mexicanus*, inhabits brush or grass areas often interspersed with *Prosopis, Larrea*, and various cacti (DAVIS, 1966). They seem to prefer sandy or gravelly soils where they can easily dig their unmarked burrows (EDWARDS, 1946). Five escape burrows were found for every main burrow, but the brood chamber was almost always constructed in side chambers at the deepest part of the main burrows. Nesting material, grasses and leaves, and twigs of mesquite were in the brood chamber during parturition and rearing of the young, but after the young had left the nest, the nesting material was taken into the sleeping chambers and the brood nest was filled with dirt, never to be used as a brood chamber again (EDWARDS, 1946).

Breeding in Texas begins the last of March or first of April and lasts for only a short period of time (EDWARDS, 1946). The males had descended testes only from the end of March until May. Females were found to have embryos as early as April 4 (EDWARDS, 1946), but parturition usually does not occur until May which infers a gestation period of approximately 30 days (DAVIS, 1966). The young at birth, 1–10 per litter with a mean of about five (DAVIS, 1966), weigh approximately 4.31 g, are covered with short fuzz, have vibrissae, are unpigmented, are blind, have their ears closed, and are relatively helpless except for the ability to cry out when handled or jostled (EDWARDS, 1946). Development proceeded quickly and by Aug. (about three months of age) the young were capable of and exhibited individual existence by occupying their own burrows and establishing their own nests. The young, although independent, do not breed during the fall and hibernate in Nov.

The spotted ground squirrel, *Spermophilus spilosoma*, preferably inhabits dry, sandy areas in the southwestern United States, but they are occasionally found in other hard surfaced areas with scattered brush, *Yucca*, and *Larrea* (DAVIS, 1966). They build their multiple entrance burrows, which terminate in a nest chamber, under plants or overhanging rocks, but their breeding and nesting behavior is not well known.

Breeding evidently begins in early Feb. and runs through June, but since these squirrels do not truly hibernate, male sexual activity may begin earlier. Young about one month old have been observed in Texas as early as April 28 and as late as Sept. This coupled with the fact that half grown young and pregnant females have been taken in June indicate that two litters per year are likely (DAVIS, 1966). Litter size is approximately 5–7 per litter, and mating behavior, gestation period, as well as growth and development are unknown.

The round-tailed ground squirrel, *Spermophilus tereticaudus*, prefers relatively flat, sandy areas of the desert extending as far north as southern Nevada in the southwestern United States. These hibernating squirrels often occur in sympatry with *S. harrisii*, but their breeding periods differ seasonally.

NEAL (1965b) found that when males emerged from hibernation in early Jan. some of them (20%) had active spermatozoa already developed in the testes, but it was not until mid Jan. through mid April that all males had active spermatozoa. Evidently, some spermatozoan development can occur while the males are presumably in hibernation. Prior to hibernation, the testes were the smallest of any time in the year and even the new males of the year did not develop prior to hibernation. NEAL (1965b) did not find the same pattern of development in the females. During non-breeding the labia were sealed, opening only during heat. The first indication of female reproductive activity was observed on March 2 when a captured female was observed to have swollen labia and an open vaginal orifice. She was kept in isolation and gave birth 25 days later which indicates that she likely mated in late Feb. This, along with laboratory data indicates a gestation period of 28–35 days. By mid March all females captured were pregnant or lactating, but after April 27, no more pregnant females were observed. Museum data indicated that pregnancy can occur in June and July but such pregnancies are likely females that lost their first litters or a second litter may occur in rare instances.

The young, 2–12 per litter with a mean of 6.2, are born naked except for vibrissae, blind, unpigmented, dependent, and the ears are closed. Development is more rapid than in *S. harrisii* although the new-born are similar, since they are weaned during the fifth week, but no sexual maturity occurs in either sex during the year in which they are born. They do not breed until 10–11 months of age (NEAL, 1965a).

Townsend's ground squirrel, *Spermophilus townsendii*, occurs in many areas of the arid southwestern United States, particularly in the northern desert areas where they occupy many types of habitats and cope with adverse environmental and food conditions by aestivating or hibernating. This burrowing animal emerges from hibernation in Feb. with the males, some of which are beginning sexual activity, coming out first. The females emerge approximately one week later and breeding begins shortly thereafter, occurring primarily between March 1–20.

The young, 5–15 per litter with a mean of 10, are born after a gestation period of approximately 24 days (HALL, 1946). At birth, the young are naked except for vibrissae, blind, have their ears closed, and are dependent. Very little is known about their maturation and development, but they live with their mother until the middle of June when they are two-thirds grown.

Thomomys bottae and *T. talpoides*

Pocket gophers, *Thomomys* spp., are common animals throughout the deserts of the southwestern United States, but due to their strict fossorial habits, they are often forgotten as desert dwellers. *Thomomys bottae* establishes runways 2–5 inches below the surface with a nesting chamber some two feet below the surface runs (DAVIS, 1966). The nest, a compact, hollow ball of dry, shredded vegetation, provides a place for the gopher to rest and also serves as the place for the female to have and rear her litters.

Breeding is thought to be continuous in the southern portions of their range (DAVIS, 1966), but in the north it is confined to the first half of the year (HALL, 1946). Although continuous in the south, there are three marked periods of increased activity. In the spring the main thrust is evidenced as in the north but there are secondary summer and early winter peaks when the females have their second litter. Young of the year are capable of reproduction.

At the onset of the reproductive season, the males leave their solidarity, breed with the female, and return to their burrows leaving the female alone to rear the young (HALL, 1946). After a gestation period of approximately 19 days, the young, 2–11 per litter with a mean of 4.8–5.4, are born. Litter size varies with the time of the year, since smaller litters

327

appear at the onset and at the end of the breeding season. The mean number of young per litter was 3.8 in Jan., whereas, in April it was 6.1 (HALL, 1946). This is likely due to the age of the female that is giving birth. Young and old females tend to have smaller litters whereas middle-aged females are the most fecund (HALL, 1946). At birth the young weigh 2–4 g, are blind, naked, unpigmented, and helpless (DAVIS, 1966 and HALL, 1946).

REFERENCES

ALCORN, J. R. 1941. Counts of embryos in Nevadan Kangaroo rats (Genus *Dipodomys*). *J. Mamm.*, 22: 88–89.

ARNOLD, L. W. 1942. Notes on the life history of the sand pocket mouse. *J. Mamm.*, 23: 339–341.

BAILEY, V. 1931. Mammals of New Mexico. N. Am. Fauna, 53, 412 pp.

BANCROFT, W. L. 1966. Reproduction, development, and behavior of the western harvest mouse, *Reithrodontomys megalotis*. Unpubl. M.S. Thesis. University of Kansas.

BEATLEY, J. G. 1969. Dependence of desert rodents on winter annuals and precipitation. *Ecol.*, 50: 721–724.

BOREL, A. E. & ELLIS, R. 1934. Mammals of the Ruby Mountains region of northeastern Nevada. *J. Mamm.*, 15: 12–44.

BLAIR, W. F. 1941. Observations on the life history of *Baiomys taylori subater*. *J. Mamm.*, 22: 378–383.

BRADLEY, W. G. 1965. Activity patterns and home range of the antelope ground squirrel, *Citellus leucurus*. Unpubl. Ph.D. Disser. Univ. Ariz.

BRAND, L. R. & RYCKMAN, R. E. 1968. Laboratory life histories of *Peromyscus eremicus* and *Peromyscus interparietalis*. *J. Mamm.*, 49: 495–501.

BRATTSTROM, B. H. 1960. Longevity in the kangaroo rat. *J. Mamm.*, 41: 404.

BUTTERWORTH, B. B. 1961a. The breeding of *Dipodomys deserti* in the laboratory. *J. Mamm.*, 42: 413–414.

BUTTERWORTH, B. B. 1961b. A comparative study of growth and development of the kangaroo rats *Dipodomys deserti* Stephens and *Dipodomys merriami* Mearns. *Growth*, 25: 127–139.

CHEW, R. M. 1958. Reproduction by *Dipodomys merriami* in captivity. *J. Mamm.*, 39: 397–598.

CHEW, R. M. & BUTTERWORTH, B. B. 1959. Growth and Development of Merriam's kangaroo rat, *Dipodomys merriami*. *Growth*, 23: 75–95.

CHEW, R. M. & BUTTERWORTH, B. B. 1964. Ecology of rodents in Indian Cove (Mohave Desert), Joshua Tree National Monument, California. *J. Mamm.*, 45: 203–225.

CLARK, F. H. 1937. Parturition in the deer mouse. *J. Mamm.*, 18: 85–87.

CLARK, F. H. 1938. Age of sexual maturity in mice of the genus *Peromyscus*. *J. Mamm.*, 19: 230–234.

COCKRUM, E. L. 1962. Mammalogy. The Ronald Press Co., New York.

DAVIS, W. B. 1966. The Mammals of Texas. Texas Parks and Wildlife Department. Austin, Texas.

DAVIS, D. E. & DAVIS, D. J. 1947. Notes on reproduction of *Peromyscus eremicus* in a laboratory colony. *J. Mamm.*, 28: 181–183.

DAY, B. N., EGOSCUE, H. H. & WOODBURY, A. M. 1956. Ord kangaroo rat in captivity. *Science*, 124: 485–486.

EDWARDS, R. L. 1946. Some notes on life history of the Mexican ground squirrel in Texas. *J. Mamm.*, 27: 105–115.

EGOSCUE, H. J. 1957. The desert woodrat: a laboratory colony. *J. Mamm.*, 38: 472–481.

EGOSCUE, H. J. 1960. Laboratory and field studies of the northern grasshopper mouse. *J. Mamm.*, 41: 99–110.

EGOSCUE, H. J. 1964. Ecological notes and laboratory life history of the canyon mouse. *J. Mamm.*, 45: 387–396.

FINLEY, R. B. 1959. Wood rats of Colorado: distribution and ecology. *J. Wldf. Mgt.*, 23: 375–376.

FORBES, R. B. 1964. Some aspects of the life history of the silky pocket mouse, *Perognathus flavus. Amer. Midl. Nat.*, 72: 438–443.

FRENCH, N. R., MAZA, B. G. & ASCHWANDEN, A. P. 1967. Life spans of *Dipodomys* and *Perognathus* in the Mojave Desert. *J. Mamm.*, 23: 339–341.

HALL, E. R. 1946. Mammals of Nevada. Univ. of Calif. Press Berkeley.

HALL, E. R. & KELSON, K. R. 1959. The mammals of North America, Vol. I. The Ronald Press Co., New York.

HAYDEN, P. & GAMBINO, J. J. 1966. Growth and development of the little pocket mouse, *Perognathus longimembris. Growth*, 30: 187–197.

HOLDENRIED, R. 1957. Natural History of the bannertail kangaroo rat in New Mexico. *J. Mamm.*, 38: 330–350.

HOOPER, E. T. 1956. Longevity of captive kangaroo rats, *Dipodomys. J. Mamm.*, 37: 124–125.

HORNER, B. E. 1968. Gestation period and early development in *Onychomys leucogaster brevicaudus. J. Mamm.*, 49: 513–515.

HORNER, B. E. & TAYLOR, J. M. 1968. Growth and reproductive behavior in the southern grasshopper mouse. *J. Mamm.*, 49: 644–660.

KING, J. A. (ed.) 1968. Biology of *Peromyscus*. Serial Publ. 2. *Amer. Soc. Mamm.*, xiii + 593 pp.

LAYNE, J. N. 1968. Ontogeny. In 'Biology of *Peromyscus* (Rodentia).' Edited by KING, J. A. American Society of Mammalogists.

MONSON, G. & KESSLER, W. 1940. Life history notes on the banner-tailed kangaroo rat, Merriam's kangaroo rat, and the white-throated wood rat in Arizona and New Mexico. *J. Wildl. Mgmt.*, 4: 37–43.

NEAL, B. J. 1965a. Growth and development of the round-tailed and Harris antelope ground squirrels. *Am. Midl. Nat.*, 73: 479–489.

NEAL, B. J. 1965b. Reproductive habits of round-tailed and Harris antelope ground squirrels. *J. Mamm.*, 46: 200–206.

PINTER, A. J. 1970. Reproduction and growth for two species of grasshopper mice *(Onychomys)* in the laboratory. *J. Mamm.*, 51: 236–243.

REYNOLDS, H. G. 1958. The ecology of the Merriam kangaroo rat (*Dipodomys merriami* Mearns) on the grazing lands of southern Arizona. *Ecol. Monographs*, 28: 111–127.

REYNOLDS, H. G. 1960. Life history notes on Merriam's kangaroo rat in southern Arizona. *J. Mamm.*, 41: 48–58.

REYNOLDS, H. G. & HASKELL, H. H. 1949. Life history notes on the Price and Bailey pocket mice of southern Arizona. *J. Mamm.*, 30: 150–156.

RUFFER, D. G. 1965. Sexual behavior of the northern grasshopper mouse *(Onychomys leucogaster). Anim. Behav.*, 13: 447–452.

SPETH, R. L., PRITCHETT, C. L. & JORGENSEN, C. D. 1968. Reproductive activity of *Perognathus parvus. J. Mamm.*, 49: 336–337.

STONES, R. C. & HAYWARD. 1968. Natural history of the desert wood rat, *Neotoma lepida. Amer. Midl. Natur.*, 80: 458–476.

SVIHLA, A. 1932. A comparative life history study of the mice of the genus *Peromyscus*. *Misc. Publ. Mus. Zool. Univ. Mich.*, 24: 1–39.

SVIHLA, R. D. 1936. Breeding and young of the grasshopper mouse *(Onychomys leucogaster fuscogriseus)*. *J. Mamm.*, 17: 172–173.

TAYLOR, J. M. 1963. Reproductive mechanisms of the male grasshopper mouse. *J. Exp. Zool.*, 154: 109–124.

TAYLOR, J. M. 1968. Reproductive mechanisms of the female southern grasshopper mouse, *Onychomys torridus longicausus*. *J. Mamm.*, 49: 303–309.

VORHLIES, C. T. & TAYLOR, W. P. 1922. Life history of the kangaroo rat. *Dipodomys spectabilis spectabilis* Merriam. U.S. Dept. Agr., Bull. 1091.

VORHIES, C. T. & TAYLOR, W. P. 1940. Life history and ecology of the white throated wood rat, *Neotoma albigula albigula* Hartely, in relation to grazing in Arizona. Tech. Bull, 86. Univ. Ariz.

WALKER, E. P. 1964. Mammals of the World. The Johns Hopkins Press Co., Baltimore.

XVI. RODENT FAUNAS AND ENVIRONMENTAL CHANGES IN THE PLEISTOCENE OF ISRAEL

by

E. TCHERNOV

Introduction

It is now generally believed that progressive desiccation has been the principal climatic trend during the late Pleistocene of Israel. This gradual shift in the climate towards a drier regime has presumably been the cause of the local extinction of the more tropical components of the Eastern Mediterranean fauna. The impact of the European glacial sequence on the one hand and of the close proximity of a great desert on the other on the evolution of the climate of Israel, and on the rodent fauna of the land through the ages, is yet to be understood in depth. An attempt has been made in the following pages to collate all the available information on the environmental changes in relation to rodent faunas in the Pleistocene of Israel.

The Main Biogeographical Changes in the Near-East since the Miocene

Towards the end of the Miocene dynamic oceanographic and geomorphological changes caused a world-wide cooling, due to what FAIRBRIDGE (1961) called 'polar topographic coincidence'. As a result the connection between the Mediterranean region of the Tethys and the Indian ocean was cut off. Following the isolation of the Vindobonian sea from the peri-Alpine depression, the Ponto-Aralo-Caspian region and the Mesopotamian bay became (during the Pontian) lagunar, and later on, continental areas. This global desiccation process was encouraged by the orogenic movements which took place during the Burdigalian, and later on by the proto-Pleistocene Levantine uplifts. While the entire region (Sarmatia and Mesopotamia) still preserved a system of closed lakes, the southern section of the Near East turned into an arid zone. This was a world wide phenomenon which affected more or less the same latitudes, to be known later on as the 'Northern World Desert Ring', or in the old word, the 'South Palaearctic Desert Belt'.

While faunal exchanges with Africa and South Asia through the Near East were effective during the Miocene, no exchanges of terrestrial forms with the regions north of the Sarmatic province was possible as late as the Pontian. Following the regression of the Tethys as a result of the later Miocene orogenic movements, the gate to the north and to the east was widely opened. During this period the developing desert belt steadily erected barriers between Africa, southern Asia and the Levant. The gradual increasing in size and aridity of the desert zone might be well demonstrated when analyzing the Near East faunas. Aquatic elements which were originated either in southern Asia or Africa were the first to be isolated in the refreshening water bodies of the Levantine regions, underwent rapid speciation, and in some cases there was sufficient time for few new genera to be evolved (like *Tristramella* spp., Cichlidae, Teleostei) (STEINITZ, 1954).

For the terrestrial faunas the desert belt was more effective as a barrier only in late Neogene and Pleistocene times. As a result, an impressive number of endemic species like *Micrelaps mülleri* and *Atractaspis engaddensis* (snakes) (HAAS, 1952, 1953); *Cinnyris osaea* and *Onychognathus tristrami* (birds) (TCHERNOV, 1962); *Procavia syriaca* were established in the Levant.

The xerotropic forms which could have better withstood desiccation have as a rule conspecific relatives in Africa south of the Sahara, as faunal exchanges could have been prolonged until a much later period (*Turdoides squamiceps, Falco concolor, Corvus rhipidurus*, (birds) or *Caracal caracal, Mellivora ratel* (mammals).

The rapid desiccation process, which was intensified with time, enforced the fauna left within the developing desert zone to undergo rapid adaptations to arid habitats.

Once the developing desert belt commenced to be an effective barrier, and no more significant faunal exchanges happened to occur between the Near-East, Africa (south of the developing Sahara) and Southern Asia, while on the other side, faunas which originated in the northern realm freely invaded (following the regression of the Partethys) the Levant and North-Africa, the main zoogeographical units were established. Compared with the Ethiopian region, the Orient shared much fewer number of common species with the Levant, like *Rhynchocalamus melanocephalum* (snake), *Ketupa zeylonensis* (fish-owl), and *Nesokia bacheri* (known as well from the late Palaeolithic of the Sudan; ROBINSON, 1966).

Of the tropical insect fauna of Israel BYTINSKI-SALTZ (1961) mentioned (Fig. 1) 80% Ethiopians, 10% Orientals and 10% Palaeotropic.

The opinions of BATE (1934), GARROD & BATE (1937), and FAIRBRIDGE (1962) that African animals invaded the Middle East only during the Aurignacian (late Palaeolithic) do not coincide with the facts that typical Palaeotropic (particularly Ethiopic) elements are found in the region since the Miocene (SAVAGE & TCHERNOV, 1968). Lower Pleistocene of Bethlehem (GARDNER & BATE, 1937; BATE, 1934, 1942, 1943; HOOIJER, 1958); or in early Middle Pleistocene of 'Ubeidiya (Jordan Valley) (HAAS, 1963, 1966, 1968; TCHERNOV, 1968); or in Middle Pleistocene of Jisr-Banat-Yakub (upper Jordan Valley) (HOOIJER, 1959) and of Jerusalem (TCHERNOV, 1968a) (Table 1). Nor does it fit

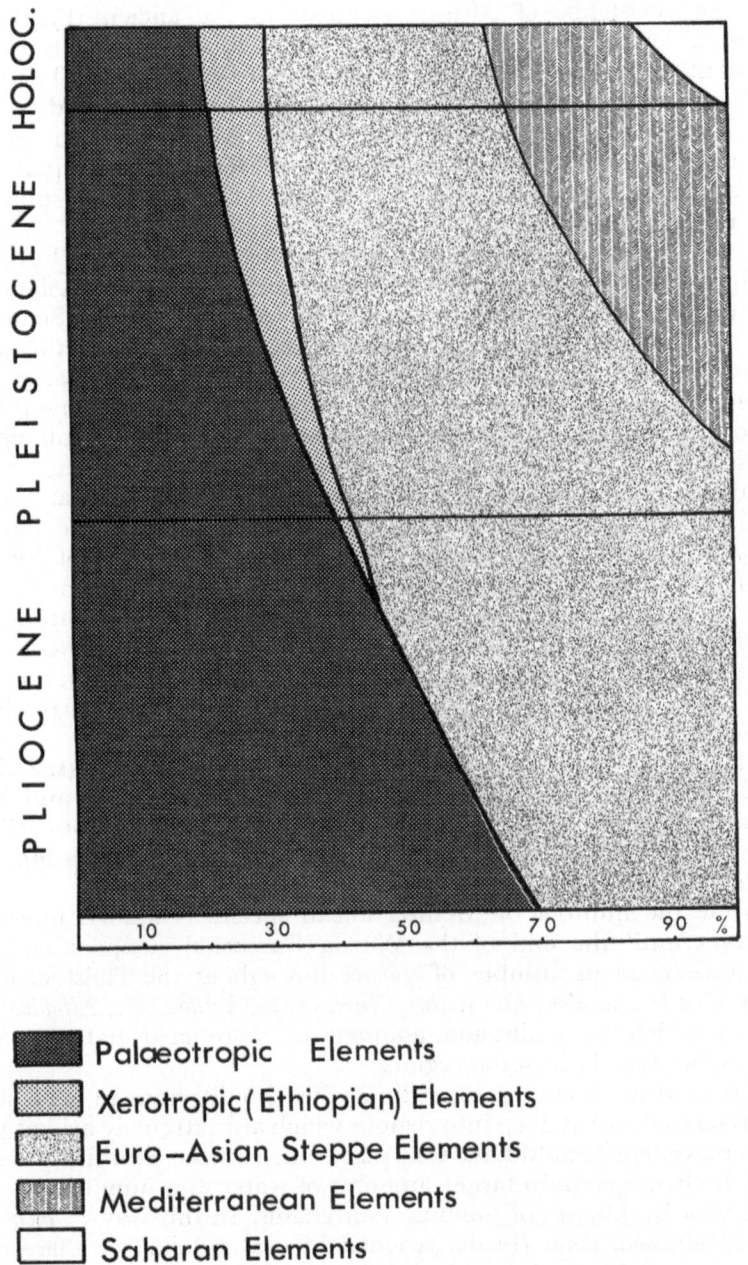

Fig. 1. Changes of the main faunal components in Israel since the Neogene time.

333

with the existence of African elements in the ancient Pleistocene of Europe. The enclosure of tropical animals in the Levant underwent a constant decline eventually since the end of the Neogene. The isolation of the tropical faunas was at first only partially effective and toward the late Pleistocene time almost totally (even with regard to few of the xerotropic species). There is no reason to believe that Ethiopian animals reinvaded Asia (save Arabian or Saharian desert species, a subject to be discussed later on).

Cooler steppe conditions prevailed during the Pliocene, humid at its beginning and more arid towards its end, enforced the Asiatic steppe biomes as far as the Northern African Atlantic shores. Semi-arid to desert conditions has no doubt been accentuated along the southern stretches of Palaearctic Asia, resulted mainly by the increasing heights of the mountainous ranges of central Asia during the Neogene (OVCHIN-NIKOV, 1946). The stamp of Irano-Turanian Vegetation is still impressive in the semi-desert regions of the Near-East. In more arid areas – Southern Sinai, southern Israel and Arabia – Irano-Turanian relics are known to occur mainly on higher altitudes amid regions which are floristically Saharo-Sindic. It must be conceded that the expansion of Euroasiatic forms in the Pliocene appeared to be a major-scale phenomenon which affected major parts of the already well determined Palaearctic region, reached the Atlantic shores of Africa where trees like *Pistacia atlantica* still grow, but never succeeded to cross the developing Saharan belt. In the Upper Pleistocene of Sudan, the central Asiatic borrowing vole *Ellobius* still occur.

Though very scattered, none of the Bethlehem remains (see above) is of European origin, but exclusively Afro-Asian. Central and Western European genera and species first appear in the Levant not earlier than early Middle Pleistocene (as found in the 'Ubeidiya formation, Jordan Valley, HAAS, 1968).

While the number of Mediterranean species remained more or less constant until the end of the Würm, Euroasian steppe animals were slowly reduced in number of species throughout the Pleistocene of the Near-East (*Lagurodon, Allocricetus, Cricetus* spp., *Jordanomys, Ellobius*). Those species which were already adapted to more arid habitats are still dominant (Fig. 1) in some regions.

It is hard to tackle the exact time when Mediterranean elements were enforced southward deep into regions which are extremely arid at present. Such movements could have take place either through gullies and streams (which always contain larger amounts of water and humidity), or along the cooler backbones of mountainous chains. In this way *P. pica asirensis* and *Dendrocopus dorae* (birds) reached Mecca and survive there as relics until present (MEINERTZHAGEN, 1954), as well as *Rana ridibunda* (HAAS, 1957). PARKER (1956) mentioned a glacial Palaearctic invasion into North Africa with regard to *Hyla arborea*. Few scattered populations of

Eliomys melanurus occupy a typical rocky desert landscapes and wadi escarpments amid the arid zone of southern Sinai and the Negev plateau of Israel. These isolated populations were probably cut off during the Pleistocene from the main normal arboreal dormice found in the north.

Animals which underwent (during the Neogene and Pleistocene) extreme adaptations to arid habitats in the Arabo-Nubian deserts (of which southern Israel and Sinai peninsula share an integral part) never invaded more humid areas, none of which is known to occur even in the most degraded Mediterranean areas of the Near-East; rarely found in semi-arid regions. These elements are as a rule limited to isohyete of 70–100 mm and less. A few species known from the Israel-Sinai fauna are: *Uromastix aegyptius, Uromastix ornatus, Agama sinaita* (Agamidae, Lacertilia); *Oenenthe leucopyga, Cercomella melanura* (birds); *Gerbillus nanus, Sekeetamys calurus* and *Acomys russatus*. Such desert animals were originated either from the ancient Palaeotropic faunas, or from the south Palaearctic stock of species which were trapped during the Neogene and Pleistocene in the slowly developing desert belt, undergoing speciation through extreme specializations to arid climates.

A swift retreat of the wood and forest dwellers (found to be at their maximum in the Upper Pleistocene of the Near-East (TCHERNOV, 1968b) took place during the Post-Glacial period, partly on the brink of historical time (apparently due to efforts of man). The dual effect of climate and man enforced few forest-dwellers (3 species of deer, or *Sciurus anomalus*) far northward. This phenomenon coincided with a rapid penetration of few typical Saharan species which invaded the stretches of dunes (TCHERNOV, 1968b) along the Mediterranean coastal plains of Israel as far as Acre in the north. Few gerbillids (ZAHAVI & WAHRMAN, 1957) underwent swift speciation during the short period parallel with the Post-Glacial time (i.e. *Gerbillus allenbyi, Meriones sacramenti*). Most of the Saharan newcomers (being well adapted to dunes) could not move eastward and join the less adapted dune dwellers of the Arava-Valley (southern section of the Israeli Rift Valley), or the dunes of central and southern Sinai, as the recent to sub-recent dunes of the Mediterranean coastal plains were never geographically connected with the much older (in origin) dunes of the Arava Valley and Southern Sinai. This new Saharan group of animals is found in Israel amid a typical Mediterranean area, still raising a kind of a paradox for the local naturalists. It demonstrates the youngest faunal assemblage – though very restricted in area – found in the Near-East and the only invasion from the African continent probably since the early Pleistocene (but never again from the Ethiopian region).

The fossil Neogene faunas of the Near-East are essentially Afro-South Asians. The earlier Pleistocene stratum (Bone Bearing Beds of Bethlehem) (Table 1) contains already close to 60% of Palaearctic elements (at a rough estimate). At the beginning of the Middle Pleistocene ('Ubeidiya formation) the percentage of Palaeotropic species decreased to 30% of the total fauna; and today is close to 20% (Fig. 1). The steady decrease of the tropical elements has taken place during the Neogene, and was accentuated throughout the Pleistocene, parallel with and as a result of the establishing of a south Palaearctic desert zone, and the total desiccation trend of the whole area.

The Composition of the Rodents Faunas in the Pleistocene of Israel

STRATIGRAPHY

An attempt was made to reexamine the species of the rodents in the different layers exposed from the main Pleistocene sites of Israel; to determine chronologically the changes which this group has undergone, and the direction in which the fauna has tended to develop during this period. A study of the faunal assemblages makes it possible to judge the nature and the trends of palaeoecological and climatic changes occurring in the different layers throughout the Quaternary. Analysis of the habitats occupied by the rodent species provided a sound basis for a reconstruction of the changing landscapes, and to draw a picture of the vicissitudes during the Pleistocene.

The material studied was uncovered from the following sites (Table 1): a) The 'Ubeidiya Formation. The site is located in the Central Jordan Valley, 3 km southward of the Sea of Galilee. The site was known for its important geological exposures since the beginning of the century when it was often used as a key to understand the stratigraphy and geology of the Jordan Rift Valley. In 1959 a prehistoric site was discovered in the same area. Following series of archaeological excavations a more accurate dating of the site could be postulated (PICARD & BAIDA, 1966, 1966a; STEKELIS, 1966; STEKELIS, BAR-YOSEF & SCHICK, 1969, HAAS, 1966; TCHERNOV, 1968). The dating of the 'Ubeidiya Formation was based on 4 criteria; tectonics, stratigraphy, fauna and prehistoric assemblages. Most scholars agree that the creation of the Jordan Valley graben was resulted by 2 main tectonic events; the first, and of the major consequence, took place in earlier Pleistocene time, while the second occurred during the post-'Ubeidiya period. The faunal assemblages of 'Ubeidiya is composed of 'Villafranchian strugglers' (HAAS, 1966, 1968), coexisting with "modern forms". The faunal assemblages of 'Ubeidiya could not have been assigned according to the early Pleistocene survivors, but rather by the post Villafranchian newcomers (TCHERNOV, 1968). The lithic assemblages of 'Ubeidiya show strong affinities with those of Olduvai Gorge (Tanzania, East-Africa), Upper Bed II, or early Middle Pleistocene (LEAKEY, 1967, STEKELIS, BAR-YOSEF & SCHICK, 1969). On the basis of those contexts it seemed appropriate to claim an early Middle Pleistocene age for the site of 'Ubeidiya. This site yielded enor-

336

Table 1. Tentative Illustration of the Stratigraphical Position of the main Vertebrate Bearing sites in the Pleistocene of Israel, with a suggested Correlation to the Glacial Chronology and Absolute Datings.

Sites and Prehistoric Cultures in Israel	Location		Absolute Dating (B.C.)	Glacial Chronology
		Holocene		
Natufian			10,000	Post Glacial
	Mt. Carmel Caves and		20,000	
Aurignacian (Upper Palaeolithich)	Hayonim Cave (Northern Israel)	Upper Pleistocene	40,000	Weichsel
Upper Mousterian			70,000	
Lower Mousterian				
Oumm-Qatafa Cave (Upper Acheulian)	Judean Hills		120,000	Eem
Jisr Banat Yakub	Upper Jordan Valley	Middle Pleistocene		Saale
Karst Fissue Filling	Give'at Shaul near Jerusalem		250,000	Holstein
'Ubeidiya Formation	Central Jordan Valley		500,000	Elster
Bone Bearing Bed of Bethlehem	Bethlehem near Jerusalem	Lower Pleistocene		

mous quantities of rodents remains which were and are studied by HAAS (see in particular 1966, 1968).

b) Karst Fissure Filling near Jerusalem. In the western suburbs of Jerusalem (Give'at Shaul) in the Judean Hills, 800 m above sea level, a Karst breccia was discovered, found to be the remains of an old cave floor deposit, composed of a fossil red soil, fragments of stalagmites and stalactites, mixed with irregularly shaped limestone pebbles (TCHERNOV, 1968). The animal remains found consisted of broken bones and numerous isolated teeth, mostly the remains of small mammals, mainly rodent. It is evident from the large numbers of teeth of the smaller, mainly nocturnal, mammals which were found in isolated accretions, that most of the material was brought from nearby by owls, whose pellets accumulated on the cave floor. The Jerusalem fauna is younger than that of 'Ubeidiya, but older than any Upper Pleistocene assemblage known from the area. It might be considered as more or less contemporary with the Holstein Interglacial. The Oumm-Qatafa beds, which represent the oldest Upper Pleistocene strata known from the Near-East have a faunal assemblage contemporary with the Eem interglacial (TCHERNOV, 1968b). The Jerusalem assemblage is definitely older than that.

337

c) The material from the Upper Pleistocene sites was excavated from the following prehistoric caves (Table 1):

1. The layers of Oumm-Qatafa which contain very plentiful faunal remains; Oumm-Qatafa is assumed to be the most ancient cave found on the Levant. RUST (1950) points out that the Oumm-Qatafa stages are at least partially older than the Jabrudian culture. Although the time range cannot be as wide as that proposed by NEUVILLE (1951), the fauna does not antedate the Last Interglacial (TCHERNOV, 1968b). WOLDSTEDT (1962) claimed a more or less maximal absolute age possible for the Oumm-Qatafa layers, assuming that they do not antedate the Eem Interglacial, which has been fixed for our purposes at 120,000 years. The cave is located in what is now the desert region of the Judean hills in the gully of Khareitoun. The cave lies about 20 km from the present shore of the Dead Sea. First analysis of the micromammalofauna of the cave was done by HAAS (1951), and later by the author (1968b).

2. Concerning the lower Mousterian fauna (cave of Tabun, Mt. Carmel), use was made of the detailed results arrived at by GARROD & BATE (1937), BATE (1942, 1943). A possible absolute age for the older Mousterian culture of the Near-East is around 70,000 years.

3. Upper Mousterian faunas were mainly uncovered from the cave of Tabun, Kebara (Mt. Carmel). Two C^{14} dates were obtained for the younger Mousterian layers of the Kebara cave, for which – 39,000 B.C. and – 42,000 B.C. were given.

4. The Upper Palaeolithic (mainly Acrignacian) faunal assemblage at our disposal was mainly uncovered from the Hayonim cave (Western Galilee, northern Israel) (BAR-YOSEF & TCHERNOV, 1966), for which a date of – 20,000 B.C. is usually given.

5. Natufian faunas were studied mainly from the caves of Wadi Fallah and Abu-Usba (Mt. Carmel) (STEKELIS & HAAS, 1952), and from the Hayonim cave (see above), for which a date of – 10,000 B.C. is assumed.

THE COMPOSITION OF THE PLEISTOCENE RODENTS (Table 2)

Remains of *Sciurus anomalus* were never found in 'Ubeidiya and Give'at Shaul (Jerusalem) Middle Pleistocene exposures, but first appear in the cave deposits of Oumm-Qatafa (early Upper Pleistocene). It showed no significant morphological changes during this period. In the layers of Oumm-Qatafa, it did not constitute an appreciable proportion of the fauna, and was much more widespread during the Mousterian, but especially in the Upper-Palaeolithic to Natufian time. Squirrels apparently existed in Israel until the Neolithic when they retreated northward, followed by the extensive propagation of the rock dwellers – *Acomys cahirinus* and *Gerbillus dasyurus* in Mediterranean regions.

The genus *Hystrix* is generally very scarce in the Pleistocene deposits of Israel, either in caves, or in open sites, except the cave of Geula

338

Table 2. Stratigraphic span of Pleistocene Rodent Species in Israel

'Ubeidiya Formation (HAAS, 1966) Early Middle Pleistocene (—500,000)	Karst Fissure Filling Jerusalem (TCHERNOV, 1968) Late Middle Pleistocene (—250,000)	Oumm Qatafa Cave (TCHERNOV, 1968b) Early Upper Pleistocene (—120,000)	Lower Mousterian (BATE, 1937, 1942) (—70,000)	Upper Mousterian (Various Authors) (—40,000)	Upper Palaeolithic (TCHERNOV, 1968b) (—20,000)	Natufian (Various Authors) (—10,000)	Recent
—	—	*Sciurus anomalus*	*Sciurus anomalus*	*Sciurus anomalus*	*Sciurus anomalus*	*Sciurus anomalus*	—
Hystrix sp.	—	*Hystrix* sp.	*Hystrix* sp.	*Hystrix angressi*	*Hystrix indica*	*Hystrix indica*	*Hystrix indica*
—	*Cryptomys asiaticus*	—	—	—	—	—	—
Myomimus sp.	*Myomimus judaicus*	*Myomimus roachi*	*Myomimus roachi*	*Myomimus roachi*	*Myomimus roachi*	*Myomimus roachi*	—
—	—	—	—	—	—	*Eliomys* sp.	*Eliomys melanurus* *Dryomys nitedula*
Spalax minutus	*Spalax ehrenbergi*	*Spalax ehrenbergi*	*Spalax ehrenbergi*	*Spalax ehrenbergi*	*Spalax ehrenbergi*	*Spalax ehrenbergi*	*Spalax ehrenbergi*
—	—	*Spalax newillei*	*Spalax newillei*	*Spalax newillei* *Spalax kerbarensis*	*Spalax kerbarensis*	—	—
Parallactaga sp.	—	—	—	—	—	—	—

Table 2. (Continued)

Progonomys spp.	—	—	—	—	—	—	—
Parapodemus jordanicus	—	—	—	—	—	—	—
Apodemus (giant sp.)	*Apodemus mystacinus*	*Apodemus mystacinus*	*Apodemus mystacinus*	*Apodemus mystacinus*	*Apodemus mystacinus*	*Apodemus mystacinus*	*Apodemus mystacinus*
Apodemus alsomyoides	*Apodemus levantinus*	*Apodemus levantinus*	*Apodemus levantinus*	—	—	—	—
Apodemus sp.	*Apodemus caesareanus*	*Apodemus caesareanus*	*Apodemus caesareanus*	*Apodemus sylvaticus*	*Apodemus sylvaticus*	*Apodemus sylvaticus*	*Apodemus sylvaticus*
—	—	*Arvicanthis ectos*	*Arvicanthis ectos*	—	—	—	—
—	—	*Rattus haasi*	—	*Rattus rattus*	*Rattus rattus*	*Rattus rattus*	*Rattus rattus*
—	—	*Mastomys batei*	*Mastomys batei*	—	—	—	—
Mus sp.	—	*Mus musculus*	*Mus musculus*	*Mus musculus*	*Mus musculus*	*Mus musculus*	*Mus musculus*
—	—	—	—	—	*Acomys cahirinus*	*Acomys cahirinus*	*Acomys cahirinus*
—	—	—	—	*Cricetulus migratorius*	*Cricetulus migratorius*	*Cricetulus migratorius*	*Cricetulus migratorius*
Allocricetus bursae	*Allocricetus bursae*	*Allocricetus jesreelicus*	—	—	—	—	—
—	—	*Allocricetus magnus*	—	—	—	—	—
Nannocricetus sp.	*Mesocricetus* n. sp.	*Mesocricetus aramaeus*	*Mesocricetus aramaeus*	*Mesocricetus aramaeus*	—	—	—

Table 2. (Continued)

	Col 1	Col 2	Col 3	Col 4	Col 5	Col 6	Col 7	Col 8
	Cricetus cricetus *Cricetus angustirostris* *Cricetus kormosi*	*Cricetus cricetus*	—	—	*Mesocricetus auratus*	*Mesocricetus auratus*	*Mesocricetus auratus*	*Mesocricetus auratus*
	Gerbillus sp.	*Gerbillus* n. sp.	*Gerbillus dasyurus**	—	—	*Gerbillus dasyurus*	*Gerbillus dasyurus*	*Gerbillus dasyurus* *Gerbillus allenbyi* *Gerbillus pyramidum*
	Meriones obeidiensis	—	*Meriones tristrami* *Psammomys obesus**	*Meriones tristrami*	*Meriones tristrami*	*Meriones tristrami*	*Meriones tristrami*	*Meriones tristrami*
	Jordanomys pusillus *Lagurodon* cf. *aranke*	*Jordanomys haasi*	—	—	—	—	—	—
	—	—	*Ellobius fuscocapillus* *Microtus guentheri*	*Ellobius fuscocapillus* *Microtus guentheri*	*Arvicola terrestris* *Microtus guentheri*	*Microtus guentheri*	*Microtus guentheri*	*Arvicola terrestris* *Microtus guentheri*

* See Text

341

(WRESCHNER, 1967) in Mt. Carmel, where a Mousterian culture was uncovered. This cave yielded numerous fragments of *Hystrix angressi* (FRENKEL, 1970), among the faunal remains. *Hystrix indica*, which is at present common in Israel, did not coexist (Table 2) with *H. angressi* in Mousterian times, but first appeared in the Upper Palaeolithic.

The existence of *Cryptomys asiasticus* (Bathyergidae) was a short episode in the Near East. It was found up till now only in the karst fissure filling near Jerusalem. Today the family is restricted to Africa, south of the Sahara, but *Gysorhychus* in Mongolia (MATHEW & GRANGER, 1923) existed in Asia (Mongolia) in Oligocene time. It is similar in certain respects to the recent African genus *Georhychus*, mainly a southern African form. After the Oligocene this group was never again found in Asia; only from Miocene times onwards did it begin to appear in Africa. At least one representative of an African genus reinvaded the Near-East, probably during the Neogene, and survived there as late as the later Middle Pleistocene period. The habitat of *Cryptomys* as known today from tropical Africa is not well defined. However, as a burrowing animal, it is more commonly found in soft soils or sands, but may also be found in more rocky areas. As to vegetation cover, *Cryptomys* prefers low bushland, large clearings or widely open forest.

The time of arrival and disappearance of several glirids in the Near-East is still an unsolved problem. The most common representative of the family is *Myomimus* which is found through the Middle and Upper Pleistocene (Table 2). An extensive and detailed study concerning this genus has been made by G. HAAS (Hebrew University) and is now in preparation (personal communication). *M. judaicus* was found in Give'at Shaul (Jerusalem) the late Middle Pleistocene karst filling, while *M. roachi* alone is known from the Upper Pleistocene of Israel. Bearing in mind that the glirids are a fairly constant group (HEROLD, 1958), and that the molar pattern of *M. judaicus* is quite different from that of *M. roachi* (TCHERNOV, 1968a) the two species do not form a chronocline. Although it has not been proved that they overlapped in time, it remains a possibility. The only living representative of this genus is *M. personatus* inhabiting eastern Europe, where large relics of Pliocene habitats still exist (KOWALSKI, 1963). This genus probably did not undergo adaptations towards arboreal life, as did most of the modern glirids, but presumably remained terrestrial. Worth noting is the fact that *M. roachi* was the predominant glirid until as late as the late Bronze (!), and then it vanished.

Strangely enough *Eliomys melanurus* is scattered today in few isolated patches in the semi-arid and arid zone of Israel and Sinai, there occupying types of rocky desert landscapes. In the northern part of the country it appears as a typical arboreal animal. The genus did not appear earlier than the Natufian age (10,000 B.C.).

Remains of *Dryomys*, which is found today only in the northern

342

mountainous region of Israel, were never recorded up till now, and might have invaded Israel very recently.

Being a terrestrial animal, the disappearance of *Myomimus* (Table 2) could not be the cause for the delayed invasions of *Dryomys & Eliomys*, primarily arboreal genera.

Many forms of *Spalax* are known to exist during the Pleistocene of Israel. According to the level of complexity of Pm_1, all of them belong to the *ehrenberg* group (= *Microspalax* Méhély, 1909) (see TCHERNOV, 1968b). *Spalax minutus* (HAAS, 1966) from the 'Ubeidiya formation is closely related to *Spalax ehrenbergi* from the Jerusalem karst filling, a species which constituted an interesting chronocline through the entire Middle & Upper Pleistocene. It showed many gradient morphological characters, which enabled us to study the evolutionary changes of the species. Two additional species were found; *S. neuvillei* appears in the Aheulian of the Oumm-Qatafa beds (—120,000 years) and retreated from the area towards the end of the Mousterian, some 35,000 B.C. *S. kebarensis*, which existed in the Mt. Carmel region from upper Mousterian times to the Natufian-Neolithic period was of relatively large dimensions, and showed a completely different and unusual construction of the coronoid process. The centre of distribution of both *S. neuvillei & S. kebarensis* was most probably far away from Israel (Central Asia?) which might explain their short episodes in the area. These blind fossorial animals occupy mainly a rather open grassland to semi arid steppe landscape, but avoid wooded or forest habitats.

Of the Dipodidae the only genus found was *Parallactaga* sp. This rather steppic form was uncovered from the 'Ubeidiya formation.

Among the Murinae, two Neogene 'stragglers' (HAAS, 1968) still survived in the Central Jordan Valley as late as during early Middle Pleistocene time. The genus *Progonomys* from which at least 2 different species could have been detected by HAAS (1966), and the genus *Parapodemus*. Israel was probably the last area of refuge for these two genera which later on were never found again.

3 species of *Apodemus* were co-existing during the whole Middle Pleistocene and major part of the Upper Pleistocene (later Acheulian and lower Mousterian). One of the species – *A. mystacinus* – continues to the present time. The other two – *A. levantinus and A. caesareanus* – spontaneously disappeared in the upper Mousterian, when *A. sylvaticus* replaced them. Today it is limited to the northern areas of the country. Owing to the large size differences in the 3 co-existing species, competition between them might well have been largely prevented. *A. levantinus* and *A. caesareanus* showed a more highly developed cheek teeth tuberculation than the recent members of 'Sylvaemus'. The more ancient populations of *A. mystacinus* showed massive proportioned mandibles, and a highly developed tuberculation of cheek teeth. At later levels it appeared in

increasing numbers of specimens, and more closely resembled the 'delicate' proportions of the recent form.

Two Ethiopian elements, *Arvicanthis ectos* and *Rattus (Mastomys) batei* (Murinae), appeared at the beginning of the Upper Pleistocene for a short period, and did not exist any more towards the later Mousterian time. *Mastomys* which was found from Neogene of the Siwalic series, rapidly retreated from Asia, whose last remnant probably occupied the Near-East in the Upper Pleistocene vanished shortly thereafter, became an exclusive member of tropical Africa.

The Near-East was probably populated by *Rattus rattus* shortly after the evacuation of the area by *Arvicanthis*, which is also restricted to Africa at present. Another species of rat – *Rattus haasi* – was uncovered only from the Oumm-Qatafa beds. According to its cheek teeth pattern, *R. haasi* is in an intermediate position between the *rattus* and *norvegicus* groups, representing the most ancient species of *Rattus* found until now in the eastern Mediterranean region. It is difficult to relate a certain biotope to this fossil species, as *Rattus* is quite euryoekous, though generally inhabiting wet areas, with higher humidity and dense vegetation.

An undetermined *Mus* is mentioned by HAAS (1966) from the 'Ubeidiya formation. The 'Ubeidiya *Mus* is different from *Mus musculus* which first appear in the Oumm-Qatafa layers, and existed in the region throughout the entire Upper Pleistocene. During this period it showed only a slight change in the tuberculation of the molars and in the proportions of the skull.

Acomys cahirinus occupies a bare rocky landscaped devoid of trees. It made its appearance in the Mediterranean region of the country during the Upper Palaeolithic and post-glacial times. At the same time few wood-dwelling species retreated northward and disappeared from Israel (*Sciurus anomalus, Cervus elaphus, Pica pica*, TCHERNOV, 1962). The wooded areas in the Near-East were degenerated slowly, a phenomenon which was sharply accentuated after the Natufian, when many more wood-dwellers disappeared from the area, or retreated into isolated 'islands' in northern Israel. *Acomys cahirinus* which probably lived in the arid zones of the Near-East throughout the Pleistocene, invaded northern areas in the wake of the declining Mediterranean woods, and the first large scale appearance of the bare rocky biotope in a formerly typical Mediterranean region. It is now occurring in the Crete & Cyprus islands which were cut off from the mainland in late Pliocene time. Populations of *Acomys* having originated in Africa should have occurred during upper Pliocene times in the Near East. Typical Mediterranean flora and fauna penetrated Israel only later, probably during Lower and early Middle Pleistocene, a period when *Acomys* has been halted up until as late as Upper Palaeolithic, when large areas began to be cleared of forest. This was the right time for *Acomys* to invade the area the Mediterranean part of the country.

While in the Lower and Middle Pleistocene, *Allocricetus bursae* inhabited

Europe and the Near East, two new forms appeared during the Upper Pleistocene in the eastern Mediterranean corner, long after the genus had vanished from Europe. These two species (*A. magnus* and *A. jesreelicus*) succeeded to survive (Table 2) until the lower Mousterian.

The genus *Nannocricetus* sp. appeared only in the 'Ubeidiya formation. The genus *Cricetus*, from which 3 species and known from the 'Ubeidiya beds (*C. cricetus*, *C. kormosi* and *C. angustirostris*); one of them *(C. cricetus)* continued until later periods of the Middle Pleistocene (Jerusalem karst filling). The genus *Cricetus* is not known from the Upper Pleistocene of the Near-East, as it probably retreated northward toward the onset of this period. *Mesocricetus* probably first appeared in the area only during the later Middle Pleistocene, having presumably been in competition with *Cricetus*. As soon as *Cricetus* retreated for good, *Mesocricetus* became quite common. A dwarf-type fossil, *Mesocricetus aramaeus*, is quite different from *Mesocricetus auratus* and apparently lived in a moister biotype (but not in the closely afforested region, as BATE propounded, 1943), from the Acheulian until the end of the Upper Mousterian. *Mesocricetus auratus* occupied more arid grassland, appeared in the Upper Mousterian, and disappeared during the Natufian-Neolithic time. Another grassland species – *Cricetulus migratorius* – appeared in the Near East along with *M. auratus*, and is now not very common in Mediterranean areas.

An undetermined species of *Gerbillus* was uncovered in the 'Ubeidiya Formation (HAAS, 1966, 1968). A much larger, and different species is known from a later stratum – the Jerusalem karst filling – both forms indicating the existence of fairly open, steppic to rocky landscapes. *Gerbillus dasyurus* lived in a biotype of bare rocks in the more arid zones of the Near-East at least during the whole Upper Pleistocene, and had manifested only a few morphological changes. It reached Mt. Carmel's Mediterranean area only as late as the Upper Palaeolithic; shortly after *Acomys cahirinus* arrived, following the wake of the declining and retreating woods and forests. Much like *Acomys cahirinus*, *Gerbillus dasyurus* is known at present from the most arid zones of Israel and Sinai, to typical Mediterranean regions of northern Israel, always occupying bare rocky slopes.

Meriones obeidiensis is known only from the Middle Pleistocene sites. Later on, during the Upper Pleistocene, *Meriones tristrami* occupied the Mediterranean regions of Israel. It has been possible to show (TCHERNOV, 1968b) that during the Acheulian *M. tristrami qatafensis* (HAAS, 1951) was not only much larger in size than the recent *M. tristrami*, but had a far closer resemblance to *M. persicus*. Furthermore, it is feasible that *M. persicus* and *M. tristrami* may have had a common ancestry, and show some affinities to *M. obeidiensis*. Oumm-Qatafa Cave is situated at the fringe of a desert area, so that animals like *Gerbillus dasyurus* and *Psammomys obesus* are expected. *G. dasyurus* appears in the Mediterranean areas of Israel only in the Upper Palaeolithic (Table 2).

The extinct genus *Jordanomys* is known till now only from the Middle

Pleistocene of Israel by two different species: *Jordanomys pusillus* from 'Ubeidiya and *J. haasi* from the Jerusalem karst filling. The genus is characterized by a very simple molar pattern. The representatives of this genus are most probably, merely stragglers from much older times, which succeeded to survive in Israel as late as the Middle Pleistocene. The genus *Jordanomys* co-existed with another extinct genus – *Lagurodon*, with rootless molars, and no cement in the re-entrant angles (HAAS, 1966). This genus is not known anymore from the Middle Pleistocene of Europe. Both extinct genera probably occupied a cooler steppic habitat, much like the closely related genus Lagurodon today (HAAS, 1968).

Arvicola jordanica – an extinct species – was found only in the 'Ubeidiya formation, while A. terrestris is known only from the Mousterian of Mt. Carmel (HELLER, 1970), and from the Hula basin today (upper Jordan Valley).

Ellobius fuscocapillus first appeared in the Acheulian of the Oumm-Qatafa beds, probably in moist steppe habitat, and disappeared toward the later Mousterian. It is known as well from the Pleistocene of North Africa, but is restricted today to Central Asia.

The genus *Microtus* is not known prior to the Upper Pleistocene of the Near East. In Israel it is known by a single species – *M. guentheri*, which forms a continuous chronocline in abundant numbers during the entire Upper Pleistocene. This is the only known vole at present south of the Syrian-Lebanon borders.

FAUNAL REACTION TO CLIMATIC FLUCTUATIONS

The difficulty in grasping an immediate, clear and definite correlation between fauna and climate is a result of the nature of the animal's reaction to these changes, especially in those regions where climatic changes were not sharp.

Israel itself is located at the fringe of a geographical region which, since the end of the Miocene, has turned into a giant arid body, the area of which is over 8 million sq. km, and its reactions to the Alpine glaciers are relatively minor. Such an arid body must have affected the south-western corner of Asia, or at least lessen the force of the polar climatic fluctuations. Even if certain climatic changes did occur in Israel, the reaction of the fauna could not have been immediate; populations might have continued to exist if ecotypes suitable to the new surroundings were left; or might have been removed to a less optimal sector of its ecological amplitude. Thus *Meriones tristrami* reaches the limit of its general distribution at around the 500 mm isohyete, while in Israel – at the 100 mm isohyete. *Spalax ehrenbergi* behaved in a somewhat parallel manner. As a rule, thanks to the gradual pace of changes, animals are preserved for longer periods, while their original biotopes have changed or disappeared.

Many investigators based their assumptions on the premise that each change in climate affects the entire fauna, and causes a complete change of its picture. When the changes are more radical, the reaction is swift, as there are no chances, or not enough time for readaptations. This is not the case when the changes are weak; then fauna shows a relative stability. And even when a change does occur, the reactions are delayed, and cannot be ascribed to any one factor. No wave of invasion was observed during the entire Upper Pleistocene in correlation with the European glacial fluctuations. Once the Palaearctic fauna was established in Israel, no gross faunal changes took place, nor did a single northern species appear in the area which could have been pushed southward resulting from the increasing of the ice caps in the north.

The paradoxical phenomenon is that the enriching or enlarging of water bodies in arid or Mediterranean areas at times parallel to cold phases, facilitated the preservation of elements of tropical origin. An increase of northern elements in epochs parallel to cold phases was often expected, but this never occurred. On the other hand, in many cases where relics of tropical fauna were found (GARROD and BATE, 1937; BATE, 1934, 1942, 1943) a hot, damp tropical climate was claimed for that period. This is also untrue as all the Ethiopian animals living at present, or during at least the later epochs of the Pleistocene in Mediterranean habitats, were long ago readapted to such conditions. The many which could not – vanished slowly throughout the Pleistocene.

The Main Ecological Changes in the Quarternary of Israel

DEFINITIONS OF THE MAIN BIOTOPES

When their biotopes are categorized, the Pleistocene rodents fall into 4 criteria: Mediterranean wood dwellers; moist steppe dwellers (BUTZER, 1966); Mediterranean open grassy to semi-arid dwellers (MOREAU, 1955) (= dry steppe dwellers); rock dwellers. Safe determinations of the various ecological requirements are not, and could not be sharply defined when fossil animals are under consideration. It must be conceded as well that some species might occasionally or regularly occur in two (or more) kinds of biotopes, or in intermediate ones. For instance, *Hystrix indica* at present in Israel occupies as a rule an evergreen open-land to semi-arid areas, but might occasionally be found in parks and wood. *Apodemus mystacinus* regularly inhabits the Mediterranean woods of Israel, but is frequently found in Maquis landscapes, unlike *Cricetulus migratorius* – normally observed in Mediterranean grassland and guarigue landscapes, but sometimes occupies woods. *Rattus rattus* and *Mus* are well adapted to varied habitats, but prefer the more humid niches. Few species (like *Acomys cahirinus* and *Gerbillus dasyurus*) are found anywhere in the country, southward in the most extreme arid zones, and continuously northward through semi-arid areas to typical Mediterranean regions with annual

347

rainfall of 1,000 mm. But both species are everywhere exclusively restricted to rocky landscapes, bare slopes of wadis and rivers, devoid of vegetation.

In order to reconstruct the main changes of biotopes and landscapes in the whole region, it is sufficient to know the general ecological requirement, and the principal kind of habitat occupied by the considered species. In cases when a whole group of species, which inhabited a certain biotope, vanished or retreated from the studied region, it certainly means that this kind of habitat was largely restricted, or disappeared. Though we will never be able to fully understand the exact habitats of extinct animals, and it will always remain highly speculative, there is general agreement as to the kind of biotopes occupied by many of the Pleistocene rodents. As the quality of the Pleistocene landscapes in the Holoarctic regions is already reconstructed into some extent, it is easier to allocate the fossil species to their appropriate landscapes.

The dynamics of the environmental changes are far more drastic in areas neighbouring arid zones, or in general, on the fringe of the Palaearctic region. The borders between arid, semi-arid and temperate zones are as a rule shifting northward and southward. These are the critical areas where the most extreme changes took place; it is true for the Levantine province, and especially Israel *a priori* its geographical situation. There are sound reasons to believe that changes in the Near-Eastern landscape during the Pleistocene were quite extreme.

Definitions of the biotopes discussed here are as follows:
1. By 'Wood Dwellers', following BUTZER (1966), we generally mean species occupying woodland or parkland landscapes, dominated by trees like *Quércus callíprinos*, *Pistácia lentíscus* and *Ceratónia silíqua*. The Mediterranean woodland zone, as mentioned before, was probably established towards the upper part of the Pliocene, and later on invaded the Levant, presumably in Villafranchian times. The temperate woodland flora and fauna which since then was predominated in the Levant, continued to spread westward along the North African coastal line.
2. The 'Moist Steppe Dwellers' (BUTZER, 1966), is a landscape with BOURLIÈRE (1963) determined as the 'Euroasian moist steppes', characterized by luxurious herbaceous vegetation, tall grasses rich with water, bogs and swamps. This kind of moist evergreen grassland supported a considerable number of northern Asiatic steppe birds (TCHERNOV, 1968) in the 'Ubeidiya Formation (early Middle Pleistocene of Jordan Valley). Such steppe floral and faunal elements were predominant especially in central and northern Asia and eastern Europe, invaded the Levant and North Africa as far as the Atlantic shores. These kinds of landscapes culminated in the Pliocene but were very characteristic in the Levant during the Lower, and at least the earliest time of Middle Pleistocene. Some of the moist steppe elements survived in the warm temperate zone of the Levant until as late as Middle Pleistocene times (*Allocricetus bursae*, *Jordanomys haasi*).

348

3. The moist steppes are gradually degraded into dry steppes in more arid areas or epochs. A clearer definition is not possible, but such drier steppe landscapes are usually inhabited by more warm-temperate species, usually originating in the south Palaearctic region, Ethiopian region, or in general, warmer biozones. We will use the term 'Grassland Dwellers' for animals occupying such landscapes, which are sometimes dominated by semi-desert xerophytic grasses, with a scattering of deciduous thorny shrubs, characteristically creating in Israel a kind of thornscrub grassland, where the dominant plant is *Potérium spinósom.*

4. The bare rocky slopes are inhabited in the Mediterranean region by *Acomys cahirinus* and *Gerbillus dasyurus.* In the arid zones of Israel and Sinai more species are found in the same biotope: *Acomys russatus, Sekeetamys calurus* and *Eliomys melanurus.* Rocky slopes as an exclusive biotope for a few species of rodents cannot be determined floristically, as it may occur everywhere in the country, and is dominated in each place by different types of plants. Rock dwellers are not known from the Pleistocene of Mediterranean Israel until as late as the late Palaeolithic age, when a severe desiccation process (contemporary with the post-Würm) resulting in the degradation of woods and forests, leaving behind gradually increasing patches of bare landscape and creating new habitats for rock dwellers like *Acomys cahirinus* and *Gerbillus dasyurus.* At the fringe of the desert areas, rock-dwellers were naturally more abundant. Oumm-Qatafa (early Upper Pleistocene, Judean hills, NEUVILLE, 1951) were by that time already located in the vicinity of the arid zone, and such rodents remains were deposited there. But northward amid the Mediterranean area, the appearance of rock dwellers is much later.

5. 'Sand-Dune Dwellers' in the Mediterranean area are found only very late in the faunal history of Israel, as this kind of biotope invaded the temperate zone of the Near-East in post-glacial ages, supporting rodents like *Gerbillus gerbillus, Gerbillus allenbyi* and *Meriones sacramenti* (ZAHAVI & WAHRMAN, 1957).

As no fossil rodents were found until now in the desert areas of Israel and Sinai peninsula, any reconstruction of the biotopes, and the changes they could have undergone through the Pleistocene is still premature.

Analysis of the Main Habitats Occupied by the Pleistocene Rodents of Israel

Most of the moist steppe dwellers (Table 3) are extinct species which were extant mainly in the Vellafranchian of Europe, but survived in the Levant until much later epochs.

The various Levantinian fossil species of the genus *Myomimus* are closely related to the still living terrestrial dormouse *M. personatus* of eastern Europe. At least 3 species lived during the Pleistocene of Israel, and, surprisingly enough, vanished during the Bronze Age!

Table 3. Distribution of Moist Steppe Species in the Pleistocene of Israel

Moist Steppe Dwellers	'Ubeidiya (−500,000)	Karst Filling Jerusalem (−250,000)	Oumm Qatafa Cave (−120,000)	Lower Mousterian (−70,000)	Upper Mousterian (−40,000)	Upper Paleolithic (−20,000)	Natufian (−10,000)	Recent
Cryptomys asiaticus	−	+	−	−	−	−	−	−
Myomimus sp.	+	−	−	−	−	−	−	−
Myomimus judaicus	−	+	−	−	−	−	−	−
Myomimus roachi	+	−	+	+	+	+	+	−
Parallactaga sp.	−	−	−	−	−	−	−	−
Arvicanthis ectos	−	−	+	+	−	−	−	−
Rattus haasi	−	−	+	−	−	−	−	−
Mastomys batei	−	−	+	+	−	−	−	−
Allocricetus bursae	+	+	−	−	−	−	−	−
Allocricetus jesreelicus	−	−	+	−	−	−	−	−
Allocricetus magnus	−	−	+	−	−	−	−	−
Nannocricetus sp.	+	−	−	−	−	−	−	−
Cricetus cricetus	+	+	−	−	−	−	−	−
Cricetus angustirostris	+	−	−	−	−	−	−	−
Cricetus kormosi	+	−	−	−	−	−	−	−
Jordanomys pusillus	−	−	−	−	−	−	−	−
Jordanomys haasi	+	+	−	−	−	−	−	−
Lagurodon cf. aranke	−	−	+	−	−	−	−	−
Ellobius fuscocapillus	−	−	−	+	−	−	−	−

Table 4. Distribution of Grassland Species in the Pleistocene of Israel

Grassland Dwellers	'Ubeidiya (−500,000)	Karst Filling Jerusalem (−250,000)	Oumm Qatafa Cave (−120,000)	Lower Mousterian (−70,000)	Upper Mousterian (−40,000)	Upper Paleolithic (−20,000)	Natufian (−10,000)	Recent
Hystrix sp.	+	—	+	+	—	—	—	—
Hystrix angressi	—	—	—	—	+	—	—	—
Hystrix indica	+	—	—	—	—	+	+	+
Spalax minutus	—	—	—	—	—	—	—	—
Spalax ehrenbergi	—	+	+	+	+	+	+	+
Spalax neuvillei	—	—	+	+	+	—	—	—
Spalax kebarensis	—	—	—	—	+	+	—	—
Cricetulus migratorius	—	—	—	—	+	+	+	+
Mesocricetus aramaeus	—	—	+	+	+	—	—	—
Mesocricetus auratus	—	—	—	—	+	+	+	—
Mesocricetus n. sp.	—	+	—	—	—	—	—	—
Meriones obeidiensis	+	+	—	—	—	—	—	—
Meriones tristrami	—	—	+	+	+	+	+	+
Microtus guentheri	—	—	+	+	+	+	+	+

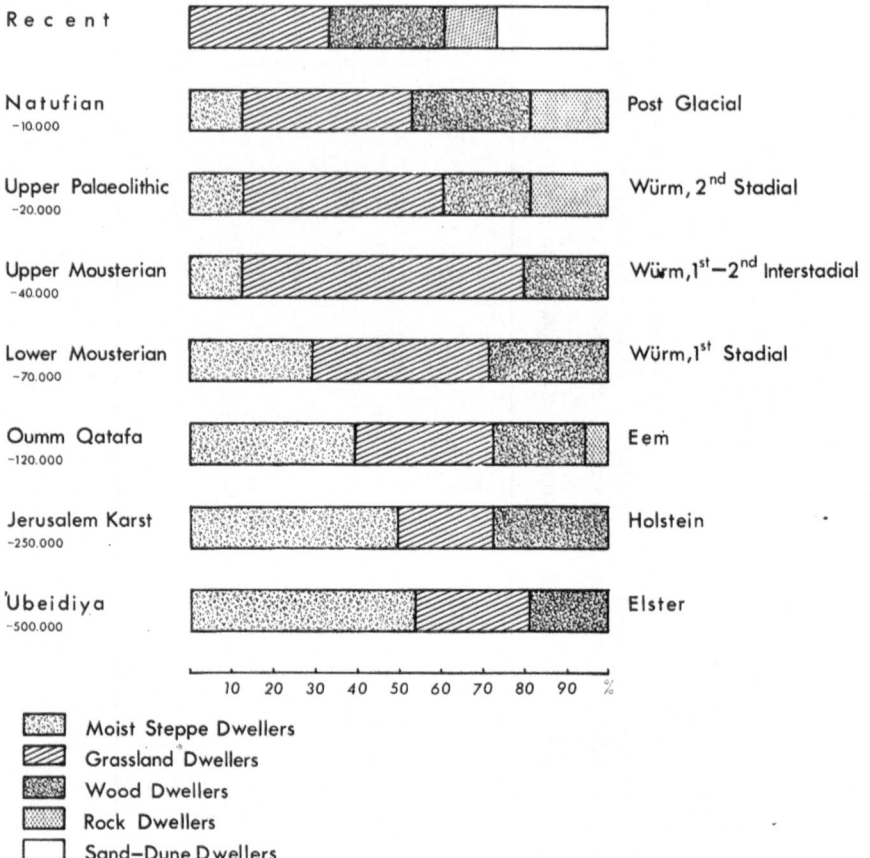

Recent

Natufian
−10.000
Post Glacial

Upper Palaeolithic
−20.000
Würm, 2ⁿᵈ Stadial

Upper Mousterian
−40.000
Würm,1ˢᵗ−2ⁿᵈ Interstadial

Lower Mousterian
−70.000
Würm,1ˢᵗ Stadial

Oumm Qatafa
−120.000
Eem

Jerusalem Karst
−250.000
Holstein

'Ubeidiya
−500.000
Elster

10 20 30 40 50 60 70 80 90 %

Moist Steppe Dwellers
Grassland Dwellers
Wood Dwellers
Rock Dwellers
Sand−Dune Dwellers

Fig. 2. Changes of the main habitats throughout the Pleistocene of Israel.

The major part of the moist steppe dwellers did not last for a long time in the Levant, and became extinct after the time of 'Ubeidiya (early Middle Pleistocene); *Parallactaga* sp. *Nannocricetus* sp; *Cricetus angustirostris, Cricetus kormosi, Jordanomys pusillus* and *Lagurodon* cf. *arankae* (HAAS, 1966, 1968a). Other species survived even beyond the 'Ubeidiya limit. In the karst fissure filling near Jerusalem (Table 2), 5 such species still occur; two of which (*Cricetus cricetus* and *Allocricetus bursae*) continued from the 'Ubeidiya times; two others belong to genera which existed in 'Ubeidiya, but appeared here as different species (*Myomimus judaicus* and *Jordanomys haasi*). The fifth one is the genus *Cryptomys* which probably occupied a kind of evergreen grassland area much like some of the blesmols which live in the plains of South Africa.

352

The genus *Allocricetus* (though appeared to be identical with *Tscherskia*, a subgenus of *Cricetulus*, we use the old term) is still found during the lower part of the Upper Pleistocene, in the Oumm-Qatafa cave, in two new species: *A. jesreelicus* and *A. magnus*. By this epoch few new genera appear for the first time, probably replacing in quite a similar biotope the above mentioned extinct forms. The new species are *Arvicanthis ectos*, *Rattus haasi*, *Mastomys batei* and *Ellobius fuscocapillus*, all of which required dense vegetation and a rich supply of water.

Dry grassland was too dry a habitat to support the following species: *A. ectos* & *Mastomys batei* (Ethiopian elements), *Rattus haasi* (probably Oriental), and *Ellobius fuscocapillus* (probably originated in the Asian steppes). Only *Mastomys batei* continued to live in the lower Mousterian age, while all the rest disappeared. The genus *Allocricetus* *(= Tscherskia)* became extinct. From later Upper Mousterian onward only *Myomimus roachi* survive and vanished in Historical time.

Mediterranean grassland dwellers (Table 4) are still few in Middle Pleistocene times, but increase in number of species (Fig. 2) during the Upper Pleistocene. In 'Ubeidiya (HAAS, 1966) 3 species were listed: *Hystrix* sp. *Spalax minutus* (which the author believes to be closely related to *Spalax ehrenbergi*) and *Meriones obeidiensis*. Porcupines are missing from the karst filling of Jerusalems' faunal assemblage, but *Spalax ehrenbergi* and *Meriones obeidiensis* are extant. The genus *Mesocricetus* (probably of a new species) appears for the first time in the Levant. At the beginning of the Upper Pleistocene (Oumm-Qatafa cave deposits) we find already 6 species living in a kind of grassland landscape, two of which (*Meriones tristrami* and *Microtus guentheri*) appear for the first time in the country, and continued until the present. Along with *Spalax ehrenbergi*, a smaller species of mole rat *(Spalax neuvillei)* coexisted. From this epoch *Mesocricetus aramaeus* was described. The two last species lasted until the Upper Mousterian, when *Cricetulus migratorius* appears for the first time, and lived until present.

A giant mole rat *(Spalax kebarensis)* and *Mesocricetus auratus*, appear in the Upper Mousterian; the first one existed for a short period and is not found anymore in post-Palaeolithic age, while the later one is quite abundant in the country at least as late as the Mesolithic (= Natufian) time, but retreat northward to the Syrian plateau probably during Neolithic age.

Wood dwellers fluctuate more or less around the same relative number of species (Table 5, Fig. 2) during the Pleistocene. No squirrels were found in Middle Pleistocene epoch. *Sciurus anomalus* first appears in the Oumm-Qatafa deposits (early Upper Pleistocene) and disappears from the area in post-Natufian times. Several species of *Apodemus* are common in the Israeli Quaternary, of which *A. mystacinus* alone is known to survive throughout the Pleistocene epoch. *A. levantinus* and *A. caesareanus* became extinct towards the Upper Mousterian, when *A. sylvaticus* first appear,

Table 5. Distribution of Woodland Species in the Pleistocene of Israel

	'Ubeidiya (−500,000)	Karst Filling Jerusalem (−250,000)	Oumm Qatafa Cave (−120,000)	Lower Mousterian (−70,000)	Upper Mousterian (−40,000)	Upper Paleolithic (−20,000)	Natufian (−10,000)	Recent
Progonomys sp.	+	−	−	−	−	−	−	−
Parapodemus jordanicus	+	−	−	−	−	−	−	−
Apodemus mystacinus	+	+	+	+	+	+	+	+
Apodemus levantinus	+	+	+	+	−	−	−	−
Apodemus caesareanus	+	+	+	+	−	−	−	−
Apodemus sylvaticus	−	−	−	−	+	+	+	+
Sciurus anomalus	−	−	+	+	+	+	+	−
Eliomys sp.	−	−	−	−	−	−	+	−
Eliomys melanurus	−	−	−	−	−	−	−	+
Dryomys nitedula	−	−	−	−	−	−	−	++

Table 6. Distribution of Rock Dwellers and Sand-Dune Dwellers in the Pleistocene of Israel

Rock Dwellers (Mediterranean Region)	'Ubeidiya (−500,000)	Karst Filling Jerusalem (−250,000)	Oumm Qatafa Cave (−120,000)	Lower Mousterian (−70,000)	Upper Mousterian (−40,000)	Upper Paleolithic (−20,000)	Natufian (−10,000)	Recent
Gerbillus dasyurus	−	−	+	−	−	+	+	+
Acomys cahirinus	−	−	−	−	−	+	+	+

Sand-Dune Dwellers (Mediterranean Region)	'Ubeidiya (−500,000)	Karst Filling Jerusalem (−250,000)	Oumm Qatafa Cave (−120,000)	Lower Mousterian (−70,000)	Upper Mousterian (−40,000)	Upper Paleolithic (−20,000)	Natufian (−10,000)	Recent
Gerbillus gerbillus	−	−	−	−	−	−	−	+
Gerbillus pyramidum	−	−	−	−	−	−	−	+
Gerbillus allenbyi	−	−	−	−	−	−	−	+
Meriones sacramenti	−	−	−	−	−	−	−	+

and since then it was only rarely recorded. It is restricted at present to the northern part of the country.

Progonomys sp. and *Parapodemus jordanicus* ('Ubeidiya Formation) considered as wood dwellers is not proved, and will remain merely a speculative possibility.

Strangely enough, arboreal dormice are never found until Natufian times. The genus *Dryomys* is known only from the recent fauna of Israel.

Rock dwellers (Table 6) (*Acomys russatus* and *Gerbillus dasyurus*) in the Mediterranean region of Israel appeared only in Upper Palaeolithic time, when desiccation and deforestation processes started to be highly effective in the Levant.

Only in post-Natufian (= uppermost Palaeolithic) did the Sand-Dune Dwellers (Table 6) move into the coastal plains of Israel, which is designated as the youngest faunal invasion to the Levant.

The most striking phenomenon in the ecological changes during the Pleistocene of Israel, is the fast and drastic elimination of the steppe animals. Except *Myomimus roachi* no more steppic species are found in the Upper Mousterian, when the grassland dwellers wholly replace them. The wood dwellers remain more or less constant in relative number of species. The severe desiccation process was critical for the moist steppe dwellers, and mainly animals living in more arid open land biozones are found from Upper Mousterian onward. The fact that during the Upper Palaeolithic (Fig. 2) there is a shrinkage of grassland elements is not a result of a more rainy period (contemporaneous with the Würm), but rather because part of the area turned to be by that epoch already too arid; bare rocks were exposed, and rock dwellers first appeared. Shortly after Natufian age Sand-Dune dwellers appear.

This desiccation phenomenon which took place in the Levant ever since the Middle Pleistocene stands with no correlation with the glacial or global climatic fluctuations. This fact was especially stressed by the author in 1968b. The fauna as a whole fails to show the small climatic fluctuations (which certainly occurred), but is highly affected by the major changing trends of the regional climate. And this idea will bring us to the next chapters.

NOTE ON THE ORIGIN OF THE ISRAELI DESERT RODENTS

Rodents of the Israel and Sinai peninsula generally belong to the Saharo-Sindien faunal region (RANCK, 1968). This faunal complex was originated either by forms invading from the Asian part, or from the African part of this region. Some other species appear to have developed *in situ*.

5 genera might be assigned as the most typical for the Saharo-Sindien region: *Meriones*, which was established in Asia and invaded North Africa across Israel and Sinai; *Gerbillus* which was originated in the Ethiopian region and invaded several times the Asiatic part of the

southern Palaearctic region, again across Israel and Sinai; the Ethiopian region is a well the cradle land of *Acomys*, which is widespread at present in several species over the more steppic and arid zones of Asia south of Caucasus. *Acomys* is more restricted in its general Palaearctic distribution than the 4th genus *Jaculus* which was developed probably in the high latitude steppes of Asia (as the rest of the Dipodidae, invaded North Africa in late Neogene or early Pleistocene times, where it is quite widespread at present north of the Sahara. The only Pleistocene record of the Dipodidae in Israel is limited to the extinct genus *Parallactaga*, described by HAAS (1966) from the 'Ubeidiya Formation.

Other Saharo-Sindien elements are to be discussed. *Tatera* is not known, and never recorded as fossil from Israel. Being absent from the Levant, North Africa and Sharan regions, but widespread in Asia and Africa south of the Sahara, raises a 'zoogeographical enigma', as RANCK (1968) pointed out. This particular pattern of distribution of a genus established in Asia, might be explained as follows: there is plenty of evidence to show that the southern corner of the Arabian peninsula (Mandab strait) was connected with the 'horn' of East Africa for several times during the Neogene and Pleistocene. On these occasions *Tatera* could have invaded Africa, avoiding the 'normal Afro-Asian route', i.e., Israel-Sinai-Egypt. This should have happened in the period when the Sahara was already too arid to be crossed by *Tatera* northward to North Africa. *Psammomys* is more restricted in distribution, and was probably created (showing affinities in many respects to *Meriones*) in the more arid zones of southern Asia. Among the jirds' species it is probably the most adapted to desert conditions, being also diurnal with pigmented skin. The most localized genus in distribution is certainly *Sekeetamys* which is endemic to Egypt, Sinai and Israel, and recently recorded from Arabia by ATALLAH & HARRISON (1967).

The genus *Gerbillus* is represented in the 'Ubeidiya formation by an undefined species, but certainly different from *G. dasyurus*. Another species was uncovered from the Karst fissure filling near Jerusalem. Appearance of *Gerbillus* in Asia (Fig. 3) took place during late Neogene time, and since then the Asiatic steppes served as a separate centre of speciation, few of which were present one at a time in the southwest corner of Asia. This is true as well for *G. dasyurus*. (Other species did arrive from Africa, see below). The Israel Middle Pleistocene gerbils occupied the moist steppe habitats, while *G. dasyurus* is a strict rock dweller. Another '*Dipodillus*' *(G. nanus)* intruded from the Arabian peninsula to a very restricted area in Southern Israel and the northern Red Sea coastal line of Sinai. The wide distribution of *G. gerbillus* in the arid zones of southwest Asia, mainly in sand biotopes its scattered pattern of distribution sometimes in isolate patches of sandy regions, demonstrate a much older invasion from North Africa. *Gerbillus pyramidum* is probably a post glacial newcomer from the Sahara. Its distribution is much more

357

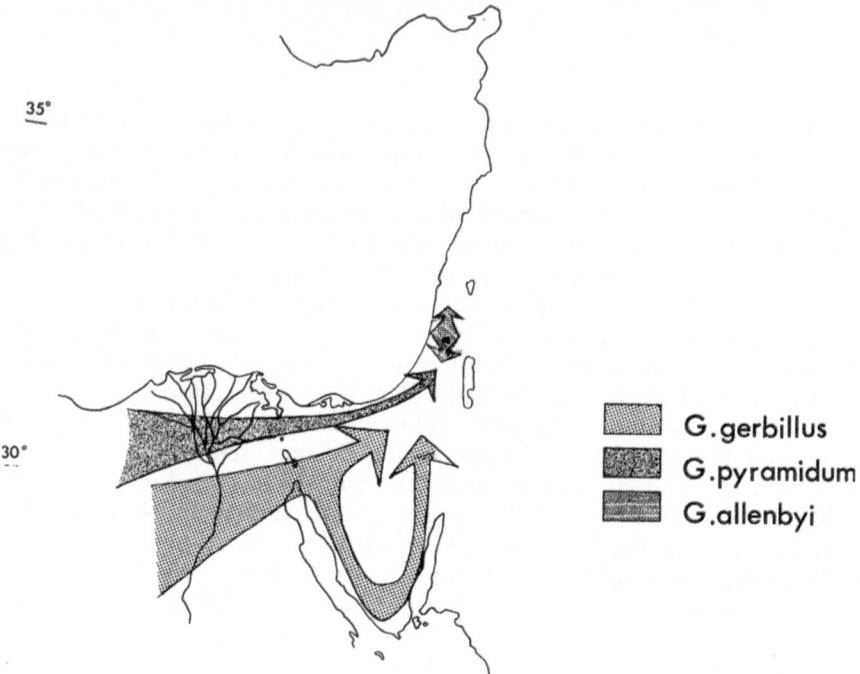

Fig. 3. A diagrammatic map exemplifying the historical spread of various Saharan species of *Gerbillus* in Israel.

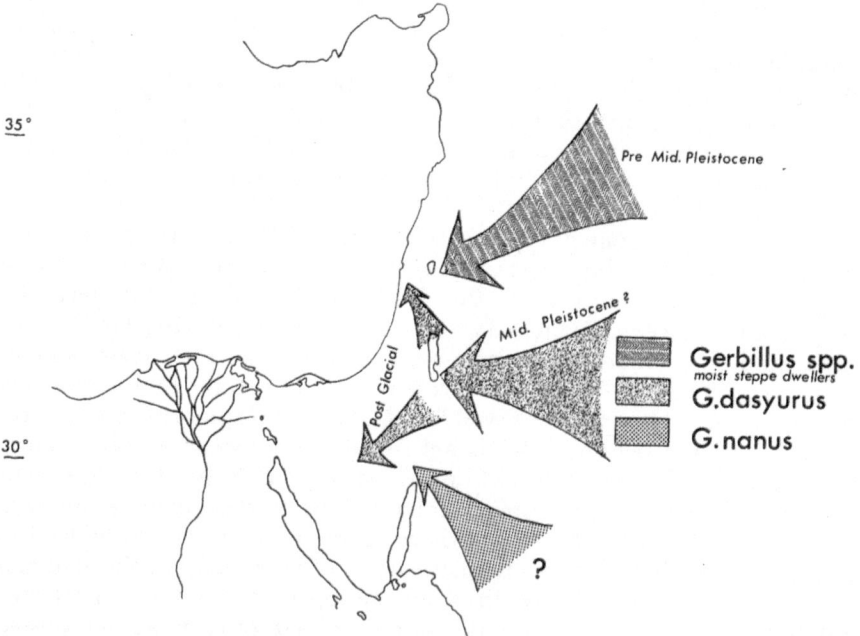

Fig. 4. A diagrammatic map exemplifying the historical spread of various species of various non-Saharan species of *Gerbillus* in the Levant.

358

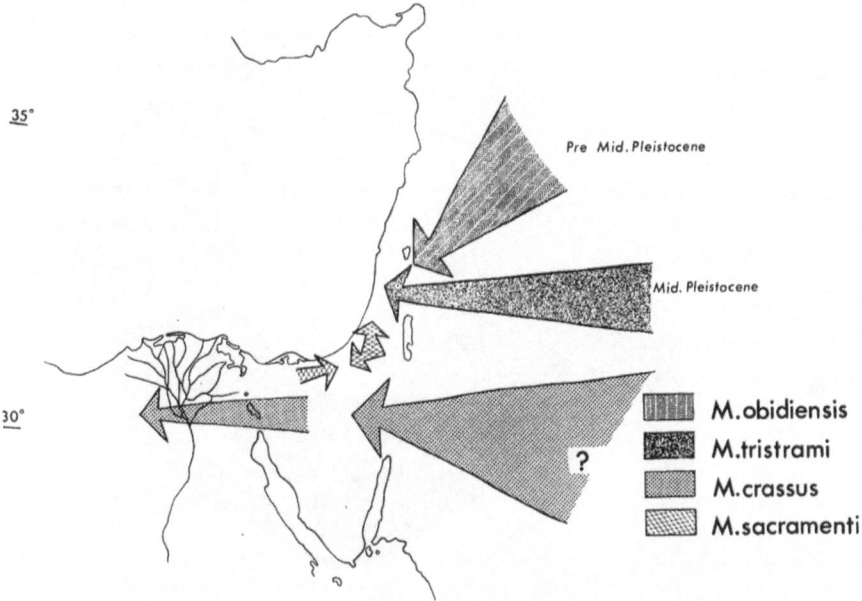

Fig. 5. A diagrammatic map exemplifying the historical spread of various species of *Meriones* in the Levant.

localized in the Levant where it is restricted to northern Sinai and the coastal sand dunes of Israel (Fig. 4). During its short history in Israel it was never able to disperse into the more and older endemic sandy areas along the Arava Valley (southern part of the Israel Rift Valley), or even to southern Sinai (HAIM, 1969), areas which are disconnected with the Mediterranean coastal dunes. An endemic sand-dune dweller – *G. allenbyi* – was established in the coastal dunes of Israel (ZAHAVI & WAHRMAN, 1957).

Meriones intruded the southwest corner of Asia probably through two different parallel lines (Fig. 5). One (demonstrated by *M. crassus*) occurred across the southern Palearctic arid belt to North Africa. Another line of dispersal took place further to the north, and is demonstrated by the more cool and moist steppe jirds, like *Meriones obeidiensis*, and *Meriones tristrami* (a grass land to semi-arid steppe animal probably a direct offspring of *M. persicus*, TCHERNOV, 1968b). *Meriones sacramenti* presents the most localized distribution of any of the Levantinian desert rodents. This endemic species is restricted to the central coastal sand-dunes of Israel (ZAHAVI & WAHRMAN, 1957). Though the genus *Meriones* was established in Asia, *M. sacramenti* is a product of a secondary invasion of a certain jird originated in the Saharan region.

The African genus *Acomys* is an old member of the Near Eastern fauna,

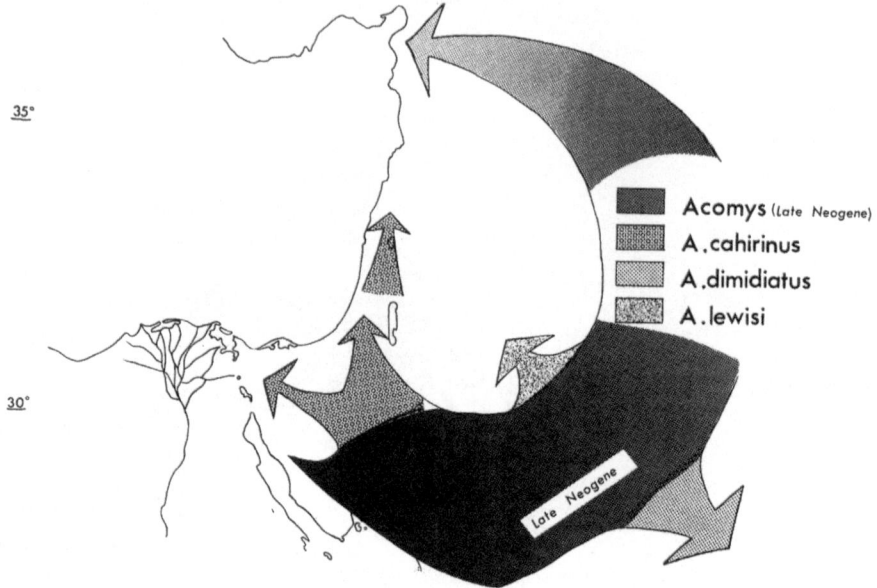

Fig. 6. A diagrammatic map exemplifying the historical spread of the genus *Acomys* in the Near East.

and existed there since Neogene times. *A. cahirinus* was differentiated in Asia, (Fig. 6) as a closely related species is found in Cyprus (BATE, 1903; VON LEHMANN, 1966) and Crete. *Acomys* should have occupied northern Syria already in late Pliocene. Such Pliocene ancestor radiated into several closely related species; *A. cahirinus, A. dimidiatus, A. levisi* (ATALLAH, 1967). No *Acomys* remains were found in the Pleistocene of Israel until as late as Upper Palaeolithic. *A. cahirinus* was probably speciated in more eastern steppic areas, and only later on invaded North Africa across Israel and Sinai. In these regions it is primarily connected with human dwellings. Where no human settlements are found in Sinai, no *Acomys cahirinus* were recorded (HAIM, 1969). This is especially notable in areas where no *Mus musculus* and *Rattus rattus* exist. *Acomys russatus* demonstrates a different line of evolution in Asia, which probably took place in the more extreme southwest Asiatic desert belt. Skull and teeth patterns are more different than members of the '*cahirinus*' group. It is well adapted to extreme arid zones, exhibiting (normally) a diurnal life, and possessing a pigmented skin. Its environmental supportable limit is, as a rule, never beyond the 100 mm isohyete.

Jaculus invaded North Africa through the Israel and Sinai peninsula, leaving behind two widespread species. *J. jaculus* and *J. orientalis*, occupying mainly Hammada to sand dune deserts.

Eliomys is the only 'stranger', being a Palaearctic element which invaded the Saharo-Sindien areas in periods when moister conditions prevailed. Later on, when aridity increased, scattered populations of *Eliomys* were left as relics, readapted to rocky landscape and terrestrial life.

REFERENCES

ATALLAH, S. I. 1967. A new species of spiny mouse *(Acomys)* from Jordan. *J. Mamm.*, 48: 258–261.

ATALLAH, S. I. & HARRISON, D. L. 1967. New records of rodents, bats and insectivores from the Arabian peninsula. *J. Zool., Lond.*, 153: 311–319.

BAR-YOSEF, O. & TCHERNOV, E. 1966. Archaeological finds and the fossil faunas of the Natufian and microlithic industries at Hayonim cave (western Galilee, Israel). *Israel J. Zool.*, 15: 104–140.

BATE, D. M. A. 1903. On the occurence of *Acomys* in Cyprus. *Ann. Mag. Nat. Hist.*, 7: 565–567.

BATE, D. M. A. 1934. Two additions to the Pleistocene cave fauna of Palestine (*Trionyx* and *Crocodilus*). *Ann. Mag. Nat. Hist.*, 10: 474–478.

BATE, D. M. A. 1942. Pleistocene Murinae from Palestine. *Ann. Mag. Nat. Hist.*, 9: 465–486.

BATE, D. M. A. 1943. Pleistocene Cricetinae from Palestine. *Ann. Mag. Nat. Hist.*, 11: 813–838.

BOURLIÉRE, F. 1963. Observations on the ecology of some large African mammals. *Viking Fund Publ. Anthropol.*, 36: 43–54.

BUTZER, K. W. 1966. Environment and archaeology. Aldine Publ. Co., Chicago, pp. 1–504.

BYTINSKI-SALTZ. 1961. The Ethiopian elements in the insect fauna of Israel. Verh. XI. Int. Kongress Entemol. Wien, 1: 457–463.

FAIRBRIDGE, R. W. 1961. Convergence of evidence on climatic change and ice ages. *Ann. N.Y. Acad. Sci.*, 95: 542–579.

FAIRBRIDGE, R. W. 1962. New radiocarbon dates of Nile sediments. *Nature*, 196: 108–110.

FRENKEL, H. 1970. *Hystrix angressi* sp. nov. A large fossil porcupine from the Levalloiso-Mousterian of the Geula cave. *Israel J. Zool.*, 19: 51–82.

GARDNER, E. W. & BATE, D. M. A. 1937. The bone bearing beds of Bethlehem; their fauna and industry. *Nature*, 140: 431–433.

GARROD, D. A. E. & BATE, D. M. A. 1937. The stone age of Mount Carmel, Vol. I, parts I, II. Oxford, Clarendon Press.

HAAS, G. 1951. Remarques sur la microfaune de mammifères de la grotte d'Oumm Qatafa. (In: NEUVILLE, R. Le palaeolithique et mesolithique du Désert de Judée). Archs. Inst. Paleont. hum., Mem. 24.

HAAS, G. 1952. Remarks on the origin of the Herpetofauna of Palestine. *Rev. Fac. Sci., Univ. Istanbul*, B 17: 95–105.

HAAS, G. 1957. Some amphibians and reptiles from Arabia. *Proc. Calif. Acad. Sci.*, 29: 47–86

HAAS, G. 1963. Preliminary remark on the early Quaternary faunal assemblage from Tel-Ubeidiya, Jorand Valley. *South Afr. J. Sci.*, 59: 73–76.

HAAS, G. 1966. On the vertebrate fauna of the lower Pleistocene site 'Ubeidiya. The Israel Acad. Sci. Human., 1–66.

HAAS, G. 1967. Some amphibians and reptiles from Arabia. *Proc. Calif. Acad. Sci.*, 29: 1–12.

HAAS, G. 1968. On the fauna of 'Ubeidiya. Israel Acad. Sci. Human.

HAIM, A. 1969. Distribution of myomorph rodents in Sinai Peninsula. Thesis submitted to the Hebrew University, Jerusalem, Israel (in Hebrew).

361

HELLER, J. 1970. The small mammals of Geula cave. *Israel J. Zool.*, 19: 1–49.

HEROLD, W. 1958. Die Variabilität der Zahwurzeln bei Schläfern (Muscardinidae). *Zool. Beitr. N. F.*, 4: 77–82.

HOOIJER, D. A. 1958. An early Pleistocene mammalian fauna from Bethlehem. *Bull. Brit. Mus. (Nat. Hist.) Geol.*, 3: 267–292.

HOOIJER, D. A. 1959. Fossil mammals from Jisr Banat Yaqub, south of lake Huleh, Israel. *Bull. Res. Counc. Israel Geo-Sc.* 8G/4: 177–199.

KOWALSKI, K. 1963. The Pliocene and Pleistocene Gliridae from Poland. *Acta Zool. Cracov.*, 14: 533–563.

LEAKEY, M. D. 1967. Preliminary survey of the cultural material from beds I and II, Olduvai Gorge, Tanzania. (In: BISHOP, W. & CLARK, J. D. (eds.)), Background to evolution in Africa. Univ. Chicago Press, pp. 417–446.

MATTHEW, W. D. & GRANGER, W. 1923. New Bathyeridae from the Oligocene of Mongolia. *Amer. Mus. Novit.*, 101: 5.

MÉHÉLY, L. 1909. Species generis Spalax. A földi Kutyàk Fajai, Budapest. pp. 1–353.

MEINERTZHAGEN, K. 1954. Birds of Arabia. Oliver and Boyd, London.

MOREAU, R. E. 1955. Ecological changes in the Palaearctic region since the Pliocene. *Proc. Zool. Soc. Lond.*, 124: 253–295.

NEUVILLE, R. 1951. Le Paleolithique et Mésolithique du desert de Judee. *Archs. Inst. Paléont. Human.*, 24: 1–270.

OVCHINNIKOV, P. N. 1946. The history of the vegetation in South Central Asia in connection with the development of Quaternary landscapes. *Trans. Inst. Acad. Sci. U.S.S.R.*, 37: 324–325.

PARKER, H. W. 1956. Species transgression in one horizon. In: SYLVESTER-BRADLEY, P. C. (ed.) The species concept in palaeontology. *The Syst. Assoc.*, 2: 9–15.

PICARD, L. & BAIDA, U. 1966. Geological report on the Lower Pleistocene deposits of the 'Ubeidiya excavations. Israel Acad. Sci. Human., pp. 1–39.

PICARD, L. & BAIDA, U. 1966a. Stratigraphic position of the 'Ubeidiya Formation. *Proc. Israel Acad. Sci. Human.*, 4: 1–8.

RANCK, G. L. 1968. Rodents of Lybia. Smithsonian Inst., U.S. Nat. Mus.

ROBINSON, P. 1966. Fossil occurence of murine rodent *(Nesokia indica)* in the Sudan. *Science*, 154: 264.

RUST, A. 1950. Die Hoehlenfund von Jabrud (Syrien). Neumunster, Karl Wacholtz, Verlag.

SAVAGE, R. J. G. & TCHERNOV, E. 1968. Miocene mammals of Israel. *Proc. geol. Soc. Lond.*, 1648: 98–101.

STEINITZ, H. 1954. The distribution and evolution of the freshwater fishes of Palestine. *Publ. Hydrobiol. Res. Inst., Fac. Sci.*, Istanbul. 1: 225–275.

STEKELIS, M. & HAAS, G. 1952. The Abu-Usba cave. *Israel Explor. J.*, 2: 15–47.

STEKELIS, M. 1966. Archaeological excavations at 'Ubeidiya, 1960–1963. Israel Acad. Sci. Human., pp. 1–32.

STEKELIS, M., BAR-YOSEF, O. & SCHICK, T. 1969. Archaeological excavations at 'Ubeidiya, 1964–1966. Israel Acad. Sci. Human.

TCHERNOV, E. 1962. Paleolithic avifauna in Palestine. *Bull. Res. Counc. Israel., Sect. B. Zool.*, 11: 95–131.

TCHERNOV, E. 1968. A preliminary investigation of the birds in the Pleistocene deposits. of 'Ubeidiya. Israel Acad. Sci. Human., pp. 5–38.

TCHERNOV, E. 1968a. A Pleistocene faunule from a karst fissure filling near Jerusalem, Israel. *Verhandl. Naturf. Ges. Basel*, 79: 161–185.

TCHERNOV, E. 1968b. Succession of rodent faunas during the Upper Pleistocene of Israel. Verlag Paul Parey, pp. 1–152.

WRESCHNER, E. 1967. The Geula caves – Mount Carmel. *Quaternaria* 9: 69–140.

ZAHAVI, A. & WAHRMAN, J. 1957. The cytotaxonomy, ecology and evolution of the gerbils and jirds of Israel. *Mammalia*, 21: 341–380.

362

XVII. PREHISTORIC RODENTS
OF THE MIDDLE EAST

by

PRISCILLA F. TURNBULL

Introduction

According to the Encyclopedia Britannica (1969), the Middle East includes the following countries: Afghanistan, Iran, Iraq, Turkey, Syria, Lebanon, Israel, Jordan, the Arabian Peninsula, Egypt, Anglo-Egyptian Sudan, Cyprus, Tunisia, Algeria, Morocco, and Libya. This paper is a survey of the prehistoric rodents found in these countries. Many more Pleistocene and subrecent sites are known, but unless rodents are contained in the faunal lists, they will not figure here. In respect of a few sites, no published literature was available.

Ubiquitous though they are, rodents are among the fossils preserved least often in terrestrial deposits. When found, they are often a mixed blessing, especially in Pleistocene caves and at prehistoric archaeological sites. On the one hand, fossilized bones and teeth of rodents will almost certainly add to the knowledge of the order's morphology, evolution, and geographic distribution; they may also aid in understanding climates of the past. Unfortunately, many remains of rodents have been present at archaeological sites but due to the type of excavation the tiny jaws, loose teeth, and broken bones were not recovered.

On the other hand, burrowing animals always pose the problem of contemporary vs. later intrusive, particularly serious in post-Pleistocene, sub-recent layers mentioned above. Both the paleontologist and the anthropologist reject as intrusive any bones of rodents that are different in appearance from other bones found at the site. However, these remains should not be wholly ignored. Owl-pellet and carnivore den accumulations can reasonably be considered to be the same age as the surrounding accumulations of cave earth. Fragmentary, disarticulated bones, single jaws and loose teeth, if stained, mineralized, and generally of similar appearance to other bone in the deposit, are probably also of similar age, but not always. White, light, unmineralized bone and articulated skeletons or groups of nestlings should be suspect and not considered part of the fossil population. C-14 and other chemical dates can be obtained, but this is expensive, may destroy the bone, and is thus seldom practical for many fossils. Recognizing that some judgement is necessary in this as in all sciences, we can proceed to survey the prehistoric rodents of the Middle East.

Prehistoric Rodents

The fossil record for rodents in the Middle East is, with few exceptions, as poor as it is for much of the rest of the world. The earliest known rodents, Paramyidae, are known from the late Paleocene of North America. Members of this group are known from Eurasia and presumably they also occurred in Africa. For details of rodent history, refer to SIMPSON (1947), STEHLIN & SCHAUB (1951), WOOD (1955) and VERESCHAGEN (1959).

A brief resume of the families that occur as fossils in the Middle East follows:

Family Aplodontidae.	Paleocene-Recent, N. Am.
	?Upper Eocene, Europe
	Lower Pliocene, Asia and prob. Africa
Family Sciuridae.	Oligocene – Recent, Europe
	Miocene – Recent, N. America
	Pleistocene – Recent, Asia, Africa
Family Ctenodactylidae.	Upper Oligocene – Mid Pliocene, Asia
	Miocene and Recent, Africa
Family Castoridae.	Lower Oligocene – Recent, N. America
	Upper Oligocene – Recent, Europe
	Upper Miocene – Recent, Asia
Family Cricetidae.	Lower Oligocene – Recent, Europe, N. Am.
	Mid Oligocene – Recent, Asia
	Pleistocene – Recent, N. Africa
Family Muridae.	Upper Miocene – Recent, Europe, Asia; Pliocene-R., N. Africa
	Recent, worldwide
Family Spalacidae.	Upper Miocene – Recent, Europe
	Pleist. – Recent, Asia, Africa
Family Rhizomyidae.	Upper Oligocene, Europe
	Pleistocene – Recent, Africa
	Upper Miocene – Recent, Asia
Family Zapodidae.	Eocene – Recent, N. America
	Oligocene – Recent, Eurasia
Family Dipodidae.	Pliocene – Recent, Asia
	Pleistocene – Recent, Europe, N. Africa
Family Gliridae.	Upper Eocene – Recent, Eurasia
Family Hystricidae.	Oligocene – Recent, Eurasia
	Pleistocene – Recent, Africa
Family Thryonomyidae.	Miocene-Recent, Africa
	?Pliocene, Asia

In early Tertiary times, the great Tethys sea still covered southern Eurasia. Not until early Miocene time did the sea withdraw and the landmasses as we know them today begin emerging. SAVAGE (1967) has written that 'the Tethyan trough, in which Mesozoic and Paleogene sediments had accumulated, was compressed'. Movements on a global scale elevated the Tethyan fringes into mountain belts; lateral forces flexed these into the complex oroclines we witness today in the Alpine-

Himalayan fold belt. The Tethys shrank, new land connections were established and intercontinental migrations became possible'.

The earliest southwest Asian rodent (gen. et sp. indet.) of record was found by KHUCHUA in 1948 at Benara, South Georgia; the deposit has been correlated with the French Quercy and other mid Oligocene beds of Khazakhstan and Mongolia.

Early Miocene (Burdigalian) rodents are known in Europe from France, Germany, Spain and Switzerland. SAVAGE (1967) gave faunal lists and maps of the sites of these Burdigalian faunas. In southwest Asia, only the Bugti Hills of Baluchistan contain early Miocene mammals and none are rodents. A site at Ismir, Turkey, produced a castoroid described by OZANSOY (1961) as *Dipoides anatolicus*.

CAUCASUS

By the mid-Miocene, the future isthmus of Caucasus had appeared as islands and was increasing in size to become a peninsula of the southwest Asian massif. Middle Miocene beds at Belomechetskaya on the Kuban River contained bones of *Paleocricetus*, used by VERESHAGEN (1959) as evidence, along with other faunal elements, that the Caucasus must have been connected with southwest Asia. In the Late Miocene (Sarmatian), the widespread *Hipparion* fauna migrated from the south into the plain and foothills of the Caucasus; this fauna was essentially the same as that of the western Mediterranean. Russian paleontologists believe (ANDRUSOV and others, as mentioned by VERESHAGEN, 1959) that the *Hipparion* fauna entered eastern Europe via the Caucasus. At the end of the Miocene, apparently, the seas surrounding the Caucasus had become shallow and less saline. The eastern gulf boundary of Caucasia had nearly dried completely and only a narrow strait of sea existed in the Manych region in southern Caucasus.

SAVAGE (1967) has speculated about possible migration routes between Europe, Asia, and Africa when the Tethys intercontinental connections arose. He suggested that one route from Asia to northern Europe ran along the Caucasus-Caspian belt, the African-Asian route perhaps passed through Israel and Arabia to Pakistan and Burma, and the African-European one led through Turkey.

According to VERESHAGEN (1959), an upper Miocene beaver was reported by EGLON in 1940 from the Kuban River near Armavir in Western Caucasia. Fragments of murid teeth were found in the Grozny region by SIZOV, along with impressions of grass and freshwater ostracods. Both of these occurrences provide supporting evidences for SAVAGE's (1967) speculation about migration routes.

During Pliocene time, the fauna of the Caucasus is mainly traceable to migrations from the north and the south. The Pontian stage was warm and became semitropical in its middle. The Kosyakin quarry near

Stavropol has produced *Amblycastor, Stenofiber, Cricetus* and *Mus*, which VERESHAGEN (1959) envisaged as inhabiting open meadows, though some mesophilous forms (mastodon, tapir, rhinoceros and warthog) also occur.

South of the Caucasus is Lake Urmia in northwestern Iran. East of the lake is Mt. Sahand, from which has been collected the Maragheh fauna of the Pontian. No rodents occur in this, but the other mammalian elements are surely indicative of the faunal ties between Asia and Caucasus. Probably some fauna adapted to cold climates migrated into the area. VERESHAGEN (1959) believed that *Mesocricetus* and *Prometheomys* are among endemic forms that evolved in Pontian time on the Caucasus.

Upper Pliocene beds on the Apsheron Peninsula (E. Caucasus) near Binagady contained bones of the castorid *Trogonotherium* and an unidentified murid. Remains of *Gerbillus* sp. were found in a diatomite near Nurus on the Zanga River, indicative that a xerophilous vegetation thrived along the southern margin of the Armenian highland. At the end of the Pliocene epoch, the hot, dry climate gave way to cooler conditions and northern faunal elements became important in Caucasia.

At the beginning of the Pleistocene epoch, the Caucasus mountains were uplifted by arching, which ultimately resulted in mountain glaciation and promoted new habitats for rodents and other mammals. The Black, Azov, and Caspian seas were correspondingly depressed. Associated with the uplift, karst lands, caves, valley terraces, volcanic action further modified the region. Pleistocene mammals in Caucasia occur in marine terraces along coasts of the Black, Azov and Caspian seas, in river and lake terraces, in loams, asphalt pools, mud flows, and in caves.

The Il'Skaya paleolithic open air site was discovered in 1818 by DEBAILLE on a tributary of the Kuban River. The tools from this site have been described as latest Mousterian or Solutrean. An oil seep from the dolomite of the area caused the bones to be well preserved. Rodents found at Il'Skaya are *Scista* cf. *caucasica* and an unknown murid. In 1957 FORMOZOV excavated a cave entrance (Mousterian) near Dakhovskaya and found bones of *Cricetus cricetus*. A list of the rodents known from the entire Caucasus is given in Table 100 of VERESHAGEN (1959).

IRAQ

A prehistoric cave site of which I have personal knowledge in the Middle East is Palegawra Cave in northeastern Iraq (BRAIDWOOD & HOWE, 1960). Based on the most recent (conventional) C-14 date, bone specimens from the deepest layer of this cave have been calculated to be as old as 14,400 ± 760 YBP[1]. The industry is considered Zarzian by the archaeologists. Rodent bones, recent at the top and fossil in lower layers,

---

[1] Y.B.P.: years before present.

occur throughout the levels. The fossil rodents have been identified (TURNBULL & REED, 1974) as:

Microtinae:	*Microtus* cf. *socialis*
	Ellobius cf. *lutescens*
	Arvicola cf. *terrestris*
Cricetinae:	*Mesocricetus* cf. *armatus*
	Cricetulus migratorius
Gerbillinae:	*Meriones* cf. *persicus*
Spalacidae:	*Spalax leucodon*

All of these forms, plus others, are known from the Zagros foothills of Iraq and Iran today (HATT, 1959; LAY, 1967).

The open site of Jarmo, east of Chemchemal, is several thousand years younger (ca. 7,000 YBP) than Palegawra cave and was worked in great detail by BRAIDWOOD, HOWE, REED and their associates from the University of Chicago for several seasons in the late 1940s and early 1950s. The only rodent remains that were collected are of the gerbil *Tatera indica*. Although REED collected live specimens of *Gerbillus*, *Cricetulus*, *Meriones*, *Microtus*, *Spalax* and *Mus* from the area around Jarmo, none of these genera appear in the faunal collection made at the site.

Shanidar cave, northwest of Jarmo, has yielded a large and varied fauna that dates back over 80,000 years with some younger occupation layers dated about 12,000 YBP. Dr. RALPH SOLECKI was largely responsible for this excavation. In his paper of 1951, he listed a specimen of *Meriones* sp., identified by SETZER. REED & BRAIDWOOD (1960) listed among the older bones (40,000 YBP) a specimen of *Ellobius*, and from the younger (12,000 YBP) a mandible of *Castor fiber*.

In a paper of 1930, describing animal remains from the Dark Cave, Hazard Merd, which lies southeast of Palegawra and Jarmo, D. M. A. BATE reported *Ellobius*, *Apodemus*, and *Spalax* cf. *ehrenbergi* (= *leucodon* ? here), dated ca. 25,000 YBP. Zarzi cave, also in northern Iraq and also reported by BATE, contained no rodents in the faunal list.

IRAN

The Warwasi shelter in Kurdestan Province of western Iran, east of and considerably higher than Palegawra, contains deposits probably somewhat contemporaneous with Palegawra and Shanidar. Three industries have been recognized: Mousterian, Baradostian, and Zarzian. The shelter was dug by Dr. BRUCE HOWE in the late 1950s (BRAIDWOOD & HOWE, 1960). In order of stratigraphic occurrence, the following rodents have been identified (TURNBULL, in press):

Zarzian layers:	*Meriones* cf. *persicus*
	Ellobius cf. *fuscocapillus*
	Microtus cf. *socialis*

Baradostian layers:	*Meriones* cf. *persicus*
	Allactaga sp.
	Ellobius cf. *fuscocapillus*
Mousterian layers:	*Mesocricetus* cf. *auratus*
	Ellobius cf. *fuscocapillus*
	Tatera indica

In 1949, CARLETON COON excavated three caves in northern Iran and in 1951 published the results. Bisitun yielded *Meriones*, 'mouse', and *Hystrix*[1]. Tamtama cave contained a gerbil and *Hystrix*. Belt Cave on the Caspian produced *Castor fiber*, *Hystrix*, a murid, and *Citellus* sp., according to COON. Radiocarbon dates on two levels indicated about 11,500 and 8,600 (HOWELL, 1959).

In 1959, COON (reported in HOWELL, 1959) wrote a preliminary report on excavations from Hotu Cave, near Belt cave, northern Iran, and recorded the rodents *Ellobius* sp., *Microtus socialis*, *Rhombomys opimus* and *Citellus fulvus*. C-14 dates for two levels in Belt Cave have been given ca. 12,000 and 9,000 YBP (HOWELL, 1959).

In southwestern Iran, Khuzistan province, HOLE, FLANNERY & NEELY (1969) investigated an early village sequence on the Deh Luran plain. FLANNERY studied the mammalian remains and reported the presence of the gerbellines *Tatera indica* and *Meriones crassus*, and murines *Nesokia indica* and *Mus*, the latter probably a recent intrusive however. The earliest C-14 date for the Deh Luran sequence was given as 8,000 B.C. and the latest, at the top of the sequence, as 3,460 B.C.

Perhaps it should be emphasized that all the materials described from prehistoric sites in Iran and Iraq are late or post-Pleistocene.

ANATOLIA

The site of Alishar Hüyük, excavated by archaeologists from the University of Chicago, Oriental Institute, in the 1920s–1930s, produced bone at all cultural levels. Only one rodent, a jaw of *Castor fiber*, was described by PATTERSON (1937) from a Hittite layer (2000–1200 B.C.).

GREECE

In 1970, FREUDENTHAL described a new *Ruscinomys* from the Pikermian (early Pliocene) of Samos. That author named the new form *R. hellenicus* and said the genus is insufficiently distinguished from *Cricetodon* of Europe.

LEBANON

The Paleolithic rock shelter, Ksār'Akil contained a rodent fauna studied

---

[1] Identified first as *Castor* and reidentified by VERESHAGEN (1959) as *Hystrix*.

first by BATE (1945), and later by HOOIJER (1961). The latter listed the following forms: *Sciurus anomalus*; *Microtus machintoni*; *M. güentheri*; *Spalax* sp.; *Apodemus mystacinus*; *A.* sp. (possibly *sylvaticus* or *caesareanus*). The rodent bones in this cave accumulation were contained in owl pellets.

The first fossil *Spalax* was collected by ZUMOFFEN from the Paleolithic Cave of Antelias in 1902. TCHERNOV placed this into the species *ehrenbergi* in 1968.

PALESTINE

Nowhere in the Middle East is the Pleistocene fauna so well studied as in Palestine. The changes that occurred in the rodent faunas of Israel and the entire Levant during the upper Pleistocene have been discussed in detail and summarized most recently by TCHERNOV (1968 and this volume). The Pliocene tropical animals of the Levant immigrated in from the north and east rather than from Africa generally. This Irano-Greek fauna was compared by BODENHEIMER (1960) to the life communities of the Ethiopian savanna. Since the early Pleistocene relatively small temperature changes have taken place, but significant changes in humidity have occurred.

HAAS (1966) studied the vertebrate remains from the early Pleistocene archaeological site 'Ubeidiya in the central Jordan valley. The rodents reported in these lacustrine deposits are listed below:

Hystricidae:	*Hystrix* sp.	
Spalacidae:	*Spalax minutus*	
Dipodidae:	*Parallactaga* sp.	
Cricetidae:	Cricetinae:	*Cricetus* cf. *nanus*
		C. kormosi
		C. cf. *angustidens*
		Allocricetus cf. *bursae*
		Nannocricetus sp.
	Gerbillinae:	*Meriones obeidiensis*
		Gerbillus sp., various
	Microtinae:	*Lagurodon* cf. *arankae*
		Jordanomys pusillus
		Arvicola? jordanica
Muridae:	*Parapodemus jordanicus*	
	Progonomys sp.	
	Apodemus mystacinus and *A.* sp.	
	Mus sp.	
	Various indet. murines	

HAAS pointed out the absence of both beaver and squirrel remains in these deposits. His collection, representing many Gerbillinae and several Cricetinae, had few large murines. The material from 'Ubeidiya is fragmentary with few complete mandibles or maxillaries and many individual teeth. The fauna, HAAS thought, was remarkable for the large number of gerbillines, especially *Meriones*, and the diversity of murines,

especially *Apodemus*. The presence of microtines has been interpreted by
HAAS as usually indicating a steppic environment and of the murines, a
woodland. *Acomys*, the African murid, immigrated into Palestine later
in the Pleistocene epoch. Today, *Apodemus* is not found in the Jordan
valley because of the extreme heat and bareness. HAAS concluded that
the incongruous mixture at 'Ubeidiya resulted from the concentration
of material from vastly different biotopes via running water in torrential
streams. The gerbillines and cricetines also are characteristic of steppe,
but the less arid steppe than the present conditions in the Jordan valley.

The 'Ubeidiya rodents have an east-European–central Asiatic character, especially the many hamsters, large *Meriones* and smaller gerbillines.
Since giraffe and hippopotamus also occur at this site, HAAS wondered
if steppe would suit these large mammals. He thought it would if enough
water in lakes was available and enough woodland close by.

Along the coast, south of Haifa, in the Kishon valley, lie the famous
caves of Mt. Carmel. GARROD & BATE (1937) excavated and described
a group of caves along the Wady el Mughara known as Tabūn, Skhūl,
Wad, and Jamal. The earliest levels contain late Acheulian tools and the
latest are associated with late Aurignacian culture; thus all finds are
related to the late Pleistocene. In general, the layers in Tabūn are lower
stratigraphically than those in el Wad. BATE worked on the fauna from
the caves, and in 1937 published the following list (combined for all
levels):

Spalax sp.
Hystrix sp.
Philistomys roachii, n. sp.
Sciurus anomalus
Ellobius pedorychus
Microtus mcCowni
M. güentheri
M. machintoni
Apodemus (4 sp.)
Arvicanthis ectos
Rattus (Mastomys) sp.
Mus carmina
Leggada sp.

TCHERNOV's detailed analysis of the succession of Upper Pleistocene
rodents of Israel (1968) charted the appearance and disappearance of
twenty-nine species during an estimated period of 160,000 years. He
followed KOWALSKI in placing BATE's *Philistomys* into the recent genus
Myomimus. TCHERNOV considered *Mus carmina* and *Leggada* sp., named by
BATE, to be forms of *Mus musculus*, a highly variable rodent. In addition
to reexamining the rodents from Mt. Carmel, TCHERNOV studied the
upper Pleistocene rodents from the caves Oumm-Qatafa, Kebara, Fallah,
and Abu Usba.

Oumm-Qatafa is the oldest cave in the Levant; it is situated in the desert area of the Judean Hills. In TCHERNOV's estimate this cave is probably no older than 120,000 years while others believe it to be 200,000–300,000 years old. All the fossil rodents found there are from Taycian and Acheulian levels, and are thus somewhat older than any of the deposits in the Mt. Carmel caves.

Kebara cave deposits were dated from two C-14 dates taken from the deep layers, 39,000 B.C. and 42,000 B.C., representing Levallo-Mousterian stage. The cave of Wadi Fallah contributed but a few fragments of *Spalax* from Naturian-Neolithic layers. The Abu-Usba fauna was originally described by HAAS (In: STEKELIS & HAAS, 1952); it is considered to be a single biologic unit of Natufian-Neolithic stage of about 10,000 B.C. Included were:

Sciurus anomalus
Spalax cf. *ehrenbergi*
Eliomys cf. *quercinus*
Microtus cf. *güntheri*
Rattus rattus
Mus musculus
Apodemus cf. *mystacinus*
A. cf. *sylvaticus*
Acomys carmeliensis (= *A. cahirinus*)
Cricetulus migratorius
Meriones cf. *shawi*
Gerbillus sp.

Of the species listed above, *Sciurus* is no longer present today except in woodlands in the hills east of the Jordan River. *Eliomys* and *Acomys* are extinct in the region. Below is TCHERNOV's list of 1968 of rodents from all Upper Pleistocene horizons in Mediterrannean Palestine.

Sciurus anomalus
Eliomys melanurus and *E.* sp. (2)
Myomimus roachi
Spalax ehrenbergi, S. neuvillei, S. kebarensis
Apodemus levantinus, A. caesareanus, A. mystacinus, and *A. sylvaticus*
Rattus batei, R. haasi, R. rattus
Mus musculatus
Allocricetus jesreelicus, A. magnus
Mesocricetus aramaeus, M. armatus
Cricetulus migratorius
Gerbillus dasyurus, G. gerbillus, G. pyramidum
Meriones tristrami
Ellobius fuscocapillus
Microtus güentheri
Pitymys sp.?
Arvicola terrestris

TCHERNOV omitted the mention of *Hystrix*, a rather common form in practically all prehistoric rodent faunas. FRENKEL (1970) recently defined *Hystrix angressi* from the Levallaso-Mousterian layers of another Mt. Carmel cave, Guela. She noted that VAUFREY (1931) first mentioned

Hystrix from Oumm Qatafa. It has, of course, also been included in Haas' fauna of 'Ubeidiya above, and in BATE's lists.

HELLER (1970) studied the other small mammals of Guela cave. This is a 'bat' cave and dated by reference with Tābun B layer at about 42,000 B.C. The rodents HELLER identified are all included in TCHERNOV's list above with a few exceptions: no *Arvicanthis*, *Rattus*, *Allocricetus*, *Cricetulus*, *Gerbillus*, *Ellobius* or *Pitymys* occur at Guela.

EGYPT

The early Tertiary of lower Egypt was very different from that of southwest Asia. WOOD (1968) stated that 'few or no connections between North Africa and Europe or, probably, Asia [existed] in the late Eoceneearly Oligocene as indicated by the striking endemism of the Fayum [rodent] faunules. The uniformity of the Fayum rodents, derivable from a single *Phiomys*-like ancestor, suggests such a form reached North Africa not long before Fayum times ... from an unknown source'. WOOD's study of a number of collections of Fayum rodents, including those found by SIMONS for Yale in the 1960s, produced the following list: From the Jebel el Quatrami formation, N.W. of Lake Qarun.

Phiomys andrewsi, *P. paraphiomyoides*, *P. lavocati*
Paraphiomys simonsi
Metaphiomys schaubi, *M. beadnelli*
Gaudeamus aegypticus
Phiocricetomys minutus

From Nubia in upper Egypt are known a few Pleistocene rodents found associated with artifacts of upper Paleolithic type. The Yale University Prehistoric Expedition to Nubia in 1962 collected an edentulus jaw of *Nesokia* cf. *indica* and a maxillary with two molars of *Psammomys obesus*, both from the archaeological site of Khor el Sil on the Kom Ombo Plain (REED & TURNBULL, 1969 and MS), about 70 km north of Aswan.

SUDAN

The University of Colorado Expedition of 1965 collected a pair of toothed jaws of *Nesokia indica* (ROBINSON, 1966) from the west bank of the Nile opposite Wadi Halfa in Sudanese Nubia. The age of these sediments has been placed at 15,000–10,000 B.C.

An indigenous African cane rat, *Thryonomys arkeli*, from a Mesolithic site near Khartoum was described and named by BATE (1947). Still another Levallo-Mousterian fauna was described by BATE (1951) from Abu Huzar, a site on the Blue Nile 225 miles south of Khartoum. BATE named *Hystrix astasobae* from amongst these remains and noted that this was a more primitive form than the modern species.

The mid Paleolithic artifact-bearing shelter Haua Fteah in Cyrenaica, northeastern Libya, described in great detail by McBURNEY (1967) contained only a few rodents, according to HIGGS who did the mammalian biology. One molar of *Microtus* sp. occurred in the early Neolithic layer. *Hystrix* sp. occurred in the Levalloiso-Mousterian, Libyco-Capsian, and Neolithic layers.

From the Hagfet ed Dabba at Wadi Derma, BATE had earlier (in McBURNEY & HEY, 1955) described the first north African microtine, *Microtus cyrenae*. *Eliomys* sp., *Apodemus* sp., and *Gerbillus* sp. were also present here. *Hystrix cristata* is recorded from the mid Paleolithic site Hagfet et Tera.

RANCK (1968) discussed the climatic fluctuations during the Pleistocene in North Africa and emphasized that during moist, cool pluvials, the rodents characteristic of the subsaharan veld and savannah moved north, while during the dry, hot interpluvials, rodents of steppe and desert extended their ranges south of the Sahara into tropical Africa. As examples of Pleistocene relicts, RANCK cited the Libyan fossil *Microtus* and a fossil *Ellobius* from Algeria, apparently the only microtines that ever entered North Africa. On the other hand, *Atlantoxerus* and *Lemniscomys* are residuals of a central African rodent fauna that advanced into North Africa during Pleistocene pluvials, though we know of no fossil forms. Thus, during warm periods of the Pleistocene, indigenous south and central African rodents intermingled with southwest Asian immigrants. During cool periods, small populations of both groups found shelters in widely separated areas in North Africa where they survived as isolated populations until the present when they are again extending their ranges.

ALGERIA

From the Pleistocene station Mechta-el-Arbi, ROMER (1928) reported the following fossil rodents: *Hystrix cristata*, Pleistocene, Recent; *Ctenodactylus gundi* and *Meriones* sp., Upper Paleolithic at Redeyef; *Gerbillus* sp., Neolithic, Recent at Saïda; *Jaculus* sp., Mousterian, upper Paleolithic, and Neolithic strata at Boui Zabouin; Recent at Redeyef; *Mus* sp., Recent, Neolithic and earlier.

In 1938, ROMER added to this list from the region of Ain Beida: *Jaculus jaculus, Meriones shawi, Eliomys quercinus*. ROMER & NESBITT (1930) named *Thryonomys logani* from a site in the Tanezrouft region in eastern Algeria, 140 km southwest of Fort Leperrine. As with another recently identified specimen of the cane rat from Mauritania (see below), this fossil indicates better watered conditions, even very locally, than occurat present.

In 1968, PETTER described a new murid, *Paraetheomys filfilae* from a late Pleistocene deposit near Philippeville.

Geographically, we are perhaps out of the bounds of the Middle East, but faunistically the southern Mediterranean is a related natural unit, and it is an important Africa-Europe and Africa-Asia migration route; thus it would seem remiss to omit it from this survey.

The shelter Mughared el 'Aliya, near Tangier, was described by HOWE (1967) and the fauna identified by ARAMBOURG (1967, appendix in HOWE, 1967). Recognized rodents were: *Hystrix cristata*, Pleistocene and sub-recent levels; *Mus* sp., Upper Pleistocene; *Tatera*, Upper Pleistocene.

At Beni Mellal, (LAVOCAT, 1967), early Pliocene (Pontian) strata have yielded a few remains of *Paraphiomys*. LAVOCAT considered *Thryonomys* and *Petromus* descendants of that form. Recently, JAEGER & MARTIN (1971) have described the earliest known Pontian murid from Africa, *Progonomys cathalai*, which these authors compared with the oldest known European murid. From these same Oued Zra (early Pliocene, but considered by most French workers to be late Miocene) beds of the central mid Atlas mountains, JAEGER & MARTIN reported *Cricetodon* sp., *Myocricetodon cherifiense*, the ctenodactylid *Africanomys*, and an indet. glirid.

MAURITANIA

Dr. P. J. MUNSON collected a small prehistoric mammalian fauna from a series of neolithic sites near Tichitt. The rodents identified in this collection (TURNBULL in MUNSON, MS) were: *Tatera* cf. *guineae* from level dated 3400–3000 B.C.; *Thryonomys* cf. *swinderianus* from level dated 3200–3000 B.C.; *Rattus (Mastomys) gambianus*; *Jaculus jaculus*; *Euxerus* cf. *erythropus*.

Discussion

From the Cretaceous to the Miocene, peneplanation of the African continent produced gently inclined plains. As Tethys disappeared during the Miocene, uplift occurred and land connections to Asia via Aden and Suez isthmus and to Europe via the Tunisian–Sicilian bridge became available. The uplift of North Africa, accompanied by rifting and volcanic action, continues today.

The invasion by Asiatic rodents that began in the Miocene in the Caucasus, in southwest Asia, and Europe became important in Africa only in the late Pleistocene. ARAMBOURG (1963) concluded that the North African fauna of the Tertiary is endemic and clearly distinguished from that of Europe. During most of the Tertiary, the mesogenic rifts were obstacles to exchange of fauna between Africa and Eurasia. The ancient Miocene rodents evolved in subsaharan areas of Africa in isolation. A few reached North Africa probably via the western coast 'road'.

The Sahara Desert has been a barrier within the continent except for occasional short periods during which wetter biotopes invaded, temporarily permitting, as an example, *Thryonomys* to find sufficient water and vegetation along the shores of lakes or streams. The connection between India and Africa established an Ethiopian-Indian biogeographic block. In the Pleistocene, certain Eurasian elements entered North Africa via the Near East and Suez.

The modern rodents of North Africa and the eastern Mediterranean are members of RANCK's Saharo-Sindian faunal zone, which extends from the Russian steppe east of the Caspian and Baluchistan to the circum-Mediterranean countries and Sinai, west across the Sahara to Morocco and Mauritania. RANCK has correlated the prevailing aridity of this region and the paucity of the rodent fauna in terms of number of genera. All prehistoric genera of Asiatic affinity are also adapted to desertic or steppic situations, though a few also lived in woodlands and savannas. Many were burrowing forms in sandy, rocky, sparsely vegetated regions. Burrowing rodents will withstand considerable change because in periods of extreme heat or cold or rain or drought, they take to their burrows until the weather takes a turn for the better. Modern *Tatera* and *Hystrix* are steppic, but they can also live in woodlands, as do *Apodemus* and *Mus*. *Pitymys* lives in open fields and tamarack swamps. *Mesocricetus, Ellobius, Meriones, Microtus, Cricetus, Allactaga, Hystrix* and *Spalax* inhabit steppe, high meadows, brush and rocky areas. *Arvicola* and *Cricetus* live near stream banks. *Gerbillus, Cricetulus, Psammomys,* and *Jaculus* thrive at the edge of the desert.

TCHERNOV has shown by his study (1968) of the rodent succession in the Late Pleistocene of Israel the gradually increasing desiccation of that country. The cave faunas of the past 60,000 years in Iran and Iraq show that steppic and savanna conditions existed in these regions. From other evidence (i.e. pollen, etc.) WRIGHT, McANDREWS & ZEIST (1967) have shown that mountain glaciers on the Zagros forced the snow line lower by 1800 meters. Recovery began in the pollen flora about 18,000 YBP, but not until about 11,500 YBP did oak and pistacio appear in the pollen flora. Woodlands appeared probably about 5,000 YBP but have subsequently been destroyed by man during the past few thousand years. The fossil rodents have been tolerant of the temperature and humidity changes. The climate today is of the Mediterranean type, with winter rain and hot, dry summers.

Thus the complex of prehistoric rodents adds to the sum of our knowledge of former times, but only to a degree. Other lines of evidence must be used along with the rodents to complete the picture of prehistoric environments. The suite of rodents reported here shows adaptability and tolerance for temperature and humidity fluctuations. Many of them are very widespread and geologically long lived. All are xeric to some degree, many to a high degree.

Acknowledgements

To my mentor and colleague, Professor CHARLES A. REED, University of
Illinois at Chicago Circle, I owe and heartily give my thanks for suggesting
that I do this interesting study and for casting his critical eye over the
manuscript, resulting in many improvements. To the librarians at the
Field Museum, I express my sincere appreciation for their skill in helping
to locate sources. My husband, WILLIAM D. TURNBULL, has patiently
and wisely advised me on many points.

REFERENCES

ARAMBOURG, C. 1963. Continental vertebrate faunas of the Tertiary of North Africa.
In: African Ecology and Human Evolution. *Viking Fund Publ. in Anthropology*, no. 36,
55–64.

BATE, D. M. A. 1930. Animal remains from the Dark Cave, Hazer Merd. *Bull. Am.
Sch. Prehist. Res.*, 6, 38–41.

BATE, D. M. A. 1937. Fossil fauna of Wady-el-Mughara caves. In: Stone age of Mt.
Carmel. Oxford Univ. Press, 1, 135–240.

BATE, D. M. A. 1942. Pleistocene murinae from Palestine. *Ann. & Mag. Nat. Hist.*, (11),
9, 465–486.

BATE, D. M. A. 1945. Note on small mammals from the Lebanon Mts., Syria. *Ann. &
Mag. Nat. Hist.* (11), 12, 141–158.

BATE, D. M. A. 1947. Extinct reed rat *(Thryonomys arkelli)* from the Sudan. *Ann. &
Mag. Nat. Hist.* (11) 14, 65–71.

BATE, D. M. A., 1950. A fossil vole from Cyrenaica. *Ann. & Mag. Nat. Hist.*, (12), 3,
no. 35, 981–985.

BATE, D. M. A. 1951. The mammals from Singa and Abu Hugar. In: Fossil mammals
of Africa no. 2. The Pleistocene fauna of two Blue Nile sites. *Pub. Brit. Mus. (N.H.)*,
Lond., 1–28.

BODENHEIMER, F. S. 1960. Animal and man in Bible lands. E. S. Brill, Leiden, 232 pp.

BRAIDWOOD, R. J. & HOWE, B. 1960. Prehistoric investigations in Iraqi Kurdistan.
Studies in ancient oriental civilization, no. 31. Univ. of Chicago, i-viii, 1–184. 29 pls.

BRAIDWOOD, R. J., HOWE, B. & REED, C. A. 1961. The Iranian prehistoric project.
Science, 133, 2008–2010.

COON, C. S. 1951. Cave explorations in Iran, 1949. Univ. Penn. (Museum), 1–96.

COON, C. S. 1952. Excavations in Hotu Cave, Iran. *Proc. Am. Phil. Soc.*, 96, 231–249.

COON, C. S. 1959. Reported in HOWELL, 1959.

FRENKEL, H. 1970. *Hystrix angressi*, sp. nov. a large fossil porcupine from the Levalloiso-
Mousterian of Geula cave. *Isr. J. Zool.* 19, no. 1, 51–82.

FREUDENTHAL, M. 1970. A new *Ruscinomys* from the late Tertiary of Samos, Greece.
Amer. Mus. Nov., no. 2402, 1–10.

GARROD, A. E. & BATE, D. M. A. 1937. Stone age of Mt. Carmel, I; excavations at
Wady el Mughara. Oxford University, 1–240.

HAAS, G. 1951. Remarques sur la microfaune de mammiferes de la grotte d'oumn-

Qatafa. In: Le paleolithique e la mesolithique du desert de Judee. *Arch. Inst. Paleo. Humaine Mem.*, no. 24, 233 pp.

HAAS, G. 1952. Fauna of layer B of the Abu Usba cave. In: STEKELIS & HAAS, *Isr. Expl. Jour.*, 2, no. 1, 35–47.

HAAS, G. 1966. On the vertebrate fauna of the lower Pleistocene site 'Ubeidiya. *Isr. Acad. Sci. & Human.*, 1966, 1–68.

HATT, R. T. 1959. The mammals of Iraq. *Mus. Zool., Univ. of Mich.*, 113 pp.

HELLER, J. 1970. Small mammals of Guela cave. *Isr. J. Zool.*, 19, no. 1, 1–49.

HIGGS, E. S. 1967. Faunal fluctuations and climate in Libya. In: Background to evolution in Africa, BISHOP & CLARK, ed. Univ. of Chicago, 149–163.

HOLE, F., FLANNERY, K. & NEELY, J. A. 1969. Prehistory and human ecology of the Deh Luran Plain, Iran. *Mem. Mus. Anthrop., Univ. Mich.*, no. 1, 488 pp.

HOOIJER, D. A. 1961. Fossil vertebrates of Ksār 'Akil, a paleolithic rock shelter in Lebanon. *Zool. Verhand.*, no. 49, 1–67.

HOWE, B. 1967. Paleolithic of Tangier, Morocco. Amer. School Prehist. Res., Peabody Museum, Cambridge, 200 pp.

HOWELL, F. C. 1959. Upper Pleistocene stratigraphy and early man in the Levant. *Proc. Am. Phil. Soc.*, 103, no. 1, 1–65.

JAEGER, JEAN-JACQUES & MARTIN, JACQUES. 1971. Decouverte au Maroc des premiers micromammiferes du Pontien d'Afrique. *C. R. Acad. Sci. Paris, Ser. D*, 272, no. 17, 2155–2158.

LAVOCAT, R. 1967. Les microfaunas du neogene d'Afrique orientale et leurs rapports avec celles de la region palearctique. In: Background to evolution in Africa, Univ. of Chicago, BISHOP & CLARK, ed., 57–72.

LAY, D. 1967. The mammals of Iran. *Fieldiana: Zool.*, 54, 282 pp.

McBURNEY, C. B. M. 1967. The Haua Fteah (Cyrenaica). Camb. Univ. Press, 387 pp. Mammalian fauna by HIGGS, E. S.

McBURNEY, C. B. M. & HEY, R. W. 1955. Prehistory and Pleistocene geology in Cyrenaican Libya. Cambr. Univ. Press, 275 pp. Appendix by BATE, D. M. A.

MOREAU, R. E. 1952. Africa since the Mesozoic. *Proc. Zool. Soc. Lond.*, 121, no. 4, 869–913.

MOREAU, R. E. 1963. Viscissitudes of the African biomes in late Pleistocene time. *Proc. Zool. Soc. Lond.* 141, no. 2, 395–421.

OZANSOY, F. 1961. Sur quelques mammiferes fossiles du Tertiaire d'Anatoli Occidentale-Turquie. *Publ. Min. Res. Inst. Turkey*, 56, 85–93.

PATTERSON, B. 1937. Animal remains from Alisher Hüyük (1930–1932). Researches in Anatolia. Univ. of Chicago, 9, pt. 3, 294–310.

PETTER, F. 1968. Un muride Quaternaire-nouveau d'Algerie, *Paraetheomys filfilae*. *Mammalia*, 32, 54–59.

RANCK, G. L. 1968. Rodents of Libya. Smith. Inst. Press, 264 pp.

REED, C. A. & BRAIDWOOD, R. J. 1960. Toward the reconstruction of the environmental sequence of N.E. Iraq. In: Studies in Ancient Oriental Civil., no. 31, Univ. of Chicago, 163–173.

REED, C. A. & TURNBULL, P. F. 1969. Late Pleistocene mammals from Nubia. *Paleoecology of Africa* 1966–1968, 4, 55–56.

ROBINSON, P. 1966. Fossil occurrence of a murine rodent *(Nesokia indica)* in the Sudan. *Science*, 154, 264.

ROMER, A. S. 1928. Pleistocene mammals of Algeria: fauna of the paleolithic station of Mechta-el-Arbi. *Logan Mus. Bull.*, 1, no. 2, 100.

ROMER, A. S. 1938. Mammalian remains from some Paleolithic stations in Algeria. *Logan Museum Bull.*, 1, no. 5, 165–184.

ROMER, A. S. 1967. Vertebrate Paleontology. University of Chicago, i-viii, 1-468, 443 figs., 4 tables.

ROMER, A. S. & NESBITT, P. H. 1930. *Thryonomys logani* from the Sahara Desert. *Ann. & Mag. Nat. Hist.*, (10), 6, 687–690.

377

SAVAGE, R. J. G. 1967. Early Miocene mammals of the Tethyan region. Systematics Assoc. Publ. #7, Ed. ADAMS & AGER. 247–282.

SIMPSON, G. G. 1947. Holarctic mammal faunas during the Cenozoic. *Bull. G. S. A.*, 58, 613–688.

SOLECKI, R. S. 1953. A paleolithic site in the Zagros mts., N. Iraq; a report on a sounding at Shanidar. *Sumer*, 9, no. 1, 60–93.

STEHLIN, H. G. & SCHAUB, S. 1951. Die trigodontie der simplicidentaten nager. *Schw. Pal. Abh.*, 67, 1–385.

STEKELIS, M. & HAAS, G. 1952. Fauna of layer B of the Abu Usba cave. *Isr. Expl. Jour.*, 2, no. 1, 35–47.

TCHERNOV, E. 1963. The successions in the rodent fauna of Israel during the later Quaternary. Heb. Univ. Res. Rept., 271 pp.

TCHERNOV, E. 1968. Succession of rodent faunas during the upper Pleistocene of Israel. *Mammalia Depicta, Berlin.* 1–152.

TURNBULL, P. F. (Ms.) A mammalian fauna from archaeological sites at Tichitt, Mauritania, W. Africa.

TURNBULL, P. F. (in press) The mammalian fauna of Warwasi rock shelter, west-central Iran. *Fieldiana: Geology*, 33, no. 8, 141-155

TURNBULL, P. F. (in press) Small birds and mammals from Jarmo, Iraq. stud. Anc. orient. civil.

TURNBULL, P. F. & REED, C. A. 1974. The fauna from the terminal Pleistocene of Palegawra cave, a Zarzian occupation site in N.E. Iraq. *Fieldiana: Anthropology*, 63, No. 3, 81–146.

VAUFREY, R. 1931. In: NEUVILLE, R. L'acheuléen superieur de la grotte d'Oumm-Qatafa (Palestine), IV – paleontologie. *L'Anthropologie*, 41, 253–263.

VERESCHAGEN, N. K. 1959. The mammals of the Caucasus. Isr. Prog. for Sci. Trans., 1968, 816 pp. (English)

WOOD, A. E. 1955. A revised classification of the rodents. *J. Mamm.* 36, no. 2, 165–186.

WOOD, A. E. 1968. The African Oligocene Rodentia in: SIMONS & WOOD, Early Cenozoic mammalian faunas of Fayum province, Egypt. *Bull. Peabody Museum Nat. Hist.*, 28, 23–105.

WRIGHT, H. E., MCANDREWS, J. H. & VAN ZEIST, W. 1967. Modern pollen rain in western Iran and its relation to plant geography and Quaternary vegetational history. *J. Ecology*, 55, 415–433.

[Ms. received March, 1972]

ADDENDA

The editors have kindly permitted me to bring this chapter up-to-date, at least to the extent of indicating the most important recent papers dealing with prehistoric rodents of the middle east.

IRAN. From Bastam, 85 km SE of Maku and NW of Aserbeidschan in northern Iran, a collection of rodents, dated between 2200–700 B.C. was described by STORCH in 1974. His list included *Cricetulus migratorius*, *Tscherskia rusa*, n. sp., and a new unnamed cricetine. From the same site, BOESSNECK (1974) described a beaver, *Castor fiber*.

TURKEY. A somewhat earlier collection of prehistoric rodents, dated 3500–1200 B.C., hailed from Vilayet Elazig in eastern Anatolia. These specimens were described in 1972 by VON D. KOCK, MALEC, and STORCH, who identified the following species: *Spalax leucodon*, *S*. sp., *Apodemus mystacinus*, *A. sylvaticus*, *Mus musculus*, *Cricetulus migratorius*, *Meriones tristrami*, *Microtus irani*.

An important southern Turkey faunule was found in caves in the southern Taurus mountains of Antalya province. From the cave at Çatallar, CORBET & MORRIS (1967) identified *Hystrix indica;* from the Cliff cave, the same workers identified *Apodemus mystacinus*, *Rattus rattus*, *Mus musculus*, *Cricetulus migratorius*, *Microtus guentheri*, and *Myomimus personatus* (which these authors synonymized with *Philistomys roachi* Bate). All of these forms, point out the authors, live today in Asia Minor, except *Myomimus personatus*, which is known from the Pleistocene and post-Pleistocene of Israel.

RHODES (Greece). So close is the Isle of Rhodes to the Turkish mainland that it ought not to be omitted from this survey. A most important contribution to the understanding of rodent prehistory was made in 1970 by DE BRUIJN, DAWSON & MEIN, who described the small mammals from a Pliocene fissure fill in Mesozoic limestone at the village of Maritsa, at the northern end of Rhodes. The quarry on Stavro Hill yielded 16 species of rodents, five of them murines. The authors emphasized that this collection, the first large late Pliocene assemblage from the eastern Mediterranean, indicates that faunal exchange between southwestern Europe and Asia was considerable. Most of the elements are Asiatic and a few are European as may be seen from the genera listed below:

Apodemus cf. *jeanteti; A.* cf. *dominous; Castillomys crusafonti; Occitanomys anomalus*, n.sp.; *Pelomys europeus*, n.sp.; *Cricetus lophidens*, n.sp.; *Mesocricetus primitivis*, n.sp.; *Cricetulus* sp.; *Calomyscus minor*, n.sp.; *Pseudomeriones abbreviatus; Spalax sotiriosi*, n.sp.; *Atlantoxerus rhodius*, n.sp.; *Spermophilinus giganteus*, n.sp.; *Myomimus maritsensis*, n.sp.; *Eliomys intermedius; Keramidomys carpaticus.*

Origin of the murines *Occitanomys*, *Castillomys*, and *Apodemus* is unclear. *Myomimus* was probably a relict of western European origin, according to DE BRUIJN *et al.*, *Pelomys* is perhaps African.

LIBYA. In his detailed and broadly conceived paper dealing with African and Indo-Australian muridae, MISONNE (1969) noted the Pleistocene occurrence of *Apodemus* sp., a genus not now present in Libya.

TUNISIA. From beach deposits on the north side of Lake Ichkeul, JAEGER (1971) described forms he believes are older than the Villafranchian and contemporary with the upper Pliocene fauna of Rhodes. An animal similar to *Ruscinomys* and another referred to *Paraethomys anomalus* were identified.

ALGERIA. The following rodents were described by JAEGER (1969) from deposits at Ternifine: *Ellobius* cf. *fuscocapillus*, *Arvicanthis* sp., *Praomys* sp., *Paraethomys* cf. *filfilae*, *Hystrix* sp., *Meriones* cf. *chowi*, *Gerbillus* cf. *campestris*. Of these, *Gerbillus* cf. *campestris* was the most abundant. Among these forms there is a total absence of European elements. *Ellobius* was an Asiatic immigrant. *Paraethomys*, *Arvicanthis* and *Praomys* are African in origin. JAEGER concluded that a barrier existed at the beginning of the mid Pleistocene, keeping out European elements. The lack of endemic Peditidae, Bathyergidae, and Thryonomidae indicates that North Africa was a different biogeographic province from the rest of Africa. This is the first reported find in the area of *Arvicanthis*, a genus usually inhabiting tropical savannah but obviously capable of adjustment to somewhat different habitats.

From Neolithique deposits at Amekni in the Ahaggar, MONOD (1970) studied specimens of *Thryonomys swinderianus*. His comparative work with other species resulted in placing *T. logani* into the former species.

MOROCCO. Three deposits, one Villafranchian and two middle Pleistocene in age, were found in caves at the western part of Jebilet, in a limestone reef structure, Jebel Irhound, on peneplaned Cambrian shist. JAEGER (1970) identified a ctenodactylid, cf. *Felovia*; sciurids, cf. *Protoxerus* and *Atlantoxerus* cf. *getulus*; the glirid, *Eliomys* cf. *quercinus*; the jaculid, *Jaculus* cf. *orientalis*; an indet. cricetodontine; the gerbillids, *Gerbillus* sp., and *Meriones* cf. *shawi*; the arvicolid, *Ellobius* sp.; and murines, *Praomys* sp., *Paraethomys* sp., *P.* cf. *filfilae*, *Mus* sp., *M. musculus*, *Arvicanthis* sp. and *Lemniscomys* sp. JAEGER indicated that *Ellobius*, *Meriones* and *Arvicanthis* were not present in the earlier Villafranchian deposits; they arrived as immigrants in the mid Pleistocene, thus changing the character of the North African fauna.

MAURITANIA. In 1970, JULIEN & PETTER described a fauna from a bat cave at the top of Guelb Moghrein, near Akjoujt. The deposit is dated 480–400 B.C. and contained the following rodents: *Meriones crassus*, *Taterillus gracilis*, *Desmodilliscus braueri*, *Gerbillus pyramidum*, *G. nanus*, and *Jaculus jaculus*.

Innumerable other prehistoric rodent finds are probably known from the middle east, and I can only hope I have not overlooked important discoveries.

ADDITIONAL REFERENCES

BOESSNECK, V. J., 1974. Ergänzungen zur einstigen Verbreitung des Bibers, *Castor fiber* (Linné, 1758). *Saugetierk. Mitt.*, 22(1), 83–88.

CORBET, G. B. & MORRIS, P. A., 1967. A collection of recent and subfossil mammals from southern Turkey (Asia Minor), including the dormouse *Myomimus personatus*. *Jour. Nat. Hist.*, 1(4), 561–569.

DE BRUIJN, H., DAWSON, M. & MEIN, P., 1970. Upper Pliocene Rodentia, Lagomorpha, and Insectivora from the Isle of Rhodes (Greece). *Proc. K. Ned. Acad. Wet.* (B) 73, 535–584.

JAEGER, J. J., 1969. Les Rongeurs du Pléistocène Moyen de Ternifine (Algerie). *C.R. Acad. Sci. Paris*, 268, D. 1492–1495.

JAEGER, J. J., 1970. Decouverte au Jebel Irhoud des premières faunes de rongeurs du Pléistocène inférieur et moyen du Maroc. *C.R. Acad. Sci. Paris*, 270, 920–923.

JAEGER, J. J., 1971. Les Micromammifères du 'Villafranchien' inférieur du lac Ichkeul (Tunesie): données stratigraphiques et biogéographiques nouvelles. *C.R. Acad. Sci. Paris*, 273, 562–565.

JULIEN, R. & PETTER, F., 1970. La faune du gisement D'Akjoujt (Mauritanie). *Bull. du Museum National D'Histoire Naturelle*, ser. 2, 41, 1290–1291.

MISONNE, X., 1969. African and Indo-Australian muridae. Musée Royal de L'Afrique Centrale – Tervuren, Belgique, 172, 1–219, pls. I–XXVII.

MONOD, T., 1970. A propos d'un Aulacode (Thryonomys) du gisement néolithique d'Amekni (Ahaggar). *Bull. de L'Inst. Fond. d'Afrique Noire*, 32, 531–550.

STORCH, G., 1974. Neue Zwerghamster aus dem Holozän von Aserbeidschan, Iran. *Senckenb. biol.*, 55, 21–28.

VON D. KOCK, MALEC, F. & STORCH, G., 1972. Rezente und subfossile Kleinsäuger aus dem Vilayet Elazig, Ostanatolien. *Zeitsch. für Säugetierk.*, 37(4), 204–229.

XVIII. DESERT RODENTS:
PHYSIOLOGICAL PROBLEMS OF DESERT LIFE

by

KNUT SCHMIDT-NIELSEN*

Introduction

The major climatic factors that shape the hot deserts of the world are high temperatures and low and irregular rainfall. The primary physiological problems of life in such areas are directly related to heat and lack of water; other aspects of animal survival, e.g. food resources, predators, cold, reproduction, etc. do not differ in principle from similar problems in non-desert areas. This chapter will therefore be addressed to the problems uniquely characteristic of the desert habitat, and furthermore, the presentation will specifically be directed to those problems that seem pertinent to rodents.

Rodents are, almost without exception, of small body size and rather uniform morphology. All the major deserts of the world are inhabited by a large variety of rodents. Even the Australian continent, which is usually thought to be inhabited primarily by marsupials, has a fauna that includes 83 native species of rodents (the total marsupial fauna of Australia numbers less than 150 species).

The success of rodents in desert areas indicates that these animals either can avoid the major stresses of heat and lack of water, or that they have successfully employed effective physiological counter-measures. The following discussion will show that the most severe heat stress of the desert is avoided by burrowing and nocturnal habits, and the major physiological problem is then reduced to how a mammal can remain in water balance in the absence of drinking water.

For an animal to remain in water balance, the total water intake must equal the total water loss. Two avenues are therefore open to the animal; it can seek to increase the water intake, or it can reduce the amounts lost. As we shall see, both avenues have been used, although the details may vary considerably from one type to another. With the limited number of possible solutions, however, there is a conspicuous functional similarity between rodents from the various major deserts, as there is in their morphology, although taxonomically they may be only distantly related. For example, in superficial appearance, in mode of locomotion, in

* Recipient of Research Career Award 1-K6-GM, 522 from the National Institutes of Health.

feeding habits, and in physiological characteristics there is a great deal of similarity between the Old World jerboas, the American kangaroo rats, and the Australian hopping mice, but they belong to three separate families, the Dipodidae, the Heteromyidae, and the Muridae.

Problems of Temperature Regulation

IMPORTANCE OF BODY SIZE

At moderate ambient temperature a mammal loses heat primarily by conduction and convection to the air, and by radiation to the environment. At high temperatures, these avenues of heat loss are eliminated, and if the ambient temperature exceeds the body temperature, heat flows from the environment into the animal. In these circumstances the body temperature can be kept from rising only by evaporation of water. The major physiological mechanisms employed in thermoregulatory evaporation are sweating (as in man) and panting (as in dogs and many other animals).

Mammals the size of rodents neither sweat nor pant. They lack skin glands in sufficient numbers to be of importance in heat dissipation, and they apparently completely lack the increase in respiratory rate during heat load which is known as panting.

The rational basis for the absence of the two major physiological mechanisms for evaporative heat dissipation in rodents is not only the scarcity of water resources, but also the generally small body size of these animals. Since small animals have a large surface, relative to their weight, they would have to respond to a given heat load with a higher relative rate of evaporation than animals of large body size. This subject has been discussed on several previous occasions (see e.g. SCHMIDT-NIELSEN, 1964).

For reasons of size alone, small rodents could not depend on the use of water for temperature regulation. Escape from the heat of the desert is therefore of greater importance to rodents than are physiological mechanisms for heat dissipation.

ESCAPE

Burrowing and nocturnal habits permit desert rodents to avoid the high daytime temperatures in the desert. In hot deserts air temperatures often exceed 40 °C, and due to the high solar radiation flux in the dry atmosphere the soil surface temperature may exceed 70 °C. In the evening the surface temperature falls rapidly because the clear atmosphere permits high levels of re-radiation, and the dryness of the soil contributes to a low heat conductance and low heat capacity, thus augmenting the rapid cooling of the surface layer. Because of these factors the nighttime surface temperature may fall far below air temperature and be more than 50 °C below the daytime surface temperature.

The fluctuations in surface temperature are rapidly damped out with increasing depth below the surface. At a depth of one-half meter the daily fluctuation is usually less than 1 °C, and therefore of no physiological significance. At this depth, which is a common depth for rodent burrows, the maximum temperature will probably not often exceed 35 °C, and thus will not be a significant physiological heat stress. There is, however, a lack of adequate and reliable information about rodent burrow temperatures under natural and undisturbed conditions, especially from the hottest desert areas of the world. (Burrow atmospheric conditions, in regard to humidity as well as to gas composition, are even less well known, but technically far more difficult to examine.)

BODY TEMPERATURE

Hyperthermia

An animal which, during heat stress, maintains by evaporation a constant low body temperature uses water far in excess of all other needs combined. If the animal instead permits the body temperature to increase (or initially has a high body temperature), the use of water will be reduced. A high body temperature can therefore be considered adaptive, and to maintain a margin of safety, a corresponding increase in the lethal body temperature would be required.

Desert rodents apparently have not, to any appreciable extent, utilized this avenue for physiological adaptation to desert life. Their normal body temperature seems to be in the same range as that of other mammals, about 36 to 38 °C. When exposed to a heat stress they become hyperthermic, but in those cases where their thermal tolerance is known, they do not seem to have higher lethal body temperatures than found in other mammals (about 42–44 °C, depending on duration of exposure). On this point more information is needed, including a larger number of rodent species than have been studied so far.

Regulation, salivation

Although neither sweating nor panting is used by rodents, some evaporative cooling can be achieved by salivation and spreading of saliva over the fur. This type of cooling is well known from marsupials, it has been described for kangaroo rats, but its quantitative aspects have been well studied only in the white rat (see HAINSWORTH & STRICKER, 1970). Apparently, the mechanism is not used to maintain the body temperature at a low 'normal' level, but is brought into effect only when body temperature rises towards near-lethal levels.

381

The role of estivation or torpor in relation to water and heat problems of desert rodents is unclear. Some desert rodents, during the hottest part of the year, enter into a state of torpor and hypothermia with greatly reduced rates of metabolism, heart beat, respiration, etc. The state is similar to the state of hibernation, and can be interpreted as a physiological escape mechanism.

Undoubtedly, the reduced metabolic rates during torpor can be interpreted as an energy saving mechanism. Furthermore, since the animals have a lowered body temperature, it has been suggested that evaporation, especially from the respiratory tract, would be lower than normal. Also, the torpid animal remains in its burrow where the atmosphere usually has a higher water vapor content than the outside desert air; the respiratory evaporation should be further reduced due to this factor.

The situation in regard to respiratory evaporation is not as simple as previously assumed. Small rodents, when breathing dry air, exhale air at temperatures far below body temperature, and even below ambient air temperature. Thus, exhaled air may be some 10 or 20° below body core temperature, and its water vapor content therefore correspondingly low (SCHMIDT-NIELSEN et al., 1970). This observation makes it desirable to re-examine whether a lowered body temperature during estivation is of major importance in water balance.

Another matter that remains inadequately understood is that some desert rodents estivate, while others do not. A classical example is that of two similar species of ground squirrels which live in the same desert habitat, one being an estivator and the other a non-estivator (BARTHOLOMEW & HUDSON, 1961). Well substantiated studies of the natural history of estivation among desert rodents should be followed by careful physiological studies in order to clarify the role of this phenomenon in desert survival.

Water Balance, Intake and Loss

For an animal to remain in water balance the total water intake must equal the sum of all water loss. Available information indicates that desert rodents normally remain in water balance for indefinite periods of time, in other words, they do not store water reserves which are slowly used during periods of drought, nor do they gradually deplete the normal body water content. Desert rodents, in all known cases, have a body water content similar to that of other mammals, about two-thirds of the body weight, and remain in a steady-state of normal water balance. In the steady state water intake equals water loss, or: Intake (Free water + Water in food + Oxidation water) = Loss (Urine + Faeces + Respiratory + Cutaneous evaporation).

Free water is available in deserts only on the rare occasions of rain. Since rodents are found in many deserts where rain occurs even less frequently than every year, drinking water cannot play any major role for these animals. Whether or not rodents will drink free water when it is available is therefore of minor physiological importance; there are reports, however, that some desert rodents will refuse to drink water (BUXTON, 1923).

A possible source of free water could be provided by dew, but many deserts have no visible dew during the major part of the year. In some areas where dew is more common and heavier than general, dew could be of importance; more information would be interesting, although probably difficult to obtain in quantitative terms. In most deserts dew is apparently not a major factor in the water intake of rodents.

PREFORMED WATER IN THE FOOD

In the absence of drinking water, free water can be obtained through the utilization of suitable food. Herbivorous animals can eat green leaves, stems, fruits, roots, tubers, and carnivorous animals obtain considerable amounts of water from the body fluids of their prey, which may contain from 50 to 80% water. Many desert rodents depend on such moist food, but others live primarily on dry seeds and other dry plant material, and their intake of preformed water is therefore quite low.

MOIST FOOD

When the food has a high water content, other interesting physiological problems may enter the picture. Many desert plants contain noxious substances, some have very high salt content, and carnivorous food yields metabolic products (primarily urea) that require water for excretion.

One toxic substance is oxalic acid, the toxicity being due to its binding of calcium. The metabolism of oxalate has received only scant attention, and many related and highly interesting problems remain unexplored.

Some desert rodents consume succulent halophilic plants whose sap has a high electrolyte content; in some cases the sap may be more than twice as concentrated as sea water. One such rodent is the sand rat (Psammomys), which has a highly efficient kidney and produces large volumes of urine with extraordinarily high electrolyte concentrations.

Carnivorous rodents enjoy a relatively high water intake, but the high protein content of their food leads to the formation of large amounts of urea as the primary excretory product. These animals generally display high urine concentrations of urea, as in the carnivorous North American grasshopper mouse *(Onychomys)* and in the Australian rodent *Notomys*, presumably primarily a plant eater (MACMILLEN & LEE, 1967). To what extent herbivorous rodents, when opportunity arises, use animal material in their diet is not well known.

DRY FOOD

When dry plant material is the primary component of the diet, such as for kangaroo rats, jerboas, and many others, the amount of preformed water is relatively small.

Air dried seeds contain several per cent of water. The amount changes with the relative humidity of the air, and thus, no exact figure can be given. The water content may exceed 10%, but is usually much lower.

In this context it is important that plant material collected at night will have a higher water content than the same plant material has during the day. Although the absolute humidity of the air usually does not change from day to night, the lower night temperature means increased relative humidity, and plant material therefore takes up more water. For this reason the nocturnal habits of desert rodents are important also to the water intake. This has been pointed out by Buxton and others (BUXTON, 1923; SCHMIDT-NIELSEN, 1964; TAYLOR, 1968). TAYLOR showed that the grass *Disperma* might change its water content 35-fold, from 2 g H_2O per 100 g dry matter to 72 g H_2O per 100 g dry matter, by being exposed to different air humidities. The principle is of importance also for those rodents that store food in their underground burrows, where the humidity is usually higher than in the outside air, and as a result, the water content of the food is increased. Further studies of this subject could contribute substantial and important information to the understanding of the ecology and physiology of desert rodents.

OXIDATION WATER

The main end products of the oxidation of foodstuffs is CO_2 and H_2O. For starch and fats these are the only end products, protein metabolism yields urea in addition to CO_2 and H_2O. Water formed in the oxidation processes can be designated as 'oxidation water', but is often called 'metabolic water'. Some authors have taken the latter term to imply that special metabolic processes can be used to produce 'metabolic' water, and the term oxidation water should serve to avoid such mistakes.

Oxidation water is derived from the hydrogen in the food, the amount formed will depend on the composition of the food, and on the rate of oxidation (metabolic rate). There is no means by which the amount of water formed from a given amount of food can be increased beyond that corresponding to its hydrogen content, and there is no way of obtaining extra water from 'unknown' metabolic processes.

Since oxygen is necessary for the formation of oxidation water, there must be a close correspondence between oxygen uptake and the amount of oxidation water formed. The rate of oxygen uptake determines the ventilation of the lungs, which in turn determines the evaporation from the respiratory tract. Thus, there is an inevitable interdependence between the two, increased formation of oxidation water is tied to increased respiratory evaporation.

The relationship between oxidation water on the one hand, and the simultaneous evaporation from the respiratory tract on the other, has been discussed on previous occasions. One aspect should be emphasized, however. The higher amount of oxidation water formed in the complete oxidation of one gram of fat, as compared to one gram of carbohydrate, is offset by the fact that fat requires a higher oxygen intake for its metabolism and therefore a correspondingly increased respiratory evaporation. Thus, the higher gain of oxidation water from fat is not a true gain, in fact, the relation between oxidation water and respiratory evaporation is slightly less favorable for fat than for carbohydrate metabolism (SCHMIDT-NIELSEN, 1964).

WATER LOSS, URINE

A high concentration of solutes in the urine means that relatively small amounts of water are required for urine production. High concentrations of urea and of electrolytes have been reported for the urine of many desert rodents. The highest urine concentrations are found in those rodents that do not depend on a high water intake. Notable examples are the North American kangaroo rats *(Dipodomys)*, Old World jerboas *(Jaculus)*, and Australian hopping mice *(Notomys)* (MACMILLEN & LEE, 1967).

Since the water loss in the urine is a major factor in the water balance of desert rodents, the renal function is extremely important. However, in principle, this situation is well understood. Numerous studies have been carried out, and further examinations of this subject are not urgently needed.

In no case has it been demonstrated that mammals can change from the excretion of urea as an end product of protein metabolism to the production of the less soluble uric acid (which therefore requires less water for its excretion). In birds and reptiles the excretion of uric acid is a primary factor in water conservation, and the excretion of uric acid has recently been described for an amphibian from the arid southeast Africa (LOVERIDGE, 1970).

WATER LOSS, FAECES

The water loss in the faeces constitutes a fairly small fraction of the total water output of desert rodents. Nevertheless, any saving of water is important, and those desert rodents that are independent of a high water intake form faeces with a very low water content. In this regard, various rodents from different continents show great similarities.

Two parameters are of importance to the amount of water lost in the faeces, the total amount of faeces produced, and their water content.

Although it is known that desert rodents produce faeces with a lower water content than, for example, white rats, no studies have been made of the mechanism involved, or of its control. It is not known whether the mechanism of intestinal water withdrawal in desert rodents differs from other rodents. A critical and careful study of this subject is much needed.

The other parameter in faecal water loss, the amount of faeces formed, is also a subject in need of investigation. It has been suggested that reingestion of faeces (coprophagy) increases the utilization of the food, thus reducing the amount of material finally excreted in the faeces, and thereby reducing their bulk. Kangaroo rats do indeed produce smaller relative amounts of faeces than do white rats on the same diet, and this is quantitatively of importance in evaluating the water balance of these two animals. However, this subject needs study on a much broader basis in order to evaluate the relative importance of food utilization, bulk of faecal material, the minimum water content of the faeces, the influence of various food material, and so on.

EVAPORATION, RESPIRATORY

The evaporation from the respiratory tract of mammals is determined by the fact that air is exhaled saturated with water vapor. Obviously, the amount of water already present in inhaled air (that is, the atmospheric humidity) will determine how much water is added from the respiratory tract before the air is exhaled. There is no indication whatsoever that any mammal can exhale air below full saturation.

Evaporation from the respiratory tract is the largest part of the total water loss from rodents. The amount of water lost by respiratory evaporation is determined by the respiratory volume, which again is determined by the oxygen consumption. Mammals in general remove approximately 5% oxygen from the inhaled air before it is exhaled again. If this amount could be increased to, say 10%, the volume of air respired would be reduced to one-half (for a given oxygen consumption), and the amount of water lost in exhaled air would be reduced accordingly. Until now there has been no indication that any desert rodent has utilized this avenue to reduce respiratory evaporation.

In contrast, an important factor in the reduction of respiratory evaporation is found in the fact that small rodents exhale air at temperatures far below body temperature. The water content of exhaled air is therefore correspondingly reduced. The mechanism is very simple: during inhalation the walls of the passageways are cooled by the air flowing in, the wall temperature decreases and may even be lower than the inhaled air because of evaporation. On exhalation, warm air from the lungs passes over the cool surface, the air is thus cooled and part of its water content is recondensed on the walls. This cooling of the exhaled air greatly reduces the water loss; the exact amount depends on ambient temperature and humidity. The phenomenon has been discussed in detail in previous publications (SCHMIDT-NIELSEN et al., 1970).

It should be noted that nasal heat exchange as described here is not unique for desert rodents. The phenomenon is a result of the geometry of the nasal passageways and physical laws of heat exchange. Thus, exhaled air temperatures in kangaroo rats and in white rats are similar. Although there is reason to believe that other rodents, desert and non-desert species alike, will behave similarly, further studies should be carried out to substantiate this assumption.

EVAPORATION, CUTANEOUS

While the evaporation from the respiratory tract of rodents is understood in principle, information about cutaneous evaporation is meager. In white rats approximately one-half of the total evaporation is supposed to be from the skin, the remainder being from the respiratory tract. By inference, it has been suggested that the cutaneous evaporation from kangaroo rats is small (SCHMIDT-NIELSEN, 1964).

Since rodents lack sweat glands, the water loss through the skin must be due to diffusion processes, a phenomenon that can be expected to take place through any structure that does not form an absolute water barrier. However, there is no clear understanding of what factors may contribute to differences between desert and non-desert species, nor is there any information about physiological factors that contribute to changes in cutaneous evaporation that may occur as animals are exposed to water deprivation.

The interesting problems of cutaneous water loss deserve further study, and it can be expected that comparison of a variety of rodents may contribute important information.

OTHER USE OF WATER, MILK

During the reproductive period a major drain on the water resources of the female rodent is in the production of milk for the offspring, commonly enhanced by a large number of young. For many rodents the reproductive period coincides with the highest availability of moisture and green vegetation, but further ecological studies of this relationship are needed.

A subject that deserves investigation is the water content of the milk. Seals and whales produce milk with a very high fat content (some 30

387

to 40%), presumably in order to transfer nutrients to the young at a particularly high rate. This can, however, be viewed in the light of the need to conserve water, which is not freely available to marine mammals.

It would therefore be interesting to know whether desert rodents produce milk with a lower water content than commonly found in corresponding non-desert species. Information in this regard would aid in the understanding of the reproductive physiology of desert rodents in relation to their water balance, and it can be expected that such studies would be highly rewarding.

REFERENCES

Since extensive bibliographies are found in several other chapters of this book, most references have been omitted from this chapter.

BARTHOLOMEW, G. A. & HUSDON, J. W. 1961. Desert ground squirrels. *Sci. Am.* 205: 107–116.

BUXTON, P. A. 1923. Animal Life in Deserts, A Study of the Fauna in Relation to the Environment. London: Arnold. 176 pp.

HAINSWORTH, F. R. & STRICKER, E. M. 1970. Salivary cooling by rats in the heat. In: Physiological and Behavioral Temperature Regulation (J. D. HARDY, A. P. GAGGE & J. A. J. STOLWIJK, eds.), pp. 611–626. Springfield, Ill.: Charles C. Thomas, Publ.

LOVERIDGE, J. P. 1970. Observations on nitrogenous excretion and water relations of *Chiromantis xerampelina* (Amphibia, Anura). *Arnoldia* 5: 1–6.

MACMILLEN, R. E. & LEE, A. K. 1967. Australian desert mice: independence of exogenous water. *Science* 158: 383–385.

SCHMIDT-NIELSEN, K. 1964. Desert Animals. Physiological Problems of Heat and Water. Oxford, England: Clarendon Press, 277 pp.

SCHMIDT-NIELSEN, K., HAINSWORTH, F. R. & MURRISH, D. E. 1970. Counter-current heat exchange in the respiratory passages: effect on water and heat balance. *Resp. Physiol.* 9: 263–276.

TAYLOR, C. R. 1968. Hygroscopic food: a source of water for desert antelopes. *Nature* 219: 181–182.

XIX. ECOPHYSIOLOGY OF WATER AND ENERGY IN DESERT MARSUPIALS AND RODENTS

by

W. V. MACFARLANE

Introduction

Small mammals and water are both difficult to find in the desert by day. Mammals are cryptozoic and largely nocturnal. By hiding under vegetation, among rocks or in burrows they avoid heat, the dryness of the air and lethal radiation from the sun. If water is to be found it is in rock pools, occasional spring, drops of dew, succulent plants or the bodies of insects and other animals. Several patterns of function for surviving the heat and drought of the desert have evolved (SCHMIDT-NIELSEN, 1964).

The seed-eating group (*Gerbillus*, *Dipodomys*, *Psammomys* and *Notomys*) combine high renal concentration powers, up to 9.3 osmol/l. (MAC-MILLEN & LEE, 1969) with low rates of transpiration and low rates of energy consumption. These animals can survive on metabolic water. Amongst the marsupial carnivores and insectivores the food supply contains more than 60% of water, but the high protein content yields nitrogenous products which have to be excreted with water. The nervous system under duress reduces metabolism by torpor or hibernation for further water conservation. Both these seed-eating rodents and insectivorous marsupials probably evolved for the desert in Pliocene times. There are however small mammals which invade desert regions from better-watered areas. They survive while there is water, but there are few physiological resources for overcoming lack of drinking water. *Sminthopsis crassicaudata* has a relatively high rate of water turnover and dies within a few days when deprived of water. But the species also can reinvade desert areas from better-watered population centres when the desert again becomes a viable habitat.

Amongst large mammals evolved for hot arid areas (camels, goats, sheep, oryx) the coat protects by reflection or insulation against the sun. There is evaporative cooling from sweat glands or the respiratory tract, water turnover is low and metabolic rate is low (MACFARLANE, 1964; MACFARLANE et al., 1971). In the camel especially, the rate of water distribution from the gut is slow, while sensitivity to vasopressin and tolerances to salt load from saline water or vegetation are very high. Even the camel is not likely to survive more than 20 days in summer without water because of the persistent evaporative cooling which ultimately dehydrates it (MACFARLANE & HOWARD, 1972).

Some of the small mammals can survive indefinitely in the desert on the other hand, without evaporative cooling and without exposure to insolation, by cryptic behaviour, torpor or renal concentration.

Methods

Rodents and marsupials were trapped in the desert or other habitat. In the laboratory, metabolic rate was measured with a Med Sci Metabolimeter. In this instrument as respiration proceeds oxygen consumption is recorded, as a continuous record. This is useful when differing levels of activity or torpor are likely to intervene during a day's observations. The temperature of the metabolic chamber was controlled to 25 °C or 31 °C for most measurements.

The turnover of water was obtained by injecting HTO intramuscularly at 200 μCi/kg. After equilibration over 2 hr, a blood sample was taken from the orbital sinus or from the tail. By using small tubes, sublimation of 50 μl of blood yielded sufficient isotope to count in a liquid scintillation system. Further samples of blood or urine were taken at 4-day intervals from the animals, which lived normally, eating and drinking as they wished in sandy enclosures at 20–28 °C.

Ecophysiology of Water and Energy

SMALL MARSUPIALS

All Dasyurids do not have large fat tails but all have some fat storage in the tail. They are mainly small animals between 10 and 200 g weight. There is evidence that the turnover rate for lipids in the tail storage is slower than in mesenteric or subcutaneous fat (SABINE et al., 1969). The tail fat is used during starvation. The advantage of local storage in the tail is that there is little additional thermal insulation when fat is concentrated in a hump (as in the camel) or in the tail of Dasyurids. Heat passes from core to shell over most of the surface, with little impedance from fat. This is useful in hot environments.

Dasyurid marsupials have uniform pin-like teeth suitable for catching insects in flight. They are very fast-moving animals, leaping after moths in flight with great agility. They also kill small mice or other mammals, then eat the viscera and muscles, leaving only the skin.

Mating of *Dasycercus cristicauda* and *Sminthopsis crassicaudata* takes place in mid-winter, the young being born within a month. Embryos migrate to the line of 8 to 10 teats and attach. The pouch is inadequate to cover the young which hang like grapes from the maternal abdominal surface, unprotected. There is only one litter in the year and the period of milk production is in the cool weather when water stress is likely to be least.

There are two groups of Dasyurids that inhabit arid areas. The desert

Table 1. Turnover/24 hr.

| Species | No. | Av. Wt. g. | Water turnover | | Oxygen turnover | | $\dfrac{H_2O}{O_2}$ Ratio |
			ml/kg	mol/kg$^{0.82}$	l./kg	mol/kg$^{0.75}$	$\dfrac{mol/kg^{0.82}}{mol/kg^{0.75}}$
Marsupials							
Dasyuroides byrnei	8	124	115	4.5	1.5	0.9	5.0
Dasycercus cristicauda	34	87	130	4.8	1.6	0.9	5.3
Antechinomys laniger	2	18	374	10.1	3.6	1.5	6.8
Sminthopsis crassicaudata	8	17	480	9.7	4.7	1.9	5.1
Rodents							
Pseudomys australis	11	50	139	4.5	2.6	1.3	3.5
Notomys cervinus	10	41	150	4.7	2.7	1.3	3.6
Notomys alexis	5	35	189	5.7	3.3	1.3	4.4

The daily use of water and oxygen is expressed in terms of body weight and as mol/metabolic kg. For water the exponent of mass is 0.82, and for oxygen 0.75, to compensate for the effects of size on rate functions.

The ratio of the mols of water consumed per metabolic kg to the oxygen used is 5.5 for the dasyurids, and 3.8 mol of water pass through for 1 mol of oxygen in the desert rodents.

Fig. 1. Desert mammals in central Australia. *a. Dasycercus cristicauda,* an insectivorous fat-tail marsupial. It burrows deeply and can live without water in the desert. Energy and water turnover in *Dasycercus* is at one third the rate of *Sminthopsis.*

Fig. 1b. Sminthopsis crassicaudata is also a Dasyurid marsupial, and insectivorous. It does not burrow though it lives in very arid areas. It has high metabolic and water turnover rates so that it dies in 4 days without water.

392

Fig. 1c. Notomys alexis is a seed-eating rodent, adapted to the desert by having low rates of energy and water turnover, but it also has a renal concentrating potential of 9 osmol/l.

forms *Dasycercus* and *Dasyuroides* make burrows which may be more than a metre deep. The diurnal temperature fluctuations are therefore small and there is shelter from the sun in a uniform environment around 28 °C. These animals have metabolic rates lower than *Dipodomys* and if deprived of food they pass into torpor. This summer torpor is associated with a fall of body temperature to ambient, though they rarely are below 20 °C. With torpor the metabolic rate is decreased and the water losses decrease proportionally. The water turnover measured by HTO varies with food supply but it is low relative to humid-zone animals of the same size. When *Dasycercus* has insect food or meat it does not require drinking water (NEWSOME & SCHMIDT-NIELSEN, 1962). The urinary concentration reaches 4 osmol/l.

Water loss through skin and pulmonary tract is low relative to other mammals. There is no evaporative cooling by sweating or panting. Thus cryptozoic behaviour and low rates of water and energy turnover, together with torpor and water-retaining membranes make these effective desert animals.

Another group of Dasyurids that live in very dry areas, is *Sminthopsis crassicaudata* and *S. larapinta* which derive from watered country. They have some fat in the tail and they pass into torpor when deprived of food (GODFREY, 1968). *Sminthopsis* differs however, in three main functions from *Dasycercus* or *Dasyuroides*. It has threefold higher rates of water turnover, metabolism and pulmocutaneous transpiration, than *Dasycercus* or *Dasyuroides*. Water loss is therefore rapid and survival even on animal food is only for 4 or 5 days in summer, if there is no water for drinking. So it is a facultative desert animal, dependent on continuous water supplies.

Sminthopsis makes nests under rock shelters or fallen *Acacia* branches.

It lives near the surface without burrows. The rate of transpiration from skin and lungs is 3 times higher than that of *Dasycercus* and it has a correspondingly higher rate of metabolism (KENNEDY & MACFARLANE, 1971).

Tropical Dasyurids such as *Antechinomys*, *Antechinus* or *Planigale* have metabolic rates and water turnovers of the same order as those of *Sminthopsis* but they do not normally venture into arid habitats. Since the large proportion of *Sminthopsis* live in well-watered country, while the fringes of their population extend into the desert, it seems that they were primarily evolved for well-watered areas and only secondarily have invaded the desert. The habitats they occupy tend to be along the river systems but they occupy some of the same territory as *Dasycercus* or *Dasyuroides* (KENNEDY & MACFARLANE, 1971).

RODENTS

Periodically rat and mouse plagues occur in low rainfall areas, generally after one or two wet years with a build-up of vegetation. Both *Mus* and *Rattus* are omnivorous but they require drinking water for survival. The rates of energy and water turnover lie between those of jungle marsupials and the desert types. They do not pass into torpor and they breed most of the year. They live like *Sminthopsis* in a cryptic fashion sheltering from the sun each day, and emerging at night. In adverse circumstances of food or water they die and reinvade from other more amenable habitats.

The native rodents of the Australian desert such as *Notomys* and *Pseudomys* lack the fat tail of Dasyurid marsupials. They have instead long thin tails, and tend to walk or jump on the hind legs. In the desert they have sheltered nests. Breeding takes place throughout the year, and they reproduce readily in captivity.

The rate of use of oxygen is low but not quite at the level of *Dasycercus* (MACFARLANE *et al.*, 1971). It appears that these desert eutherians have higher rates of metabolism than marsupials of the same size (MAC MILLEN & LEE, 1970). The rate of pulmocutaneous evaporation is relatively low and is proportional to the overall low water turnover rate. *Notomys* concentrates urine to an average of 6 and maximum of 9.3 osmol/l. *Pseudomys* as well as *Notomys* turnover less water than marsupials of the same size, so the ratio of water to oxygen used (3–4) is lower than among Dasyurids (5–6). The desert survival of *Notomys* while eating dry seeds arises from low rates of metabolism, and unequalled powers of water reabsorption in the long renal papillae.

Discussion

The ratio of water to energy consumption is relatively constant amongst small mammals. The ratio is not quite the same in each group but it appears that as energy consumption rises so water use increases and these functions change in parallel for animals adapted to either wet or to dry habitats. Amongst Dasyurids about 5 mol water are used to 1 mol oxygen. This applies to both desert animals like *Dasycercus* and jungle genera such as *Antechinus* or *Sminthopsis*. Even when the rate of metabolism is 3 times greater in one genus than the other the water: energy turnover rates maintain the same ratio (MACFARLANE *et al.*, 1971).

Amongst the rodents there is a rather lower ratio, so that 3 to 4 mol water pass through these animals for 1 mol oxygen consumed.

It is not clear how water and energy relations are linked (LIFSON & McCLINTOCK, 1966) but it seems likely that the rate at which cells move electrolytes determines the rate of water movement through cells and tissues. Some 60 to 70% of the energy of cells (ISMAIL-BEIGI & EDELMAN, 1970) is consumed in pumping electrolytes, so that the rate at which this occurs is likely to be related to the greater water turnover. A link then between oxygen use and water use would occur at the cell membrane. Hormones regulate this rate and in particular thyroxine can modulate electrolyte pumps.

Evolution of animals in the desert seems to have led to selection for reduced energy consumption, low water turnover and storage of fat against emergency. Desert Dasyurids not only have lower metabolic rates than eutherians of the same size, but they also have a lower oxygen turnover than non-desert marsupials. In addition they reduce energy expenditure and water use further by torpor in summer. Sometimes the renal concentrating power has been highly developed as in *Notomys* but in other genera a low rate of urine output through control of glomerular filtration takes place, as in the camel. The pulmonary and cutaneous surfaces are relatively impermeable to water, proportionally to the low water turnover and metabolic rate.

Animals evolved in wetter climates with high metabolic rates and rates of water use can invade the desert and survive there while there are adequate food and water supplies. When drought strikes, however, the species is likely to be eliminated locally with the possibility of reinvasion from the heartland when next fertility returns to sections of the desert.

REFERENCES

GODFREY, G. K. 1968. Body temperatures and torpor in *Sminthopsis crassicaudata* and *S. larapinta* (Marsupialia-Dasyuridae). *J. Zool. Lond.* 156, 499–511.

Ismail-Beigi, F. & Edelman, I. S. 1970. Mechanism of calorigenic action of thyroid hormone: role of sodium transport. *Fed. Proc.* 29, 582 Abs (No. 1881).

Kennedy, P. M. & MacFarlane, W. V. 1971. Oxygen consumption and water turnover of the fat-tailed marsupials *Dasycercus cristicauda* and *Sminthopsis crassicaudata*. *Comp. Biochem. Physiol.* 40A, 723–732.

Lifson, N. & McClintock, R. 1966. Theory of the use of the turnover rates of body water for measuring energy and material balance. *J. theoret. Biol.* 12, 46–74.

MacFarlane, W. V. 1964. Terrestrial animals in dry heat: ungulates. In Handbook of Physiology, 4, 509–530. Environment. Ed. D. B. Dill, E. F. Adolph & C. G. Wilber. Amer. Physiol. Soc., Washington, D.C.

MacFarlane, W. V. & Howard, B. 1972. Comparative water and energy metabolism of wild and domestic animals. In Comparative Physiology of Desert Animals. Ed. G. M. O. Maloiy & W. V. MacFarlane, Academic Press, London.

MacFarlane, W. V., Howard, B., Haines, H., Kennedy, P. J. & Sharp, C. M. 1971. The hierarchy of water and energy turnover of desert mammals. *Nature. Lond.*,234, 483–484.

MacMillen, R. E. & Lee, A. K. 1969. Water metabolism of Australian hopping mice. *Comp. Biochem. Physiol.* 28, 493–514.

MacMillen, R. E. & Lee, A. K. 1970. Energy metabolism and pneumocutaneous water loss of Australian hopping mice. *Comp. Biochem. Physiol.* 35, 355–369.

Sabine, J. R., Wigley, G. R., Brewer, N. A. & MacFarlane, W. V. 1968. Fat metabolism of several members of the family Dasyuridae. Proc. Int. Soc. Fat Research 1–17. 9th Congress, Rotterdam.

Schmidt-Nielsen, K. 1964. Desert Animals. The Clarendon Press, Oxford.

Schmidt-Nielsen, K. & Newsome, A. E. 1962. Water balance in the mulgara *(Dasycercus cristicauda)*, a carnivorous desert marsupial. *Aust. J. biol. Sci.* 15, 683–689.

396

XX. THERMO-REGULATION AND WATER ECONOMY IN INDIAN DESERT RODENTS

by

PULAK K. GHOSH

Introduction

As in all hot deserts, rodents in the Great Indian Desert (24.5° to 30.5°N, 60° to 70°E) face two serious hazards, viz. intense heat and lack of water. The basic problems of mammalian physiology under conditions of heat and water stress and details of the compensatory mechanisms in several species of animals have been extensively reviewed in the past (see, for example, CHEW, 1961, 1965; HUDSON, 1964; MACFARLANE, 1964; SCHMIDT-NIELSEN, 1964a, b). PUROHIT (1967) has reviewed the work done on the physiology of small desert mammals, including rodents, of the Great Indian Desert. A recent symposium of the Zoological Society of London has also brought out a mass of interesting information on the general topic of desert animal physiology (MALOIY, 1972).

The rodents of the Indian desert

PRAKASH *et al.* (1971) have reported the occurrence of rodents of 17 species and subspecies, belonging to the families *Sciuridae*, *Muridae* and the sub-family *Gerbillinae*, in the desert areas of Rajasthan. In their survey of the region, these workers found *Meriones hurrianae* (26.0 percent) and *Tatera indica indica* (24.7 percent) as the most abundant rodent species in this desert. Others which follow are: *Rattus meltada pallidior* (13.1 percent), *Rattus cutchicus cutchicus* (12.0 percent) and *Gerbillus gleadowi* (11.1 percent). *Mus platythrix* and *Funambulus pennanti* also occur here in recognisable numbers while the other rodent species and subspecies are rather thinly distributed.

Behavioural and physiological studies on Indian desert rodents have been almost completely restricted so far to four species, viz. 1) *M. hurrianae*, 2) *T. indica indica*, 3) *G. gleadowi* and 4) *F. pennanti*. Of these four species again, *M. hurrianae* has received the maximum attention. Most of the material in this chapter will, therefore, relate to this species, commonly referred to as the Indian desert gerbil or simply the gerbil.

Methods

All animals used in these studies were captured from different locations

397

Table 1. Rodent species of the Rajasthan desert and their activity patterns (I. PRAKASH – personal communication).

Diurnal	Nocturnal	Nocturnal, but active during part of the day
Funambulus pennanti	*Hystrix indica*	*Rattus rattus rufescens*
Meriones hurrianae	*Gerbillus nanus indus*	*Mus musculus bactrianus*
Golunda ellioti gujerati	*Gerbillus gleadowi*	
	Tatera indica indica	
	Rattus cutchicus cutchicus	
	Rattus meltada pallidior	
	Rattus gleadowi	
	Mus booduga booduga	
	Mus cervicolor phillipsi	
	Mus platythrix sadhu	
	Nesokia indica indica	
	Bandicota bengalensis	

Table 2. Water evaporated during heat stress in *M. hurrianae* (GHOSH & PUROHIT, 1964).

Air temp.	Exposure time (hr)	Water evaporated (gm)	Evaporation	
			gm/hr	gm/hr/100 gm body weight
40 °C	4.5	2.8	0.62	1.46
42 °C	3.5	2.4	0.68	1.82
44 °C	2.5	4.65	1.99	3.04

within the Rajasthan desert and then brought to the laboratory at Jodhpur for experimentation. The methods employed were all of a standard nature and have been fully described in the relevant publications cited in this review.

Thermo-Regulation

TEMPERATURE TOLERANCE

The normal rectal temperature of the gerbil, *M. hurrianae*, is 37.7 °C. When subjected to different thermal environments in the laboratory, with a view to determine the gerbil's tolerance limit for high ambient temperature, it was found that the rodent can effectively tolerate temperatures upto 40 °C, but beyond that there is a sharp decline in survival percentage (GHOSH & PUROHIT, 1964). The maximum and minimum periods of survival at 42 °C were 320 and 120 minutes and at 44 °C, 165 and 120 minutes respectively. When subjected to heat stress, the body temperature rises and death supervenes at a body temperature of about 44 °C. A marked fall in the body weight of the animals has been consistently found in all cases of heat exhaustion. There was conspicuous wetting of the fur under the chin and throat, presumably due to copious salivation. The cooling effect of this reaction is, in the opinion of SCHMIDT-NIELSEN (1956) who observed this phenomenon in kangaroo rats, not of much significance except as an emergency measure lasting for a short period. A similar response occurs in the ground squirrel (HUDSON, 1962) and in the Egyptian rodent – *Dipus* (KIRMIZ, 1962).

The amount of water evaporated from the body of the gerbil at these temperatures has been indicated in Table 2. From the corresponding data on the kangaroo rat (SCHMIDT-NIELSEN, 1964a, p. 155), it would appear that the gerbil loses much less water per 100 g body weight at the very high temperature of 44 °C ambient, the figure for the kangaroo rat being 10.3. Again, none of the kangaroo rats in SCHMIDT-NIELSEN's study could survive for more than $1\frac{1}{2}$ hr at an air temperature of 43 °C, whereas the maximum survival period of *M. hurrianae* at 44 °C air temperature was 2 hr and 45 minutes. Even then, the gerbil's performance under conditions of thermal stress, as imposed in these studies, does not indicate any unusual thermolytic efficiency of this animal. It became apparent in the course of these studies that the important component of the gerbil's strategy for survival must be its behavioural adaptations.

BEHAVIOURAL ASPECTS OF TEMPERATURE REGULATION

The Rajasthan desert, in common with the other hot deserts of the world, has the characteristic diurnal temperature regime of wide amplitudes. Moreover, for nearly nine months during the year, this desert remains a hot inferno with howling, sand-laden winds desiccating everything exposed to the elements. The rodents having a much larger body surface relative to mass, compared to the larger animals, have a low 'thermal inertia'. Therefore, it is impossible for the rodents to afford using water to keep the body temperature from rising in the hot environment of their habitat, assuming that metabolic heat production (or oxygen consumption) is proportional to the surface area of the body. This theoretical consideration alone would lead us to expect that the primary thermoregulatory strategy evolved in desert rodents would be based more on behavioural modifications rather than on physiological adjustments. The more important behavioural aspects, in this context, are as follows:

Nocturnality

The general activity pattern of the rodents of this desert conforms to the pattern observed by other workers in deserts elsewhere (e.g., MAC-MILLEN, 1972). As much as 82.35 percent of the rodent species occupying the Rajasthan desert are nocturnal in their habit (Table 1). Apparently, the rewards of nocturnality are ample in the sense that the evaporative heat loss from the body does not lead to a situation where the critical water balance of these creatures goes out of gear. The Indian desert gerbil, *M. hurrianae*, the most predominant rodent species here, is, however, diurnal in its habit, but it generally restricts its surface activities to the cooler twilight hours of the early morning and evening. An exception must be made here of an important group of Rajasthan desert rodents – the palm squirrel *(F. pennanti)* which is a strictly diurnal animal. However, these animals are generally arboreal and the thick foliage and the moisture-laden air of gardens and orchards offer a particularly protective microclimate to the squirrel and, hence, this rodent does not have to bear the full impact of the hot desert day. Moreover, *F. pennanti*, though not a strictly aestivating animal, is known to limit its foraging activities to the minimum during the summer, thereby avoiding body heat gain to a considerable extent.

Fossoriality

Next to nocturnality, the most conspicuous behavioural adaptation in desert rodents is their fossorial habit. The burrow in the soil is the desert rodent's most ideal haven. SCHMIDT-NIELSEN & SCHMIDT-NIELSEN (1950), KENNERLY (1964) and PRAKASH et al. (1965) have reported on the generally uniform temperature and humidity conditions within the burrows of various species of desert rodents. In their study of the microclimate of the *M. hurrianae* burrow, PRAKASH et al. (1965) observed that in summer, when the soil surface temperature was as high as 55.5 °C, the temperature inside the burrows at all depths (50–200 cm) fluctuated within the narrow range of 33.6° to 37.6°C. Even during the monsoon, the burrow environment was found to be cooler than the soil surface by 11.4°C. These workers further observed that the amplitude of temperature variation within the burrow environment was of a small order almost throughout the year: 1.2 to 1.5°C in winter, 1.1 to 2.4°C in summer, 2.9 to 8.2°C in monsoon and 0.8 to 1.1°C during the post-monsoon period. These temperature variations are indeed small in comparison to those prevailing either on the soil surface or in the air in the desert through the year. PRAKASH et al. (1965) had concluded that 'the burrows have a relatively constant temperature throughout the year and consequently impose no temperature stress on the animals.' It is apparent, therefore, that by intermittently retiring to the more hospitable environs of their burrows, the desert gerbil not only passively shakes off some portions of its body heat gained during its sojourn on the surface, but it also adds to the moisture status of the burrow air. The entire arrangement would, therefore, seem to be excellent from the point of view of desert survival.

Shift in circadian rhythm

PRAKASH (1962) observed that during summer, *M. hurrianae* comes out of its burrow just after dawn and retires only after a few hours before it gets too warm. It again comes out at dusk and returns to its burrow after about two hours of foraging. In winter, however, it is out of its burrow throughout the day but not during mornings and evenings when it is quite cold. PRAKASH (1962) considered this shift in diel activity an adjustment to the fluctuating heat and cold of the desert. Large-scale summer migrations of colonies of *M. hurrianae* have also been observed by PRAKASH (1962). Such migrations, presumably in search of food, are likely to be of considerable importance from the point of view of the water balance (and heat tolerance) of the animals also.

The behavioural adaptations of the gerbil are not peculiar to this species. These are a common feature of the 'beat the Sun' kit of the majority of desert-dwelling rodents everywhere and these instinctive 'escape' mechanisms are meant to supplement their physiological resources. As MACMILLEN (1972) has pointed out, these rodents' existence under the extreme conditions of the desert 'depends upon a continual interplay between physiology and behaviour'. A similar view has also been held by BARTHOLOMEW (1964).

Water Economy

WATER REQUIREMENT

The average daily water consumption rates in *M. hurrianae* and *G. gleadowi*, under laboratory conditions, are 0.57 and 3.28 percent of body weight respectively (GHOSH & GAUR, 1966). In Nature, these rodents do not, probably, drink water for the simple reason that their habitat does not provide easily-accessible sources of free water. PRAKASH (1962) is of the view that *M. hurrianae* depends chiefly on the preformed water of its food, and also on the metabolic water produced in its body for meeting its water requirements. It may be of some relevance to note that locusts comprise a major portion of the food of *M. hurrianae* during summer (PRAKASH, 1962). Insects and their larvae are known to have a high body water content (46 to 92 percent of body weight, – ROBINSON, 1928). The summer preference for locusts may, therefore, be a part of the gerbil's adaptive strategy for maintaining water balance. *M. hurrianae* is decidedly at a more advantageous position with regard to its water requirement than *G. gleadowi*. This phenomenon is partly explicable in behavioural terms. Although both these desert rodents are adapted for fossorial life, *G. gleadowi* is strictly nocturnal in its habit while *M. hurrianae* is a diurnal animal. Being comparatively less exposed to the desiccating daytime desert atmosphere, *G. gleadowi* are presumably less efficient in economizing on water expenditure in comparison to *M. hurrianae*. Moreover, *G. gleadowi* being of a relatively smaller size, the gap between its total water loss and metabolic production of water is theoretically greater than that of *M. hurrianae*. The latter species has the advantage of its body size to place primary dependence on metabolic water for homoeostatic purposes.

PHYSIOLOGICAL RESPONSES DURING WATER DEPRIVATION

Body weight

The importance of metabolic water for desert survival has been dramatically shown in a laboratory study involving *M. hurrianae* maintained without water on a diet of moisture-free grains. The gerbils survived and maintained body weight for 16 months on this regime (GHOSH & PUROHIT, 1964). In the field *M. hurrianae* maintain a total body water content of 68 percent throughout the year. When kept on a laboratory diet that included water, no significant change in the gerbil's body water content has been observed. It is doubtful, therefore, if the gerbil is capable of storing water in its body. When water-deprived in the laboratory, seasonal fluctuations in the body weight of the gerbils have been noted, with low values during the extreme hot, dry periods. The lowered atmospheric humidity at such periods is likely to have affected the moisture status of the feed and, consequently, the water balance of the gerbil. The gerbil cannot possibly withstand water deprivation, at least under laboratory conditions, beyond sixteen months when its weight falls rapidly and the animal shows signs of general weakness. It should, however, be pointed out that in Nature, these burrowing rodents hardly ever face the severity of the laboratory conditions. When the gerbil apparently reaches its limit of tolerance for water deprivation in the laboratory, provision of water produces a dramatic and quick recovery of the animal, the body weight increases and food consumption returns to normal.

It would seem that under natural conditions the gerbil is capable of withstanding several droughts in succession.

Body composition

Table 4 indicates the changes in gross body composition of *M. hurrianae* brought about by water deprivation for 60 days.

Table 3. Seasonal fluctuations in the average body weight of water-deprived *M. hurrianae* (GHOSH & PUROHIT, 1964).

			Body weight in gm.		
	Winter	Spring	Summer		Autumn
			Dry	Monsoon	
First year	41	43	24	64	50
Second year	58	51	32 (On the verge of death)	57 (water provided *ad lib.*)	

Table 4. Effect of water-deprivation on body composition in *M. hurrianae* (GHOSH et al, 1962).

Condition of animal	Whole body			
	Water %	Fat %	Cholesterol %	Phospholipid %
Freshly-captured	70.17	4.04	0.158	0.359
Dry-fed for 60 days	59.99	11.62	0.190	0.343

The total body water content of the freshly-captured animals was found to be significantly higher than that of the water-deprived gerbils while there was a significant increase in the total body fat content in the latter group of animals. No significant variation was observed in the whole body cholesterol and phospholipid contents between the two groups of animals.

Body fat dynamics

The significant increase in total body fat of water-deprived *M. hurrianae*, as indicated in Table 4, is likely to be a physiological adaptive mechanism. As reported by STROHL (1929), there are localised or general fat accumulations in most of the desert mammals at the beginning of the dry period. These fat deposits are likely to act not only as energy stores, but probably also as potential water stores, as 106 parts of water can be obtained from 100 parts of fat by oxidation. However, the SCHMIDT-NIELSENS (1952) believe that the increased pulmonary ventillation

Table 5. Body composition of normally hydrated and water-deprived *M. hurrianae* as affected by two levels of dietary fat (GHOSH & PUROHIT, 1964).

Group	Treatment	Average body weight in gm		Body fat %	Body water %
		Initial	Final		
1	2% fat diet with water	48.0	55.6	8.49	68.1
2	2% fat diet without water	49.7	42.7	16.77	61.3
3	15% fat diet with water	52.0	82.3	30.45	49.9
4	15% fat diet without water	50.0	48.5	14.08	65.1

needed for increased oxidation, with the resultant loss of moisture through the respiratory passage, would offset any advantage of the gain in metabolic water. Although it would be difficult, in view of theoretical considerations, to establish a causal relationship between water deprivation (or desert habitation) and body fat accumulation, yet it is surprising that a number of small desert mammals, e.g. the African rodent, *Pachyuromys steatomys* and the marsupials from the arid regions of Australia, *Antechinus maculatus* and *Sminthopsis crassicaudata* have fat tails (SCHMIDT-NIELSEN, 1964a). Certain desert lizards, e.g. *Uromastix* and *Heloderma* also have fat tails.

Table 5 gives the findings on the effect of dietary fat load on the body composition of water-deprived gerbils (GHOSH & PUROHIT, 1964). All the groups of animals in this study received isocaloric diets over a period of two months. The pair-feeding technique was employed between groups 1 and 2 and between groups 3 and 4.

While the differences between groups 1 and 2 in respect of body fat and water percentages just reached the level of statistical significance, the differences between groups 3 and 4 were highly significant in both these respects. Since the body fat percentages in the water-deprived gerbils of groups 2 (2% fat diet) and 4 (15% fat diet) were not significantly different from each other, it would seem that dietary fat level does not influence the laying down of body fat when water is not available to the animals. It would also appear that the body fat in the desert gerbil serves a dual purpose, viz., as energy and water stores. It may be speculated that when the gerbil faces water crisis on a predominantly carbohydrate diet, its metabolism is directed towards establishing a fat bank to serve primarily as a water reservoir, but when its water needs are fulfilled and it is subjected to a high dietary fat load, it stores the extra fat in the body in the form of an energy bank. It should, however, be

404

pointed out that as water deprivation is prolonged beyond three months, the gerbil's body fat store is slowly exhausted and the animal eventually becomes quite lean.

An interesting feature of the gerbil's body fat dynamics under conditions of water stress is the gradual unsaturation of its body fat as the period of water deprivation is prolonged, with the iodine number of the body fat of such animals steadily increasing from the normal value of 58 to the considerably high value of 135 after 15 months of water deprivation (GHOSH & PUROHIT, 1964). There is a report that in tropical Merino sheep, water deprivation in summer for two days is followed by a five-fold rise of plasma unsaturated fatty acids (MACFARLANE, 1963). It is, however, difficult to explain the significance of the increased unsaturation of body fat in water-deprived animals on the basis of our present understanding of fat dynamics *vis-a-vis* water metabolism.

Haemoconcentration

When *M. hurrianae* are dry-fed for a period of two months, considerable haemoconcentration results (GHOSH *et al.* 1962). This is apparently due to a reduction in plasma volume consequent upon a general reduction in body water content. The results of blood analyses done on freshly-captured and water-deprived gerbils have been presented in Table 6.

Although haemoconcentration, presumably due to a shrinkage in plasma volume, is apparent in the dry-fed animals, it is, however, difficult to account for the sizable increase in the number of red blood cells without assuming an increase in the rate of erythropoiesis due to some unknown factors. Interestingly, when water is withheld from the gerbil for more than six months, a normal blood picture prevails. The normal blood specific gravity of the gerbil is around 1.065. This may go up to 1.075 during short-term water deprivation, but the normal value is regained as the period of deprivation extends beyond six months. These findings are in accord with the observations of HENSCHEL (1964) in man.

Organ weights

As the maintenance of the normal fluidity of the blood is of primary importance from the homoeostatic point of view, it is to be expected that even vital body organs will contribute towards this end by releasing a part of their intracellular water as necessity arises. Since, however, the gerbil is known to suffer haemoconcentration during the initial stages (up to two months) of water restriction, it is apparent that the rodent utilises plasma water during water crisis, in preference to the intracellular water contained in its vital organs. This will be further indicated from the data presented in Table 7.

Table 6. The blood picture in normally hydrated and water-deprived *M. hurrianae* (GHOSH *et al.*, 1962).

Condition of animals	Red blood cells (millions/cu.mm)	White blood cells (thousands/ cu.mm)	Haemoglobin (g/100 ml)
Freshly-captured	2.352	4.91	12.66
Dry-fed for 60 days	7.29	6.62	16.83

Table 7. Weight of different body organs in g in normally hydrated and water-deprived *M. hurrianae* (GHOSH & GAUR, 1966).

Organ	Normally-hydrated	Water-deprived for 60 days
Adrenals	0.0197	0.0513
Brain	0.8787	0.957
Kidneys	0.3398	0.3618
Liver	1.8528	1.5263
Heart	0.1141	0.1096
Spleen	0.0902	0.0935
Lungs	0.3377	0.3406
Pancreas	0.2301	0.2879
Reproductive tract	0.3230	0.1537
Alimentary tract	1.6487	1.2157

The results presented in Table 7 indicate that the weight of most of the vital organs in the gerbil remain unaffected when the animals are subjected to water restriction for two months. Apparently, the imposed stress, although causing significant negative water balance and haemo-concentration in the animal (GHOSH *et al.*, 1962), failed to cause any desiccation of the organs studied.

There were significant increases in the adrenal and brain weight in the water-deprived animals. The increase in adrenal weight is likely to be due to hypertrophy of the gland in response to the imposed stress. Actual hyperplasia of the zona fasciculata in the adrenals of water-deprived gerbils has been reported (PUROHIT & GHOSH, 1963). Although aldosterone synthesis is believed to occur only in the zona glomerulosa, there are evidences to suggest that corticosterone, also involved in water metabolism, is produced from both the glomerulosa and the two inner zones of the cortex (TURNER, 1961). Reduction in intravascular volume after dehydration is believed to be a strong impetus for increased secretion of aldosterone (WILLIAMS, 1962). Therefore, the fact that the gerbil

Table 8. Faecal water and dry matter excretion in normally-hydrated and water-deprived *M. hurrianae* (GAUR & GHOSH, 1971).

Condition of animals	Faeces, g dry matter/ 100 g food	Water, g/100g dry faecal matter	Water, g lost with faeces/100 g food
Normally-hydrated	12.05	38.97	4.69
Water-deprived (2–2½ months)	7.80	22.71	1.76

experiences haemoconcentration during the initial stages of water restriction suggests enhanced aldosterone, and possibly corticosterone, mediation and, consequently, a hypertrophied gland.

The significant increase in the brain weight in the water-deprived animals cannot be explained except by assuming that increased neuro-secretory function of the hypothalamus in elaborating antidiuretic hormone has been the cause of this increase.

CONTROL OF WATER LOSS

The major avenues of water loss are faecal, urinary and pulmo-cutaneous.

Faecal water loss in water-deprived M. hurrianae

When *M. hurrianae* are deprived of water for 2 to 2½ months, they appear to conserve moisture through the faecal route by a) excreting less dry matter per 100 g of food consumed and b) by voiding faeces with a lowered water percentage, in comparison to normally-hydrated gerbils (GAUR & GHOSH, 1971). These will be indicated by the data presented in Table 8.

The higher efficiency of food utilisation together with lowered water loss through faeces would indicate that considerable water economy is achieved by the gerbil during water stress.

Renal concentrating mechanism in M. hurrianae

By far the most important means of reducing body water loss in the gerbil appears to be its capacity to concentrate urine when subjected to water stress (GHOSH *et al.*, 1962; GHOSH & PUROHIT, 1964). The renal excretory mechanism of this species seems to be similar in many ways to that described by SCHMIDT-NIELSEN (1964a) for the kangaroo rat. Normally, the gerbil voids about 1.5 ml of urine in 24 hrs. When water is denied for a period of two months, the urine volume gets considerably reduced and, consequently, the concentrations of urinary metabolites (chloride, total electrolytes, urea and total nitrogen) are increased. The results reported by GHOSH et al. (1962) indicate that the efficiency of the kidney of the Indian desert gerbil is very high indeed, when compared to data presented and/or reviewed by ADOLPH (1943), BODENHEIMER (1957), IRVING et al. (1935) and SCHMIDT-NIELSEN (1948), regarding different species of animals.

In the light of SPERBER's (1944) observations on the typically long kidney papillae of desert rodents, it would be expected that the functional efficiency of the gerbil kidney may be related to its structure. However, *M. hurrianae* does not possess any unusually long papillae (PUROHIT & GHOSH, 1963). A few other rodents of the Rajasthan desert, viz., the antelope rat, *Tatera indica indica* (PUROHIT & GHOSH, 1963), the hairy-footed gerbil, *Gerbillus gleadowi* and the mouse, *Mus platythrix* (PUROHIT, 1967) possess elongated renal papillae characteristic of desert rodents. It is possible that a specialised renal architecture alone may not determine renal concentrating ability, but the overall reabsorptive efficiency at different levels of the nephron may also be an equally important factor in this respect.

Pulmocutaneous water loss

No attempt has yet been made to quantify pulmocutaneous water loss in any rodent species of the Indian desert. However, it seems a fair guess that these rodents, in common with a large number of nocturnal desert rodents from other parts of the world (MACMILLEN, 1972) are able to compensate entirely for pulmocutaneous water loss by metabolic water production.

TOLERANCE OF SALINE WATER

The effect of a high-salt diet on the metabolism of *M. hurrianae* at different stages of dehydration have been studied by GHOSH et al. (1962). It was observed that the excessive demands on the excretory capacity of the kidney induced by salt-loading did not leave the animals totally incapacitated as shown by the slow gain in lost body weight when the animals were transferred from the salt-grain diet to the basal dry diet. The longer the period an animal was on the dry diet, the greater was the percentage loss in body weight when it was subjected to the heavy salt load for a short period. These observations indicate that the gerbil kidney is highly efficient in filtering out a large excess of salt even when the body is subjected to severe water stress. This renal mechanism of excreting urine of a very high osmotic ceiling is apparently aimed at conserving the maximum body water under xeric conditions. Another rodent species of this desert, *G. gleadowi*, would appear to have even

Table 9. Effect of saline drinking on plasma chloride level (mEq/1) in *M. hurrianae* and *G. gleadowi* (GHOSH & GAUR, 1966).

Species	Tap water treated	1.2 M NaCl treated
M. hurrianae	107.8	143.5
G. gleadowi	106.4	114.4

Table 10. Effects of various concentrations of NaCl in drinking solution on fluid intake and salt ingestion in *M. hurrianae* and *G. gleadowi* (GHOSH & GAUR, 1966). (Results are expressed as percentage body weight/day).

	Conc. of NaCl in drinking soln., M						
	0.0	0.2	0.4	0.6	0.8	1.0	1.2
NaCl soln., mg/ml	0.00	11.70	23.40	35.10	46.80	58.50	70.20
M. hurrianae							
Fluid intake	0.57	0.63	0.59	0.51	0.49	0.29	0.18
NaCl ingestion	0.00	0.007	0.013	0.017	0.021	0.016	0.012
G. gleadowi							
Fluid intake	3.28	3.96	3.72	3.71	3.85	3.40	2.66
NaCl ingestion	0.00	0.046	0.082	0.130	0.170	0.199	0.187

more powerful kidneys than *M. hurrianae* (GHOSH & GAUR, 1966). Concentrations of NaCl solutions utilised freely for drinking by *M. hurrianae* and *G. gleadowi* have been found to be 0.4 and 1.0 M respectively. *G. gleadowi* drinks more of NaCl solution up to a concentration of 1.2 M than *M. hurrianae* with a consequent increase in salt ingestion. The chloride content of urine of both the species increased with increasing concentration of the drinking solution. In general, while drinking NaCl solutions of concentration up to 0.8 M, the mean urine chloride levels in *G. gleadowi* were well above the chloride levels of the respective drinking solutions. *M. hurrianae*, however, excreted urine of less chloride content than the particular drinking solutions.

The normal plasma chloride levels in *M. hurrianae* and *G. gleadowi* are not significantly different from each other when provided with tap water. When 1.2 M NaCl was provided, the plasma chloride content significantly increased in *M. hurrianae* but not in *G. gleadowi* (Table 9).

The actual salt ingestion data for both *M. hurrianae* and *G. gleadowi* while on various drinking regimes have been summarized in Table 10.

In *G. gleadowi* salt ingestion was found to be directly related to concentration of drinking solution upto 1.0 M, while the same is true for *M. hurrianae* only up to a concentration of 0.8 M. Beyond these limits the two species tended to consume less fluid and, consequently, less salt. On an average, salt ingestion (and fluid intake) by *M. hurrianae* was about one-sixth to one-seventh that of *G. gleadowi* on unit body weight basis.

409

The maximum salt intake by *M. hurrianae* was 0.021 percent of body weight per day while drinking 0.8 M NaCl solution. *G. gleadowi* ingested a maximum of 0.199 percent of body weight per day on 1.0 M saline, its salt ingestion tending to level off at the next higher concentration. It has, however, been observed in this laboratory that when force-fed, individual *G. gleadowi* can tolerate salt ingestion of upto 0.6 percent of body weight per day for 3 to 4 days. *M. hurrianae* can, likewise, tolerate salt intake of upto 0.22 percent of body weight per day for a short period. It is apparent, therefore, that significant species difference exists with regard to salt tolerance. *G. gleadowi* is more tolerant than *M. hurrianae* and the former species is comparable to the kangaroo rat in this regard (SCHMIDT-NIELSEN *et al.*, 1948). PRAKASH & RANA (1973) have reported a preponderance of *G. gleadowi* over *M. hurrianae* on the sand dunes occurring in the extreme arid, 100 mm rainfall zone of the Rajasthan desert. Interestingly, this region is marked by the presence of vegetation like *Salvadora oleoides, Dipterygium glaucum, Cyperus rotundus, C. conglomeratus* and *Haloxylon salicornicum*, indicating the somewhat saline nature of the soils of the region. It would also seem possible that the dissolved salt content in these halophytes would be considerable. This ecological situation would, therefore, lend support to the experimental observations of GHOSH & GAUR (1966) regarding the relatively superior salt tolerance of *G. gleadowi*.

Conclusion

The physiological capacity to withstand water stress, developed in *M. hurrianae* and several other rodent species of the Rajasthan desert, becomes meaningful when considered along with the behavioural adaptive measures evolved in these animals for the purpose of a desert existence. Apart from the gross observations made on these rodents' wisdom in respect of their selection of microhabitats and food, and in regulating their diel activities, very little is known about the subtleties of inter-specific differences in behavioural patterns that enable some species like *M. hurrianae* to make the most of the resources of their niche in the desert. Such knowledge is obviously needed for a clearer understanding of the rodents' overall strategy for desert survival.

REFERENCES

ADOLPH, E. F. 1943. Do rats thrive when drinking sea water? *Amer. J. Physiol.*, 140: 25–32.

BARTHOLOMEW, G. A. 1964. The roles of physiology and behaviour in the maintenance of homeostasis in the desert environment. *Symp. Soc. exp. Biol.*, 18: 7–29.

BODENHEIMER, F. S. 1957. The ecology of mammals in arid zones. Pp. 100–137, in Human and Animal Ecology. Reviews of Research. UNESCO, Paris.

CHEW, R. M. 1961. Water metabolism of desert-inhabiting vertebrates. *Biol. Rev.*, 36: 1–31.

CHEW, R. M. 1965. Water metabolism of mammals. Pp. 43–178, in Physiological Mammalogy, Vol. II (W. V. MAYER & R. G. VAN GELDER, eds.), Academic Press, New York.

GAUR, B. S. & GHOSH, P. K. 1971. Effect of water deprivation on faecal water loss in the Indian desert gerbil *Meriones hurrianae*, Jerdon. *J. Animal Morph. Physiol.*, 18: 121–126.

GHOSH, P. K. & GAUR, B. S. 1966. A comparative study of salt tolerance and water requirements in desert rodents, *Meriones hurrianae* and *Gerbillus gleadowi*. *Indian J. exp. Biol.*, 4: 228–230.

GHOSH, P. K. & PUROHIT, K. G. 1964. Effects of water stress on gross body composition and renal function in small desert mammals. Proc. UNESCO/India Symp. on Problems of Indian arid zone, Jodhpur, Pp. 298–304.

GHOSH, P. K., PUROHIT, K. G. & PRAKASH, I. 1962. Studies on the effects of prolonged water deprivation on the Indian desert gerbil, *Meriones hurrianae*. In Environ. Physiol. Psychol. in Arid Conditions. Proc. Lucknow Symp., UNESCO, Paris, Pp. 301–306.

HENSCHEL, A. 1964. Minimal water requirements under conditions of heat and work. In Thirst: 19–30 (M. H. WAYNER, ed.), Pergamon Press, Oxford.

HUDSON, J. W. 1962. The role of water in the biology of the antelope ground squirrel, *Citellus leucurus*. *Univ. Calif. Publ. Zool.*, 64: 1–56.

HUDSON, J. W. 1964. Water metabolism in desert mammals. In Thirst: 211–235 (M. H. WAYNER, ed.), Pergamon Press, Oxford.

IRVING, L., FISHER, K. C. & McINTOSH, F. C. 1935. The water balance of a marine mammal, the seal. *J. Cell. comp. Physiol.*, 6: 387–389.

KENNERLY, T. E., Jr. 1964. Microenvironmental conditions of the pocket gopher burrow. *Texas J. Sci.*, 16: 395–441.

KIRMIZ, J. P. 1962. Adaptation to desert environment. A study on the jerboa, rat and man. Butterworths, London.

MACFARLANE, W. V. 1963. Endocrine functions in hot environments. Pp. 153–222, in Environ. Physiol. Psychol. in Arid Conditions. Reviews of Research. UNESCO, Paris.

MACFARLANE, W. V. 1964. Terrestrial animals in dry heat: ungulates. Pp. 509–531, in Handbook of Physiology – Environment (D. B. DILL, ed.), American Physiological Society, Washington, D.C.

MACMILLEN, R. E. 1972. Water economy of nocturnal desert rodents. Pp. 147–174, in Comparative physiology of desert animals, (M. O. MALOIY, ed.). Academic Press, London.

MALOIY, M. O. 1972. Comparative physiology of desert animals. Symp. Zool. Soc. Lond. No. 31. Academic Press, London.

PRAKASH, I. 1962. Ecology of the Gerbils of the Rajasthan desert, India. *Mammalia*, 26: 311–331.

PRAKASH, I., GUPTA, R. K., JAIN, A. P., RANA, B. D. & DUTTA, B. K. 1971. Ecological evaluation of rodent populations in the desert biome of Rajasthan. *Mammalia*, 35: 384–423.

PRAKASH, I., KUMBKARNI, C. G. & KRISHNAN, A. 1965. Eco-toxicology and control of Indian desert gerbil, *Meriones hurrianae*. Pt. IV. Burrow temperature. *J. Bombay nat. Hist. Soc.*, 62: 237–244.

PRAKASH, I. & RANA, B. D. 1973. A study of field population of rodents in the Indian desert. III. Sand dunes in 100 mm rainfall zone. Zeitschrift fur Angewandte Zoologie, 60: 31–41.

411

Purohit, K. G. 1967. The Great Indian Desert. Perspectives in the ecology and physiology of small desert mammals. *Mammalia*, 31: 28–49.

Purohit, K. G. & Ghosh, P. K. 1963. Histological and histochemical studies on the tissues of two desert rodents, *Meriones hurrianae* and *Tatera indica*. *Annals of Arid Zone*, 2: 26–34.

Robinson, W. 1928. Water conservation in insects. *J. Econ. Entomology* 21: 897–902.

Schmidt-Nielsen, B. & Schmidt-Nielsen, K. 1950. Evaporative water loss in desert rodents in their natural habitats. *Ecology*, 31: 75–85.

Schmidt-Nielsen, K. 1956. Animals and Arid Conditions. Arid Zone Res. XI. Climatology and Micro-climatology. Proc. Canberra Symp., 217–224.

Schmidt-Nielsen, K. 1964a. Desert animals: Physiological problems of heat and water. Oxford Univ. Press, London.

Schmidt-Nielsen, K. 1964b. Terrestrial animals in dry heat: desert rodents. Pp. 493–507, in Handbook of Physiology – Adaptation to the environment, (D. B. Dill, ed.), American Physiological Society, Washington, D.C.

Schmidt-Nielsen, K. & Schmidt-Nielsen, B. 1952. Water metabolism of desert mammals. *Physiol. Rev.*, 32: 135–166.

Schmidt-Nielsen, K., Schmidt-Nielsen, B. & Schneiderman, H. 1948. Salt excretion in desert mammals. *Amer. J. Physiol.*, 154: 163–166.

Sperber, I. 1944. Studies on the mammalian kidney. *Zool. Bidr. Uppsala*, 22: 249–431.

Strohl, J. 1929. Wasserhaushalt und Fettbesttand bei Steppen- und Wustentieren. *Verh. naturf. Ges. Basel*, 40: 422–440.

Turner, C. D. 1961. General endocrinology. W. B. Saunders Co., Philadelphia.

Williams, R. H. 1962. Textbook of endocrinology. W. B. Saunders Co., Philadelphia.

XXI. THE PHYSIOLOGICAL ADAPTATIONS OF DESERT RODENTS

by

L. I. GHOBRIAL & T. A. NOUR*

Introduction

The word desert denotes areas characterised by high temperatures and low irregular precipitation, resulting in extensive drought and scarcity of vegetation. Animals living in such an environment are faced with two major physiological problems: obtaining sufficient water for the needs of the body, and keeping the body temperature at a level compatible with life. It is evident that the high temperature of the environment often imposes the problems of additional water being required for use in heat regulation.

Animals inhabiting deserts have, in the course of evolution acquired morphological and physiological characteristics which enable them to live and thrive in an environment which is hostile and uninhabitable to other closely related families or species. Among rodents there are many well adapted desert forms which are found in all the major deserts around the world. What are the physiological adaptations which enable these desert rodents to live in such a hostile and uninhabitable environment?

Desert rodents, whether diurnal like pocket mice (*Perognathus flavescens, P. fasciatus, P. baileyi*) and round tailed ground squirrels (*Citellus tereticaudus, C. spilosoma macrospilotus, C. leucurus, C. nelsoni, C. mohavensis*) or nocturnal rodents like kangaroo rats (*Dipodomys merriami, D. spectabilis, D. agilis*), jerboa (*Jaculus jaculus, J. orientalis*), gerbils (*Gerbillus gerbillus, G. pyramidum, G. dasyurus*) and the pack rat (*Neotoma* spp.) are all well adapted to their desert environments.

The present data suggest that all desert rodents escape the excessive heat that is imposed by high solar radiation, air temperature and high ground surface temperature, by remaining in their underground burrows or in the shade. They need small amounts of moisture with their food even if they are seed eaters. They have efficient physiological means for regulating their body temperature and conserving their water balance inspite of high temperature and shortage of water.

In deserts, the temperature of the ground surface shows extreme variations. During the day the air temperature may be 45 °C and the

---

* Present address: Dept. of Chemistry, University of Alger, Algerie.

ground heated by sun, may attain a surface temperature of 70 °C or more; at night the surface temperature may fall far below the cool night air. Thus the diurnal variations in the temperature of the soil surface may be two or three times as great as the variation in air temperature.

The tremendous variations in soil temperature diminish rapidly with increasing depth. The extent of damping depends on the nature of the soil and its water content. GEIGER (1957) found that variations of 3 °C per 20 cm depth was representative of the effect of damping with diurnal variations almost non existent at 80 cm.

Measurements in Arizona showed that during the year, the surface temperature varies by more than 80 °C, but the annual fluctuations at 1 metre depth are reduced to about 12 °C (TURNAGE, 1937). This is a common depth for many rodent burrows.

The burrow temperature normally does not exceed 31 °C (VORHIES & TAYLOR, 1946) even on the hottest day. The temperature in burrows of the round tailed ground squirrel *(Citellus tereticaudus)* was recorded for an entire year in South Arizona. In June, when the maximum air temperature increased to over 40 °C and the soil surface temperature was 70 °C, the deep burrow temperature remained below 27 °C. Even when the soil temperature reached a maximum of 75 °C the deep burrow temperature did not exceed 29 °C and the ground squirrel's microclimate was made less severe by the presence of shade from vegetation and shrubs near its burrow. Although they are diurnal, they avoid the serious heat load by being inactive during the hottest part of the day (VORHIES & TAYLOR, 1946).

The jack rabbit shelters in depressions under bushes in such positions that it is never exposed to the sun. SCHMIDT-NIELSEN (1964) suggested that the ground temperature of the depression would be something of the order of 35 °C while the exposed surface temperature would be 75 °C. Although the jack rabbit is diurnal, it avoids heat stress by being active mainly at night or in the early morning and evening when the air is cool and moist.

The microclimate of the nocturnal kangaroo rat's *(Dipodomys merriami)* burrow shows a temperature of 28 °C and it never exceeds 33 °C, indeed temperatures above 35 °C have been proved to be fatal. The microclimate of the burrows of *Dipodomys spectabilis* shows the lowest temperature of 9 °C which was observed in January, and the highest, 31 °C, observed in July (VORHIES & TAYLOR, 1946).

Pack rats provide themselves with partial shade by building a shelter of debris. They site their nests in the trunks of the mesquite bush which gives additional shade. Measurements in the nest showed a soil temperature of 46 °C while the temperature in the open desert was 75 °C (VORHIES & TAYLOR, 1946), and although the nest was rather shallow the maximum air temperature in the nest did not exceed 31 °C. Since the

414

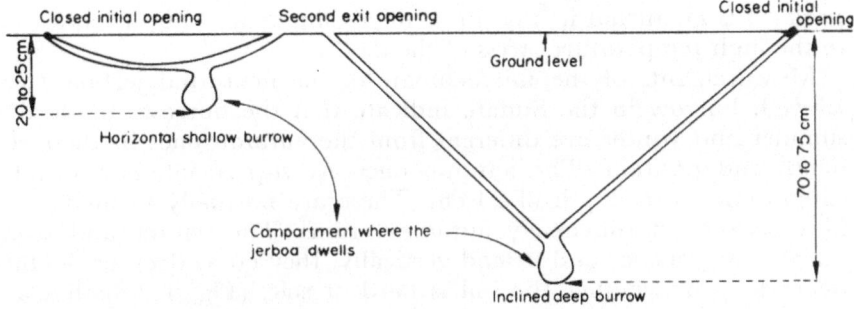

Fig. 1. Burrows of *Jaculus jaculus* during summer and winter.

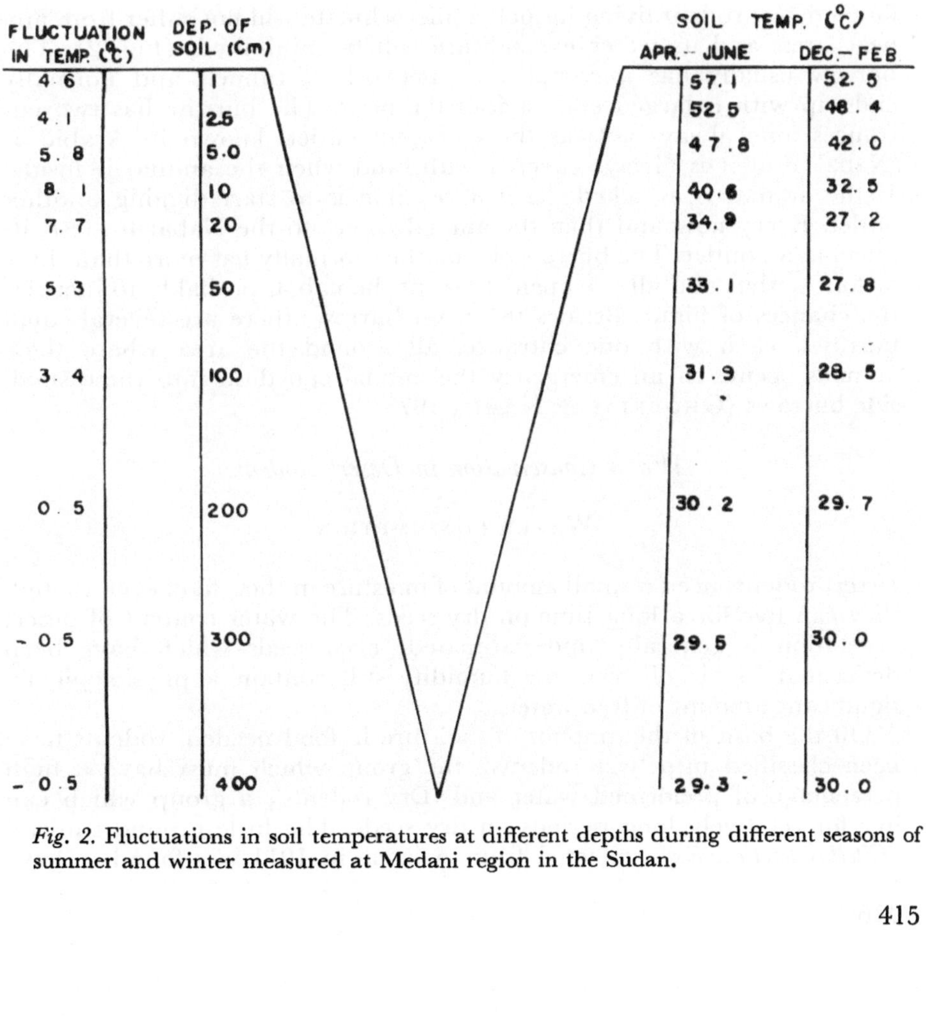

FLUCTUATION IN TEMP. (°C)	DEP. OF SOIL (Cm)	SOIL TEMP. (°C)	
		APR.– JUNE	DEC.– FEB
4·6	1	57.1	52.5
4.1	2.5	52.5	48.4
5.8	5.0	47.8	42.0
8.1	10	40.6	32.5
7.7	20	34.9	27.2
5.3	50	33.1	27.8
3.4	100	31.9	28.5
0.5	200	30.2	29.7
- 0.5	300	29.5	30.0
- 0.5	400	29.3	30.0

Fig. 2. Fluctuations in soil temperatures at different depths during different seasons of summer and winter measured at Medani region in the Sudan.

415

pack rat is nocturnal it stays in its microclimate and never exposes itself to the high temperature stress of the day.

Measurements of the microclimate of the nocturnal jerboa *(Jaculus jaculus)*, burrow in the Sudan, indicate that the burrows made during summer and winter are different from the autumn ones in their shape, depth and position. The autumn ones are superficial, horizontal and extend only to a depth of 20 cm. These are normally formed on high hills because of the heavy autumnal rains. The winter and summer burrows are deeper and extend vertically, they go as deep as 50–80 cm, depending on whether the soil is hard or soft, (Fig. 1). Such seasonal variations in the depth, length and positions of the burrows help in achieving more or less constant temperature and humidity conditions within the burrows. Summer burrows are deep and extend vertically and when the ambient temperature outside is as high as 55 °C and the relative humidity is very low, these burrows have temperature in the range of 28–30 °C and relative humidity in the range of 68–80%. Therefore a desert rodent living in such a microclimate will not suffer from any heat stress and its water expenditure will be minimised, (Fig. 2). The burrow usually has a complicated network of tunnels and normally ends up with enlarged end cavities the nests. The burrow has two entrances, one always acts as the emergent outlet, known in Arabic as 'Nabal' and it is always covered with sand when the animal is inside. If the animal is attacked, its first reaction is to start digging another outlet. Every now and then the animal comes to the Nabal to 'test' its safety as an outlet. The burrow of a mother normally has more than three or four outlets and all are opened except the nabal, probably to increase the chances of flight. Besides the main burrow, there are several small burrows, each with one entrance, all around the area where these animals occur. In an emergency the animal can dash into these small side burrows (GHOBRIAL & KAMIL, 1973).

Water Conservation in Desert Rodents

WATER CONSUMPTION

Desert rodents need a small amount of moisture in their food even though they can live for a long time on dry seeds. The water content of desert vegetation is generally underestimated; even seeds which have been desiccated in air of very low humidity still contain a physiologically significant amount of free water.

On the basis of the amount of moisture in food needed, rodents have been classified into 'wet rodents', the group which must have a high percentage of preformed water and 'Dry rodents', a group which can live for relatively long periods on dry seeds. The little ground squirrel *(Citellus tereticaudus)* collected from Arizona in 1911 was found to have a stomach half full of moist food consisting of about 90% green vegetation, 5% pulp of tuberous or bulbous roots and 5% small insects. The antelope squirrel, in Arizona, lives largely (especially during dry winters) on seeds with 34% moisture and pulps of cactus fruit and the big yellow pear shaped capsules of visnaga with a moisture content of 85% moisture (BAILY, 1923). When round-tailed squirrels *(Citellus tereticaudus)* were kept on a dry grain diet in Arizona, they lost weight and died of haemoconcentration through lack of water (SCHMIDT-NIELSEN, 1964). Some specimens of antelope ground squirrels *(Citellus nelsoni)* prefer insects and flesh which form 90% of their diet in general, and obtain all their water and food supply from their prey. In winter, however, when few insects are available, green plants dominate their diet.

Another desert rodent which needs moisture in its food is the pack rat *(Neotoma)*. Like other desert rodents, it does not drink free water, but eat moistened food. The principal food of this rat is green and succulent vegetation which furnishes an abundant supply of water. Distribution of these animals is greatly affected by the distribution of food. Pack rats are found in abundance where cactus is common. Cactus contains nearly 90% water and forms half of their food. Consumption of cactus increases during the dry season, but during the rainy season green leaves and pods of the mesquite *(Prosopis)* are important, also fresh-green grass is preferred when available.

Apparently the pack rat is well adapted to arid conditions, not by any special tolerance but by living where the juiciest food is available, preferring this even to the more nutritious seeds. If the pack rat is given only dry food its ability to survive is lowered. Pack rats fed on dry grains, died within 4–9 days while white rats survived 15–21 days. The rate of weight loss was similar in the two animals, but the pack rats died when they had lost about 30% while the white rats tolerated 50% loss (SCHMIDT-NIELSEN, 1948).

'Dry rodents' or the extreme desert rodents can survive and live for long periods entirely on dry food and even eat dry food by preference. Members of this group belong to the family Heteromyidae and are the most common rodents in Arizona and the Sahara desert. The length of time they can survive exclusively on a diet of dry grains seems to be unusually long. The Merriam Kangaroo rat *(Dipodomys merriami)* eats dry food by preference and can live for an indefinite period without an additional supply of moisture (SCHMIDT-NIELSEN, 1964). Similar reports have been obtained for other species of Kangaroo rats, e.g. *D. deserti* and *D. spectabilis. D. merriami* lives in the hottest part of the United States, in the lower Sonoran zone from El Paso, Texas, West through the Gila and Colorado valley of Arizona to eastern California. These rodents are found in the dry valley south of Tuscon, Arizona and in a dry winter following a dry summer, they were found flourishing in their normal condition. When caught and instantly killed, it was found that their pockets were filled with little juicy tubers of a small *Portulaca* whose moisture content is estimated to be 75–80 per cent and which is generally abundantly available an inch or two below the surface of the desert even during the drought season. Some of them dig their burrows to reach one of these underground Portulaca *(Talinum angustissimum)*. Thousands of little pits were found over the surface of the ground, which had been dug by rodents to extract roots, bulbs and tubers which serve as food and drink. There is also an abundance of small bushes of *Hymenoclea monogyra* which bring up moisture from great depths with their long roots. Many Kangaroo rat burrows lead to the roots of these bushes which have an estimated water content of 90% (BAILEY, 1923). *Dipodomys spectabilis* are unique; they inhabit a particular mound house and store large quantities of food to last through the winter and through long periods of drought and scarcity of vegetation. Their stores mainly consists of seeds and dried vegetation, but in all seasons they can find moist vegetation on the ground or under the ground although they occupy the driest areas of southern Arizona, New Mexico and western Texas. In capacity they require either some green or juicey vegetation or considerable water to drink and they evidently suffer very much if deprived of both.

Jaculus jaculus in the desert and semi desert regions near Khartoum, Sudan feed and rely mainly on *Cyperus* sp. and on the green parts above ground during the rainy season and on the underground bulbs (during the dry season) which have a moisture content estimated to be 58 to 74% (GHOBRIAL & KAMIL, 1973) and 8% protein content although KIRMIZ (1962) reported that the jerboa has frequently been found to get along well in captivity without water. However, in the Sudan it was found that *Jaculus jaculus* could not survive on dry Sorghum (with a moisture content of 8.5%). The animal lost weight steadily and stopped eating completely when the body weight loss was 15%. A comparison was made with the Nile rat *(Arvicanthis niloticus)* which lives in fields of beans, wheat and cotton near the banks of the river Nile. The jerboa lost body weight at the same rate as the water dependent Nile rat; when both animals were fed dry millet for 3 weeks, the Nile rat stabilised itself and started to feed normally and to recover its original weight within a period of 2 weeks if supplied with free water, while the jerboa continued to lose weight and would only flourish if its food was supplemented with cucumber (Figs. 3a–3d) (GHOBRIAL & AWAD, 1974).

However, it seems that there is a good deal of adaptational variations within the same family and species depending on the conditions the animal is forced to live in. CHEW (1961) suggested that physiological adaptations in mammals may be subspecific. LINDEBERG (1952) found that the minimum amounts of free water needed for survival by deer mice, *Peromyscus* sp., from different population groups are in accordance with the subjectively judged aridities of the respective habitats. LEE & MACMILLEN (1959) studied the wood rat, *Neotoma lepida*, from the coastal

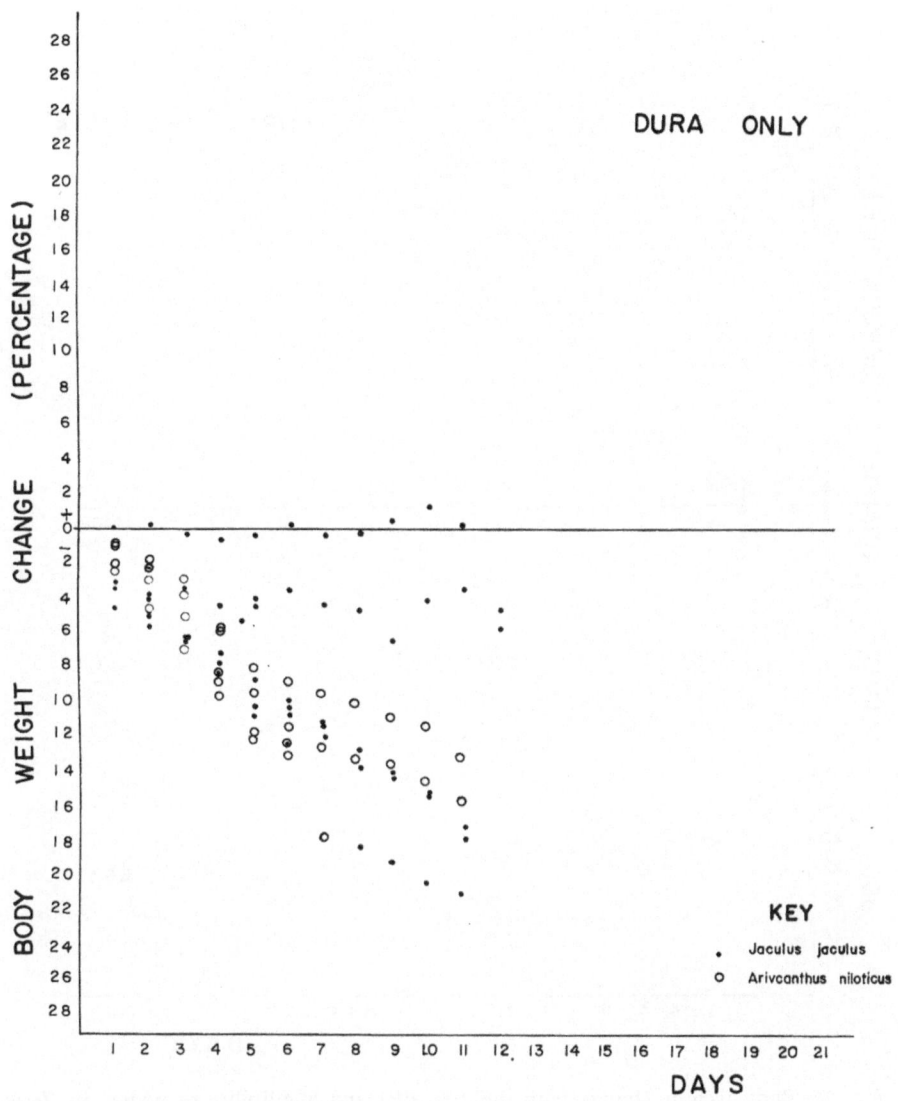

Fig. 3a. Body weight change with different diet and availability of water in *Jaculus jaculus* and *Arvicanthis niloticus*, Dura only.

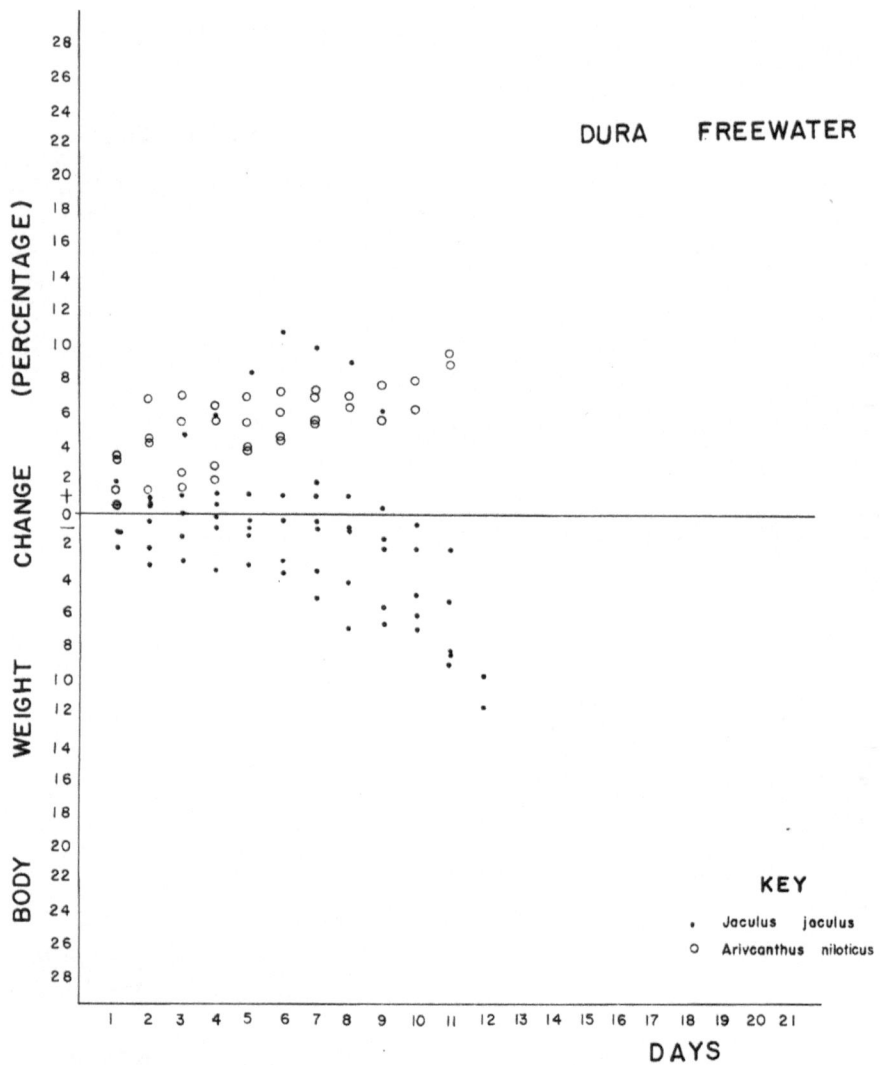

Fig. 3b. Body weight change with different diet and availibility of water, in *Jaculus jaculus* and *Arvicanthis niloticus*, Dura + Free water.

420

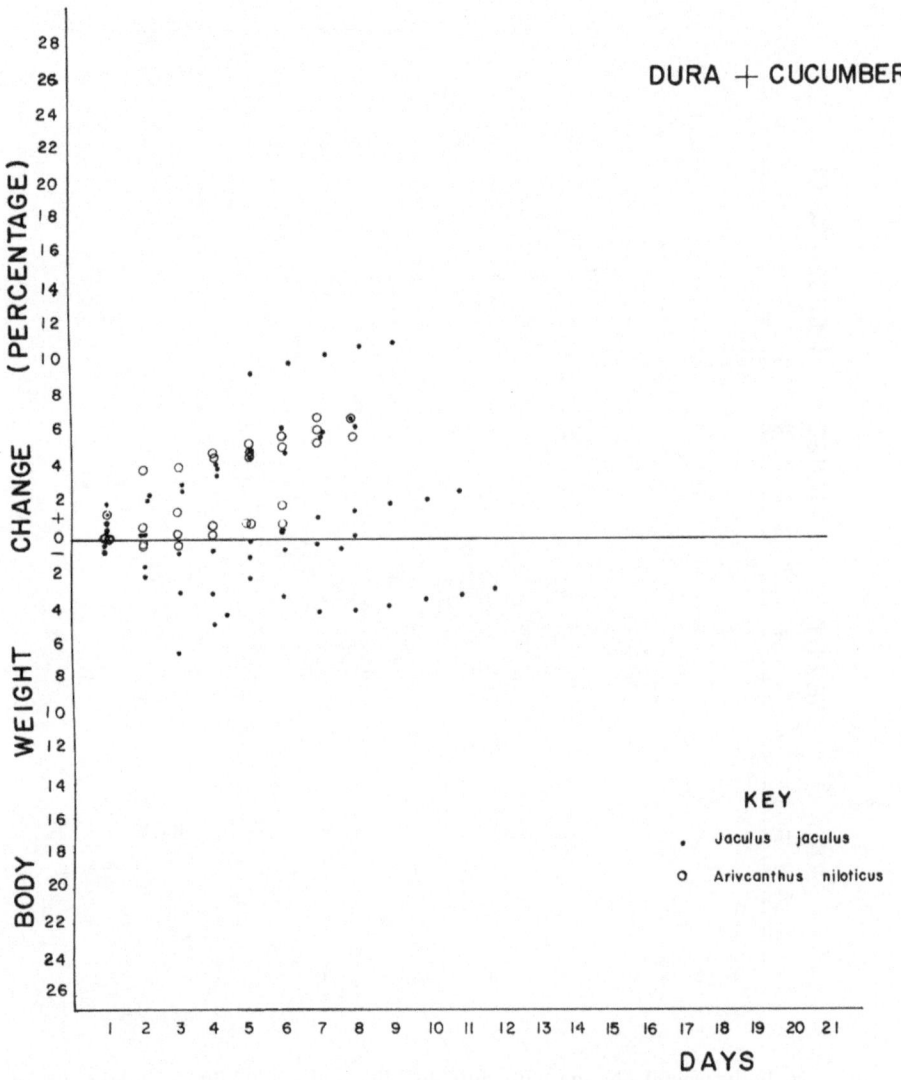

Fig. 3c. Body weight change with different diet and availability of water, in *Jaculus jaculus* and *Arvicanthis niloticus*, Dura + Cucumber.

421

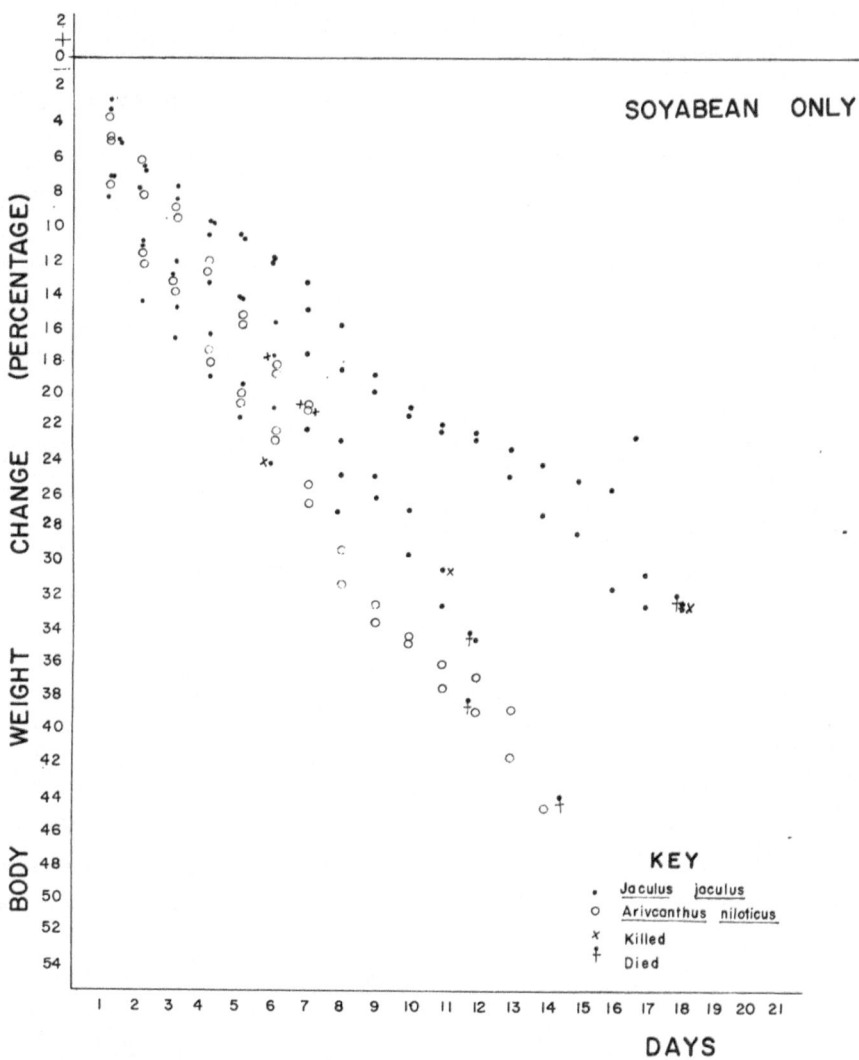

Fig. 3d. Body weight change with different diet and availability of water, in *Jaculus jaculus* and *Arvicanthis niloticus*, Soyabean only.

population in California living in areas where cacti are abundant. They found that these animals were able to survive on only *Opuntia* joints (8% moister). On the other hand, individuals from inland populations, living in areas dominated by the desert shrub-creosote *(Larrea)*, did less well. This is because *Larrea* provides less water than *Opuntia* and because *N. lepida* eat cactus leaves inefficiently. When the two populations were dehydrated, the inland population lost a smaller percentage of initial body weight.

BAILEY (1923) reported that Kangaroo rats kept for a week on dry food were extremely thirsty and drank eagerly and repeatedly as if famished and some of them died after two or three weeks without water or green moist food. HOWELL & GERSH (1935) reported that several species of *Dipodomys* which were kept on a diet of dry rolled oats, without water for a period of three months, lost little weight and were fairly ravenous for moisture at the end of the trial period.

Of the 'Dry rodents' group, the pocket mouse seems to be the member most adapted to survive without a supply of moisture. The pocket mouse *(Perognathus fallax)* lived for more than three years on dry rolled oats without showing any ill effect (STEPHENS, 1906) and even when the pocket mouse *Perognathus bailey* was offered dry rolled oats and barley plus water melon it never displayed any interest in the water melon, while Kangaroo rats *(Dipodomys merriami* and *D. spectabilis)* were fairly eager for it. Wistar rats *(Rattus norvegicus)* and Neotomas *(Neotoma albigula)* were found to consume even larger amounts of water melon than the Kangaroo rats every day (SCHMIDT-NIELSEN, 1948).

BAILEY's famous pocket mice maintained body weight for six weeks on a diet of dry weed seeds without any water supply. At the end of the sixth week they started to lose weight. When provided with free water they refused to take any, but when given fine snow, they ate it and replaced all their losses, and even managed to suck moisture from soaked cotton. KUTSCHER (1968) measured plasma volume in control and water deprived rats *(Rattus norvegicus)*, hamsters *(Mesocricetus auratus)*, gerbil *(Meriones unguiculatus)* and guinae pigs *(Cavia porcellus)*. The mean plasma volume declined with water deprivation in all four species but this change was not statistically significant in desert gerbils.

The antelope jack-rabbit *(Lepus alleni)* lives on dry valley slopes of Southern Arizona, distant from water sources. However, even if water is available, they will not go near it. The greater part of their food is green grass and growing vegetation and the many abundant species of cactus. Also they feed on pulps and pods of *Opuntia engelmanni* more than any other species. Dissected stomachs of the jack-rabbit were found to contain considerable amounts of the moist mucilagenous pulps of cactus pads with a moisture content of 78 per cent. They also feed eagerly on the common 'Visnaga' cactus *(Echinocactus wislizeni)*. The crisp juicy flesh of this cactus has been shown to have 94% moisture. Also they feed on underground tubers of *Talinum angustissimum* with a moisture content of 70 per cent.

WATER EXPENDITURE

Water is lost from the organism by three routes, in the urine, in faeces

423

and by evaporation from the lungs and skin. Another important route through which rodents lose water is by the copious salivation that cool their body surface if they are forced to face stresses of heat at lethal and sublethal temperatures.

It seems clear now that desert rodents are capable of reducing all their water losses to a minimum. Measurements of the water expenditure through urine, faeces and evaporation give the capacity of the water conservation mechanisms. Urine formation varies with the need for excretion of urea and electrolytes and the efficiency of the kidney while surface evaporation depends on the water content of the atmosphere and ambient temperature.

Evaporative Water Loss

Loss of water by evaporation from the animal body takes place from the skin, as well as from the respiratory tract. Even if no detectable sweating occurs, and even in the absence of sweat glands there is a small amount lost from the surface of the skin in the form of insensible prespiration. The water loss from the moist surface of the respiratory tract is considerable and depends upon the amount of air passing over the surface, the water content of the inspired air and the temperature of the expired air.

The ways in which desert rodents reduce both respiratory and skin water losses are uncertain. Different workers have reported different mechanisms.

HOWELL & GERSH (1935) reported that the reduction of water loss by respiration may be correlated with the amount of moisture available. The rate of respiration in undisturbed *Dipodomys mohavensis* on a moist diet was found to vary from 80–106/min; after being on dry diet for 10 days, the average was 53 and after 17 days the average was 44/min. The same individual was then fed on a moist diet once more and after 17 days the average rate increased to 130/min.

SCHMIDT-NIELSEN et al. (1950b) reported a lower evaporative water loss for the Heteromyides than the white rats. The average water evaporated from Heteromyides breathing dry air was 0.53 mg H_2O/ml O_2 while for the white rat it was 0.93 mg. They suggested that the mechanism responsible for this reduction in water loss involves changing of the pattern of breathing, a decrease in the ventilation of lungs and an increase in the utilization of alveolar oxygen (through an increase in the ratio of alveolar air to clean air). However, this hypothesis could not be confirmed by GJONNES & SCHMIDT-NIELSEN (1952) who found that the respiratory characteristics of Kangaroo rat blood were similar to those of white rat blood.

The lower temperature of the nasal mucosa is probably another important factor affecting the reduction of pulmonary water loss in rodents in general. Measurements of the nasal mucosal temperature in white rats and Heteromyides showed that in the two groups the nasal mucosa temperature was 6–7 °C lower than the rectal temperature. At an air temperature of 24 °C the Kangaroo rat was found to have a nasal mucosal temperature of 24 °C while its body temperature ranged between 29–30 °C (SCHMIDT-NIELSEN et al., 1950a). This difference of temperature permits the expired air to be saturated at a temperature lower than the body temperature, and these relatively cool surfaces condense considerable moisture from the warmer saturated air coming from the lungs before this water is expired. Since the water content of saturated air rises sharply above 30 °C, keeping the expired air temperature below this level results in a considerable saving of moisture.

If the air contains some water, it obviously takes less water to bring the expired air to saturation. Consequently the evaporation from the animal depends on the humidity of the atmosphere, for every milligram of water in the inspired air means a corresponding reduction in water loss. Since warm air at a certain relative humidity contains more water than cold air at the same relative humidity, the respiratory evaporation will be higher in warm air. It is the absolute humidity of the air which determines how much additional water must be added by the animal to bring the air to saturation. Thus the Heteromyides decrease moisture evaporation from their lungs by staying in their more humid underground burrows which have a higher absolute humidity than the outside. This reduces evaporation loss through the lungs to about 75% of the amount that would have been lost had the animals been exposed to the dry desert air outside.

Although the white rat has a low nasal mucosal temperature, it evaporates more water than the Kangaroo rat, because the white rat, in addition to evaporation from the lungs evaporates about an equal amount from its skin (TENNENT, 1946). However, the jerboa (Jaculus jaculus) completely lacks sweat glands (GHOBRIAL, 1970) which results in a complete absence of surface evaporation. Mice and Kangaroo rats, too, lack sweat glands and this is probably related to their relatively larger surface area/volume ratio. These animals are so small that they can not afford to utilize water for temperature regulation (EDNY, 1960). Studies on the sensitivity of skin water loss to rate of metabolism, excitement of the animal and anasthesia (TENNENT, 1945, CHEW, 1955) suggest that the loss through this route is limited physiologically. Studies on the skin of jerboa (Jaculus jaculus) reveal a thin epidermis and a horny layer of 0.02 mm, in comparison to an epidermal layer of 0.06 mm in a diurnal animal like the camel (GHOBRIAL, 1970) (Figs. 4, 5). Such thin layers may permit considerable losses of water by transpiration in dry air. The jerboa overcomes this difficulty by adopting nocturnal habit, and by staying in their humid burrows during the hours of heat stress; this results in a reduction of losses by evaporation through the skin. MALI (1956), working with isolated pieces of human skin, detected a

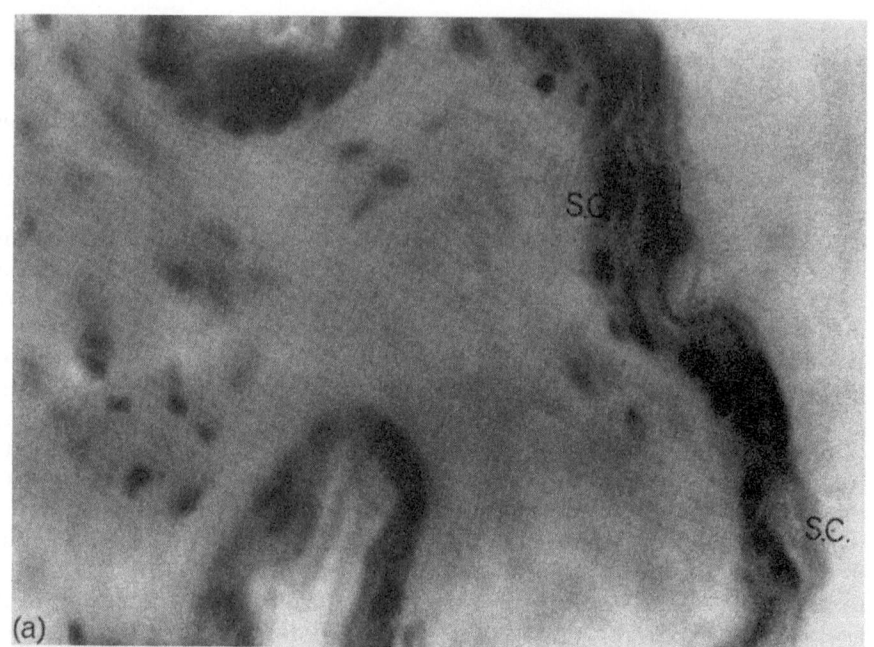

Fig. 4. Different strata of the epidermis of the jerboa (**X** 1000) S.C. Stratum corneum, S.G. Stratum germinativum.

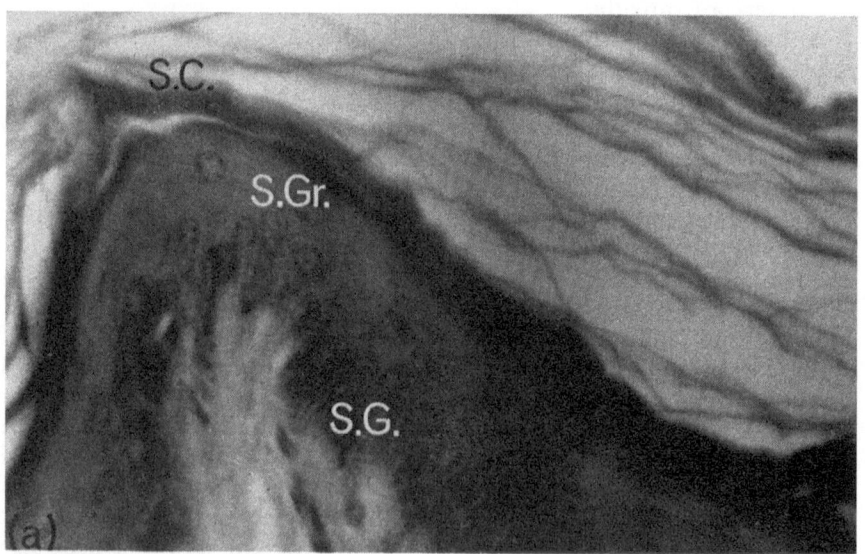

Fig. 5. Different strata of the epidermis of the camel (**X** 1000). S.C. Stratum corneum, S.G. Stratum germinativum, S. Gr. Stratum granulosum.

426

physical vapour barrier at the junction of the stratum granulosum and stratum corneum. Study of the skin of desert rodents might reveal a similar or even a more effective, physical barrier than that found in human skin. The jerboa skin shows a highly compact dermis with densely packed collagene fibres and striated muscles. RIAD (1960) reported that the jerboa dermis exhibits seasonal variations; at the beginning of winter it is formed wholly of adipose tissue which undergoes gradual transformation to fibrous connective tissue as the fat is used up and so by early summer the dermis is largely devoid of fat. This probably indicates that the dermal fat serves as a store of energy on which the animal depends when it faces food scarcity in nature.

Water losses through salivation

In rodents the water lost through the saliva may constitute a major component of evaporative water loss. Such losses occur only when the animal is exposed to lethal or near lethal temperatures. When the Kangaroo rat *(Dipodomys merriami)* was exposed to an air temperature of 43 °C the fur under the chin and throat became wet and completely soaked with moisture from a copious flow of saliva (SCHMIDT-NIELSEN, 1964). A similar response occurs in ground squirrels (HUDSON, 1962) and in the Egyptian rodent *Dipus* sp. (KIRMIZ, 1962). In the Sudan when a large number of jerboa *(Jaculus jaculus)* were placed in one cage and exposed to temperature regimes of 40 °C–43 °C, not only regions under the chin or throat but the whole animal became wet. A similar phenomenon has been reported for many other animals e.g. cat, rabbit, etc.

Water Used For Urine

Urine is one of the main routes through which the animal loses water. The more concentrated the urine the animal produces the less water is used for the elimination of metabolic wastes. The mechanisms of urine concentration is highly developed in small desert rodents as minimum water loss through the urine is of vital importance for their existance. The Kangaroo rat, *Dipodomys merriami*, has been found to achieve a urine concentration of 3840 mM urea/l, or 23% of urea, which is almost 5 times as high as the most concentrated urine produced by man (SCHMIDT-NIELSEN *et al.* 1948).

SCHMIDT-NIELSEN (1964) has reported that the Kangaroo rat produces a very concentrated urine which crystallises as soon as it is passed out of the body, crystallization taking place even when the urine was collected in a pipette.

In our laboratory, the jerboa *(Jaculus jaculus)* produced an average of 5.5 ml urine/24 hours when kept on cucumber (moisture content 85%). When fed on dry millet *sorghum* (moisture 8.9%) the urine production decreased to an average of 1.5 ml/24 hours of a very concentrated solution in the form of drops which crystallised as soon as these were excreted, so that for further chemical analysis it was necessary to dissolve the crystals in distilled water. Peculiarly, however, the urine remained in soluble form as long as it was inside the bladder, crystallising immediately it was released. What controls the solubility of this very concentrated urine inside the body is not known and further investigation on this point should be of value.

Desert rodents are so well adapted to their environment that even when loaded with water they find 'difficulty' in eliminating the excess, presumably because of factors associated with their adaptation to xeric conditions. Hofman (1956) reported that when the Kangaroo rat, *Dipodomys merriami*, was given 5 ml distilled water/100 gm body weight by stomach tube it eliminated only 40% of the load after 367 minutes, while the white rat did so in 84 minutes. When loaded with water intraperitoneally the Kangaroo rat showed still more pronounced water retention ability (Rollason, 1964). *D. spectabilis* excreted the same percentage of loaded water (85–88%) within 6 hours as did the white rat, but only after a lag of 3 hours (Cole *et al.* 1963). Hummel (1963) reported that the gerbil, *Meriones shawii*, cannot excrete orally administered tap water by means of a hypotonic diuresis when kept on dry food.

Desert rodents seem to be capable of adapting their physiological mechanisms to the harsh environments they are compelled to inhabit, and their degree of adaptation is proportional to the degree of aridity of their habitat. Chew & Mitchell (1965) compared the rate of excretion of loaded water in different rodent species (*Dipodomis merriami*, *D. agilis* and *Meriones unguiculatus*) living in different habitats of different aridity. They reported that all these animals excrete an orally administered water load at a rate slower than those reported for the white rats. Moreover, the rate of water excretion in these species was inversely related to the degree of aridity of their respective habitats.

Structure and Concentrating Mechanisms in the Desert Rodent Kidney

The production of very concentrated urine by desert rodents it is brought about by the peculiar structure and concentrating mechanisms of their kidneys. Schmidt-Nielsen, working in the Arizona desert, on the water metabolism of desert rodents (Heteromyidea), found that these animals can achieve urinary concentrations of 1.2 N electrolytes and 3.8 M urea, almost twice the values found in the urine of white rats. An even higher urea concentration of 4.3 M/l, i.e. 24% was recorded for the jerboa (*Jaculus jaculus*). The Egyptian gerbil (*Gerbillus gerbillus*), when kept on dry food, produced urine with a urea concentration of 3.4M/l (Burns, 1956).

During dehydration the antelope ground squirrel can produce urine with an osmotic pressure excess of 3500 mosm., 1200 of which are due to electrolytes (BARTHOLOMEW & HUDSON, 1959). The urine concentration in this species may be 9.4 times as high as the plasma concentration, urine/plasma ratio being higher than any recorded for other animal species (HUDSON, 1960, 1962). Such high urine concentrations were found in both animals trapped from their natural environments and in animals forced to eat excessive amounts of salt and protein. All these findings have attracted many workers to investigate the structure and mechanisms of the powerful desert rodent kidneys.

Studies of mammalian kidneys by SPERBER (1944) showed that there is a remarkable correlation between the length of the loop structure and the collecting tubules, that is the relative thickness of the renal medulla, and the aridity of the habitat in which the animal lives. He found that out of thirty-four rodent species studied, seven from desert regions had such long papilla, extending beyond the pelvis of the kidney into the ureter.

HOWELL & GERSH (1935) had demonstrated the increasing concentration of urine as it passed along the tubules, by injecting a solution of sodium ferrocyanide into the Kangaroo rat *Dipodomys agilis*. The concentration gradient of sodium ferrocyanide was demonstrated along the renal tubules, in the loop of Henle and in the collecting tubules. Micropuncture studies showed that the glomerular filtrate is isotonic with plasma. As the filtrate passes down the renal tubules through zones of increasing hypertonicity in the medullary and papillary tissues, the concentration of the urine increases (WIRZ, 1957, GOTTSCHALK, CARL & MYLLE, 1959). All this indicates that both the outer and the inner zones of the medulla act as a countercurrent multiplier system.

Further evidence for the counter-current hypothesis was furnished by the work of SCHMIDT-NIELSEN & O'DELL (1961) through their studies of the distribution of urea and electrolytes in different kidney types during antidiuresis. They found the beaver to have 100% short-looped nephrons, the rabbit, 44% long-looped nephrons and the desert dwelling *Psammomys*, 100% long-looped nephrons. They also found that during antidiuresis all kidneys have the same sodium and urea concentrations in the outer zone of the medulla. In kidneys with inner zones, the sodium and urea concentrations continued to increase throughout the inner zone, reaching the highest values in kidneys with the thickest medulla. These findings indicate that there is no important qualitative difference in the functions of short and long-looped nephrons in the kidneys. These findings, as well as the above mentioned correlation between concentration ability and medullary thickness, indicate that the inner zone of the medulla also functions as a counter-current multiplier system.

Histological studies of Heteromyid kidneys showed that the distal convoluted tubules follow a markedly tortuous course. The lumen is of considerable width, and the linings of epithelial cells are dome shaped and thus somewhat protruding into the lumens, thus increasing the surface area available for absorption. In this respect the Heteromyid kidney is different from that of other rodents.

The rate of diffusion of urea from the kidney tubules into the blood is greater in desert rodents than in man. The urea clearance in these animals was shown to be only 25–30% of the creatinine clearance. The reabsorption of water was considerable, sometimes the creatinine in the urine would be 300–400 times as concentrated as in the serum. Urea diffuses freely throughout body fluids and cells, and the plasma concentration represents its concentration in all body fluids; therefore, if urea is to be retained in the body, its concentration in the blood plasma increases. High concentrations of urea in the body weight lead to the question whether desert rodents conserve water by excreting less urine than required to eliminate the waste products from the body and therefore store their waste products. Evidences furnished by SCHMIDT-NIELSEN et al. (1948) indicate that desert rodents do not, in fact, conserve body water at the cost of waste elimination. However when *Jaculus jaculus* was maintained on Soyabeans only or millet only, the total food intake decreased, the body weight decreased but the urea excreted was far less than the amounts retained in the plasma. (GHOBRIAL & AWAD, 1974).

On the other hand, the wood rat, *Neotoma*, and the white rat are unable to excrete their waste products efficiently when kept on dry food. Their plasma concentration of urea under such circumstances often increases to a level that they are not able to cope up with.

It is well known that the volume of urine formed and its concentration are under endocrine control. An increased release of the antidiuretic hormone (ADH) from the neurohypophysis leads to an immediate renal response with increased tubular reabsorption of water resulting in the production of highly concentrated urine. Could production of highly concentrated urine by desert rodents be caused by unusual amounts of circulating ADH?

The kidney has a characteristic concentrating ability which depends on its structure and which differs from species to species. The amount of ADH produced in the hypophysis and released into the blood stream determines to what extent the concentrating capacity of the kidney is utilized at a given moment, but when the concentrating mechanism works at its maximal capacity, the release of more hormone should have no further effects. The administration of exogenous ADH to *Gerbillus gerbillus* (BURNS, 1956) did not result in higher levels of urine concentration.

Desert rodents in general show a high concentration of plasma ADH. *Dipodomys merriami* has a high plasma ADH level of 19.3 mU/ml compared to the hydrated white rat with an ADH level of zero, and a maximum of 6 mU/ml of plasma in dehydrated white rats (AMES & VAN DYKE, 1950; DICKER & NUNN, 1957). Such a chronically high concentration of ADH in desert rodents not only accounts for their usually high urine concentration, but this may also be a measure of the concentrating ability and efficiency of their kidneys and may also be the explanation for the fact that desert rodents take a long time to excrete a water load, and often show kidney damages after water loading.

The high-powered kidneys of desert rodents enable them not only to survive without access to free water, but even to drink solutions of higher electrolytic concentration than sea water, extract the amount of water required from such solutions and excrete all excess electrolytes.

When the Egyptian gerbil *(Gerbillus gerbillus)* was kept on dry barley and an extra salt load it was found that the concentration of electrolytes in its urine was 1600 m Eq/l which is about three times as high as sea water (BURNS, 1956), while the Kangaroo rat *(Dipodomys merriami)* can produce urine with a maximum electrolyte concentration of 1200 m Eq/l, which is about twice as concentrated as sea water.

Experiments on ground squirrels *(Citellus leucurus)* showed that they can drink solutions of NaCl more concentrated than sea water (BARTHOLOMEW & HUDSON, 1959). If free water is replaced by increasing concentrations of NaCl, they are able to utilize concentrations as high as 800 mM, i.e. 1.4 times as saline as see water. On 1000 mM (5.8 per cent NaCl), they lost weight. Ground squirrels maintained on sea water have been known to survive for double the period that water-deprived animals of this species live.

Pocket mice, *Perognathus*, maintained on dry grains plus 10% NaCl solution, were capable of excreting electrolytes in the urine up to a concentration of 1088 mN when the same diet was supplied to *Dipodomys* the average concentration of urinary electrolytes was 865 mN (SCHMIDT-NIELSEN *et al.* 1948).

SCHMIDT-NIELSEN *et al.* (1950c) induced negative water balance in the Kangaroo rat *(Dipodomys merriami)* by feeding it on dry soya beans, which have a high protein content. The result of this was that larger than usual amounts of urea were excreted. On a diet of soya beans with sea water to drink the Kangaroo rats lost weight for the first 2–3 days and suffered from diarrhoea due to the appreciable amounts of magnesium and sulphates in sea water, but afterwards they recovered and resembled the control animals which drank fresh water, and even gained weight. The animals were capable of excreting all excess urea and salt and managed to keep a constant body fluid composition.

Animals kept only on dry soya beans did not survive, they lost weight and died in 8–16 days. But, although, in general desert rodent kidneys are capable of producing urine with very high concentrations of electrolytes, variations occur depending on factors such as ambient temperature, relative humidity, dietary constituents and moisture content of the diet.

431

The work of SCHMIDT-NIELSEN and his colleagues (1948) showed that there is a seasonal fluctuation in the amount of electrolytes in the urine of some rodents, a pronounced decrease often indicating a hydrated state of the body.

MEFFERD et al. (1958) found that, in rats, urinary Na+, K+ and Ca+ tend to vary inversely with the ambient temperature.

The nature of the diet is known to have a great controlling influence on the concentration of urinary electrolytes. For example, on a protein-rich diet *Dipodomys merriami* were found to excrete electrolytes in concentrations of up to 12 N (7%), while on a diet containing sufficient moisture these animals tended to excrete much less electrolytes than when on a dry diet (SCHMIDT-NIELSEN et al., 1948).

GABRE & SHALABY (1964) examined the total urine electrolytes in both sexes of jerboa *(Jaculus jaculus)* and under different conditions of temperatures. They found that for animals maintained on a normal diet, the average urinary electrolyte concentrations in summer and winter were 36.8 and 28.1 gm/l respectively for females and 38.8 and 25.4 gm/l respectively for males. When fed on dry barley, the females excreted 46.8 and 25.5 gm total electrolytes/l of urine and the males, 42.2 and 26.5 gm total electrolytes per litre of urine during summer and winter respectively.

For animals fed on moist food, the average urinary total electrolyte values were 29.9 and 17.1 gm/l for the females in hot and cold seasons respectively, while the corresponding values for males were 36.6 and 17.3 gm/l. These authors have reported a higher total sodium concentration in the urine of males than in that of females.

Faecal Water Loss

Considerable amounts of water are normally lost with the faeces, as well as in the urine. HOWELL & GERSH (1935) noted that the phenomenon of low water excretion through faeces was common to most desert animals and they considered it a physiological adaptation for arid environments. SCHMIDT-NIELSEN et al. (1951) reported that conservation of water in the heteromyids is accomplished by a highly developed ability to excrete a concentrated urine and dry faecal pellets. They reported that faeces deposited by the Kangaroo rat *Dipodomys merriami* was drier than that of the white rat. The water content of the faeces was found to be 45% for the Kangaroo rat and 68% in the white rat; following the ingestion of 100 gm pearled barley, the amount of dried faeces eliminated by the Kangaroo rat was found to be half that eliminated by the white rat, and the amount of water lost with the faeces for the Kangaroo rat was found to be about one fifth of that for the white rat (2.54 gm and 13.5 gm water/100 gm dry faecal matter respectively).

FARGHALY & SHALABY (1956) found that the faeces of *Gerbillus pyramidium* has a lower water content than that of the white rate, the values being 51.5% and 69.4% respectively. After feeding on dried barley, the water content of the faeces of *Gerbillus pyramidium* showed a pronounced decrease, while the white rats showed hardly any difference in the water content of their faeces. At the same time, *Gerbillus pyramidium* eliminated almost the same amount of faeces as was eliminated by the white rat, after consuming the same amount of dried barley.

BAILEY (1923) reported that *Dipodomys spectabilis* produce pellets as dry as the dry food, but produce copious urine if supplied with moist food. In our laboratory we found that if the animal was provided only with millet (8.9% moisture content) the animal excreted very dry pellets and very concentrated urine which crystallised as soon as it was produced. When the animals were maintained on millet plus cucumber they continued to produce dry pellets, but the urine volume rose to about 5.8 ml/day (GHOBRIAL & AWAD, 1974).

HOW LONG DESERT RODENTS SURVIVE WITHOUT ACCESS TO FREE WATER?

Desert rodents can be classified into 'Dry' and 'Wet' types, on the basis of their primary reliance on 'dry seeds' and on 'succulent vegetation'. Wet rodents are adapted to arid conditions by inhabiting regions rich in succulent cacti ensuring that their water relation is always balanced. On the other hand 'Dry rodents' face the difficulty of water shortage; how can they survive indefinitely on dry food? Is the small amount of water available to them from the moisture in dry seeds and water produced by oxidation balanced by reduced water losses through evaporation, urine and faeces? A full answer to these questions can hardly be given at this stage because of insufficient information, except for the hypothetical assumption of SCHMIDT-NIELSEN *et al.*, (1951) regarding the water balance of the Kangaroo rats when maintained on dry food, under different humidity conditions and his experimental verification of it.

We decided to use jerboa *(Jaculus jaculus)* in further experimental work designed to verify SCHMIDT-NIELSEN's hypothetical assumptions. The results obtained showed that animals kept on dry Sorghum (millet) tended to lose weight at humidities of 10 to 15% and temperatures 30 to 35 °C, but at higher humidities of 40% to 60% at 20 to 25 °C, the animals lost weight during the first 3 to 4 days but regained their normal weight and even gained more weight (GHOBRIAL & AWAD, 1974). This is probably because millet tends to absorb moisture and increases its preformed water content. The normal preformed water content is of this grain of the order of 0.8 gm/100 calories under 30% relative humidity conditions (GHOBRIAL, 1968).

The oxidation water obtained from the metabolism of dry seeds is supplemented by increasing amounts of absorbed water at higher humidities.

433

The humidity inside the rodent burrows is usually much higher than the determined critical limits. The dry desert rodents being usually nocturnal in habit, escape the critical hours of the day in their burrows and the seeds that they store in these burrows absorb moisture there. This seems to be explanation for the physiological ability of these desert rodents to survive periods of water scarcity.

TOLERANCE TO HEAT AND THERMO-REGULATION OF DESERT RODENTS

The quality that characterises homeothermic animals is their ability to regulate the production and dissipation of heat so as to maintain constant body temperature under different environmental temperatures. This is done mainly by mobilisation of water and therefore thermal regulation is inconceivable without the adjustment of body fluids.

Tolerance to high temperatures differs significantly from species to species and even from individual to individual and varies largely with the size and natural habitat of the animal and its requirements. The physiology of animals living in hot dry deserts enables them to tolerate heat much better than their relatives living in cooler, more humid regions. The camel which has an average body temperature of 34.5 °C, is capable of increasing this temperature by 6.2 °C in the course of 24 hours. Such variations enable the animal to store excess heat during the hot hours of the day, and this can be dissipated during the cool hours of the night without expenditure of water. At the same time the high body temperature reduces the passage of more heat from environment to the body, due to the small gradient difference (SCHMIDT-NIELSEN et al., 1957). In comparison, man can increase his body temperature up to only 2 °C. The medium size desert gazelle *(Gazella dorcas dorcas)* cannot fluctuate its body temperature by more than 2 °C even when deprived of water and the maximum rectal temperature which can be tolerated by them is 41 °C. The animal avoids the stresses of the hot hours of the day by restricting its surface activity to the periods between 0600 and 1000 hours and between 1400 and 1800 hours i.e. during the cooler hours of the day (GHOBRIAL, 1974; GHOBRIAL & CLOUDSELY-THOMPSON, 1974).

Desert animals are faced with the difficulty of heat exchange with their surroundings resulting in a net movement of heat load passing from the surroundings to the body through conduction from hot air, radiation from the sun, radiation from the hot ground and heat generated by food metabolism. If the animal is to tolerate heat stresses, all these excess loads must be lost through radiation and evaporation from the body surface, and therefore the animal must be able to afford the expenditure of large volumes of water for heat regulation.

Rodents, being small in size, and having a large surface area in proportion to volume, accumulate heat at a higher rate than larger animals and therefore must use larger volumes of water for heat regulation.

Desert rodents are adapted to their arid life both behaviourally and physiologically. They create their own microclimate in their burrows, which have a far lower temperature than the outer atmosphere. Small animals living in such a microclimate do not face any heat loads or any excess losses of water through evaporation. The microclimate of the pack rat's nest indicates that this nocturnal animal is never exposed to serious heat stresses, the nest has large piles of debris to shelter it and has an average temperature of 31 °C, while the soil temperature in the open desert is 75 °C and the soil temperature near the nest is 46 °C. The burrow temperature of the Kangaroo rat *(Dipodomys spectabilis)* has a highest recorded temperature of 31 °C while the outer atmospheric temperature was 45 °C and the soil temperature 70 °C.

DAWSON (1955) reported that when ground squirrels are exposed to high ambient temperature in the laboratory their body temperature becomes elevated, indicating that there is no physiological mechanism which can effectively counteract the rise in body temperature. Ground squirrels *(Citellus leucurus)* when exposed for 3½ hours to an ambient temperature of 39 °C became heated to 43.1 °C and died shortly afterwards. The animals had salivated profusely near the end of three hours, but this reaction did not seem to be sufficient to save it from heat death. However, when animals were exposed to a temperature of 38 °C they survived. It seems that ground squirrels can not usually stand exposure to heat for prolonged periods but are able to tolerate shorter periods of heat exposure. KIRMIZ (1962) compared the effects of external temperatures on the jerboa *(Dipus aegypticus)* and the rat with respect to body temperature, energy metabolism and insensible perspiration. He reported that generally all these parameters are lower in the jerboa than in the rat: The average body temperature of the jerboa is 37.0 °C, while in the rats it is 37.5 °C. The Merriam Kangaroo rat *(Dipodomys merriami)* has a body temperature usually around 36–38 °C at an atmospheric temperature of about 20 to 25 °C. Exposure to different temperatures show that the animal is able to maintain its body temperature up to an air temperature of 37 °C. A further increase in ambient temperature would lead to a corresponding increase in body temperature, and there is no apparent physiological reaction that prevents the body temperature from rising. SCHMIDT-NIELSEN (1964) reported that four animals out of a group of 5 died after 3–5 hours exposure to an air temperature of 39 °C, when they had rectal temperatures of 40.5 to 41.1 °C. However, when the body temperature was proceeding towards the lethal limits (around 41 °C), the Kangaroo rat's mechanism of thermoregulation came in force and there was copious secretion of saliva, wetting the fur under the chin and the throat and leading to evaporation of moisture which keeps the body temperature from rising further. This is an 'emergency' heat regulation mechanism which apparently comes into operation only when the conditions are critical for survival. SCHMIDT-NIELSEN

435

demonstrated this phenomenon in an experiment in which he exposed Kangaroo rats to an air temperature of 43 °C. All the animals died between 45 minutes to $1\frac{1}{2}$ hours. They showed a continuous body temperature rise until the lethal level 41 °C was reached, when there was a break in the curve indicating an active resistance to the rising temperature and for the next 40 minutes there was no further rise. This plateau part of the curve coincided with copious salivation. The reactions of the white rat when exposed to similar heat loads, are different; there is no apparent ability in these animals to lower the body temperature even by salivation, and death occurs at a much lower air temperature (about 39 °C) than that for Kangaroo rats although the lethal body temperature is nearly the same in the two species.

McManus et al. (1969) reported that the Mongolian gerbil *(Meriones unguiculatus)* is able to maintain its body temperature near 38.3 °C for extended periods at ambient temperatures ranging from —23 °C up to 40 °C, the higher limit being tolerated for up to 3 hours. Saliva spreading occurs under heat stress conditions and the lethal body temperature for this species is near 44 °C.

Aestivation and hibernation

Aestivation and hibernation are other physiological adaptive mechanisms which enable desert rodents to escape critical periods: high atmospheric temperatures (above 25 °C) during the summer, and cool temperatures during winter (below 10 °C). During these periods the animal enters into a state of lethargy or torpidity. BARTHOLOMEW et al. (1956) and TUCKER (1962) reported that there is no real physiological difference between aestivation and hibernation, in both conditions there is a general reduction in all physiological processes to the minimum. Body temperature is reduced to approach that of the animal's surroundings and metabolic rates are also reduced. Moderate energy reserves are extended over long periods, and the ventilation of the lungs is reduced thereby diminishing the pulmonary evaporation of water.

Extensive studies have recently been carried out into the physiology of rodents entering into these states of torpidity, and the significance of such states to the survival of animals living in regions of excessive heat and shortage of water and food.

In the hibernating American marmot both CO_2 and O_2 contents of the blood were found higher than in active animals (RASMUSSEN, 1915). The higher blood CO_2 and O_2 contents might be explained by the relatively greater number of erythrocytes per volume of blood, due to the withdrawal of the liquid serum of the blood as a consequence of the minimization of the amount of fluid needed by the body during this dormant period.

The recent studies of RUSEV & STEFANOV (1969) on changes in the metabolism of the ground squirrel *(Citellus citellus)* during hibernation

436

Fig. 8. The Sudanese Jerboa, *Jaculus jaculus.*

Fig. 9. The Sudanese Rat, *Arvicanthis niloticus.*

437

showed that hibernation is characterised by a generalised lowering of the oxidative processes with the highest depression occurring in the kidney and the liver; only the heart muscle retained its oxidative capacities. During arousal, the most quickly activated organ is the kidney; this is probably meant to get rid of the accumulated metabolic end-products indicating that during aestivation and hibernation water loss through the kidneys is reduced to the minimum.

Inspite of the reduction in the general metabolic rate during the dormant period, the brown fat increases in mass and weight, becomes much more vascularised and exhibits a higher rate of O_2 consumption during this period. CHAFFEE *et al.* (1966) reported that at an ambient temperature of 2 °C the hibernating ground squirrel *(Citellus lateralis)* had a 20% decrease in body and liver weight while there was an increase in the weight of the brown fat by 40% and that in respiration rate by 36% over normal.

REPRODUCTION IN DESERT RODENTS

PRASAD (1954a & b) analysed the influence of annual variation in day and night length, temperature and in the type and quantity of food available on the reproductive cycle in the gerbil *(Tatera indica)*. His data showed that light, temperature and rainfall exert an indirect influence on the reproductive cycle of these rodents and he considered that this effect may be brought about by increased physical activity during periods of long nights and by an increase in food supply with the onset of the rainy season.

Desert animals are subjected to considerable variations in environmental temperature and the availability of food and water. In general, desert rodents breed towards the end of the rainy season and during the first few months afterwards, regardless of when the rainy season occurs. MEASROCH (1954) showed that different species of gerbils live in regions which experience different types of summer and winter rains. Breeding is minimal in *Tatera brantsi* and in *Mastomys* or ceases entirely in *Elephantulus* at the end of the dry season and the beginning of the rainy season.

PRASAD (1954b; 1956) reported that spermatogenic activity in the male gerbil decreases towards the end of the dry season and ceases about the time that the rains begin, but the period of quiescence is short and both males and females become reproductively active again before the end of the rainy season. HAPPOLD (1966) reported that desert rodent species, in the Sudan, breed at the end of the rainy season, while wild rodent species found along the Nile and the commensal species probably breed throughout the year. He suggested October to be the month of the greatest breeding activity and found no evidence for any breeding activity from March to May, the hottest and driest part of the year.

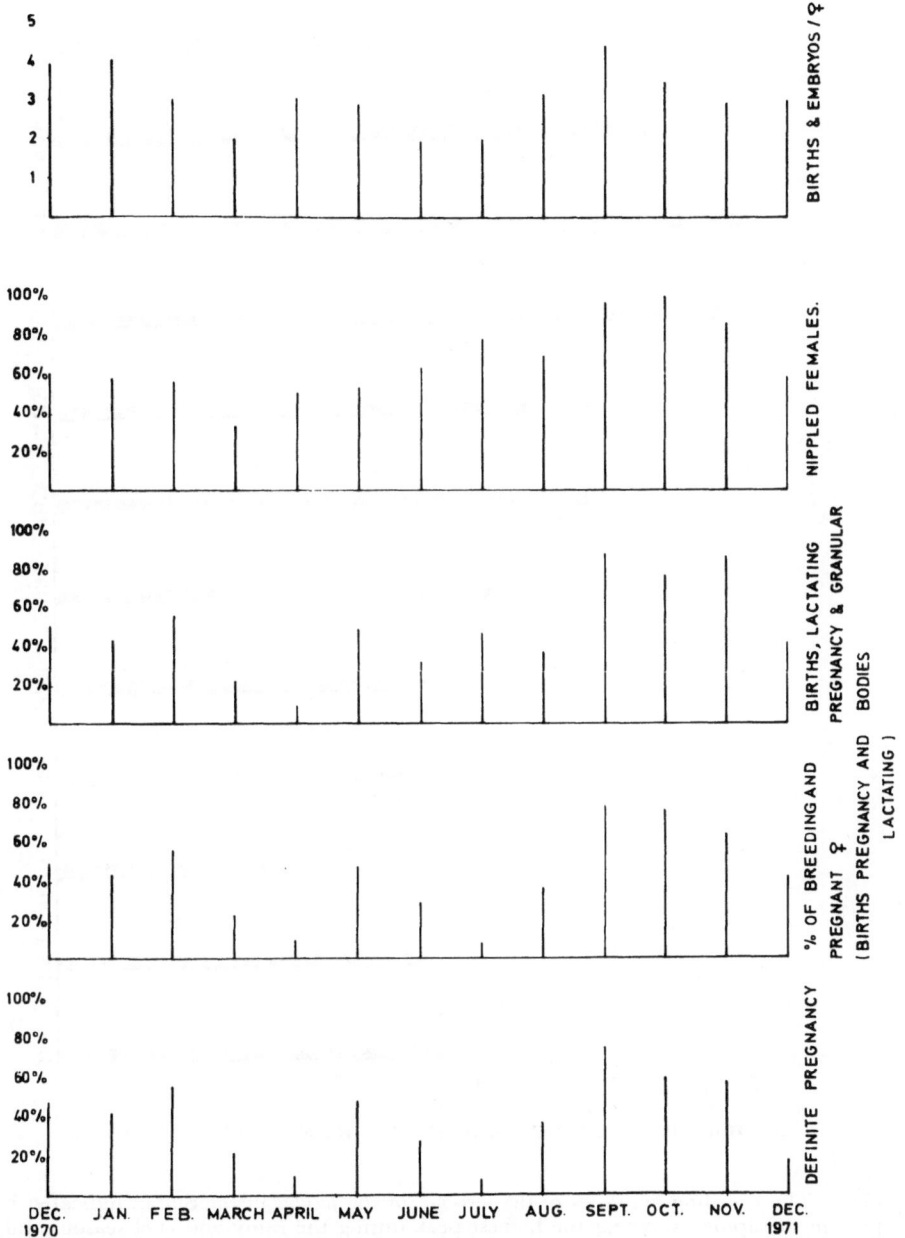

Fig. 6. The breeding season of the gerboa, *Jaculus jaculus* in the Sudan. Breeding activity is indicated from evidence of pregnancies, birth, young, placental scars and prominent nipples. The rains occur in July, August and September.

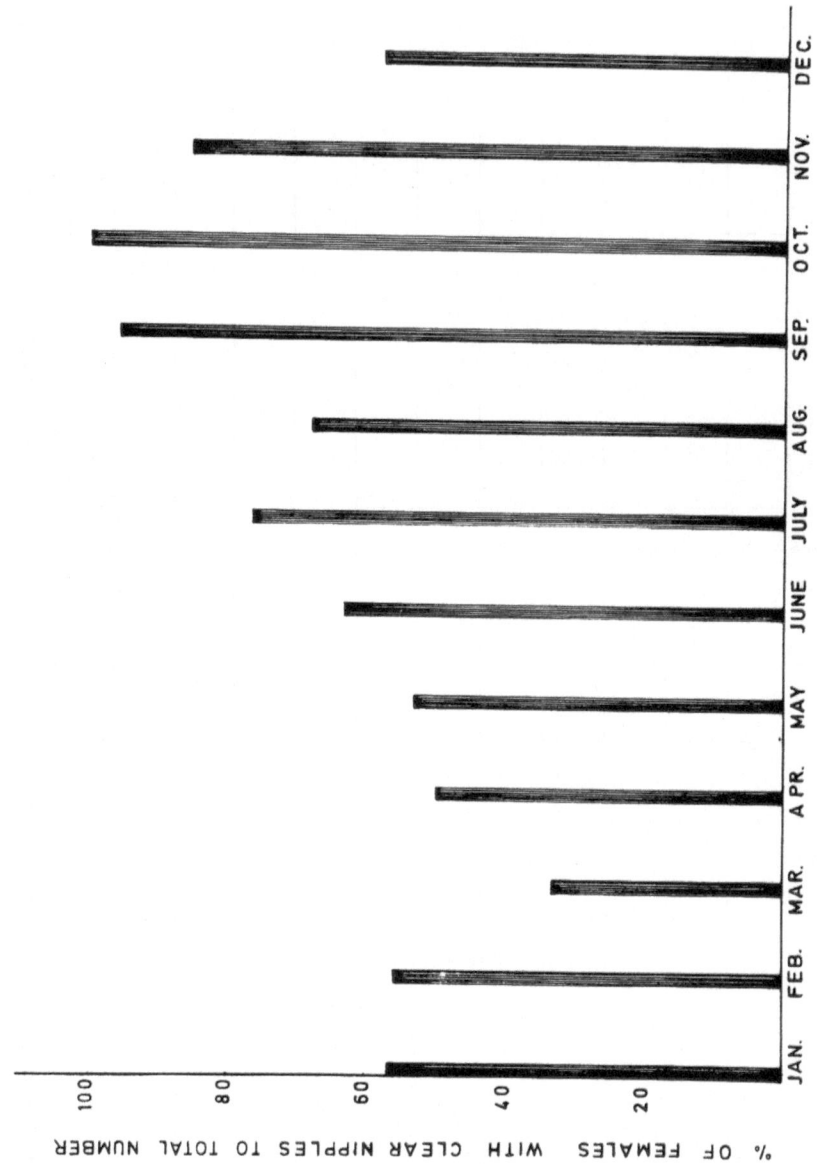

Fig. 7. The breeding activities of the jerboa, *Jaculus jaculus* in the Sudan indicated by prominent nipples, showing the highest peak during the rainy and cool seasons, July to February and lowest from March to June.

440

Studies on breeding and reproduction of the Sudanese jerboa *(Jaculus jaculus)* presently being carried in our laboratory suggest (GHOBRIAL & KAMIL, 1973) that these desert rodents breed throughout the year (Fig. 6), but show a single breeding season of eight months starting at the beginning of the rainy season, in July, when most of the females show either definite pregnancy or the presence of granular bodies in the uteri. September and October are the two months for high pregnancies and births, and all females are either pregnant or with prominent nipples in these months (Fig. 7). HAPPOLD (1970) suggested two breeding seasons for the Sudan desert rodents viz. one just after the rains in October-November and the other between February-March. The highest peak in reproductive activity occurs during the rainy and cool season when the air temperature ranges from 15–30 °C and the relative humidity from 40–70%. The lowest breeding activity occurs during the dry and hot season between March and May, when the air temperature ranges from 35–55 °C and the relative humidity from 0–20%. The number of litters produced varies in different seasons, there being an average of 2 per female during the dry season and 5 per female during the breeding season.

The males also show seasonal variations in breeding activity during the breeding season, the testes are large, reddish, highly vascularised and scrotal with an average weight of 0.5–0.55 gm, while during the dry season, they are abdominal with an average weight of 0.2–0.25 gm. Spermatogenesis also shows prominent seasonal variation with a higher cell division and concentration of sperms in the vesicular seminalis during the wet and cool seasons but a marked reduction during the dry season. ALLANSON (1933, 1958) reported similar results for males of the gerbil *(Tatera afra)*. During periods of complete anoestrus in females lasting from two to four months no males with spermatic testes were found. The mean testis weight of adult animals during the height of the breeding season varied between 1.5–4.8 gm, but during the anoestrus period this decreased to 0.9 gm. The male Indian gerbil *(Tatera indica)* shows an annual breeding season of ten months and a short period of prohibited spermatogenic activity of two months (PRASAD, 1956).

REFERENCES

ALLANSON, M. 1933. Changes in the reproductive organs of the male grey squirrel *Phil. Trans.* (B). 222: 79.
ALLANSON, M. 1958. Growth and reproduction in the males of two species of gerbil, *Tatera brantsii*, and *Tatera afra. Proc. Zool. Soc. Lond.* 130: 373–396.

AMES, R. G. & VAN DYKE, H. B. 1950. Antidiuretic hormone in the urine and pituitary of kangaroo rat. *Proc. Soc. Exp. Biol. N.Y.* 75: 417–420.

BAILEY, V. 1923. Sources of water supply for desert animals. *The Scientist Monthly.* 17: 66–86.

BARTHOLOMEW, G. A. & CADE, TOM, J. 1956. Temperature regulation, hibernation and aestivation in the little pocket mouse *(Perognathus longimembris). J. Mammal.* 38: 60–72.

BARTHOLOMEW, G. A. & HUDSON, J. W. 1959. Effects of sodium chloride on weight and drinking in the antelope ground squirrel. *J. Mammal.* 40: 354–360.

BURNS, T. W. 1956. Endocrine factors in the water metabolism of the desert mammal, *Gerbillus gerbillus. Endocrinology.* 58: 243–254.

CHAFFEE, R. R. J., PENGELLEY, E. T., ALLEN, J. R. & SMITH, T. E. 1966. Biochemistry of brown fat and liver of hibernating golden mantled ground squirrel *(Citellus lateralis). Can. J. Physiol. Pharmacol.* 44(2): 217–223.

CHEW, R. M. 1955. The skin and respiratory water losses of *Peromyscus maniculatus sonoriensis. Ecology.* 36: 463–467.

CHEW, R. M. & MITCHELL, O. G. 1965. Response of Xeric-adapted rodents to water loading. *Proc. Soc. Exp. Biol. Med.* 120: 336–338.

COLE, P. M., CHESTER-JONES, I. & BELLAMY, D. 1963. *J. Cell. Comp. Physiol.* 38: 165.

CRAIG, H. H. & POULSON THOMAS, L. 1970. Circadian rhythm. II Endogenous and exogenous factors controlling reproduction and hibernation in chipmunks *(Eutamias)* and ground squirrels. *Comp. Biochem. Physiol.* 33(2): 357–383.

DAWSON, W. R. 1955. The relation of oxygen consumption to temperature in desert rodents. *J. Mammal.* 36: 543–553.

DEANESLY, R. & PARKES, A. S. 1933. The oestrous cycle of the grey squirrel *(Sciurus carolinensis). Phil. Trans.* (B) 222: 47.

DEANESLY, R. & PARKES, A. S. 1935. Growth and reproduction in Stoat *(Mustela erminea). Phil. Trans.* (B). 225: 459.

DICKER, S. E. & NUNN, J. 1957. The role of the antidiuretic hormone during water deprivation in rats. *J. Physiol.* 136: 235–248.

EDNEY, E. B. 1960. The survival of animals in hot deserts. The Smithsonian Report for 1959. 407–425.

FARGHALY, A. M. & SHALABY, A. A. 1956. Water economy of the Egyptian gerbil *(Gerbillus pyramidum).* M. Sc. thesis Zool. Dept. Alexandria Univ. Egypt.

GABRE, M. E. A. 1962. Glucose content of the blood of the desert mammal *Jaculus jaculus* in relation to its body water balance. *J. Zool. Soc. Egyp. Bull.* No. 14.

GABRE, M. E. A. & SHALABY, A. A. 1962. Glucose content of the blood of the desert mammal *Jaculus jaculus* in relation to its body water balance. *J. Zool. Soc. Egypt. Bull.* No. 17: 17–20.

GABRE, M. E. A. & SHALABY, A. A. 1964. Studies on the concentration of the total electrolytes and some inorganic ions in the urine of the desert mammal *Jaculus jaculus* in relation to its body water economy. *J. Zool. Soc. Egypt. Bull.* 19.

GEIGER, R. 1957. The climate near the ground. Harvard University Press Cambridge, Mass. (Rev. ed., 2). XXI–494 pp.

GHOBRIAL, L. I. 1968. Physiological adaptations of some desert mammals. Ph.D. Thesis, Zool. Dept. Khartoum Univ. Sudan.

GHOBRIAL, L. I. 1970. A comparative study of the integument of the Camel, Dorcas gazelle and jerboa in relation to desert life. *J. Zool. Lond.* 160, 509–521.

GHOBRIAL, L. I. 1974. Water relation and requirement *of* the Dorcas gazelle in the Sudan. *Mammalia,* 38, No. 1.

GHOBRIAL, L. I. & CLOUDSLEY-THOMPSON, J. L. 1974. Climate, distribution and the behaviour of the Dorcas gazelle in the Sudan. *Ecology* (in press).

GHOBRIAL, L. I. & KAMIL, A. S. 1973. Climate and Seasonal Variations in the breeding of desert jerboa *(Jaculus jaculus)* in the Sudan. *J. Reprod. Fert. Suppl.* 19.

GHOBRIAL, L. I. & AWAD, A. I. 1974. A comparative study of water consumption and conservation in the Sudanese jerboa *(Jaculus jaculus)*, gerbil *(Gerbillus gerbillus)* and the brown rat *(Arvicanthis niloticus)* *J. Zool. Lond. Tran.* (in press).

GOTTSCHALK, C. & MYLLE, M. 1959. Micropuncture study of the mammalian urinary concentrating mechanism, evidence for the counter current hypothesis. *Amer. J. Physiol.* 196: 927–936.

GOTTSCHALK, C. & MYLLE, M. 1960. Osmotic concentration and dilution in the mammalian nephron. *Circulation* 21: 861–868.

GJONNES, B. & SCHMIDT-NIELSEN, K. 1952. Respiratory characteristics of Kangaroo rat blood. *J. Cell. Comp. Physiol.* 39: 147–152.

HAPPOLD, D. C. D. 1966. Breeding periods of rodents in the Northern Sudan. *Rev. Zool. Bot. Afr.* LXXIV: 3–4.

HAPPOLD, D. C. D. 1970. Reproduction and development of the Sudanese jerboa *Jaculus jaculus*. *J. Zool. Lond.* 162: 505–515.

HOFMAN, F. G. 1956. Hormones of water and salt-electrolyte metabolism in vertebrates (ed. JONES, C. & ECKSTEIN, P.). Cambridge Univ. Press. 38–40 pp.

HOWELL, A. B. & GERSH, I. 1935. Conservation of water by the rodent *Dipodomys*. *J. Mammal.* 16: 1–9.

HOWELL, A. B. & GERSH, I. 1962. The role of water in the biology of the antelope ground squirrel *Citellus leucurus*. *Univ. Calif. Publ. Zool.* 64: 1–56.

HOWELL, A. B. & GERSH, I. 1964. Temperature regulation in the round-tailed ground squirrel *Citellus tereticaudus*. *Annales Academiae Scientiarum Fennicae* Series A, IV Biologica 71: 15.

HUDSON, J. W. 1960. Water requirements and thermoneutrality in the antelope ground squirrel *Citellus leucurus*. *Anat. Rec.* 138: 357–358.

HUDSON, J. W. 1962. The role of water in the biology of the antelope ground squirrel *Citellus leucurus*. *Univ. Calif. Publ. Zool.* 64, 1–56.

HUMMEL, V. R. HELV. 1963. *Physiol. Acta* Suppl. XIV, I.

KIRMIZ, J. P. 1962. Adaptation to desert environment. Butterworths, London.

KUTSCHER, C. L. 1968. Plasma volume changes during water-deprivation in gerbils, hamster, guinea pigs and rats. *J. Comp. Biochem. Physiol.* 25: 929–936.

LEE, A. K. & MACMILLEN, R. E. 1959. Utilization of vegetation as a water source by coastal and inland populations of desert wood rat. *Bull. Ecol. Soc. Amer.* 40: 48.

LINDEBERG, R. G. 1952. Water requirements of certain rodents from xeric and mesic habitat. *Contr. Lab. Vertebr. Biol. Univ. Mich.* 58: 1–32.

MALI, J. W. 1956. Transport of water through the human epidermis. *J. Invest. Derm.* 27: 451–469.

MARSHALL, A. J. 1951. The refractory period of testic rhythm and its possible bearing on breeding and migration. *Wilson. Bull.* 63: 238.

MCMANUS, J. J. & JOSEPH, A. M. 1969. Temperature regulation in the Mongolian gerbil, *Meriones unguiculatus*. *Bull. N.J. Acad. Sci.* 14(2): 21–22.

MEASROCH, V. 1954. Growth and reproduction in the females of two species of gerbil, *Tatera brantsi* and *Tatera afra*. *Proc. Zool. Soc. Lond.* 124: 631–658.

MEFFERD, R. B., HALE, H. B. & MARTENS, H. M. 1958. Nitrogen and electrolyte excretion of rats chronically exposed to adverse environments. *Amer. J. Physiol.* 192: 209–218.

MORHARDT, J. E. 1970. Body temperature of White-footed mice *(Peromyscus* sp.) during daily torpor. *Comp. Biochem. Physiol.* 33(2): 423–439.

PRASAD, M. R. N. 1954a. Natural history of the South Indian gerbille *Tatera indica cuvierii*. *J. Bombay nat. Hist. Soc.* 52: 184.

PRASAD, M. R. N. 1954b. Food of the Indian gerbille, *(Tatera indica cuvierii)* *J. Bombay nat. Hist. Soc.* 52: 321.

PRASAD, M. R. N. 1956. Reproductive cycle of the male Indian gerbille *(Tatera indica cuvierii)*. *Acta Zool. Stockholm* 37: 1.

PENGELLEY, E. T. & CHAFFEE, R. R. J. 1966. Changes in plasma Mg concentration

443

during hibernation in golden mantled ground squirrel *(Citellus lateralis)*. *Comp. Biochem. Physiol.* 17(2): 673–681.

PENGELLEY, E. T. & KELLY, K. H. 1967. Plasma potassium and sodium concentrations in active and hibernating golden mantled ground squirrel *(Citellus lateralis)*.

PENGELLEY, E. T. & FISHER, K. C. 1968. Ability of the ground squirrel *(Citellus lateralis)* to be habituated to stimuli while in hibernation. *J. Mammal.* 49(3): 561–562.

RASMUSSEN, A. T. 1915. *Amer. J. Physiol.* 39: 20.

RIAD, Z. M. 1960. Integumentary peculiarities related to life in the desert. *Proc. Egypt. Acad. Sci.* 15: 37–43.

ROBERTS, JANE, C. & SMITH, R. E. 1967. Effect of temperature on metabolic rates of liver and brown fat homogenates (Ground squirrel, rat). *Can. J. Biochem.* 45(II): 1763–1771.

ROLLASON, H. D. 1964. *Amer. Zoologist.* 4: 305.

RUSEV, G. & STEFANOV, S. 1969. About some changes in the ground squirrel *(Citellus citellus)* metabolism in respect to hibernation. *God Sofiiskiya Univ. Biol. KNI Zool. Biokhim Zhiovtn.* 61: 197–208.

SCHMIDT-NIELSEN, B., SCHMIDT-NIELSEN, K., BROKAW, A. & SCHNEIDERMAN, H. 1948. Water conservation in desert rodents. *J. Cell. Comp. Physiol.* 32: 331–360.

SCHMIDT-NIELSEN, B., SCHMIDT-NIELSEN, K., BROKAW, A. & SCHNEIDERMAN, H. 1950a. Evaporative water loss in desert rodents in their natural habitat. *Ecology,* 31: 75–85.

SCHMIDT-NIELSEN, B., SCHMIDT-NIELSEN, K., BROKAW, A. & SCHNEIDERMAN, H. 1950b. Pulmonary water loss in desert rodents. *Amer. J. Physiol.* 162: 31–36.

SCHMIDT-NIELSEN, B., SCHMIDT-NIELSEN, K., BROKAW, A. & SCHNEIDERMAN, H. 1950c. Do Kangaroo rats thrive when drinking sea water? *Amer. J. Physiol.* 160: 291–294.

SCHMIDT-NIELSEN, B., SCHMIDT-NIELSEN, K., BROKAW, A. & SCHNEIDERMAN, H. 1951. A complete account of the water metabolism in Kangaroo rats, experimental verification. *J. Cell. Comp. Physiol.* 38: 165–181.

SCHMIDT-NIELSEN, K., SCHMIDT-NIELSEN, B. & BROKAW, A. 1948. Urea excretion in desert rodent exposed to high protein diet. *J. Cell. Comp. Physiol.* 32: 361.

SCHMIDT-NIELSEN, B. & O'DELL, ROBERTA. 1961. Structure and concentrating mechanism in the mammalian kidney. *Amer. J. Physiol.* 200: 1119–1124.

SCHMIDT-NIELSEN, K., SCHMIDT-NIELSEN, B., JARNUM, S. A. & HOUPT, T. R. 1957. Body temperature of the camel and its relation to water economy. *Amer. J. Physiol.* 188: 103–112.

SCHMIDT-NIELSEN, K. 1964. Desert Animals. Physiological problems of heat and water. Clarendon Press, Oxford.

SPERBER, I. 1944. Studies on the mammalian kidney. *Zoologiska bidrag Fran Uppsale.* 22: 249–431.

STEPHENS, S. F. 1906. California mammals. San Diego, California: West Coast Publ. Co. pp. 351.

TENNENT, D. M. 1946. A study of water losses through the skin in the rat. *Amer. J. Physiol.* 145: 436–440.

TUCKER, VANCE, A. 1962. Diurnal torpidity in the California pocket mouse. *Science.* 136: 380–381.

TURNAGE, W. V. 1939. Desert Sub-soil temperature. *Soil. Sci.* 47: 195–199.

VORHIES, C. T. & TAYLOR, W. P. 1940. Life history and ecology of the white throated wood rat *(Neotoma albigula albigula)* in relation to grazing in Arizona. *Univ. Agric. Exper. Sta. Tech. Bull.* No. 86: 455–529.

VORHIES, C. T. & TAYLOR, W. P. 1946. Water requirements of desert animals in the South-West. *Univ. Arizona Agric. Exper. Sta. Tech. Bull.* 107: 487–525.

WIRZ, H. 1957. The location of antidiuretic action in the mammalian kidney (From the neurohypophysis. ed. H. HELLER) London. Butterworth Sci. Publ.

XXII. NEMATODE PARASITES OF THE INDIAN DESERT RODENTS

by

SYLVESTER JOHNSON

Introduction

BAYLIS' (1936, 1939) classical account of the nematodes occuring in India laid the foundation of a long series of studies in this field. His survey was, however, limited to the confines of the then British India and did not cover areas which were formerly under princely rule. The Indian desert or the Thar desert represents one of the important areas thus left out.

From the zoogeographical point of view, the Indian desert is a continuation of the Saharo-Iranian desert. By far a majority of rodents inhabiting the Indian desert have palaearctic affinities although, geographically, the desert is situated in the Oriental realm. A few rodent species have, however, migrated into the desert from the east as well. I have had an opportunity of examining nematode parasites of the rodents having 'western' affinities, like *Tatera*, as also of those with 'eastern' relationships, e.g. *Funambulus*, *Rattus*, etc. I have endeavoured here to present the systematics of the nematode parasites, new forms in particular, their population pattern, and to discuss if there is any similarity among the nematodes parasitising rodents of different origin.

Material and Methods

The rodents were trapped either in the field or from domestic dwellings depending upon their habitat. These were dissected and the various organs were examined for possible helminth infection. The parasites were killed by hot 70% alcohol and then stored in 4% formaldehyde. The standard and well-known techniques were followed for studying the nematodes. All drawings were made by the help of a camera lucida. In those cases where a population study was undertaken a careful count was kept of the number and the sex of the hosts dissected as well as the number and the sex of the parasites recovered from them.

Nematode Parasites of the Indian Desert Rodents

The rodent families Sciuridae and Muridae, with its two subfamilies Gerbillinae and Murinae, are well represented in the Indian desert. All

445

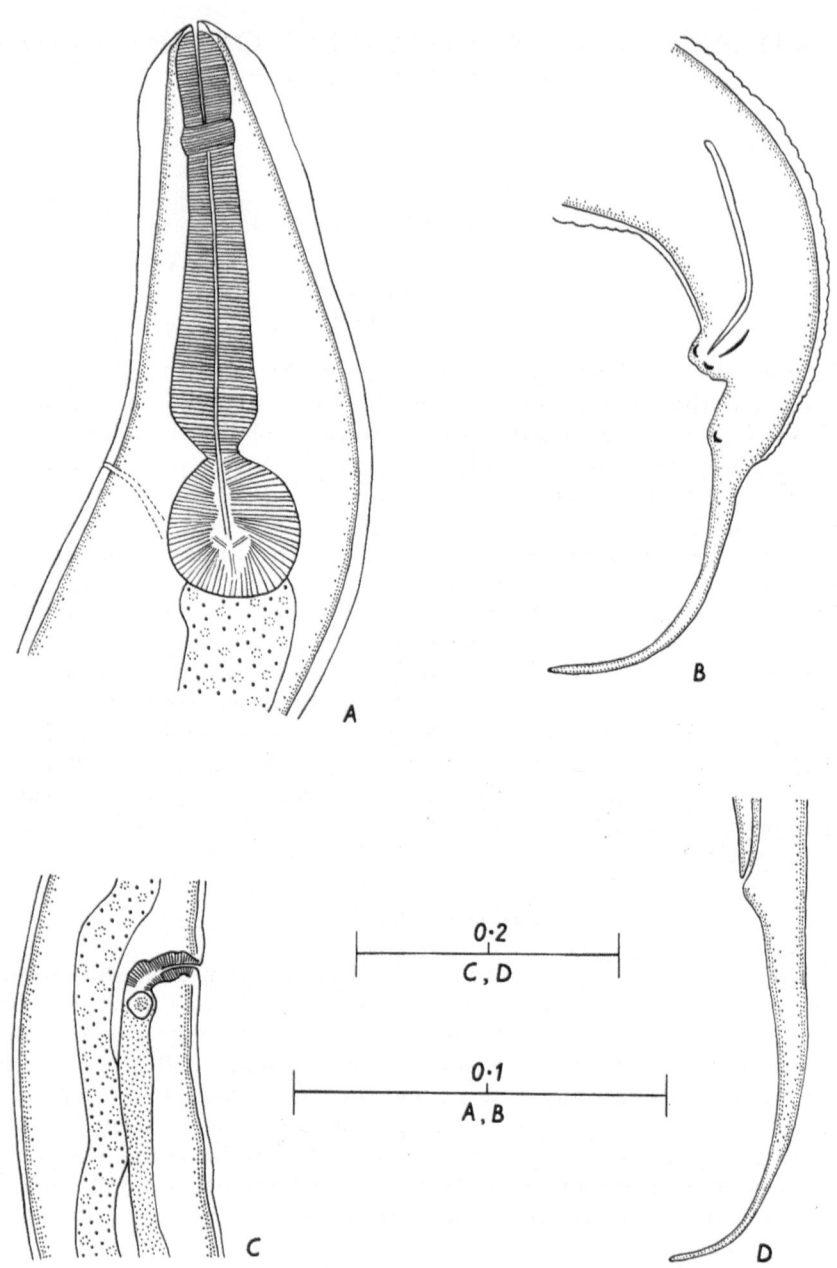

Fig. 1. Syphacia funambuli Johnson, 1967; A. Anterior region of male; B. Caudal region of male, lateral view; C. Vulvar region, lateral view; D. Posterior region of female; lateral view. (All scales are in mm.)

of these have been subjected to preliminary nematological investigations and the important results are summarised below.

The Indian palm-squirrel *Funambulus pennanti* Wroughton is one of the most common rodents found in this region. Our studies with it have yielded so far a new oxyurid and an interesting rictularid nematode.

Syphacia funambuli

The oxyurid *Syphacia funambuli* Johnson, 1967 (JOHNSON, 1967) was recovered from the rectum of *Funambulus pennanti*. The male worm possesses two ventral mamelons or bosses, and a poorly chitinized and unbarbed gubernaculum. The female has a tail with a rounded tip and a post-oesophageal vulva (Fig. 1).

Another species of the genus *Syphacia* Seurat, 1916 described from *Sciurus palmarum* (? *Funambulus pennanti*) from India is *S. sciuri* Mirza and Singh, 1934 (MIRZA & SINGH, 1934). This species is characterised by having in the male a dorsal mamelon also in addition to the two ventral mamelons. In *S. tineri* Khera, 1954 (KHERA, 1954), reported from the brown rat *Rattus norvegicus* from Lucknow, India, the male possesses three ventral mamelons. *S. funambuli* comes closest to *S. eutamii* Tiner, 1948 (TINER, 1948a), described from *Eutamias minimus* from U.S.A., but differs from it in having an unbarbed and much smaller guberna-culum.

Rictularia sp.

Only one male nematode, belonging to the spirurid genus *Rictularia* Froelich, 1802, was once collected from the stomach of the common Indian squirrel *Funambulus pennanti* (JOHNSON, 1969a). It is characterised by a dorsally inclined oral opening; 58 combs; 4 fans; 2, unequal spicules; 9 pairs of caudal papillae, arranged in three groups of three each – first group is precloacal, second postcloacal and the third further behind (Fig. 2).

The worm resembles very much the 5-fanned *Rictularia ratti* Khera, 1954 (KHERA, 1954), described from *Rattus norvegicus* from Lucknow, India, except that whereas the former has 9 pairs of caudal papillae, in the latter all the 6 pairs are postcloacal, since the number of fans on the rictularid male appears to be variable as demonstrated by TINER (1948b) in *R. citelli* McLeod, 1933. *Rictularia onychomis* Cuckler, 1939 (CUCKLER, 1939), recorded amongst others from *Sciurus niger*, has the oral opening more anterior than dorsal as compared to the distinctly dorsally inclined oral opening in our specimen.

447

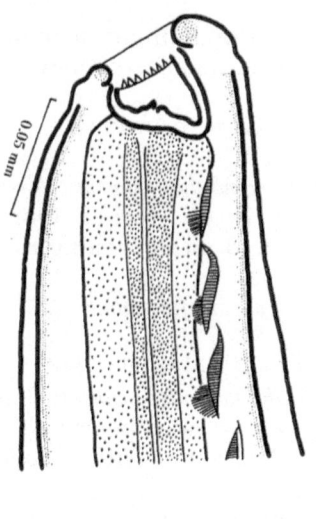

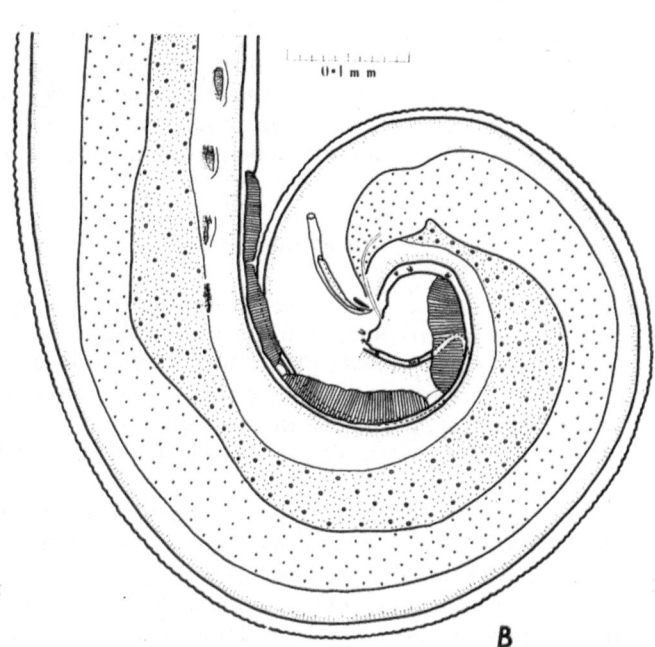

Fig. 2. Rictularia sp. Male (After JOHNSON, 1969); A. Anterior region; B. Posterior region.

Table 1. Helminth infestation of *Tatera indica indica*

	Host			Parasite	
	Examined	Infected	% of infected hosts	No. recovered	Intensity per 100 hosts
Total	64	56	87.5	590	1053.5
♀	37	30	81.08	338	1126.6
♂	27	26	96.2	252	969.2

Thus, although it was not possible to identify this worm with any known species of the genus, it was not considered desirable to assign a specific status to it on the basis of a solitary male alone.

FAMILY MURIDAE

Members of the Family Muridae contribute a great deal to the rodent population of the Indian desert. Both the subfamilies Gerbillinae and Murinae are very well represented.

SUBFAMILY GERBILLINAE

The Indian gerbil or the Antelope-rat *Tatera indica indica* was subjected to a rather intensive investigation. The study yielded three new nematodes, two spirurids and one trichurid, and some interesting data regarding their incidence and population patterns (Tables 1 and 2).

Streptopharagus indicus

The spirurid nematode *Streptopharagus indicus* Johnson, 1969 (JOHNSON, 1969) belonging to the subfamily Ascaropsinae Alicata and McIntosh, 1933, is a fairly common parasite of *Tatera indica indica*. The worms are characterised by the pharynx describing a half-spiral turn at about its middle. In the male the posterior region is coiled spirally and possesses broad, transversely striated caudal alae. Caudal papillae comprise 5 pedunculate pairs – 4 precloacal in 2 groups of 2 each, and one post-cloacal at about the middle of the tail, and 5 pairs of sessile papillae arranged in a group close to the tail tip. There are two unequal spicules and a gubernaculum. In the female there is a small median tesselated area just in front of the anus, and the slightly pre-equatorial vulva is inconspicuous (Fig. 3).

So far *Streptopharagus indicus* appears to be the only record of the genus *Streptopharagus* Blanc, 1912 from rodents in India. It is interesting to note that subsequent to its report BARUŠ *et al.* (1970) reported *S. kutassi*

449

Table 2. Helminth parasites of *Tatera indica indica*

Species	Host			Parasite		
	Examined	Infected	% of infected hosts	No. recovered	Intensity per 100 hosts	% of total helminth infestation
1. *Streptopharagus indicus*	64	29	45.3	316	1089.6	53.5
2. *Trichuris barusi*	64	31	48.4	175	564.5	29.6
3. *Rictularia taterae*	64	18	28.1	54	300.0	9.15
4. *Rodentolepis fraterna*	64	24	37.5	45	187.5	7.2

Table 3. Incidence of *Streptopharagus indicus* in *Tatera indica indica*

	Host			Parasite	
	Examined	Infected	% of infected hosts	No. recovered	Intensity per 100 hosts
Total infection	64	29	45.3	316	1089.6
Sex ratio					
1. Parasites: ♀	64	29	45.3	218	751.7
♂	64	28	43.7	98	350.0
2. Host: ♀	37	13	35.1	180	1384.6
♂	27	16	59.2	136	850.0
3. Parasite-host:					
♀ parasite & ♀ host	37	13	35.1	118	907.6
♀ parasite & ♂ host	27	16	59.2	100	625.0
♂ parasite & ♀ host	37	13	35.1	62	479.6
♂ parasite & ♂ host	27	16	59.2	36	225.0

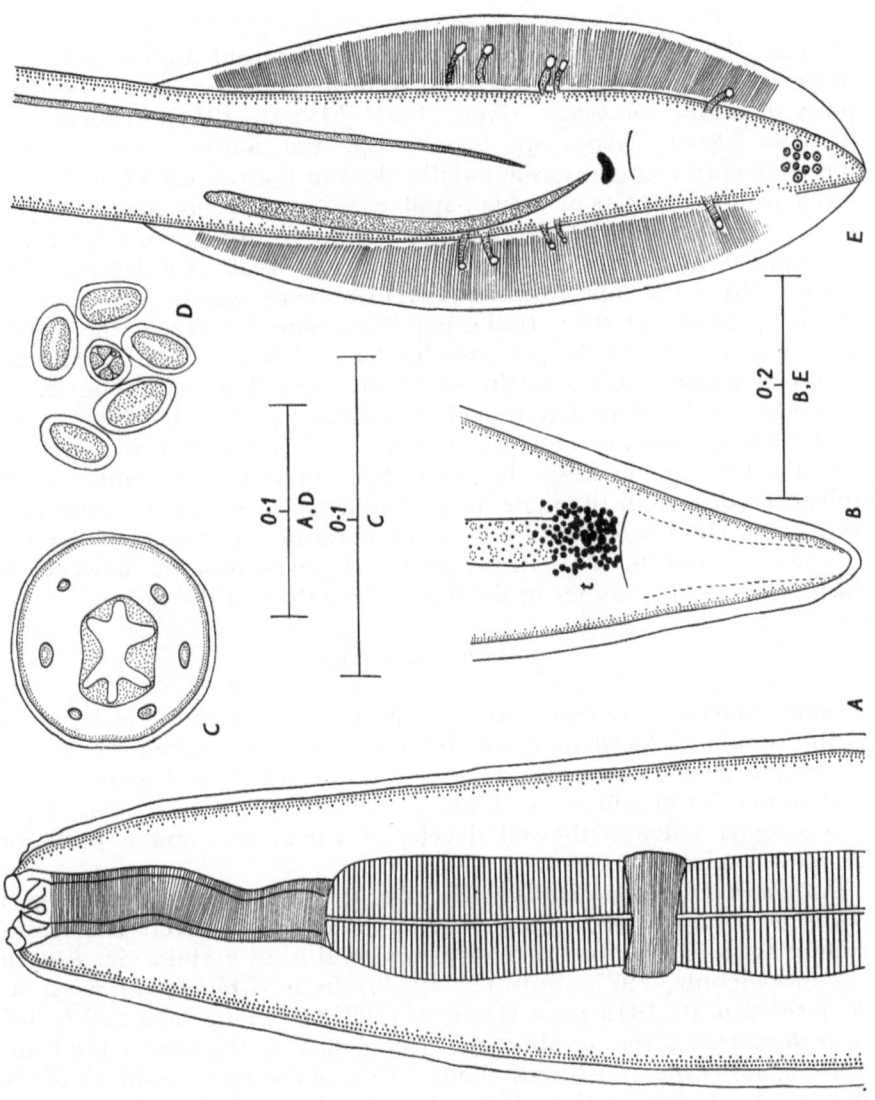

Fig. 3. Streptopharagus indicus Johnson, 1969; A. Anterior region of female, ventral view; B. Posterior region of female, ventral view (t – tesselated area); C. *en face* view; D. Eggs; E. Posterior region of male, ventral view. (All scales are in mm.)

(Schulz, 1927) from the rodent *Marmota caudata* from Afghanistan. Other species of the genus described so far from the rodents belong to different parts of Africa. *S. kuntzi* Myers, 1954 (MYERS, 1954), reported from *Meriones libycus, Acomys* sp. *Gerbillus* sp. and *Rattus* sp. from Egypt, possesses only a single sessile papilla close to the tail tip while *S. indicus* has a group of 5 pairs of sessile papillae. *S. lerouxi* Quentin, 1964 (QUENTIN, 1964), described from *Tatera lobengulae* from Congo, has only the left caudal ala, while *S. geosciuri* Le Roux, 1930 (LE ROUX, 1930), reported from the cape ground squirrel *Geosciurus capensis*, is characterised by the presence of three sessile papillae – one median at the anterior border and a pair at the posterior border, which are absent in *S. indicus.*

Streptopharagus indicus constitutes slightly more than half, about 53.5%, of the total helminth infestation of the Indian gerbil (Table 3). It appears that male gerbils are more susceptible to *S. indicus* infection than the female. The female hosts, however, harbour a larger number of this spirurid worm than the male hosts. The female parasites far outnumber the male. The incidence of the female parasite is not much different in the two sexes of the host. In the case of the male parasite, however, the incidence is much higher in the female host than in the male.

Rictularia taterae

Another spirurid commonly met within the stomach of the Indian gerbil is the rictularid *Rictularia taterae* Johnson, 1969 (JOHNSON, 1969c). It is characterised by a dorsally inclined oral opening, 63–64 pairs of combs and spines out of which 44–45 are prevulvar and 19 postvulvar, a post-oesophageal vulva with well developed vulvar lips and a posteriorly directed vagina (Fig. 4).

Rictularia taterilli Baylis, 1928 (BAYLIS, 1928), described from the gerbil *Taterillus gracilis angelus* from Nigeria, is a comparatively shorter and stouter form. Two other rictualrids reported from gerbils, viz. *Rictularia caucasica* Schulz, 1927 (SCHULZ, 1927b) from *Gerbillus meridianus* and *R. proni* Seurat, 1915 *sensu* GENDRE (1921) from *Gerbillus emini*, differ from *R. taterae* in the number and arrangement of the combs and spines.

Rictularia taterae constitutes about 9.15% of the total helminth infestation of the Indian gerbil (Table 4). The *Rictualria* infection in *Tatera indica indica* is, apparently, not very severe. Although no male parasites were found, it is clear that the incidence of the female parasites is much more in the male host, and, thus, these seem to be more susceptible to *Rictularia* infection than the female gerbils.

Trichuris barusi

The Indian gerbil *Tatera indica indica* usually harbours yet another nematode *Trichuris barusi* Johnson, 1973 (JOHNSON, 1973) located in the

452

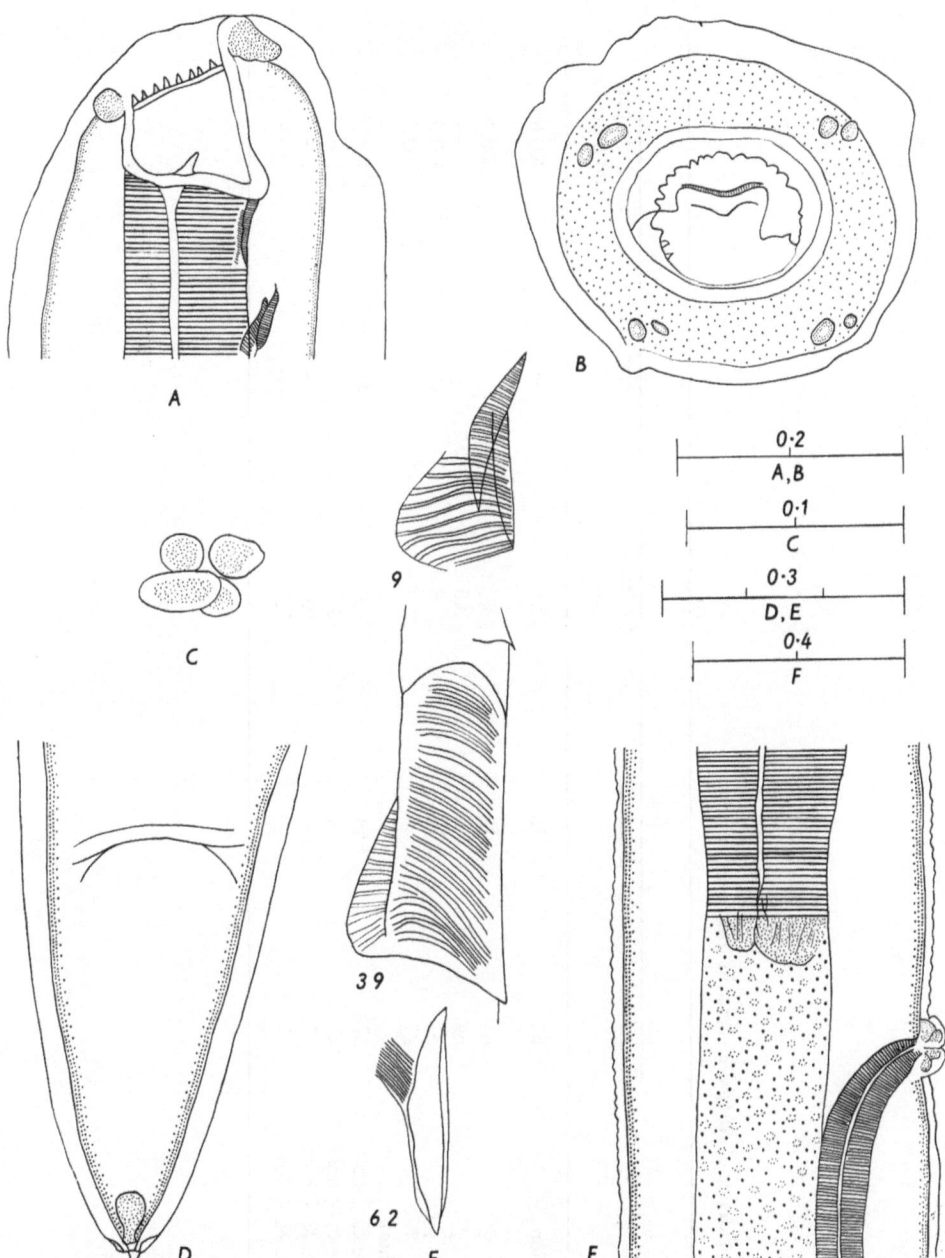

Fig. 4. Rictularia taterae Johnson, 1969. Female; A. Anterior region, lateral view; B. *en face* view; C. Eggs; D. Posterior region, ventral view; E. Combs and spines (numbers refer to the serial number of combs and spines); F. Vulvar region, lateral view. (All scales are in mm.)

Table 4. Incidence of *Rictularia taterae* in *Tatera indica indica*

	Host			Parasite	
	Examined	Infected	% of infected hosts	No. recovered	Intensity per 100 hosts
Total infection	64	18	28.1	54	300.0
Sex ratio					
1. Parasite: ♀	64	18	28.1	54	300.0
♂	64	—	0.0	—	—
2. Host: ♀	37	14	37.8	46	328.5
♂	27	4	13.7	8	200.0
3. Parasite-host					
♀ parasite & ♀ host	37	14	37.8	46	328.5
♀ parasite & ♂ host	27	4	13.7	8	200.0
♂ parasite & ♀ host	37	—	0.0	—	—
♂ parasite & ♂ host	27	—	0.0	—	—

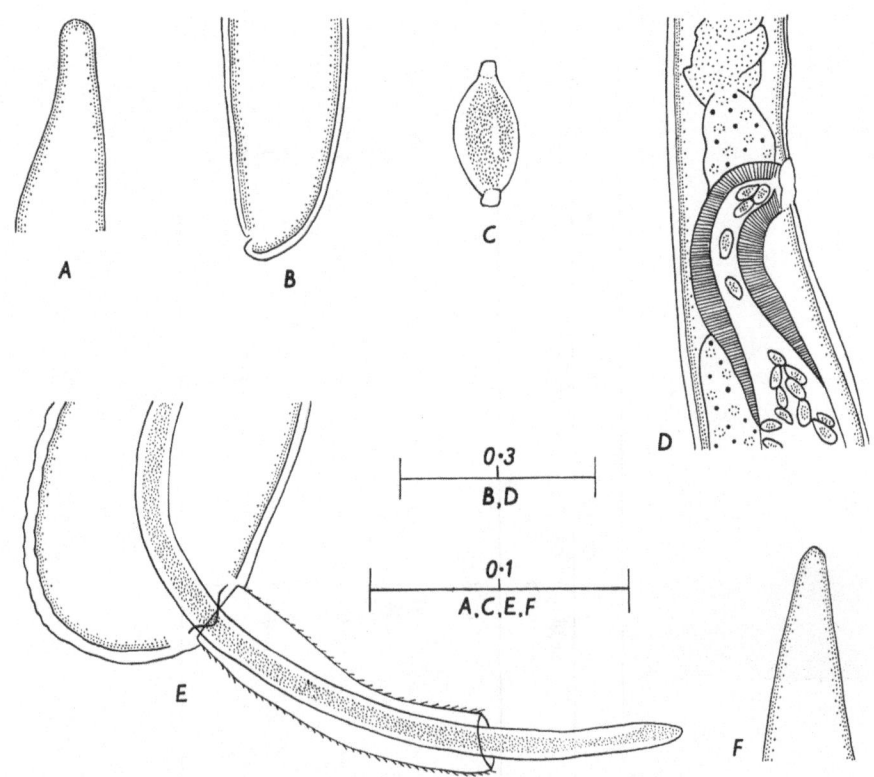

Fig. 5. Trichuris barusi Johnson, 1973; A. Anterior extremity of female; B. Posterior extremity of female, lateral view; C. Egg; D. Vulvar region, lateral view; E. Posterior region of male, lateral view; F. Anterior extremity of male. (All scales are in mm.)

ileum. This trichurid worm is characterised by the absence of any vesicular swelling at its anterior extremity and in having the thin anterior and the thick posterior portions of the body almost equal in length in both the sexes. In the male there is a single spicule which is strongly chitinized except at the rounded tip. The spicular sheath is well developed and provided with small spines. There is also present one pair of caudal papillae situated laterally. The massive testes are divided into rounded parts. The cloaca is shorter than the ejaculatory duct and the cloacal orifice is subterminal. The female has a muscular vagina and a sub-terminal anus (Fig. 5).

Two other trichurid species have been reported from the rodent genus *Tatera. Trichuris carlieri* Goedoelst, 1916, re-described by QUENTIN (1964) from the African gerbil *Tatera lobengulae*, differs from *Trichuris barusi* in having very different dimensions for the spicule, spicular sheath, eggs and the female worms. *Trichuris bahanus* Tenora, 1969 (TENORA, 1969),

455

Table 5. Incidence of *Trichuris barusi* in *Tatera indica indica*

	Host			Parasite	
	Examined	Infected	% of infected hosts	No. recovered	Intensity per 100 hosts
Total infection	64	31	48.4	175	564.5
Sex ratio					
1. Parasite: ♀	64	30	46.8	120	400.0
♂	64	30	46.8	55	183.3
2. Host: ♀	37	14	37.8	82	585.7
♂	27	17	62.9	93	547.05
3. Parasite-host					
♀ parasite & ♀ host	37	14	37.8	54	385.7
♀ parasite & ♂ host	27	17	62.9	66	388.2
♂ parasite & ♀ host	37	14	37.8	28	200.0
♂ parasite & ♂ host	27	17	62.9	27	158.8

Table 6. Incidence of *Rodentolepis fraterna* in *Tatera indica indica*

| | Host | | | Parasite | |
	Examined	Infected	% of infected hosts	No. recovered	Intensity per 100 hosts
Total infection	64	24	37.5	45	187.5
Sex ratio: Host – ♀	37	14	37.8	26	185.7
♂	27	10	37.03	19	190.0

Table 7. Distribution of Helminth parasites in *Tatera indica indica*

Location	Parasite	% of total helminth infestation
Stomach: Pyloric	*Rictularia taterae*	9.15
Ileum: Proximal part	*Streptopharagus indicus*	53.5
Distal part	*Trichuris barusi*	29.6
	Rodentolepis fraterna	7.2

reported from *Tatera indica* from Afghanistan, is a stout and large form with a very long spicule and a comparatively longer spicular sheath. A pair of caudal papillae present on the male of *Trichuris barusi* has not been reported either for *T. carlieri* or *T. bahanus*. Moreover, according to TENORA (1969), *T. bahanus* appears to parasitise only synanthropic populations of *Tatera indica* and, unlike *T. barusi*, is not found in gerbils living in free nature. *T. barusi* differs from *T. muris* Schrank, 1788 (SCHRANK, 1788), reported also from the genus *Meriones*, by its much longer spicule.

Trichuris infection was found in the largest number of hosts although the total number of worms recovered and the intensity of infestation is much lower as compared to *Streptopharagus indicus* Johnson, 1969. *Trichuris barusi* contributes about 29.6% of the total helminth infestation of the Indian gerbil (Table 5).

The male host appears to be more susceptible to *Trichuris* infection than the female. The female parasites number more than double the male parasites. Whereas the former were almost evenly distributed in the two sexes of the host, the latter were more in the male host than in the female.

Rodentolepis fraterna

The hymenolepid cestode *Rodentolepis fraterna* Stiles, 1906 (STILES, 1906) completes the helminth fauna of *Tatera indica indica* in the Indian desert.

457

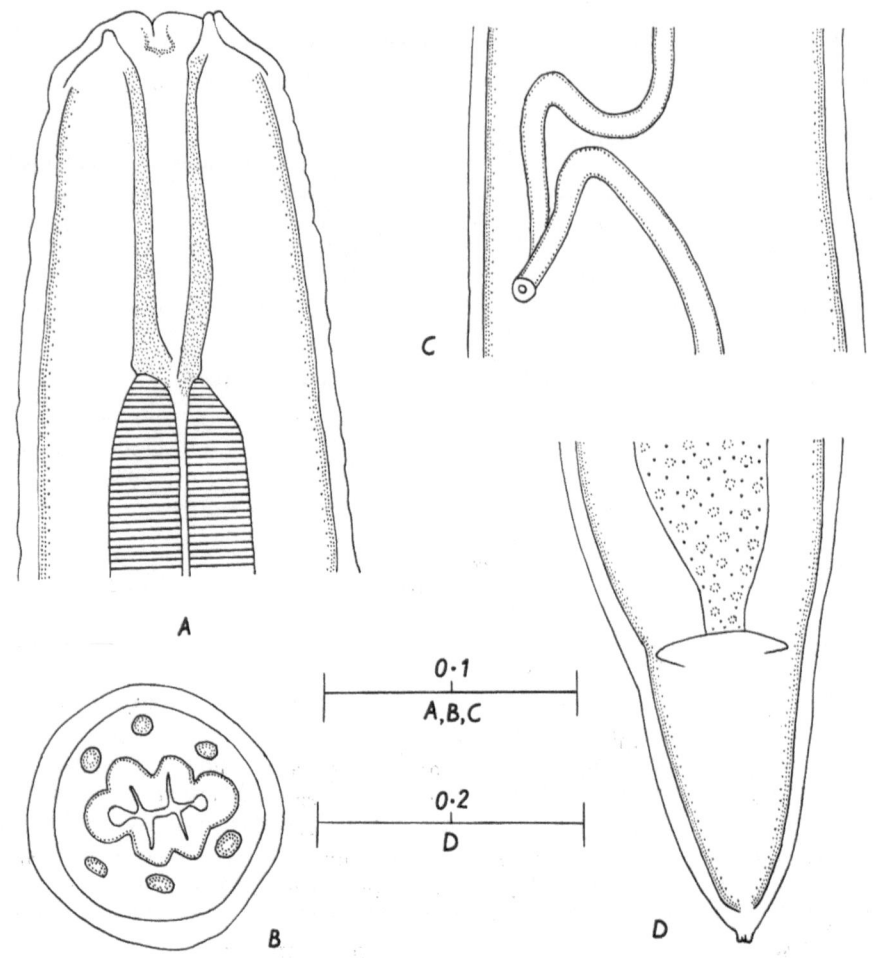

Fig. 6. *Neospirocerca rajasthanensis* Johnson, 1968. Female. A. Anterior region, ventral view; B. *en face* view; C. Vulvar region; D. Posterior region, ventral view. (All scales are in mm.)

It contributes about 7.2% of the total helminth infestation of the Indian gerbil (Table 6). The cestode infestation is the smallest and both the sexes of the host appear to be equally susceptible to it. Although more worms were obtained from the female hosts, the intensity of infestation is slightly higher in the male host.

Distribution of parasites in Tatera indica indica

9.15% of the total helminth infestation of the Indian gerbil, contributed

458

by *Rictularia taterae*, is confined to the pyloric stomach. The heaviest infestation 53.5%, represented by *Streptopharagus indicus*, is harboured by the proximal portion of the ileum. To the distal part of the ileum is confined the residual infestation made up by *Trichuris barusi* (29.6%) and *Rodentolepis fraterna* (7.2%).

SUBFAMILY MURINAE

Two species of the common genus *Rattus*, viz. *R. meltada pallidior* and *R. rattus*, were examined for nematode infection.

Rattus meltada pallidior
Neospirocerca rajasthanensis

The soft-furred field rat or metad *Rattus meltada pallidior* Gray, 1837 harbours a spirurid worm *Neospirocerca rajasthanensis* Johnson, 1968 (JOHNSON, 1968). The genus, akin to *Spirocerca* Railliet and Henry, 1911, is characterised by an hexagonal mouth without definite lips; six cephalic papillae; subsidiary papillae and branches of pulp masses absent; long cylindrical vestibule devoid of spiral thickenings; and an undivided oesophagus. The female has a digitate tail and vulva at the anterior third of the body (Fig. 6). It is interesting to note that the members of the closely related genus *Spirocerca* parasitise carnivora causing tumours in or on oesophagus, stomach, blood vessels and lungs, while *Neospirocerca rajasthanensis* is found free and without the characteristic tumours in the metad.

Rattus rattus
Aspiculuris (Paraspiculuris) ratti

An oxyurid nematode *Aspiculuris (Paraspiculuris) ratti* Johnson, 1969 (JOHNSON, 1969d) is found in the rectum of the common house rat *Rattus rattus*. The worm has a cephalic bulb and four submedian papillae. The cervical alae extend upto the end of oesophagus, terminate into sickle-shaped margins and then continue as narrow lateral alae. There is a conical and tapering tail. The inconspicuous vulva is pre-equatorial, and the vagina runs anteriorly for a short distance before taking a posterior course (Fig. 7).

The oxyurid genus *Aspiculuris* Schulz, 1924 (SCHULZ, 1927a) is a fairly wide spread group of nematode parasites of rodents. It has been reported from Europe, Siberia, China, Japan, India, Pakistan, U.S.A. and Argentina. The very extensive host list includes various species of the genera *Mus*, *Rattus*, *Cricetus*, *Apodemus*, *Microtus*, *Chiromys*, *Arctomys*, *Jaculus*, *Clethrionomys*, *Peromyscus*, *Rhombomys*, *Neotoma* and *Caviella*.

AKHTAR (1955) divided the genus *Aspiculuris* into five subgenera

459

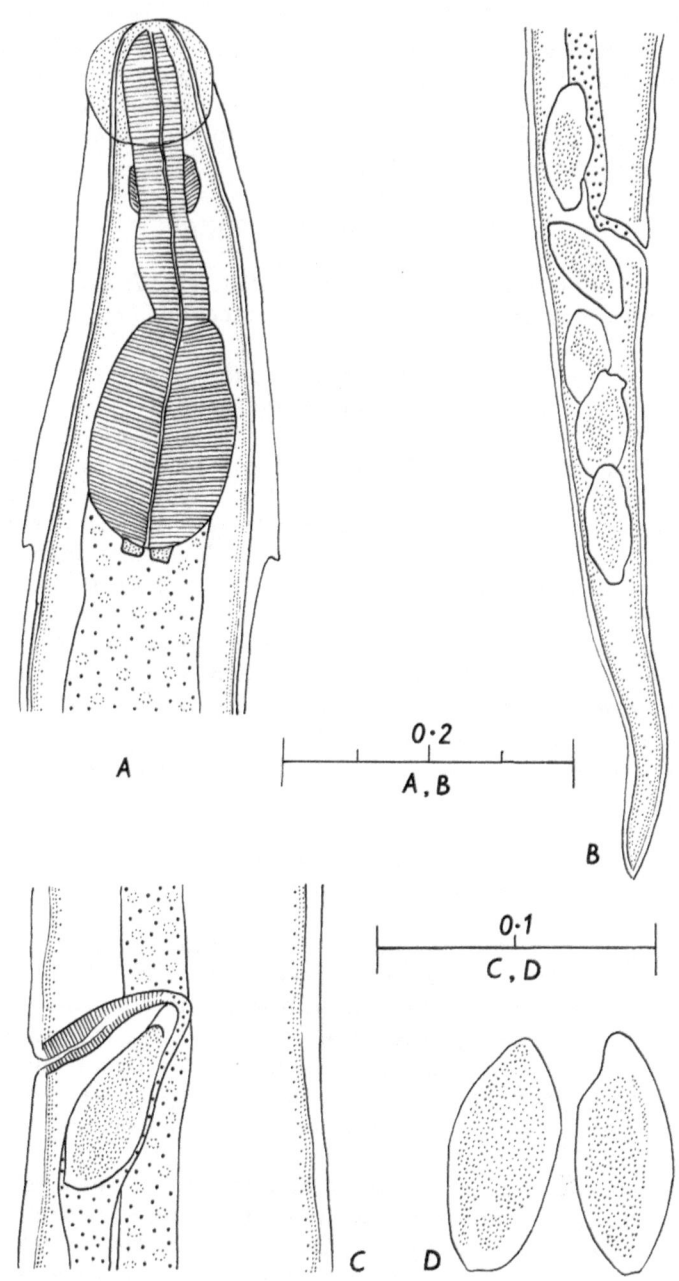

Fig. 7. *Aspiculuris (Paraspiculuris) ratti* Johnson, 1969. Female. A. Anterior region; B. Posterior region; C. Vulvar region; D. Eggs. (All scales are in mm.)

460

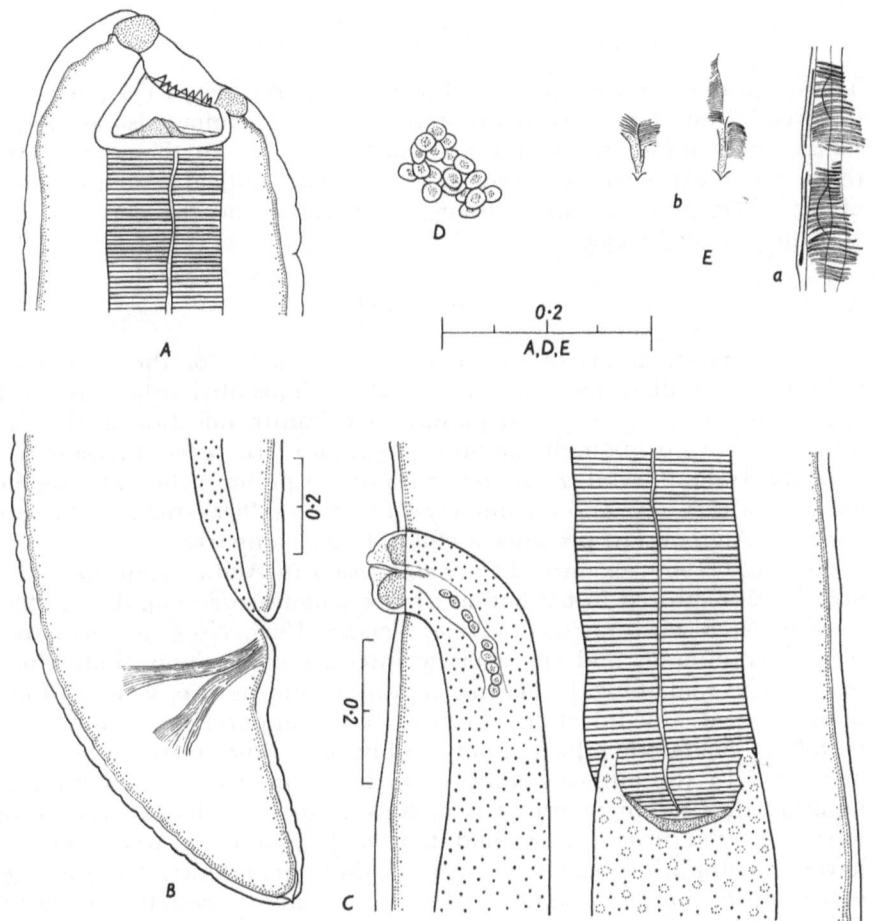

Fig. 8. *Rictularia ratti* Khera, 1954. Female. A. Anterior end; B. Posterior end; C. Vulvar region; D. Eggs; E. Combs and spines: a. prevulvar, b. postvulvar. (All scales are in mm.)

according to the presence or absence of the cephalic bulb and the lateral alae, and the nature of the termination of the cervical alae and their relationship with the lateral alae. In the subgenus *Paraspiculuris* Akhtar, 1955 the cervical alae have sickle-shaped termination and are continuous posteriorly with the lateral alae. *Aspiculuris (Paraspiculuris) pakistanica* Akhtar, 1955 *(l.c.)*, described from *Rattus rattus rufescens* from Lahore, Pakistan, compared to *A. (P.) ratti*, is a stouter and larger form and possesses a comparatively posterior vulva and nerve ring, smaller eggs and a tail which narrows down abruptly to a point near the apex.

461

Rictularia ratti

The spirurid worm *Rictularia ratti* Khera, 1954 (KHERA, 1954), originally described from *Rattus norvegicus* from Lucknow, India, also parasitises *Rattus rattus* found in the Indian desert (Fig. 8). In the worms from these two hosts a slight variation is, however, noticable in the length of the worm and the tail, position of the vulva and the nerve ring, as also the size of the eggs.

Discussion

Detailed investigations of the helminth infestation of the rodents inhabiting the Indian desert are sadly lacking. This obviously restricts the emergence of any generalised pattern of helminth infection in this very important group of small mammals. Author's (In press) investigations with the Indian gerbil *Tatera indica indica* are perhaps the only detailed analysis of this type and fortunately TENORA's (1969) study of the same gerbil in Afghanistan provides useful data for comparison.

TENORA (1969) examined 53 *Tatera indica* in Afghanistan. Out of 41 gerbils, all captured in tea-houses from a synanthropic population, 29% were infected with *Trichuris bahanus* Tenora, 1969. No other nematodes were encountered, and the average intensity of invasion (individuals) was 3. The remaining 12 gerbils, captured from nature, were all found to be free of any infection. TENORA *(l.c.)* apparently examined the gerbils for nematode infection only and hence did not mention any other helminth. In the author's study carried out on *Tatera indica indica* inhabiting the Indian desert, it is interesting to note that all the gerbils were collected from nature and that 87.5% of these were found to harbour helminth infection comprising three nematodes, *Streptopharagus indicus, Rictularia taterae* and *Trichuris barusi*, and a cestode *Rodentolepis fraterna*. Mostly the infection was multiple with two or more species parasitising the same individual and only sometimes a single helminth species was encountered in one gerbil.

It appears, therefore, that the helminth fauna of the Indian gerbil depends a great deal on the ecological conditions of the host. In Afghanistan only the synanthropic population of the gerbils is infested and that too only by the trichurid worms, while the gerbils in the natural pastures are, apparently, altogether devoid of any helminth infestation. In the Indian desert, on the other hand, the gerbils in free nature are very heavily infested with helminths comprising four species.

On the basis of our studies on *Tatera indica indica* in the Indian desert, it can be concluded that the helminth infestation is very common in the Indian gerbil and the helminth fauna consists of three nematodes and one cestode. Although the male gerbils are more susceptible to helminth infection than the females, the intensity of infestation is, however, higher in the latter. It is clear that more than half of the total helminth infestation is confined to the proximal part of the ileum. While both the sexes of the gerbil are more or less equally infested with the cestode *Rodentolepis fraterna*, the male gerbils are more susceptible to infection by *Streptopharagus indicus* and *Trichuris barusi* and the female gerbils are more prone to *Rictularia taterae* infection. The intensity of infestation of the various helminth species, beginning with the highest, is in the following order: *S. indicus*, *T. barusi*, *R. taterae* and *Rodentolepis fraterna*. Finally, the Indian gerbil inhabiting different ecological localities harbours helminth fauna different both in nature and degree.

There can be little doubt regarding the importance of detailed helminthological investigations of the various rodents found in the Indian desert. Unless and until such studies are undertaken and sufficient data is available many important and pertinent questions like that of the host specificity, the relationship of nematode parasites to the zoogeography of the hosts, the possible influence of the nematode numbers on the mortality rate of the rodents, etc., must remain unanswered. Thus, for instance, the distribution pattern of the various nematode genera discussed here would strongly suggest that some of these at least are restricted to hosts which inhabit the deserts, viz. *Tatera*, *Meriones*, etc., whereas the members of the subfamily Murinae, which are wide-spread over the world, appear to harbour nematode genera quite cosmopolitan in distribution. Such tempting conclusions, however, can and should be drawn only if supported by sufficient data, which, in turn, will accumulate only as a result of intensive and extensive helminthological investigations.

REFERENCES

AKHTAR, S. A. 1955. On nematode parasites of rats and mice of Lahore, with some remarks on the genus *Aspiculuris* Schulz, 1924 and two new species of the genus. *Pakistan J. Sc. Res.*, 7(3): 104–111.

BARUŠ, V., KULLMANN, E. & TENORA, F. 1970. Neue Erkenntnisse über Nematoden und Acanthocephalen aus Nagetieren Afghanistans. *Věst. Čs. spol. zool.*, 34: 263–277.

BAYLIS, H. A. 1928. On a collection of nematodes from Nigerian mammals (chiefly rodents). *Parasit.*, 20: 280–304.

BAYLIS, H. A. 1936. The Fauna of British India, including Ceylon and Burma. Nematoda, Vol. 1. (Ascaroidea and Strongyloidea). London, 408 pp.

BAYLIS, H. A. 1939. The Fauna of British India, including Ceylon and Burma. Nematoda, Vol. 2. (Filarioidea, Dioctophymoidea and Trichinelloidea). London, 274 pp.

CUCKLER, A. C. 1939. *Rictularia onychomis* n.sp. (Nematoda: Thelaziidae) from the grasshopper mouse *Onychomis leucogaster* (Weid.). *J. Parasitol.*, 25: 431–435.

GENDRE, E. 1921. Sur deux espèces de Nématodes africains. *Actes Soc. Linnéenne Bordeaux*, 73: 28–36.

JOHNSON, S. 1967. A new nematode of the genus *Syphacia* (Oxyuroidea) from the squirrel *Funambulus pennanti* from Rajasthan, India. *Proc. zool. Soc., Calcutta*, 20: 83–85.

JOHNSON, S. 1968. A new spirurid genus from the soft-furred field rat *Rattus meltada* (Nematoda, Spirurinae). *Rev. Brasil. Biol.*, 28(2): 111–114.

JOHNSON, S. 1969a. On some nematodes belonging to the genus *Rictularia* (Nematoda: Spiruroidea). *Rev. Biol. Trop.*, 15(2): 289–297.

JOHNSON, S. 1969b. A new nematode of the genus *Streptopharagus* (Spiruroidea) from the Indian gerbil *Tatera indica indica*. *J. Helminth.*, 43(3/4): 343–346.

JOHNSON, S. 1969c. A new nematode of the genus *Rictularia* (Spiruroidea) from the Indian gerbil *Tatera indica indica*. *Folia parasit. (Praha)*, 16: 371–374.

JOHNSON, S. 1969d. On a new oxyurid nematode of the genus *Aspiculuris* from the common house rat *Rattus rattus*. *Indian Jour. Helminth.*, 21(2): 147–149.

JOHNSON, S. 1973. A new trichurid nematode from the Indian gerbil *Tatera indica indica*. *Folia parasit. (Praha)*, 20: 275–277.

JOHNSON, S. Helminth infestation of the Indian gerbil *Tatera indica indica*. *Věst. Čs. spol. zool.* (In Press).

KHERA, S. 1954. Nematode parasites of some Indian vertebrates. *Indian Jour. Helminth.*, 6(2): 27–133.

LE ROUX, P. L. 1930. A spirurid (*Streptopharagus geosciuri* sp. nov.) from the stomach of the cape ground squirrel (*Geosciurus capensis*). *Rept. Director Vet. Services Animal Ind., Onderstepoort*, 16: 201–204.

MIRZA, M. B. & SINGH, S. N. 1934. *Syphacia sciuri* n.sp., a new oxyurid worm from *Sciurus palmarum*. *Current Sc.*, 2(9): 345–346.

MYERS, B. J. 1954. Helminth parasites of reptiles, birds and mammals in Egypt. 1. *Streptopharagus kuntzi* sp. nov. from rodents with a review of the genus. *Can. J. Zool.*, 32: 366–374.

QUENTIN, J. C. 1964. Nématodes parasites de rongeurs du Congo. *Expl. Parc. natu. Upemba Miss.* G. F. de Witte, Brussels, 69: 73–91.

SCHRANK, F. P. 1788. Verzeichniss der bisher hinlänglich bekannten Eingeweidewürmer, nebst einer Abhandlung über ihre Anverwandtschaften. München, 116 pp.

SCHULZ, R. E. S. 1927a. On the genus *Aspiculuris* Schulz, 1924, and two new species of it – *A. dinniki* and *A. asiatica* from rodents. *Ann. Trop. Med. Par.*, 21(2): 267–275.

SCHULZ, R. E. S. 1927b. Zur Kenntnis der Helminthenfauna der Nagetiere der Union S.S.R. II. Spirurata Railliet et Henry, 1914. *Trudy. Gosudarstv. Inst. Eksper. Vet.*, 4(2): 36–65.

SEURAT, L. G. 1915. Sur les Rictulaires des Carnivores du Nord-Africain et les affinités du genre *Rictularia*. *C.R. Soc. Biol.*, 78: 318–322.

STILES, C. W. 1906. Illustrated key to the cestode parasites of man. *U.S. Pub. Health Service, Hyg. Lab. Bull.*, 25: 1–104.

TENORA, F. 1969. Parasitic nematodes of certain rodents from Afghanistan. *Věst. Čs. spol. zool.*, 33(2): 174–192.

TINER, J. D. 1948a. *Syphacia eutamii* n.sp. from the least chipmunk, *Eutamias minimus*, with a key to the genus. *J. Parasit.*, 34: 87–92.

TINER, J. D. 1948b. Observations on the Rictularia (Nematoda: Thelaziidae) of North America. *Trans. Am. Micr. Soc.*, 67: 192–200.

XXIII. ECOLOGY OF DESERT RODENTS OF THE U.S.S.R. (JERBOAS AND GERBILS)*

by

N. P. NAUMOV & V. S. LOBACHEV

Introduction

The rodents are an integral component of the desert biogeocenose. Soviet zoologists have been conducting considerable research work on the rodent fauna of the U.S.S.R. for the last several decades. In this communication, we have made an attempt to collate the available information on the ecology of two groups of rodents of the Russian deserts, viz., the Jerboas and the Gerbils. Under every rodent species, as far as possible, we have included its distributional range, habitat preference, activity pattern, association and interaction with other rodent species, food, various aspects of reproduction, hibernation and predators. The species** which are included in this Chapter are:

Family   DIPODIDAE

Subfamily   Cardiocraniinae

      Cardiocranius paradoxus
      Salpingotus crassicauda

Subfamily   Allactaginae

      Allactaga euphratica
      A. elater
      A. sibirica
      A. major
      A. severtzovi
      A. bobrinskii
      Alactagulus pygmaeus
      Pygeretmus platyurus

---

* This article was originally written in Russian and later translated by the authors into English. The authors' English text has, subsequently, been subjected to extensive editorial modifications. We thank our friend and colleague Dr. R. P. DHIR who willingly helped us in finalising the legends of the figures. Because of the language barrier, necessitating handling of the article by various persons at various stages, and in view of the perpetual race against time, we are painfully aware that much literary deficiency is persisting in this final form of the article. We owe a special debt of gratitude to our Publishers, Dr. W. Junk B.V. for their understanding and cooperation in including this chapter even at the very late page-proof stage of the book. – Eds.

** The nomenclature, sequence of arrangement and classification of the various species adopted by the authors do not always conform to ELLERMAN & MORRISON-SCOTT (1951).

Subfamily Dipodinae

    Dipus sagitta
    Scirtopoda telum
    Jaculus turcmenicus
    Eremodipus lichtensteini
    Paradipus ctenodactylus

Family MURIDAE

Subfamily Gerbillinae

    Meriones persicus
    M. vinogradovi
    M. tamariscinus
    M. blackleri
    M. unguiculatus
    M. meridianus
    M. libycus
    M. crassus zarudnyi
    Rhombomys opimus

The arid zone of the U.S.S.R.

In the Soviet Union, desert and semidesert regions make up almost 12 per cent of the country's total area. The proportion rises to between 50 and 60 per cent in the Central Asian Republics of Uzbekistan, Kazakhstan and Turkmenistan. The territory includes: 1. the dry steppes, semi-deserts and deserts of the Armenian and Azerbaidjanian uplands; 2. the semi-deserts of Azerbaidjan and the North-Caspian plains; 3. the Volga-Ural sands; 4. the semi-deserts of the North-West Caspian region; 5. the plateaus of Usturt, Mangishlack and deserts of Betpackdala; 6. the deserts of Kazakhstan and Middle Asia; 7. the upland steppes and semi-deserts of Tian-shan and Pamir; and 8. the upland steppes and semi-deserts of South TransBaikal.

The arid territory of the U.S.S.R. belongs to the category of Moderate Zone deserts. The region has a strong continental climate with very low temperature regime: 25–30 °C (upto 50°) in summer, and from —37° to —45 °C in winter. Annual precipitations are limited to 70–250 mm per year. Summers are very hot and the rate of evaporation is very high. Annual and even daily fluctuations of the atmospheric humidity and of air and soil temperatures are very high. The deserts of the U.S.S.R. differ from those of Central Asia and the Mediterranian region in respect of the seasons of precipitation. Rain and snow fall in the deserts of the U.S.S.R. during the cold period of the year, chiefly in spring and autumn. The seasonal distribution of precipitation is more even in the northern deserts. Usually the evaporation of moisture exceeds precipitation by 7 to 8 times.

The intensity of solar radiation on the slopes of the Middle Asian deserts is of the order of 140 Kcal/cm²/year and it is quite sufficient for the growth and development of desert plants.

466

The most pronounced continentality of the climate exists in the north-eastern deserts, where the difference between the summer maximum (+50°) and the winter minimum temperature (—37°) is about 87°C. The temperature of the sand surface reaches 80°C on clear summer days.

In the U.S.S.R. there are different types of deserts viz., sandy, clayey, rocky and saline.

The sandy deserts may be of two types, according to the thickness of the sand stratum and the forms of relief, viz. ridgy and hilly. They differ also in respect of the density of the cover and biomass of plants.

The large sandy deserts of the Volgo-Ural Sands are located between the Volga and the Ural rivers. Large territories in Kazakhstan, Uzbekistan and Turkmenistan are occupied by the sandy terrains of Kizil-Kum (near Aral Sea), North and South Kara-Kum, Mujum-Kum and some other sandy patches like those of Sary-Ishik-Otrau (near lake Balkhash), Great and Little Barsuki (near Aral Sea) and many other smaller areas. For the most part these sandy deserts emerged in the Pleistocene. The characteristic plants of the sandy deserts are ephemeral and xerophytic, such as white saksaul *(Haloxylon persicum, H. aphyllum)*, *Calligonum, Salsola richteri, Ammodendron*, etc.

Clayey or ephemeral deserts are located in alluvial depressions alongside the valleys of large rivers or around salt marshes. These are prominent features of the landscape in the Turgai depression and in the ancient valleys of the Amu-darya and the Syr-darya rivers. A broad strip of such deserts, a few hundred kilometers in length is located along the northern side of the Kopetdag and the Alatau mountains. This type of desert has more surface water resources than the sandy deserts. Wormwoods *(Artemisia* sp.) and ephemeral species of the Graminae family *(Poa bulbosa, Bromus*, etc.) are the most important plants of these terrains. Only lichens and some blue-green algae inhabit the 'takirs' – which are the bare, flat, asphalt-like surfaces of the shallow depressions.

Animals of the clayey deserts are represented by many specialized species, such as *Phrynocephalus helioscopus, Eremias arguta* (lizards), *Allactaga major, A. sibirica, Alactagulus pygmaeus* (jerboas), *Gazella subgutturosa* and *Saiga tatarica* (antelopes), *Meriones libycus erythrourus, M. c. zarudnyi, Rhombomys opimus* (gerbils), *Microtus afghanus, M. socialis* (voles) and some snakes *(e.g. Ancistrodon halys)*.

The rocky deserts were more widely distributed in the past. They chiefly occupy the plateaus of Betpackdala, Usturt and the small mountains in Kizil-Kum; these deserts are located also along the foothills of the Kopetdag and the Gissar mountains. These are called 'rocky' or 'gypsum' deserts, as gypsum crystals occur on their surface. This type of desert is especially characterised by deficiency of ground water. The central parts of these deserts do not usually have any water resources; only in spring some ephemeral ponds may be found there. The sub-soil waters are located very deeply.

467

The plant cover of the rocky deserts is very thin and occupies only 1/10 or 1/5 of the soil surface. Usually only one species, such as *Salsola arbuscula*, *Anabasis salsa* or *Nanophyton erinaceum* dominates. Ephemeral plants like *Tulipa*, *Carex physodes*, *Salsola rigida* and *Reaumuria fruticosa* are scanty. Large numbers of antelopes *(Saiga, Gazella)* graze on these deserts seasonally. The jerboas – *Allactaga bobrinskii*, *A. elater*, *A. major*, *A. severtzovi*, *Alactagulus pygmaeus* and *Pygeretmus* are numerous here. The desert dormouse, *Selevinia betpackdhalensis*, is an endemic species of this type of desert.

The saline-marsh deserts are situated on the sides of the old sea transgressions. The larger deserts of this type are located along the shores of the Caspian and the Aral seas (Tentjak-sor, Tshelkar-Tengize and Tshubar-Tengize in the north-west of Kazakhstan; Barsa-Kelmes in Usturt, Sarykamish to the south-west of the Aral sea, saline-marshes at the lower part of the Chu river, etc.). The area occupied by the saline-marshes is not large and these are often located as comparatively smaller patches in the other types of deserts. The sub-soil water is at a relatively higher level in the saline-marshes and after rains or after the thawing of snow these tracts are transformed into shallow ponds or lakes. The intensive evaporation of moisture through the surface of these deserts makes their microclimate more moderate in comparison to the climates prevailing in the other types of deserts. The depressions in the saline marshes do not usually have any plants, these being covered with layers of salts. But on areas surrounding the depressions many halophytes such as *Salicornia herbacea*, *Salsola*, *Halocnemum strobilaceum*, *Kalidium caspicus*, *Petrosimonia*, etc. are found.

The saline-marshes usually have a small number of vertebrate species. Rodents and small carnivores are not found here as the habitat is not fit for making burrows and other shelters. The antelopes can only graze for a limited time upon the halophytic vegetation. Only the eurytopic jerboas, such as *Alactagulus pygmaeus* and *Pygeretmus* can exist here more or less continuously. Because of its characteristic features the saline-marsh desert has, practically, no narrowly specialised species of vertebrates. After the breeding season large numbers of larks *(Melanocorypha, Calandrella)*, snipes *(Charadrius asiaticus, Burhinus oedicnemus)* and some other small birds exploit the saline-marshes. The animal population is richer in places where the desert shrubs *(Tamarix, Haloxylon)* occur.

Jerboas

Satunin's pygmy jerboa

Cardiocranius paradoxus Satunin, 1903

This species was previously known only from the Mongolian deserts (the region of Nan-Shan, Northern and Central Gobi). Then it was found in Tuva in 1962 (BERMAN, 1962), and in 1971 – 1600 km westwards – in Kazakhstan (the north foothills of Bektauati mountains). Recently KAPITONOV (1972) has reported its occurrence in two strictly localized areas each of 1–2 km in length. Undoubtedly, then, it is a specialized form, adapted to rocky desert.

This species of jerboa is characterised by its large head as in *Salpingotus*, with long thick vibrissae and a long, thin, fur-trimmed tail. Its body shows conspicuous fat deposits in summer. It has greatly developed bullae tympani and smaller os interparietale in comparison to other rodents. As in Zapodinae, the metatarsus bones of *Cardiocranius* and *Salpingotus* are not arranged in a way suitable for jumping, thereby distinguishing them from other jerboa species. This character also points to the antiquity of the subfamily *Cardiocraniinae*.

The abundance of this thinly distributed rodent not only depends on the availability of its preferred habitat, viz., gravelly plain along with cereals to feed upon but also on the presence of burrows of other rodent species where it may hibernate during the severe winter months.

On a 224 km motor journey through Kazakhstan 154 jerboas were encountered of which 9 (5.8%) were *Cardiocranius paradoxus* (1 animal per 27 km of auto-route).

The mode of living of *Cardiocranius* has not been studied in any detail. The first specimens from North Gobi were obtained by Kasakevich in the sands with the burshwoods of *Nitraria schoberi*, but, considering the morphology of its extremities, it was concluded that this species is an inhabitant of clayey deserts. The areas of dwelling of *Cardiocranius* in Kazakhstan are the same viz. terrains of gentle slope (5–15°) in the foothills.

On the plane surface of this sandy terrain there are some thinly growing plants like *Nanophyton erinaceum*, *Artemisia*, *Stipa* and *Salsola*, (covering not more than 30%), the height of the plants being 8–25 cm. Apparently, the presence of the plane surface, strewn with pebbles and having some species of Graminae, is the necessary condition of living for this jerboa. In the opinion of BERMAN (1962), *Cardiocranius* rarely digs burrows (refuges) itself, but makes use of the burrows of other jerboas, *Lagurus* and field-voles *(Microtus socialis)*.

Obviously, *C. paradoxus* is a seed-eater. Seeds of feather-grass *(Stipa)* were found in the stomach of 5 jerboas from Kazakhstan, and when in captivity the animals ate cereal grains with thin seed-coats and soft scales like *Festuca sulcata*, *Poa bulbosa*, *Setaria* and, especially *Echinochloa grus* (galli) and refused to eat the green parts of the plants, beetles (Coleoptera), locust (Acrididae), flies (Diptera) and spiders (Aranei). Like many other jerboas, they do not drink water.

Towards the middle of summer *Cardiocranius paradoxus* grow considerable amount of body fat (on the tail – upto 8 mm, on the stomach – upto 3.5 mm).

In the Kazakhstan desert the testicles of male *C. paradoxus* are found to be conspicuously developed even in July and a female with four enlarged nipples was encountered on July 18 (KAPITONOV, 1972). Thus, only one spring reproduction is likely to take place in this species.

The shedding of hair in the younger as well as in the grown-up animals seems to begin from the head. In captivity the thick and fluffy fur of *Cardiocranius* has been found to turn greasy and the animals have always been found to 'powder' its coat in the dust.

Observations in nature (18–25 July) revealed that it forages from 2200 hours till about dawn (0400 hours). The leaner individuals may remain active for a longer time (till 0600 hours in the morning) and these may also start their surface activity somewhat earlier.

The animals keenly react to sounds but less to smells. In captivity their low sqeaking voice can be heard during fights. While attacking each other, the animals jump up high, pushing each other in the air, or bite each other at the root of the tail. During feeding *C. paradoxus* move by little jumps touching the ground with the ends of its long vibrissae directed downwards. The rodent searches for seeds in the sand and beneath pebbles with their hind legs, leaning on their forelegs and tail. Like *Salpingotus* and *Pygeretmus* this species also faces danger and rarely ever runs away.

A small number of lice (Anoplura) and chiggers (Trombidiiformes) have been found to occur as parasites on *Cardiocranius* in Kazakhstan.

THICK-TAILED PIGMY JERBOA

Salpingotus crassicauda Vinogradov, 1924

A very rare species, it is distributed in western Mongolia, north-western China and the U.S.S.R. The species has been reported from the Zaisan depression valley along both banks of the river Black Irtish, the lower reaches of the left bank of the Irtish, the southern territory of Balkhash lake and, in the recent past, from the Aralskie Kara-Kum to the northeast of the Aral sea. Its subspecies, initially described as a new species, *S. geptneri*, is known to inhabit the north-western regions of Kizil-Kum.

The biology of *Salpingotus crassicauda* has not been studied to any great extent. It has been mainly found in the *Stipa*-sandy steppe in the hillocky sands and in sandy areas overgrown with vegetation. In the Zaisan-lake depression it dwells in the sands in association with *Calligonum* and *Caragana* (ELIZARIEVA, 1949; PARASKIV, 1960). It principally occurs in stabilised steppes (VORONTZOV *et al.*, 1969) as well as in the semi-stabilised sandy plains (SHUBIN & ISMAGILOV, 1969). In the south Balkhash-lake region it is found on the gray alluvial plain with sandy ridges (TRUKHACHEV, 1965). In Aral Kara-Kum it is found on the sandy plain in association with *Agropyron sibiricum* and *Artemisia terrae alba* (LOBACHEV, 1971). In Mongolia it has been captured from the desert with loose sands (BANNIKOV, 1954). The species is quite common in localised areas in Kazakhstan.

Thus, this jerboa seems to avoid the hilly and unstabilised sandy terrains and can be found only at the outskirts of the big sand massifs on the plains. The hair on its hind legs is less coarse and thicker, than that in *Dipus*.

It digs either simple and temporary or composite and permanent types of burrows, the length of which often reaches 3 m. The entrances to the burrows are plugged. It feeds on complex animal and vegetable food. Insects and arachnids with no hard chitin integuments, are much preferred. Its relationships with other small rodents have not been studied in detail. Only in the region near the Balkhash lake it has been caught near the burrows of great gerbils.

In the Zaisan depression VORONTZOV and his colleagues have noted breeding activity only in Spring. But SHUBIN & ISMAGILOV (1969) have reported two consecutive peaks in breeding activity, May–June and June–July. The number of young born per gravid female is small and almost constant, viz., 2.7 ± 0.07. The young do not attain sexual maturity in the year of their birth.

EUPHRATES JERBOA

*Allactaga euphratica** Thomas, 1881

This species is widely distributed in the mountains, semi-deserts and desert steppes of Asia Minor (north-western Iran, Turkey, eastern and north-eastern Zaravkzjie). In the U.S.S.R. in the north, this rodent has been observed along the beach of the Caspian Sea upto the town of Hatchmas and in the west upto the town of Leninkan. This is the only form of Dipodidae which has penetrated by several, relatively isolated populations found in the caves, in mountains, of Azerbaidjan and Armenia.

This rodent lives in the semi-deserts and in mountain steppes upto a height of 2500 m above sea-level (e.g. the Ararat mountain). Its favourite habitats are the rocky deposits and virgin and permanent fallow lands in the mountain steppes. *A. euphratica* live in small groups on the sandy, clayey and slightly hilly plots measuring 10–20 hectares, particularly in places of intensive agricultural operations (e.g. in the environs of Jerevan).

The living habits of this rodent have not been studied in detail. At

---

* *A. williamsi = A. euphratica* (ATTALAH & HARRISON, 1968).

471

some places (e.g. the Apsheron peninsula) this species occurs in large numbers. In contrast to *A. elater* it prefers the more rocky and hard surfaces of the foothills or the mountain valleys having xerophilous vegetation. Although in the Ararat valley *A. euphratica* are not found at altitudes higher than 1000 m, it has been found to occur elsewhere upto a height of 2500 m (POGOSJAN, 1955). Interestingly the populations found at the higher altitudes are darker in colour than those living on the lower mountains and in the plains. *A. euphratica* do not have any conspicuous morphological adaptive characteristics for high altitude living. The digits on the hind legs of this species are somewhat longer than those of *A. elater*.

The density of *A. euphratica* in the Talish mountains (the Diabar hollow) is of the order of 1 animal per hectare. This jerboa digs simple superficial burrows on gentle slopes (20–25°) without any sharp bends. The burrows are usually 120–170 cm long and 45–70 cm deep. The only inlet, which is usually not oval in shape as in other Dipodidae burrows, is 9–12 cm in diameter and is closed by the animals with a soil plug. Because of the hardness of the soil, the number of burrows found in an area usually is not large and the burrows of individual animals are situated quite apart from one another. The structure of the burrows on the slopes and on the hilly terrains depends on the level of the subsoil water.

The mountain jerboa feeds on seeds and on the green and underground parts of plants. According to the observations made on captive animals, these rodents, in contrast to other *Allactaga* species, preferred to eat the ears of cereals to the juicy plants and tulip bulbs. It carefully dug holes in melons and ate the seeds after splitting them apart.

In the Ararat valley these rodents are known to hibernate from the end of October upto the end of February. The jerboas come out on the surface when there is still some snow here and there. It appears that in many populations of this species hibernation does not take place at all in warm winters (for instance, in Talish).

According to POGOSJAN (1955) pregnant females were found in Armenia from May to October (mainly from the end of April upto the end of May), indicating that the breeding season in this species is considerably long. This is likely to be one of the traits of adaptation to the severe mountain conditions. The litter size varies from 2 to 8. During 1960–1972, in the Diabarskaja depression (Talish) out of 29 females caught in August, one was found gravid (with 3 embryos) and another nursing. Among 4 females captured in March, 8 in April, 22 in May and 34 in June none were found to be pregnant. In the Ararat valley a grown-up female was caught that gave birth to 3 young ones on May 3.

In the Ararat and Diabarskaja depressions the jerboas regularly take advantage of the burrows of the Middle-Asian gerbils *(Meriones blackleri)*, and on the Leninakan upland, those of the sousliks *(Citellus c. xantho-prymnus)*.

In Zakavkazje the flea, *Mesopsylla apscheronica*, was found on this jerboa.

SMALL FIVE-TOED JERBOA

Allactaga elater Lichtenstein, 1825

This widely distributed species inhabits the clayey-saline deserts of the U.S.S.R. (from the eastern Predkavkazje and Zakavkazje up to the Zaisan depression), the north-western part of China, western Mongolia, eastern Turkey, Iran, Afghanistan and western Pakistan. Two isolated pockets of their distribution are situated in Armenia and Azerbaidjan.

The northern border of the range of this species touches the Black Jar town on the Volga and the Volga-Ural sands, the whole of the Aral Sea region upto the Irgise river and the towns of Karsakpai, Betpackdala, the north Balkhash lake region and the Zaisan and Alakol depressions. The species is common in the whole of Kazakhstan and Middle Asia upto the foothills of the Tian-Shan and the Kopet-Dag mountains.

It prefers to inhabit relatively soft loamy soils with wormwood cover. Like most other jerboas it does have a special liking for areas of minimal vegetation cover, but, nevertheless, it is also adapted for living in areas of thin grass cover and bushy vegetation. It does not live in the sand though it is often capable of crossing great sandy spaces. It is the only jerboa species to penetrate into the outskirts of desert river forests and territories overgrown with *Haloxylon aphyllum*. It is also commonly found in the irrigated valley zones and in kitchen gardens and melon fields. It has been found to occur on mountains at a height of 1200 m above sea level. Rodents of this species are found in relatively large numbers all over the semidesert and desert zones. But, although the rodent can be found almost everywhere, it seems to attain the maximum population density in the zones of irrigated agriculture. Upto 6 or 7 (mean 5) individuals were observed per km, with the aid of headlights, in the vicinity of kitchen gardens in Chelkar town (Actubinsk district), while the density was as low as 0.6 individuals per km in the nearby wormwood associations on the sandy soils. On clayey plots with *Anabasis salsa*, *Nanophyton erinaceum* and wormwood the density of population was 0.02 per km. In the clayey desert between the Terek and Kuma rivers, TEMBOTOV (1972) obtained 15 individuals on a 6 km route between 20 and 21 hours in April, 1968.

In the valley of the Syr-Darya river this jerboa is especially numerous and lives in a variety of habitats ranging from clayey hills, with *Poa bulbosa* and *Poa*-wormwood associations (upto 8.5 individuals per km of foot route; upto 7 individuals per hectare), to the wet low lands, thickly covered with wormwood, *Eurotia* and *Lasiagrostis splendens*.

In the Aral Kara-Kum, during 1969–70, a mean density of 0.02–0.06 animals per hectare was observed on the clayey, wormwood and wormwood-*Anabasis salsa* associations. On the outskirts of the wavy sand plains in places with thinned out vegetation cover the density was the same but *A. elater* has not been found to occur very deep in this biotope. Along the banks of the saline flats, however, a relatively high density of 4.9 individuals per km has been observed, while in the *Poa-Artemisia* associations on the loams 3.8 individuals per km (8–12 per cent of these have been usually captured when traps were set in line near the holes) and in regions of *Anabasis salsa* cover 0.7 individuals per km have been observed.

In the north-western Kizil-Kum *A. elater* also prefers to inhabit plots with comparatively thick vegetation (for other jerboas) and subsandy soils. On the wide space of the takir-plains of the old valley of the Amu-Darya river with exceptionally thin vegetation cover the number of this gerbil was observed as 0.05–0.3 individual per km. On the subsandy plots with wormwood and *Haloxylon aphyllum* cover, situated on the borders of the takirs and sands, these gerbils were found in more numbers, about 0.5 individuals per km. On neglected irrigated fields, overgrown with tamarisk bushes (*Tamarix* sp.) and *Peganum harmala* the density reached 1.0 per km. On saline clayey tracts overgrown with Salsola there were 5.0 animals per km. In the Zaunguz Kara-Kum also the maximum number of these animals was observed on the edge of the sands. On the takir-like clayey plains its population was much less.

The burrows of *A. elater* are simpler in structure than those of most other jerboas. The main entrance to the burrow is not usually closed with soil plugs. The burrow is about 1.5 m long, the nest chamber is situated at 50–60 cm depth and usually there are 2–3 blind holes on the surrounding surface. The nests are made by the animals with soft vegetation. A detailed description of the burrows, as observed in Kizil-Kum, has been made by SABILAEV (1971a). The burrows that are used temporarily usually have a 120–350 cm long passage below the surface of the soil. Usually, the passage from which the animal had started digging gets closed with soil, and the jerboa comes out through the short lateral passage and keeps it closed with soil plug during day time.

The permanent burrow is the transitional form between the temporary burrow and the nest burrow (Fig. 1). It has no nest chambers. The closed main passage leads to a passage at deeper levels through a sharp turn of the burrow (the 'jerboa turn' according to ARGIROPULO, 1939). This deep passage has several broadenings and ends in a chamber.

The littering burrows always have a nest chamber and are somewhat complicated in structure. The nest proper is 10–12.5 cm in diameter, and is usually found to contain dried stems of plants and sometimes feathers, rags, paper and wool.

The contacts of this with other species have not been studied in detail. In the Usturt region (SABILAEV, 1970a) an average of 6 per cent of the

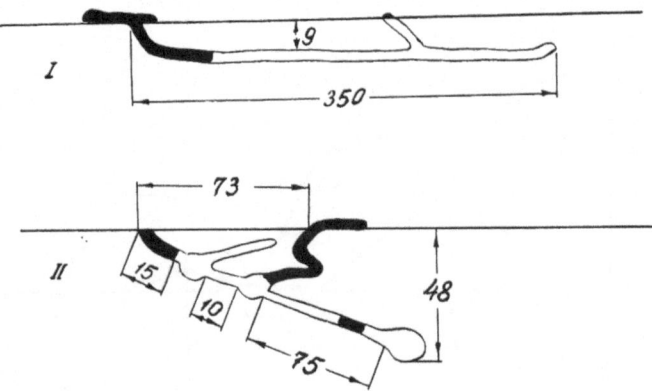

Fig. 1. Summer burrows of *A. elater*. I. Temporary burrow; II. permanent burrow (after SABILAEV, 1971a). (Distances are shown in cm).

colonies (burrows) of great gerbil are inhabited by *A. major*. In the Aral Kara-Kum region, this species can be more frequently captured in the burrows of the great gerbil than any other species (Table 2).

Along the coast of the Aral Sea (Koktem), *A. elater* were found to occur in 97% of the 32 colonies of *Rhombomys opimus* that were observed. It intrudes into the burrows of the great gerbil situated on hummocks among thick bushes of the cough-grass *(Agropyron sibiricum)* and over-growths of wormwood and *Calligonum* sp.

The food of *A. elater* is considerably varied. The vegetative juicy parts of plants are eaten as well as seeds, bulbs, roots and sometimes, insects. In Usturt and Kizil-Kum, it has been found to eat 40 species of plants of which about 80 per cent were ephemerals and annual grass crops. It readily feeds on *Anabasis salsa* seeds and on those of *Aeluropus literolis*, *Bromus tectorum*, *Eremopyrum triticeum* and *Malcomia africana*. In summer it frequently eats the seeds of *Lepidium perfoliatum* and of *Polygonum* sp. The significance of the roots of perennial herbs in the ration of the jerboas increases in spring and that of green feeds in spring and autumn (SA-BILAEV, 1971).

In Kizil-Kum *A. elater* comes out of its burrows earlier than other rodent species. One can observe these animals during summer 2–3 hours before sunset, but their major foraging activity on the surface takes place in the twilight hours of the dawn and the dusk. The young ones return to the burrows by 3–4 o'clock in the morning. Sometimes, individual animals of the species continue feeding till 9–10 in the morning.

These animals go into hibernation during October-March. In Kizil-Kum and in the Usturt deserts, hibernation takes place from the third ten-day-period of November to the first ten-day-period of March. As in other jerboa species, the adult fatty males are the first to lie down and

475

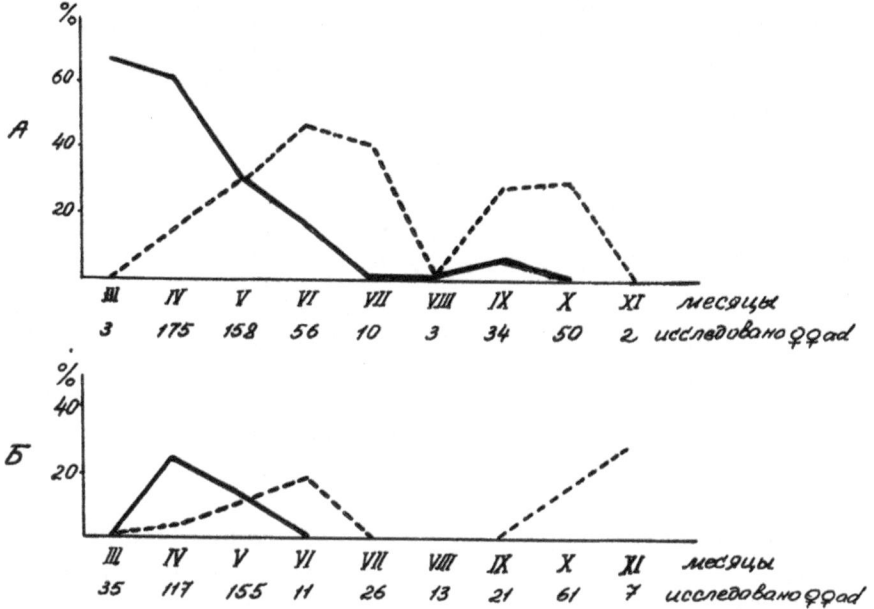

Fig. 2. Reproduction in *A. elater.* A. in the Aral Kara-Kum; B. in northern Turkmenia. (The continuous and the dotted lines denote percentages of pregnant and lactating females respectively. Months have been shown in roman numerals (e.g. March – III). Arabic numerals shown below each month indicate total number of adult females captured in that month).

the last to do so are the young females when the ambient temperature drops to 2–3 °C or even lower. The winter sleep of *A. elater* is interrupted at the time of the thaws. In Kizil-Kum RUDENTCHIK & SABILAEV (1971) repeatedly observed active animals in the end of January (ambient temp. +2.6 °C) and in the second part of February.

No clear information is available on the reproductive biology of this species. KOLESNIKOV (1934) reported the production of 3 litters in a year in the foothills of the Karatau mountains, each litter corresponding to a reproductive peak in spring, autumn and summer. VINOGRADOV (1937) and OGNEV (1948) on the other hand, considered that there is only one prolonged period of reproduction in this species. ISMAGILOV (1948) found its period of reproductive activity to last from April till June (the spring-summer period of reproduction) on the Barsa-Kelmes island in the Aral Sea. In Zacavkazje (ALIEVA, 1965) the jerboas bred from April till autumn, and in the valleys of the river Ural and in Turkmenia – only during spring. ISMAGILOV (1961) observed two littering activities in a year in *A. elater* in western Kazakhstan and in Mujunkum respectively. An autumn (August) litter, together with a spring or a spring-

476

Table 1. Fecundity of *A. elater*

Location	Number of females having litters of various sizes: Number of implanted embryos								Mean embryo No. per gravid female	Source
	1	2	3	4	5	6	7	8		
South Kazakhstan	1	4	2	12	13	2	—	—	4.1 ± 0.21	KOLESNIKOV, 1934
South-west Kizil-Kum	—	6	15	29	19	16	2	1	4.4 ± 0.14	ORLOV, 1957; SABILAEV, 1967a; Our data
Usturt desert	—	3	12	45	56	22	5	5	4.7 ± 0.10	SABILAEV, 1967a
North Turkmenia	—	2	5	13	9	13	5	1	4.9 ± 0.20	Our data
Aral Kara-Kum	—	3	17	56	50	15	4	2	4.5 ± 0.09	Our data
Gurjev district	—	1	3	8	14	5	1	1	4.8 ± 0.21	ORLOV, 1957

The biological significance of the shorter period of reproduction in the north is not quite clear. It is probably connected with the markedly higher longevity of the jerboas of the north presumably due to a longer hibernating phase resulting in a shortening of the active period of life in this species.

summer one, has been observed in north Tajikistan (DAVIDOV, 1964), in south Uzbekistan (LOBIZOVA, 1971) in south-western Kizil-Kum (ISHUNIN & PAVLENKO, 1966) in Kirghizia (TOKTOSUNOV, 1958) and in Semiretschje (Sevenrivers region, SHNITNIKOV, 1936). In north-western Kizil-Kum (SABILAEV, 1971) *A. elater* generally breeds only once in April-May although some females have been found gravid in June also. ORLOV (1957) has reported about a third (September-October) littering in this species for this territory.

In Kara-Kum near the Aral sea gravid females of *A. elater* are found now in March–June and in September, and nursing females in April–July and in September–October. In north Turkmenia gravid females have been observed in April–May and in October–November (Fig. 2).

Summing up these informations we come to the conclusion that 3 litters are dropped in a year by *A. elater*, in spring, summer and autumn, although the second littering may not always be a regular feature.

The fecundity of *A. elater* is subjected to geographical modifications (Table 1). As in other species, these animals tend to have a lowered breeding potential in the north.

JUMPER JERBOA OR FIVE-TOED SIBIRICA JERBOA

Allactaga sibirica Forster, 1778

The wide area of habitation of this species occupies the north desert, the semi-desert and the steppes of Zabaikalje, Mongolia and northern

China. It is distributed from the lower reaches of the Ural river including Usturt and Aralian deserts, Betpakdala and the northern Balkhash sea region upto the foothills of the Tchu-Ili mountains and Ili Alatai mountains. In eastern Kazakhstan the species is found as far in the north as Pavlodar town. Besides that, *Allactaga sibirica* dwells in the Chuisky steppe on the Altai, South Tuva and the South Zabaikalje, to the east of Selenga river, penetrating there from Mongolia. In Mongolia it inhabits nearly the whole of the country, except the taiga regions of Kentay and Prikosogolje, the Alpine zone of Khangaj and the Mongolian Altai, and also, except the fiercely desertic terrain, towards the south comprising the Zaaltai Gobi and the Bordzon Gobi (BANNIKOV, 1954).

A. sibirica is one of the most versatile species among the jerboas as it is known to inhabit practically all biotopes. It penetrates through the clayey parts of the river valleys and reaches upto 3560 m on mountains (TOKTOSUNOV, 1958). It is widely spread in the dry rocky steppes of Kazakh upland and the desert parts of Mongolia. In the steppe zone of Zabaikalje it lives only on plots with scant vegetation cover. In the sandy desert of western Kazakhstan it is ordinarily found on the dunes. In south-eastern Zabaikalje it has been found in ilm groves on the sandy soil (SARZINSKI, 1963) and in the tape pine forests (NEKIPELOV, 1940). In the north Aral sea region, according to our observations, the jumper is much more versatile than *Allactaga major* and the former dwells practically everywhere, avoiding only the larger takirs and the sand massifs where it can be found only on the outskirts. Its number sharply reduces also in the river valleys where at places both species occur. *Allactaga severtzovi* is more common. *Allactaga sibirica* is quite numerous on the banks of the saline plots.

There are occasional pockets of high population density of this species to the north of the Aral Sea, along the borders of the saline plots. For example, an average of 1.4 jumper was observed per km during the night with the aid of a head-light. On the sand dunes 0.1–0.3 specimen per hectare was noted under the above conditions. On most parts of the Usturt region the number of the jumper is generally low in comparison with other species of jerboas, and is found in appreciable numbers only at certain pockets of the South Usturt (SABILAEV, 1970b).

The permanent burrow of this species seems to be simpler than that of *Allactaga major*. The burrow is almost horizontal to the surface, is usually 5 m in length and has 3 holes on the outside (FORMOZOV, 1929). At a depth of 20–65 cm below the point of divergence of the holes there is a nest chamber. There may be 2–3 of these chambers and sometimes a vertical passage may lead to them (NEKIPELOV, 1935). The burrows are sometimes broad and have air-holes presumably for purposes of ventilation. The inlets are not always closed with soil plugs, usually made of loose soil. The nests are covered with dry grass and sometimes with sheep wool.

Table 2. Occurrence of different jerboa species in burrows of the great gerbils in the Aral Kara-Kum, according to observation made on more than 5000 burrows from 1958 to 1968.

Species	Per cent occurrence in great gerbil burrows	Approximate species composition of the jerboas in the Aral Kara-Kum, in per cent
A. major	2.9	5.1
A. sibirica	4.3	8.4
A. elater	17.4	10.2
A. pygmaeus	3.7	27.6
P. platyurus	0.2	0.1
S. telum	70.9	31.6
D. sagitta	0.6	16.5
E. lichtensteini	—	0.5

The temporary burrows are shorter, 60–120 cm long and 20–30 cm deep, sometimes they are closed with soil plugs. These burrows end in a broad extremity and do not have any nest chamber (BANNIKOV, 1954). The number of the temporary burrows may be quite considerable. These are utilized intermittently and the animals often dig new burrow.

In Kazakhstan the jumper frequently, but not always, encounters the great gerbil by visiting the latter's burrows and sharing its food. Of the jerboa species obtained in the burrows of the great gerbil in the north Aral territory, *A. sibirica* has been found to constitute 4.3 per cent of the total number of species (Table 2) while it formed 8.4 per cent of the jerboa species found in nature in that season.

In the deserts bordering the Aral sea, the jumper seems to compete with *Allactaga major* and the latter species seems to have forced out the former to places with thicker vegetation covers.

Allactaga sibirica feeds on a variety of food items. It seems to be one of the most carnivorous among the jerboas. NEKIPELOV (1940) found in Zabaikalje, insects and beetles (mostly Neodorcadion) and ants (Formicidae) and other larvae in 80 per cent of the stomachs and vegetable remains (in most cases the seeds of *Setaria viridis* and *Astragalus* in 36 per cent). In the Chuiski steppe KOLOSOV (1939) found in the stomachs of these animals the seeds of sedge *(Carex)*, cereals (Gramineae) and *Salsola* and only in one case the remains of a beetle (Coleoptera). In Mongolia the stomach contents of these animals showed remnants of greens, the pulp of bulbs, seeds and larvae (FORMOZOV, 1929). Seeds of leguminous plants and dark brown beetles (Tenebrionidae) have also been reported to occur in the stomach of this species (BANNIKOV, 1954). In Usturt, the wormwood *(Artemisia)*, keyrek *(Salsola rigida)* and tulip bulbs *(Tulipa sogdiana)* are among the preferred food. In captivity they

479

eagerly take to tulip bulbs and the green as well as dry branches of wormwood, leaving the roots. The roots of *Anabasis salsa* are also devoured. In nearly every stomach there were remains of insects (SABILAEV, 1969a). The jumper catches butterflies (Lepidoptera) and beetles in the air perfectly easily by jumping to heights of 1 m or even more. The species apparently has little trap shyness and can be comparatively easily captured in traps using bread and vegetable butter as baits.

The night activity of the jumper in early spring and in autumn is greatly supplemented by its activity during morning and evening.

Specific information on hibernation in this species is not available. Usually animals of this species go into hibernation from September–October upto March and more often upto April. Some observations indicate that the jumper appears even when there are patches of snow (March 21) and it has been found to be active under conditions of $-17\,°C$ of frost (Sept. 25, Mongolia; BANNIKOV, 1954). But these must be exceptional cases as even with the first frost the jumper is almost certain to disappear from view.

In Zabaikalje (NEKIPELOV, 1940; BOGORODSKI, 1967; CHABAEVA, 1968), in Tuva (FLINT, 1970), and in Mongolia (BANNIKOV, 1947; TCHUGUNOV, 1962) there is one prolonged period of reproduction for the jumper, usually lasting from May to the beginning of June. In the North Priaralje, according to our observations, gravid females were encountered from April till June, and nursing females were found during the whole summer and upto the beginning of September. In Usturt SABILAEV (1970a & b) found gravid females (3 out of 8, the remaining 5 in heat) in April and one of the 5 in June. From this observation probably originated the speculation, apparently baseless, that young jumpers also take part in reproductive activity.

The number of the young per litter of this species is from 1 to 7 with a mean of 3–4 (Table 3). The species seems to maintain its number by means of its long period of reproduction, the preservation of the population during hibernation and the comparatively longer duration of life of its individuals as observed in the Aral. Among 31 individuals 41.9 per

Table 3. Fecundity of the jumper jerboa.

Locality	Number of females having litters of various sizes							Mean embryo number per gravid female	Source
	Number of implanted embryos								
	1	2	3	4	5	6	7		
Zabaikalje	1	5	9	3	—	—	—	2.8 ± 0.19	BOGORODSKI, 1967
Tuva	1	10	11	7	3	5	3	3.7 ± 0.26	FLINT, 1970
Aral Kara-Kum	—	1	4	3	1	—	—	3.4 ± 0.29	Present authors

cent of the animals were found to be 2 years old and 3.3 per cent 3 years old.

At some places the jumpers have been known to cause damage to kitchen gardens and melon fields. BANNIKOV (1954) has reported a case of total destruction of a cucumber plantation in the course of a single night.

GREAT JERBOA (EARTH HARE)

Allactaga major Kerr, 1792

This species is widely spread in the forest-steppe, steppe zones and in the northern semi-desert areas. It is one of the largest of the jerboa species. It is found as far as the towns of Nikolaev and Crivoi Rog, in the region of Uman, on the right bank of the Dnieper river, in the region of Cherkassi and in the towns of Kiev and Novozibkov in the west. In the north it occurs upto the southern part of Briansk district on the right-bank of the rivers Oka and Volga, almost near to the latitude of Kazan, and upto the left bank of the rivers Kama and Belaja. It is found on the north foothills of the Big Caucasus in the south and along the beach of the Caspian sea as far as upto the town of Makhatshkala. In Asia the northern border of its habitat touches Verchne-Uralsk, Chelabisnk, Shadrinsk, Jalutorovsk, Novosibirsk and Barnaul towns. The southern border passes by the Usturt region, the valley of the river Syr-Darya (an isolated population occurs to the south of this river), the foothills of the Pamiro-Alay and Alakolian regions and the Zaisanian depressions and upto the western foothills of the Altai mountains.

In the clayey deserts it inhabits places with few bushes and prefers the wormwood-*Salsola* associations with thin grass cover, although they may also be found on the bare takir. Towards the south it may occur in irrigated fields particularly melon fields. It is widely spread in the steppe and in the forest-steppe, but it generally keeps to the roads, the curves of the fields, pastures and permanent fallow lands without in-filtrating into the virgin lands. At places where the soil is compact and the vegetation scarce, its population density increases. In the forest steppe it inhabits patches covered with bulbous plants and almost no other vegetation. It also occurs on the podsol soils, on the sand dunes, especially near the pine forests. It is found on the Alpine steppes upto a height of about 1,100 m (in North Kirghizia).

Towards the north, in the loamy terrain with wormwood-*Salsola* association from 0.8 to 1.7 animals of this species have been observed per km at night time.

These jerboas are known to live solitary lives. Their burrows are dug in the packed soil. In the digging of their burrows, as in other Dipodidae, chiefly the cutting teeth (SAMSONOV, 1953) forming an obtuse angle

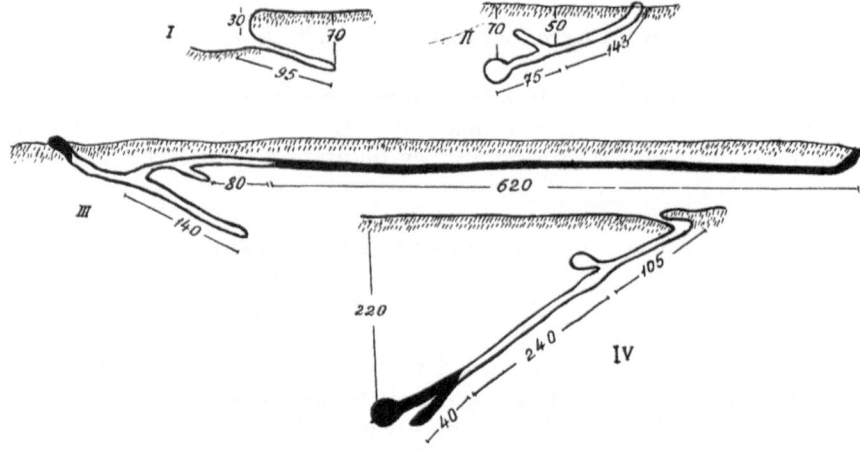

Fig. 3. Burrows of *A. major*. I temporary burrow; II–III. summer permanent burrows of females with young ones; IV. winter permanent burrow of a male (after FENJUK, 1928). (Distances are shown in cm).

with each other, are used. This arrangement of the cutting teeth in this species as in *A. elater* enables it to dig up even very compact clayey soils and to make absolutely smooth surfaces (VINOGRADOV, 1937; EISENBERG, 1967).

The structure of its burrow is similar to that of other five-fingered jerboas, and differs only in the size of the diameter of the passages and that of the inlet which may reach from 7 to 9.5 cm (Fig. 3). In any individual plot there may be from one to four permanent burrows, each more than 2 m deep and having 1–2 chambers covered with bedding of dry grass. Each burrow has one, seldom two inlets. The entrance to a burrow is usually protected with an earthen plug. Along the passage of the burrow also 1–2 earthen plugs may be found. The animals make temporary burrows in great numbers in different parts of their individual plots and at places where they generally congregate. The entrance to a temporary burrow is generally kept open and is more horizontal than in a permanent burrow. In the temporary burrow there is a typical 'Starting pad' and a landing ground at a distance of 1.2 to 1.7 m from the burrow. The temporary burrows do not always have chambers and are usually not deep. The animal knows perfectly well the arrangement of all burrows and covers in the vicinity and quickly takes shelter at times of danger. Sometimes it uses the burrows of *Citellus fulvus, C. pygmaeus, Rhombomys opimus* and other species.

This species does not seem to have wide contacts with animals of other species. Only 3 individuals of *A. major* were found in a total of 1355 dug up burrows of the great gerbil in the Usturt desert region (SABILAEV,

482

1970a). In 1966 in the lower reaches of the river Turgai, we could observe considerable intrusions of this species into the burrows of *C. fulvus*. When traps were set near gopher burrows, 25 *A. major* were captured in 4 nights. The contacts of *A. major* with sousliks were also observed by BONDAR (1956) in the Betpakdala desert region. *A. major* generally feed on the underground parts of plants (bulbs, tubers, etc.), rarely on the green sprouts. The gopher and this jerboa dig almost identical feeding pits, the latter's dug out being characterised by its more gentle slope on one side and by the impressions of its tail and footprints along the edges. During periods of scarcity, seeds form the largest part of its feed. Insects and Arachnidae are usual components of the food of *A. major*. At night when seen with the aid of a car's head-light the rodents are found to be eagerly catching moths and beetles and are the least bothered about the presence of observers. In the small feeding pits the remains of insects are not usually uncommon. Frequently, the jerboas destroy melon fields and crops of foodgrains.

As in all jerboa species, this one also is nocturnal in its habit, but its period of surface activity is usually longer than that of the others. Though the time of coming out on the surface varies greatly, most of these animals come out in the twilight, an hour or an hour and a half after sunset. One can mark two peak activity periods in summer – one in the beginning and the other at the end of the night. In captivity, however, the animals generally pass both the day and the night by dozing off in different poses (Fig. 4). Its feeding activity is the most intensive during the first part of the night. In autumn one can usually find *A. major* even in the day-time. It generally hibernates from October to March, although this period may vary somewhat in different geographical regions.

According to observations made by FENJUK (1928) in the Lower Volga Valley, the breeding season of *A. major* is spread from April to July. In the foothills of Karatau (South Kazakhstan) 3 litters have been known to be born in a year: the first – in April and the beginning of May; the second – in the end of June and the beginning of July and the third – in September. The conclusion regarding the presence of the third litter was made on the ground that a nest with 3 young ones of 30–40 days age together with a fully grown-up nursing mother was found on October 22.

Most other authors have, however, reported only one season of peak reproductive activity, e.g. – in the south of the European part of U.S.S.R. – in May–June (BARABASH-NIKIPHOROV, 1957), in western Kazakhstan – in May (FOKANOV, 1954), in the Usturt region in April (SABILAEV, 1968), in the Kazakhstanian plateau in May and in the north of Kirghizia – in April–May (TOKTOSUNOV, 1958). However, most of the authors mentioned above lacked full information in this respect.

In the Aral Sea Kara-Kum region we observed gravid females from April to June and nursing ones in April–July and in September–October.

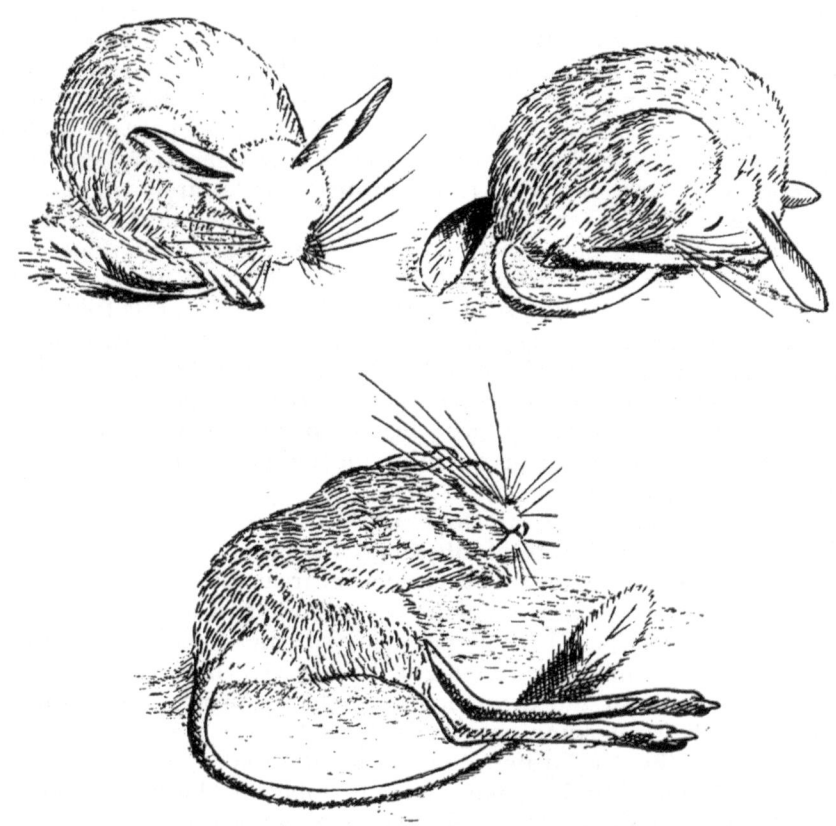

Fig. 4. Postures adopted by *A. major* during sleep and rest (after FENJUK, 1928).

On June 20, 1970, in the Aral Sea desert region, one female *A. major* with 4 imperfectly developed embryos along with 3 conspicuous placental spots was captured. Thus, along with the spring reproductive activity, a second breeding phase in the autumn for *A. major* cannot be ruled out.

The size of the litter, according to the number of the embryos, varies from 2 to 7. However, in comparison to other jerboas, the number is somewhat less variable in this species. In south Kazakhstan, according to KOLESNIKOV (1934), the number of the embryos is 3.0 ± 0.16 (n = 8, min. 2, max. 4), in Aral Kara-Kum, according to our observations it is 4.6 ± 0.38 (n = 3, min. 3, max. 6) while in the Guriev region, according to ORLOV (1957) it is 4.1 ± 0.49 (n = 9, min. 4, max. 6).

FIVE-TOED SEVERTZOV'S JERBOA

Allactaga severtzovi Vinogradov, 1925

This species is distributed over different types of desert terrains in Middle

Asia and South Kazakhstan. To the west it goes upto the beach of the Caspian Sea and to the north – the northern Usturt desert, the northern Aral and Sir-Darja deserts, south Betpackdala, south Balkhash lake region upto the foothills of Tian-Shan mountain. It is widely spread in Kizil-Kum. To the south its habitat extends to south-western Turkmenia in the region of Sarikamishian lakes. In Turkmenia, the southern border of its range goes along the river bed of Usboi. It has been found near Termez, in south-western Tajikistan and in the western part of Fergana valley.

It is an inhabitant of clayey, saline loess and gypseous deserts. It prefers comparatively soft, clayey and loess soils with bushes and rich ephemeral and *Salsola* vegetation.

At Usturt it has been found in association with *Haloxylon* and along the banks of saline flats and down the slopes of the large saline cavities, while in the south it occurs in association with *Anabasis salsa*. It avoids places with thick covers of *Salsola laricifolia*, rocky or hard clayey plots, big takirs and loose sands. In north-western Turkmenia it is found on sandy loam soil and in plains, with thin vegetation covers. It is rare on the takir like plains and on packed smooth sands. In Kizil-Kum it is found on clayey soils with rich *Salsola* vegetation, on saline flats and in the ravines. It is very numerous on permanently fallow lands and on takir like plots in the plateau and on the sands with overgrowths of *Haloxylon* and of bulbous ephemerals. In the south and southwestern Kizil-Kum it occurs on compact gypsum gray-brown soils having *Salsola laricifolia* and *Salsola rigida* association. In south-western Tajikistan the species occurs in the valleys with gray soils and in the foothills and in the north Mujun-Kum in plots of *Haloxylon* and in the vicinity of the saline flats. As one can see, the stenotopic habit of this species increases towards the north and to the south of the area.

These animals are found in any significant number (2–4 specimens per km of motor route) only in the valleys of the rivers, on gray soils, on permanently fallow lands, in association with overgrowths of *Haloxylon* and on the fringes of saline flats.

As with *Allactaga elater* this species does not avoid thick bushy and grassy areas (with the grass cover not exceeding 10–30%) and regions of irrigated agriculture. On the takir-like plains overgrown with *Anabasis salsa* in north-western Kizil-Kum the density of this species was noted as 0.07 individuals per km in July 1971 while the density was almost 10 times more in plots of *Haloxylon aphyllum*.

Allactaga severtzovi digs its burrows in the loose soil on open places, usually on the higher plots. According to SABILAEV (1969a) the temporary burrows for the summer season usually found on the road sides generally have one shallow passage (Fig. 5, I). These are used for one, and rarely for several days' rest. These temporary protective burrows are similar to the ones reported for *Allactaga major* (FENJUK, 1928). The

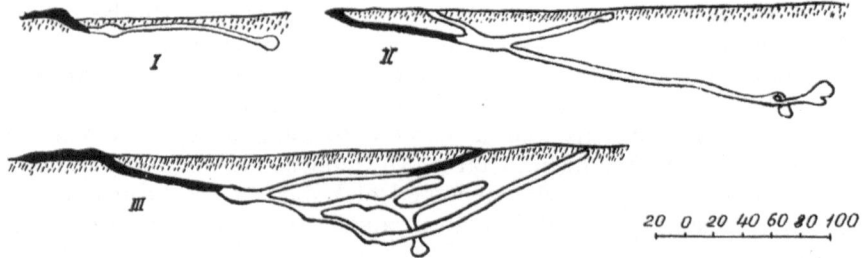

Fig. 5. Burrows of *A. severtzovi* (after SABILAEV, 1969a) I. temporary burrow; II. summer breeding burrow; III. winter burrow. (The scale denotes distances in cm).

difference is in the comparatively smaller size (6 × 9 cm) in the usually opened oval inlet and in the presence of 2–3 broad chambers along the passage of the burrow. The summer temporary littering burrows are deeper (Fig. 5, II), more complex and have chambers 12–17 cm wide with beddings of small roots and stems of plants. The length of the passages may reach upto 5.3 m. This jerboa can dig a passage 30 cm in length and 10 cm deep within 38 minutes. The winter burrows (Fig. 5, III) have 1–4 reserve outlets. The wintering chambers are at depths of 0.5–0.55 m.

It is characteristic for this rodent to have at least several protective burrows with the typical long winding path. Besides that *Allactaga severtzovi* often uses bushes and burrows of other rodents for covering. The individual animal may change its dwelling several times during the summer season.

Although it has been generally held earlier that *Allactaga severtzovi* feeds primarily on the underground parts of plants (ANDRUSHKO, 1939), subsequent observations made in Kizil-Kum indicated the great role of vegetative and generative parts of plants in their ration. SABILAEV (1969a) reported on the basis of 126 stomach studies, the occurrence of juicy shoots and leaves of *Salsola*, ephemerals and tulpan *(Tulipa)* bulbs. In all, 22 species of plants could be identified. The seasonal nature of feed preferences is well-expressed in this species. The seeds and fruits of ephemerals are eaten by these animals beginning from the second half of May upto July, later the fruits of *Anabasis salsa* and of *Salsola rigida* and the flowers and seeds of wormwood are consumed. Ordinarily and especially during the summer, this jerboa eats invertebrates like beetles (Coleoptera), wood-lice (Isopoda), butterflies (Lepidoptera), ants (Formicidae) and flies (Diptera). The jerboas catch the insects by jumping in the air.

It is active, like other jerboas, at night. With the onset of spring this species comes out of its burrows 45–50 minutes after sunset and in summer 10–20 minutes after sunset. With the shortening of the day in autumn, the picture of surface activity is reversed (SABILAEV, 1969). There are records of two peaks of activity during the night, with the maximum in the first part of it.

Allactaga severtzovi has a longer winter hibernation than other Dipodidae. Observations made in Kizil-Kum indicate that it hibernates from the beginning or middle of October upto the beginning to middle of March, that is for nearly 5 months. Its waking up is associated with an ambient temperature of 3–6 °C.

Besides spring reproduction as observed by many workers in *Allactaga severtzovi*, FENJUK & KAMNEV (1957) have suggested the existence of a second summer breeding phase in the Usturt region. These authors reported the finding of 4 new-born *A. severtzovi* along with a lactating female in a dug-up burrow in July. A few sub-adult animals were also observed. ORLOV (1957) considered that there may even be three littering periods for *A. severtzovi*. In the western Kizil-Kum he obtained gravid females in April and young animals in the second ten-day-period of June and in the first and third ten-day-period of October. He obtained a lactating female in the first week of November. SABILAEV (1969a) observed that in Usturt and in the northern Kizil-Kum the second littering (June) is irregular and is dependent on the prevalence of favourable climatic conditions (precipitation) in May. In the south-western Kizil-Kum in October one female out of four captured was found lactating (ISHUNIN & PAVLENKO, 1966). In the foothills of the Nuratan ridge, PAVLENKO & HUBAIDULINA (1970) found two generations of *A. severtzovi* within a year, one in spring (March–April) and the other in autumn (end of August–September).

From 1 to 7 embryos (mean 4–5) have been found to occur in *A. severtzovi* (Table 4).

In common with other jerboa species, the sex ratio in this species appears to be very variable. In Tajikistan more (53–60%) males have

Table 4. The fecundity of *A. severtzovi*.

Place	Number of females having litters of various sizes Number of implanted embryos								Mean embryo number per gravid females	Source
	1	2	3	4	5	6	7	8		
South Uzbekistan	—	—	+	+	+	+	—	—	4.0	LOBIZOVA, 1971
Tajikistan	+	+	+	+	+	+	+	+	3.5	NERONOV et al., 1964
South-west Kizil-Kum	—	1	11	22	7	7	+	—	4.1 ± 0.14	SABILAEV, 1967
North Turkmenia	—	1	—	4	5	1	3	—	5.0 ± 0.38	Our data

been noted (NERONOV *et al.*, 1964). We obtained 8 grown-up males and only 3 females in the north-western Kizil-Kum in the summer of 1971.

FIVE-TOED BOBRINSKI'S JERBOA

Allactaga (Allactodipus) bobrinskii Kolesnikov, 1937

This species inhabits the southern deserts. It is rather restricted in its occurrence. It is found in south and west Kizil-Kum (from the lower reaches of the Amu-Darya river upto the region of its middle streams), in the northern periphery of the Nukus latitude and upto the Bukhar oasis to the south. It has been found in the east of Zaunguz Kara-Kum (in north-eastern Turkmenia). It is interesting to note that the present course of Amu-Darya has not been an obstacle to the spread of *A. bobrinskii* as it lives on both its banks.

KOLESNIKOV was the first to describe *A. bobrinskii* obtained on the borders of sandy and clayey deserts. He considered this animal to be a psammophile, because of the presence of the typical hair on its hind legs as in the case of the three-fingered Dipodidae. This view has not found support in different later reviews. LOBIZOVA (1971) reportedly captured these animals on metalled roads. SAPOZHENIKOV (1963) and STALMA-KOVA (1957) believed that this animal, like *Jaculus turcmenicus* makes burrows in compact soil and feeds in the sand. There can hardly be any doubt that this jerboa is an inhabitant of hard and loamy plains having *Salsola rigida*, *Anabasis salsa*, *Haloxylon aphyllum* and *Ceratocarpus turkestanicus* covers. It rarely occurs in the sandy deserts. The hair on the fingers of its hind legs is most likely to be the result of its evolutionary development (subgenus *Allactodipus*) in the conditions of Kizil-Kum where large sand flats are rare. Apparently this species developed during the destruction of the ancient mountains in central Kizil-Kum and the appearance of the gravelly plains there. In the Zaunguz Kara-Kum it lives on the crusty saline flats strewn with pebbles and overgrown with the juicy *Salsola*.

The characteristic habitats of *A. bobrinskii* (SABILAEV, 1968; FOKIN, 1969) are the saucer-like depressions cleared with stones or gravel (Fig. 6) which these animals seem to use as sand baths.

The number of this relativelys stereotype species is rather high in its typical locations – ranging from 2 to 5 (mean 2.3 individuals) per km in the environs of Nukus in summer 1971 according to our data. The highest trapping success has been achieved with the Gero trap (3–6% of hit) on gravelly soils (SABILAEV, 1968).

The burrows of *A. bobrinskii* are situated on slightly raised grounds free from vegetation. In the open burrows jerboas rarely occur during the day time; the occupied burrows are usually closed with soil plugs.

Fig. 6. Entrance of burrow of *A. bobrinskii* (photo by G. SHENBROT).

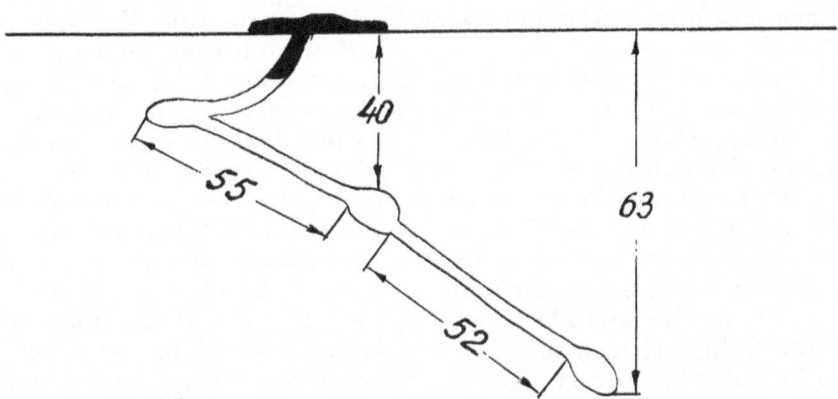

Fig. 7. Summer burrow of *A. bobrinskii* (after SABILAEV, 1968). (Distances are shown in cm).

The structure of these burrows (Fig. 7) is the same as that of the burrows of *A. elater*. In the permanent burrows usually one or two chambers are found with nesting materials comprising dry roots and stems, and sometimes rags, wool and feathers.

The flowering branches of the grey wormwood and shoots of *Anabasis salsa*, *Alhagi pseudoalhagi* and *Halimocnemis villosa* are its preferred feeds.

489

30–50% of the stomachs of captured animals contained seeds. It sometimes eats insects too. It is remarkable that the vegetation of the sandy terrains is rarely eaten by this species (FOKIN, 1969).

Its hibernation characteristics are not properly understood, but in Karakalpakija country *A. bobrinskii* were trapped as early as on the 29th of March, and as late as on the 14th of October (ambient temperature +5 to +1 °C).

Very little information is available about its reproduction. In western Kizil-Kum and in north Turkmenia it breeds in April–June and in September–October, the yearlings also taking part in the autumn reproduction. It appears that *A. bobrinskii* has two (spring-summer and autumn) and probably three (spring, summer and autumn) peak breeding seasons in a year. The number of embryos in this species, in north Turkmenia has been found to range from 3 to 7, the mean per gravid female being 5.3 ± 0.22 (n = 32, FOKIN, 1969).

LITTLE EARTH-HARE JERBOA

Alactagulus pygmaeus Pallas, 1778

This is one of the most common and widely spread species. Its habitat range includes the Caspian region in the west upto the Don river (the Ilovlinskaya station), the towns of Rostov-on-Don, Stavropol and Machatshkala, in the south upto the river along the western beach of the Caspian Sea (the only jerboa species to penetrate into the centre of the Volga delta – 'the Ber hillocks'), Usturt, Aral Sea region, south and central Kazakhstan and the Balkash Lake region upto the Alacol depression. In the north it is spread upto Volsk town on the Volga and the middle reach of the Ural river, in the east from Asktubinsk town and the Turgai river region to all over Kazakhstan and the environs of Semipalatinsk town. In the south it occurs in the whole of Middle Asia, Iran, the south-eastern beach of Caspian Sea and Mongolia (the foothills of the Gobi Altai and the central region of north China – Alashan and Ordos). It inhabits exclusively the clayey and gravelly parts of the semi-deserts in the north and south of Kazakhstan, the Middle and Central Asia. It is found in the south steppes only along the roads and in places grazed by cattle. In the semi-deserts it lives on the saline vegetation and in the deserts – on the thin *Salsola* growths on firm soil. Its most typical habitats in the deserts are the hard, saline and gravelly plots overgrown with *Anabasis salsa*, and plains and takirs of various sizes. Dried-up beds of lakes are also inhabited by this species. It exploits areas least inhabited by other species of desert animals. In the foothill regions, it inhabits only the gravelly and saline patches (Fig. 8). It does not occur on mountains but is found in small numbers in the river valleys and on loessed soils.

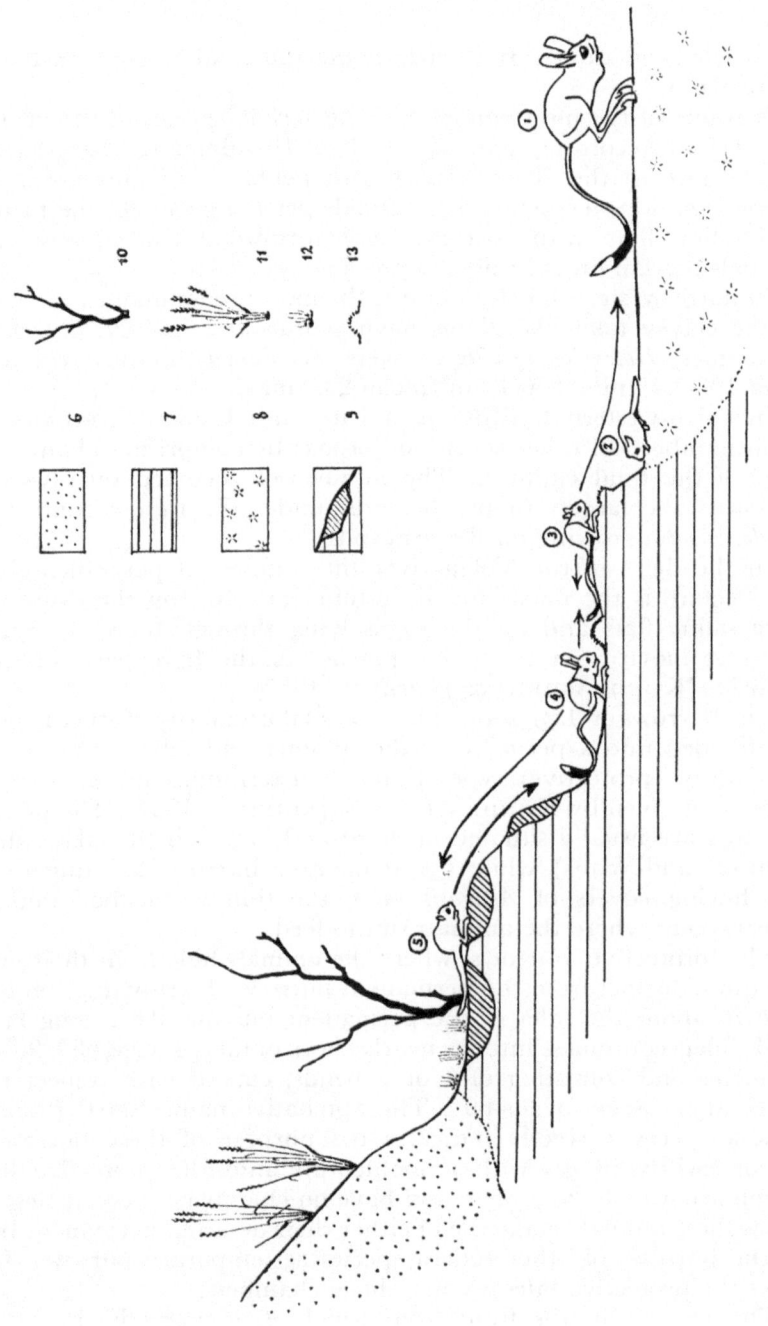

It avoids sand dunes. It is rare in the inner takirs of Kizil-Kum and Kara-Kum.

It is one of the most common of the Jerboa species in the north Aral Sea region. According to our data, about 10 animals of this species occur per hectare in the clayey desert with pebbles and hummocks having *Anabasis salsa* cover, while 5–6 animals per km occur on the firm takir-like saline flats with *Anabasis salsa*-wormwood associations and 0.3 animals per km on the saline depressions.

In north-western Kizil-Kum it is the most predominant rodent species of the clayey takir-like plains having *Anabasis salsa, Salsola rigida* and, sometimes, *Haloxylon aphyllum* covers. Its density in this region varies from 1.3–2.7 animals per km (mean 2.0; maximum 10.0).

In north-western Kizil-Kum and in south Usturt (SABILAEV, 1967) it outnumbers all other species of jerboas by comprising about 30.5 per cent of the total captures. The number of occupied burrows in that region is the largest (5 per hectare) under the thin ground cover of *Anabasis salsa* growing on the firm soils.

In the delta of the Volga river the number of permanent burrows per hectare is the maximum in autumn (12–16), on the dried up and bare saline flats and on the roads lying through them. In spring *A. pygmaeus* move from the saline patches to the bare parts of the 'Ber hillocks' (KONDRASHKIN & JEDIKINA, 1957).

The burrows of *A. pygmaeus*, like those of the majority of other Dipodidae, are divided into a permanent (the summer and hibernating ones) and temporary (protective) type. Detailed descriptions of these structures have been given by KONDRASHKIN & JEDIKINA (1957). The permanent burrows are usually situated on hard and bare soil (in takirs, denuded pastures and roads) while the temporary burrows are dug on softer soil having covers of *Anabasis salsa*, the thin wormwood and *Salsola* brushwoods, where the animals run to feed.

The 'protective' burrows, where the animals hide from their enemies, are quite distinct from the permanent burrows. Freshly dug soil is found thrown about the inlet of the permanent burrow like a long fan. The oval inlet continues into a nearly horizontal passage, 30–90 cm in diameter and consisting of 1 or 2 bends, curved with respect to each other at an angle of 30–60°. The apt native name 'start' ('razgon' in Russian) very correctly indicates the purpose of these burrows. The jerboa swiftly jumps with a momentum into the protective burrows which are usually kept open and have no chambers. Nevertheless, sometimes these animals make use of either their deserted permanent burrows or the burrows of other rodent species as temporary burrows. In such cases the protective burrows may have chambers.

The passage in the front from which *A. pygmaeus* begins to dig its permanent summer burrow, later becomes obstructed with soil*. At first this thrown out soil is easily noticeable on the surface as a heap, but later on it is usually washed away and cannot be noticed. The initial passage is usually upto 6 m long (frequently less), 3–5 cm wide and with several broadenings of upto 6–8 cm diameter. It turns into a permanent passage, consisting of two main parts.

One part of the permanent passage serves as the inlet ending in the narrow round outlet gnawed through on the surface. The outlet is closed during the daytime with a small plug of soil. This soil plug not only has a protective function but has also thermoregulatory significance. The increase in the number of the open burrows is directly proportional to the increase in summer temperature (KONDRASHKIN & JEDIKINA, 1957). Besides that, during the hot season (July–August) the length of the plug increases upto 8–17 cm (from 2–3 cm in spring and autumn).

The other part of the passage (or passages) of the permanent burrow leads downwards and ends in a chamber. The nests contain split stems of *Poa* sp., *Ceratocarpus turkestanicus* and other plants, and also wool, feathers, paper etc. These nests are made by *A. pygmaeus* at depths of 50 cm by males and non-lactating females and in depths of upto 70 cm by gravid and nursing females. The size of the nest in the latter case is usually 2–3 times larger – upto 1400–2000 cm³, and the burrow itself is more complex – the special type of the 'littering burrows'. The structure of these burrows is continuously modified to suit the time of the year (for purposes of thermoregulation) and the physiological state of the animals. As the mean monthly air temperature increases, the number of open burrows gets reduced. This is particularly evident during the period June–August. The length of the summer burrow is longer for females (306–348 cm as per measurements made on 107 burrows) and particularly so for the gravid and nursing ones (average 425 cm as per 83 burrow measurements). The usual length of the passages in the burrows of male animals does not exceed 308 cm (average 256 cm according to measurements made on 109 burrows). While 83–86% of the nest chambers of the males and the dry females are situated at depths of not more than 50 cm, those of gravid and nursing females are found anywhere between 30 and 70 cm deep in the soil, the depth of occurrence of the nest being quite constant for nursing females.

In the burrows used during hibernation the chambers and passages are situated at greater depths. The jerboas do not enter into hibernation in the nest, but in the so-called 'chambers of hibernation' (KONDRASH-KIN & JEDIKINA, 1957) containing no bedding material and situated at depths of 22 to 32 cm in the case of males and 40 to 60 cm in the case of females. At places where the soil temperature goes much below the

---

* In Kizil-Kum the summer burrows with the animals estivating in them do not usually have the front passages blocked with soil. There are some suppositions that the presence of such a blocked passage is connected with the nature and density of the soil.

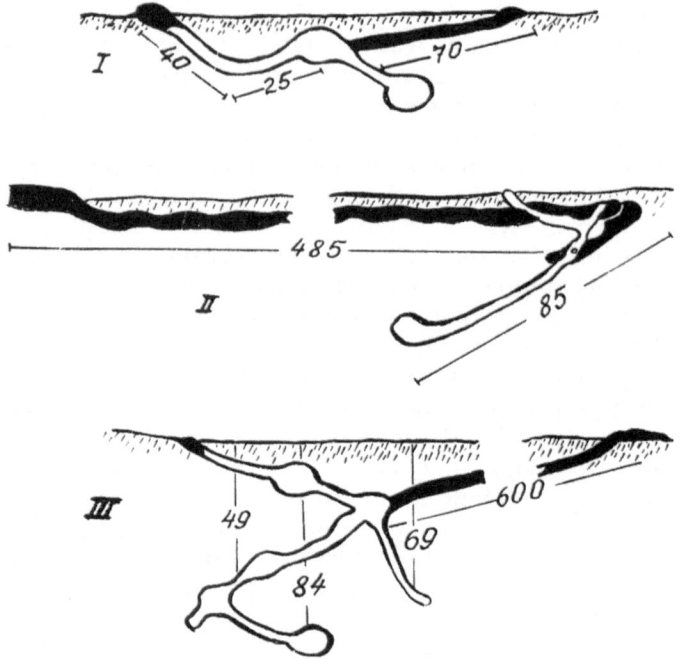

Fig. 9. Burrows of *Alactagulus pygmaeus*. I, II. summer burrows; III. winter burrows (after SKVORTZOV, 1955). (Distances are shown in cm).

freezing point, the chambers may be situated at greater depths, while on wet loam soil the chambers may be situated at higher levels. The increased humidity of the soil near the foothills of 'Ber hillocks' (in the lower Volga region) makes *A. pygmaeus* dig new chambers of hibernation in the offshoots of the burrows nearer the surface.

In Turkmenia during December–February, (SKVORTZOV, 1955) in none of the 34 burrows examined any *A. pygmaeus* was found hibernating in the nest chambers, while such animals were found in the hibernating burrows (Fig. 9) at mean depths of 68 cm. Hibernating *A. pygmaeus* have been found at 41.5 cm depth while awakened ones have been found in the chambers of hibernation and in the broadenings of the passages having no bedding materials. It has been shown that the process of hibernation is not interrupted by fluctuations in the temperature of the surrounding air between 1.6 and 4.6 °C. The animals are aroused at ambient temperatures between 6.8 and 7.0 °C. It seems that the chambers of hibernation not only play a thermoregulatory role, but also physically save the animals during times of flodding of the burrows after the winter thaws in the same way as the burrows of some ground squirrels function (KALABUKHOV, 1956).

494

Thus, the relative degree of efficiency of the thermoregulatory mechanisms and differences in the availability of suitable energy resources for different species of hibernating rodents are behind the geographical modifications in burrow structure and the animals' choice of the most suitable depth for construction of the hibernation chambers. *A. pygmaeus* predominantly subsists on green vegetation and its caecum is almost twice the size of that of *A. elater* (130–170 mm against 70–90 mm; ISMAGILOV, 1961). It feeds on the green parts of plants in spring and on a mixed food of green stems, flowers seeds, fruits and especially the flowers of *Anabasis salsa* in summer. The role of insects in its dietary is negligible.

In the southern part of Usturt and in Kizil-Kum *A. pygmaeus* feeds principally on *Anabasis salsa* and somewhat on the succulent *Salsola*. It does not relish *Anabasis brachiata* and *Nanophyton erinaceum*. In captivity it is known to consume *Tulipa sogdiana* and the fruits of *Malcomia* (SABILAEV, 1967, 1971).

In the Volga delta KONDRASHKIN & JEDIKINA (1957) observed two distinct periods characterised by differences in the type of food consumed by these rodents. In spring *A. pygmaeus* mostly prefer the bulbs of *Poa bulbosa* and the tulips *(Tulipa)* as well as the seeds and stems of *Poa* and ephemers; in autumn it eats the seeds of *Ceratocarpus turkestanicus*, *Bromus* sp., *Salsola* and cultivated cereal grains.

It is characteristic of *A. pygmaeus* to enter into hibernation relatively late, and to have a rather interrupted hibernation (particularly in the southern deserts) of short duration. It is, therefore, understandable that the animal does not accumulate much body fat. In fact, in certain cases *A. pygmaeus* have been known to be hibernating without any trace of subdermal or deep body fat layer. As in other species, the young animals of this species also have shorter periods of hibernation.

In the Usturt region (SABILAEV, 1971) this species enters into hibernation in November and awakens in February–March. In some winters (1962–63, for instance) the length of the period of hibernation was of 50 days only and even within this period a part of the population awakened during the thaws.

The hours of surface activity for *A. pygmaeus* are regulated by seasonal, climatic and physiological parameters. In Usturt in the beginning of April these animals appear on the surface just after the sunset while in May they emerge 40 minutes later and in June after a lapse of 20–30 minutes after sunset. The two peak periods of surface activity (March–April and October–November) are connected with reproduction and weight gain before the hibernation.

The young of *A. pygmaeus* born during the summer and the winter remain in hibernation long after their growth ceases and hence their body development lags behind that of the young born later in spring (Fig. 10). Only by next June do the former groups of animals attain

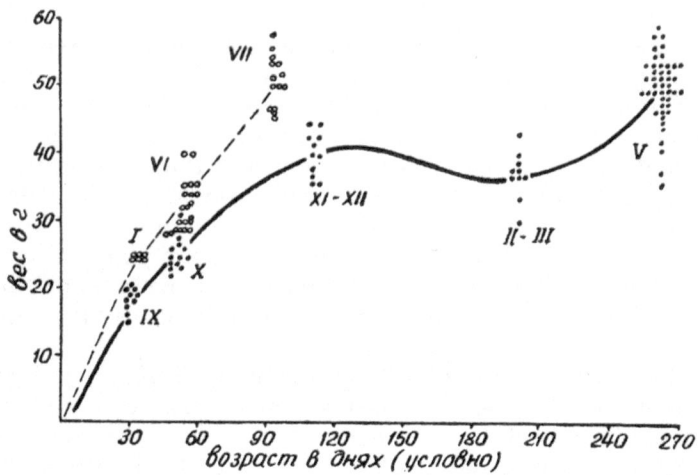

Fig. 10. Body weight in relation to age in *Alactagulus pygmaeus* of spring–summer and autumn litters. The roman numerals indicate the months of collection of the samples (after KONDRASHKIN & JEDIKINA, 1951). (Ordinate: weight of the animals in g; abscissa: age in days).

sexual maturity and begin to propagate themselves when the spring-born animals of the same year have already become sexually active. The reproductive period in this species is, thus, considerably prolonged.

The majority of workers (KONDRASHKIN & JEDIKINA, 1957; SABI-LAEV, 1971; LOBIZOVA, 1971; KAMNEV *et al.*, 1959) have described the spring or the spring-summer peak in reproductive activity for *A. pygmaeus*.

Table 5. The fecundity of *A. pygmaeus*.

Place	Number of females having litters of various sizes No. of implanted embryos								Mean embryo No. per gravid female	Source
	1	2	3	4	5	6	7	8		
Lower Volga region	2	43	99	65	18	4	—	—	3.3 ± 0.06	KONDRASHKIN & JEDIKINA, 1957
Gurjev district	1	12	43	15	—	—	—	—	3.0 ± 0.08	ORLOV, 1957
Aral Kara-Kum	—	2	9	10	4	2	1	—	3.9 ± 0.22	Our data
Usturt region	—	11	14	11	5	2	—	—	3.4 ± 0.17	SABILAEV, 1967
North-western Kizil-Kum	1	23	42	21	13	4	—	—	3.4 ± 0.11	SABILAEV, 1967 and Our data

496

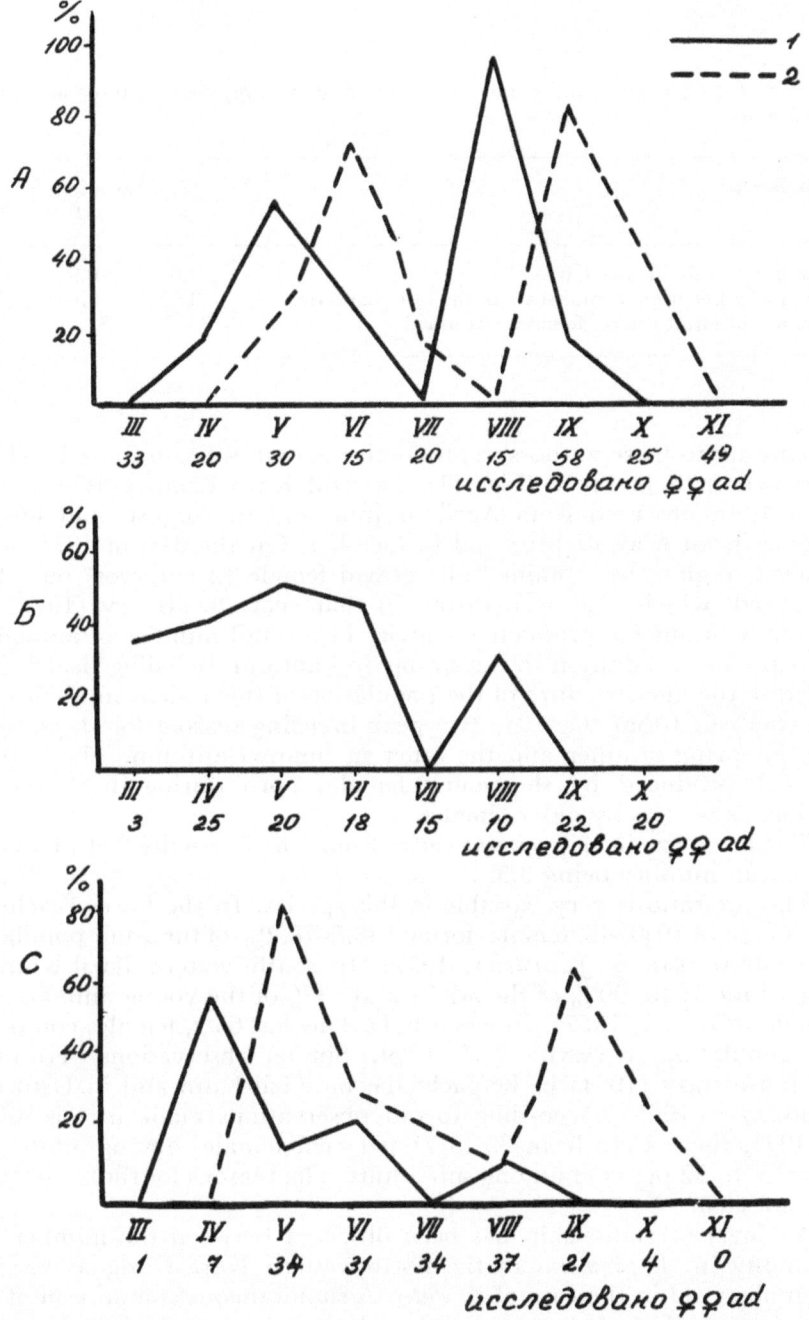

Fig. 11. Reproduction in *A. pygmaeus*. A. in the lower Volga (after KONDRASHKIN & JEDIKINA, 1957); B. in north-western Kizil-Kum (after SABILAEV, 1971a); C. in Aral Kara-Kum (our data). 1. (continuous line) – percentage of pregnant females; 2 (broken line) – percentage of lactating females. In the abscissa the roman numerals indicate the months of sampling and the arabic numerals indicate the number of adult females examined.

Table 6. Comparison of the number and fecundity of *A. pygmaeus* at three localities in Kizil-Kum.

Populations	Locations		
	I	II	III
Number of animals per km	2.6	0.9	0.6
Mean number of placental scars of the last generation	3.17	3.45	4.0
Number of adult gravid females examined	23	31	7

According to these authors, reproductive activities begin in early March and end in September (Fig. 11). In Aral Kara-Kum, gravid females have been observed from April to June and in August, and nursing females from May to June and in October. On the 31st of 1971 in the Chelcar region, we obtained one gravid female (2 embryos) out of six catpured, which was born earlier in that year. SABILAEV (1971) observed a summer reproductive activity in a small number of animals of this species, in addition to the spring and autumn breeding peaks. Considering the age structure of the population of this rodent in Kizil-Kum we concluded that there are two peak breeding seasons for *A. pygmaeus*, one in spring-summer and the other in summer-autumn. The autumn litter is produced by the young females born during the preceding spring, as well as by older females.

The litter size in *A. pygmaeus* varies from 1 to 7, usually 2–4 (Table 5), the mean number being 3.5.

The sex ratio is very variable in this species. In the lower reaches of the Volga in 1944–46, females formed 48.5–78.2% of the adult population (KONDRASHKIN & JEDIKINA, 1957). In south-western Kizil-Kum, in May–June 1946, 90% of the adults and 63% of the young animals were females (ORLOV, 1957). In western Turkmenia, 65% females comprised the population (KAMNEV *et al.*, 1959). Similar observations were made by ISMAGILOV (1961) in Betpackdala, in Kizil-Kum and in Usturt by SABILAEV (1965). According to our observations made in Kizil-Kum in 1971, there were from 33 to 71 per cent females among adults and from 30 to 52 per cent among sub-adults. The reasons for these variations are not clear.

An inverse relationship has been observed between the number and fecundity in *A. pygmaeus* in the north-western Kara-Kum, as has also been observed in the case of *A. elater*. A simultaneous comparison of two populations of *A. pygmaeus* of different sizes, located 50–100 km away from one another has indicated that this relationship is statistically significant (r = 0.32) (Table 6). Probably, shifts in the sex ratio also plays a determining role in regulating the number of these animals.

Pygeretmus platyurus Lichtenstein, 1823

This rare species is endemic to the Aralo-Caspian deserts and semi-deserts. For about 40–50 years it was known only through a single specimen captured. It is now believed that it inhabits a wide area and is quite numerous in pockets within the semi-arid regions. From the northern shores of the Caspian Sea and the right-bank of the Ural river near Uralsk town, the range of this species extends to northern Usturt and Usturt and to the south upto the Mangishlask peninsula and includes the desert regions to the north from the Aral Sea upto the Sari-Su river. The northern border of the recent distribution of this species coincides with the southern border of the Festuca-wormwood steppes in the environs of Chelcar station (Asktubinsk district) and the middle reaches of Irgiz and Sari-Su rivers (SILVERSTOV *et al.*, 1969). *Pygeretmus platyurus* were earlier distributed more intensively in the north and the south. It has now yielded much of its older habitat to *P. zhitkovi.*

An isolated relict population of *Pygeretmus* has remained in northern Kizil-Kum probably since the change in the course of the Syr-Darya river took place there. Undoubtedly this species is well adapted for the semi-deserts and the northern deserts. It rarely occurs in the southern and sandy deserts. This is clearly corroborated by the fact that out of 251 *Pygeretmus platyurus* captured between the Ural and Emba rivers from 1957 to 1964, 155 specimens were found in the semi-desert zones, 92 in the desert zone and only 4 on the border between these two zones.

In spite of its preference for the semi-desert and the desert niches, *P. platyurus* is fairly wide-spread. It has been captured from almost all the bio-climatic zones examined although nowhere has it been found to be as numerous as *A. elater* or *A. pygmaeus.*

P. platyurus is mostly found on clayey plains with thin covers of *Salsola.* Its habitat is mostly of the nature of takirs, with covers of *Anabasis salsa* and other annuals of the *Salsola* family. These places are comparatively poorly inhabited by other rodent species. Hence, the number of openings of burrow of other species of jerboas and of the small ground squirrel in these places have been found to vary widely (from 2 to 20 per hectare) from region to region (the mean for Aral Kara-Kum being 7). This is in sharp contrast to the 300 odd burrows per hectare found on more vegetated habitats, particularly with wormwood-*Anabasis salsa* associations. The plant association of its habitat consists of the genera *Salsola* viz. *S. crassa* and *S. rigida*, and others such as *Kalidium caspicum, Sueda* sp. and *Nanophyton erinaceum* with an occasional plant of *Halocnemum strobilaceum,* here and there. On hummocky terrains with *Anabasis salsa* and on small saline plots this rodent is rare. More than any other jerboa species, *P. platyurus* avoids places of thick vegetation cover, particularly where the vegetation is 10 to 15 cm tall. *Allactaga elater* also prefers similar situations while *Alactagulus pygmaeus* keeps in the periphery of these regions. In the region between the Ural and Emba rivers both *P. platyurus* and *Scirtopoda telum* are found to occur in the same habitats. Incidentally, the abundance of both these species is more marked towards the south, indicating thereby their preference for the semi-deserts and dry steppes, and not for the deserts. Feeding on halophytic plants, *P. platyurus* dig their burrows near the takir-like slopes of gravelley sandy hillocks. Night observations revealed that these rodents usually live away from the main roads.

Relatively high densities of *P. platyurus* population have been found to occur at the following places: Donghuztau (north Usturt), the Tassai valley (north Usturt), the environs of the settlements along the Irgiz river, the environs of Chelcar town and at or around the Copmulla and Saksaulskaya stations. This jerboa is numerous at places in the extreme south of Urals district. In the course of night countings during summer, 1974, a total of 152 jerboas, including 17 *P. platyurus* were observed on a 100 km length of road.

The population density of this species in the environs of Dossor between Ural and Emba rivers, at Bytschunas, Dangar and Karatal, is very high. 3 specimens of *P. platyurus* were trapped from each of four 0.25 hectare plots in one of these regions in late April, 1957. During the autumn of 1960, 4 specimens of *P. platyurus*, 4 of *Alactagulus pygmaeus*, 3 of *Allactaga major* and 4 of *Allactaga elater* were caught at night on a 20 km route in such a region. In the Aral Kara-Kum 7 specimens of this species were caught on 4 plots of 5 hectares each in its typical habitat. In the north Aral region, these jerboas are found in larger numbers on the borders of the ancient valley of Turgai and on the south-western beach of the Aral Sea, as evidenced by the occurrence of their remains in the pellets of birds of prey.

P. platyurus is particularly numerous in the region of the station Chelcar, where on the clayey plains it is the most numerous Dipodid species with a mean density of 3 to 5.2 animals per km. This species inhabits there not only on clayey, but also on gravelly wormwood plots, although its number at the latter sites is small. Apparently, there is an ecological optimum density level for this species in such places. In the pellets of *Athene noctua bactriana* from the north-western Aral sea region the remains of bones of *P. platyurus* comprised from 0.09 to 2.0% and in the pellets of the eagle-owl – from 0.4 to 0.5% of all mammalian remains. Out of 13 samples of the owl's food examined the remains of *P. platyurus* could be detected in only 5. Remains of both *A. elater* and *A. pygmaeus* have been found to occur in the owl's pellets in a similar way. This points to the lower density of *P. platyurus* in this region in comparison to the western part of the area between the Ural and Emba rivers. In a single pellet of the eagle-owl in Usturt the remains of 7 specimens of this jerboa were observed (3 to 5 specimens are usually found). The remains of *P. platyurus* are also frequently found in the pellets of the brown owl. Obviously, because of its comparatively slow speed, the density of *P. platyurus* population is more likely to be influenced by its predators in comparison to the populations densities of other jerboa species.

The data of awakening from the state of hibernation for *P. platyurus* are not exactly known. According to our data, the earliest date on which an adult male of this species was observed, in 1959, in the Uralo-Emba district, near a burrow of *Citellus pygmaeus*, was April 12. In Aral Kara-Kum the first post-hibernating *P. platyurus* were caught on 25 and 28 April. It would, therefore, appear that these animals resume their surface activity in the second ten-day-period of April. The last animal seen before the onset of hibernation was caught on 10th October, 1950 in Gurjev district. The accumulation of body fat begins in these animals somewhat earlier than in other jerboa species and in some years this may occur even from the beginning of June when adult males with thick tails are observed.

P. platyurus, unlike many other jerboa species (e.g. *Scirtopoda telum. Allactaga major*), are apparently active only during the night. The first animals to appear on the surface have been marked at about 22 hours. During daytime these animals speedily dig in the soil and take cover in their burrows. Their nocturnal habit is also proved by the fact that upon examination of a large number of pellets of birds of prey (*Buteo buteo, Buteo rufinus, Falco tinnunculus*, the eagles) the body parts of *P. platyurus* have never been found, although the remains of other jerboas do infrequently occur in such pellets.

In the Ural-Emba region the breeding season of this species begins at the end of April and lasts till May or the beginning of June. Towards the end of April the size of the testicles of the males increases. Young animals of this species have been captured from the end of May to the end of June. In the northern part of its habitat (Turgai river region) a female with 4 recent placental scars was found on June 30, 1966. More to the south, in north Usturt, a nursing female with 6 bare youngs with unopened eyes was obtained on the 17th of May, 1950, and in the Aral Kara-Kum, a gravid female with 5 embryos (12 mm length) was obtained on the 18th of May, 1962. Further to the south, in Karakalpac Usturt, two lactating females and three young ones were obtained at the end of May, 1960, while two young animals were found at the end of the second ten-day-period of May.

Thus, this species seems to have only one peak reproductive season in spring. The males predominated in a sample of the population of *P. platyurus* from the Ural-Emba region (39 specimen), the male:female ratio being approximately 3:2.

While other species of jerboas freely move in the paths of other animals, *P. platyurus* tends to avoid such paths and prefers to move on the virgin

Fig. 12. Typical postures of *P. platyurus*. (Drawings by V. M. Smirin based upon photographs by V. B. Silverstov & V. S. Lobachev).

soil. This rodent is fearful, but, at the same time, a curious animal. When faced with any unwelcome noise, it shows nervousness and tenseness. The jerboa stands on its hind legs, stretches its body, straightens up its ears and turns its head towards the side of the noise (Fig. 12). It sometimes runs hither and thither and sometimes stands erect like a column or it hides itself lying flat on the surface, and after sometime it goes away with its usual hobbling step. When caught in the path of a light at night, the jerboa usually jumps away slowly, stopping several times and then takes cover in any secluded place available. Its usual hiding places are near the bushes of *Anabasis salsa* or the depressions in the ground. In the beam of a spot light one can see this rodent moving with its ears held closely back as in the small gerbil *(Meriones)* or the *Cricetulus*, but unlike in other jerboas. The careful movements of its hind legs resemble the characteristic stealthy movements of a cat. The running animal can be easily caught. When it is followed closely, the animal hops on its hind legs and usually changes the direction of its movement abruptly. In one jump it may clear 40–50 cm.

Observations made at night on *P. platyurus* with the help of infra-red light and other suitable devices indicate that it does not stay at one place too long and slowly moves all the time. We did not find any permanent burrows of these animals. One of the marked rodents changed burrows 4 times within 3 days. At daytime the jerboa, when released from captivity, does not run away but begins to dig a burrow. There are no reports in the literature about the way *P. platyurus* digs its burrow. When released in *Anabasis salsa* woodlands, this jerboa does not run away but inspects and sniffs at the ground around it till it finds a proper place. It then scrapes the ground with its upper cutting teeth. The rodent pushes under itself the loosened lumps of soil with its forelegs and then energetically throws these out with the hind legs. When the hole becomes 4–6 cm deep the jerboa changes the way of throwing out the soil. The dug out soil is deposited with the help of its hind legs at the edge of the outlet of its future burrow, and then the rodent turns and quickly throws the soil out of the hole, with the help of its breast and nose while supporting its hind legs against the wall. When a nearly straight passage, 20 cm long at an angle of 30–45°, has been dug, the animal makes a small chamber of 10–12 cm wide where it hibernates. The depth of this chamber may be 10–15 cm. All its burrows of daily occupation have the same simple structure. The inhabited burrows are closed from the outside with soil plugs, 5–8 cm in length and it takes nearly 20 minutes to throw this plug out. It digs its burrows of day-to-day occupancy from 3.05 hours till 4 in the morning. It apparently does not use any strange burrow for its day's rest.

The contents of 14 stomachs of *P. platyurus* from the Ural-Emba territory and of 8 from the Aral Kara-Kum, examined by us, comprised of only green mass. 7 animals living in captivity, when offered a variety of plants, chose and ate the most juicy parts of *Anabasis salsa* only.

In the typical habitats of *P. platyurus* bushes of *Anabasis salsa* are frequently present, the stem and some parts of the root of which are dug to a depth of 5–8 cm. We had a chance to observe one *P. platyurus* near a bush of *Anabasis salsa*. The animal stretched its hind legs and reached for the most green and juicy parts of the stems. Hearing a sharp rustle the jerboa recoiled into a hole, dug out near the stem, and hid itself there, hugging closely to the ground. In captivity in the Aral Sea region, *P. platyurus* was found to eagerly eat animal food (spiders, little insects). In July they greedily fell upon the stems, splashed on with water, and did not eat the tulip bulbs *(T. schrenki, T. biflora)*, rhizomes and large beetles that were offered to them.

It is interesting to note that among all the jerboas this species is least afflicted with fleas. In the Aral Kara-Kum none of the 18 animals examined had ectoparasites. The flea *Mesopsylla tuschkan* was found on the body of this rodent in the southern part of Ural district, while 2 species of fleas, viz. *M. tuschkan* and *M. lenis* were found on 7 animals catpured in Usturt. In the semi-desert zone along the Ural-Emba river 4 species of fleas were found to occur on this rodent, viz. *Ophtalmopsylla volgensis, Neopsylla setosa, M. hebis* and *M. tuschkan*.

From 2.1 to 18 per cent of the animals captured in the Uralsk district in the Ural-Emba rivers in 1964 were found to be infested with fleas. Fleas counted on 68 animals gave an average figure of 0.3 flea per animal.

The flea, *M. hebis*, was found to occur on 21 *P. platyurus* captured from the desert areas of the Ural-Emba region in 1963. The mean index of abundance of this flea slightly exceeded 0.9 specimen per rodent. Sometimes the Ixodidae, in most cases belonging to the genus *Rhipicephalus*, are found on this jerboa.

P. platyurus have almost no contact with colonies of the great gerbil, at least in the north Aral territory. In the whole of north Priaralje (including Usturt, Aral Kara-Kum and the right bank of the middle portion of the Syr-Darya river) only one specimen of this rodent was found in any burrow of the great gerbil in the course of 20 years of extensive field studies.

Among the thousands of subfossil bony remains gathered from the vicinity of burrows of the great gerbil in one particular area for ten years (1947–1957), not a single bone of *P. platyurus* was included. Later on, however, the remains of two *P. platyurus* could be detected among 1452 specimens of animal remains collected from the colonies of the great gerbil. In our studies in Aral Kara-Kum, the remains of only one *P. platyurus* could be detected out of 1240 specimens of bony remains of small animals found in the burrows of the great gerbil.

In the Ural-Emba region, however, these animals apparently, have frequent contacts with the great gerbil and the small ground squirrel *(Citellus)*. It seems that in the northern and western parts of its habitat the contacts of this species with other rodents are considerable. But over a considerable part of its habitat in the north Priaralje area its contacts with other animals, unlike in the majority of the jerboa species (e.g. *Scritopoda telum*) are almost negligible.

Dipus sagitta Pallas, 1773

It inhabits the sandy landscapes of south-eastern Europe, Kazakhstan, Middle and Central Asia. It is spread from the western and north-western Caspian Sea (Tersko-Kum sands, Kuma river, lower-Volga and Volga-Don sands upto the mouth of the Medveditsa river in the north), north Aral Sea and Balkhash Lake to the north along the Ural river up to Inderborks town, then the Temir sands, Bolshie Barsuki, Sari-Suiskie and Pri-chuiskie Mujunkumi, along the valley of Irtish river upto 52° of the north latitude, and to the east upto the tape pine-forests of Prialtaiskaya steppe (Rubcovsk town district). It is also widely spread in Mongolia (Gobi) and in north-western China and north Afghanistan.

Its range includes the steppes, semi-deserts, and the north and south deserts. But in these limits *Dipus sagitta* can be found in separated populations, sometimes in a relative isolation. It inhabits the sands with different relief including the dune sands, overgrown with pine forests (tape pine forests along the Irtish river), and in thin sands in the depressions of the modern and ancient lakes of Mongolia. It avoids long chains of bare barkhans, and keeps to their outskirts, where vegetation grows. *Dipus sagitta* can be more often met on the tops of the semi-anchored sandy ridges and hillocks, where it dwells with *Paradipus ctenodactylus*. Stretches in the clayey deserts, several km wide, do not obstruct its distribution as it also dwells in the sandy islands in the middle of the clayey plains. In the Mongolian Altai it is also found on the crop areas, together with the jumper, and in the brushwood of *Caragana* on sands. It penetrates the mountains upto the height of 3088 m (the Altai ridge), and in the valleys of the rivers and in the depressions of the lakes on the thin sands and sandy hillocks; in Tuva – it occurs even in the sandy tracts of considerable grass covers and *Potentilla*-wormwood steppes. It is one of the most numerous and common jerboa species. The changes in its number during the year are characterized (KAZANTZEVA & FENJUK, 1937;

Table 7. Burrow and population density of *D. sagitta* per hectare in the semi-stabilised sands of north and south Kizil-Kum (after SABILAEV, 1971).

Type of land form	North		South	
	burrows	animals	burrows	animals
Sandy terrain with small hillocks	2.3	2.6	4.1	4.2
Medium-sized hillocks	3.2	3.2	5.3	5.5
Large hillocks	1.5	2	4	4.1

SABILAEV, 1971) by a peak in July–August (in Kizil-Kum-mean 7–9 specimens per hectare, maximum upto 21). A vast majority of one year old animals (about 4–6 specimens per hectare) perish by autumn, and in spring, the lowest number is observed (2–3 specimens per hectare).

The yearly fluctuations of the rodent have not been worked out but it is observed that their density remains relatively stable. In the southern deserts this species is more numerous than in the northern ones (Table 7).

In the outskirts of the Nucus town in 1971, its density was 0.6 specimen per km as observed by foot counting but the rodent was absent on the plots with thick cover of slag. In the period of re-establishment, the young males (2 per km of the motor route) in the north Priaralje were found in the areas where *Agropyron sibiricum* was growing in abundance and on wormwood sandy plain. On the clayey plain in north-western Kizil-Kum, however, its population was very low: 0.03 specimens per km of the motor route.

Unlike the *Allactaginae*, *D. sagitta* digs burrows with the forelegs, using the teeth only for the removal of obstacles. It throws the sand out with its hind legs and pushes it out with the nose. From the strong motions of the hind legs the sand scatters and forms a typical fan-like mound. The animal can dig about 45 cm deep within 10–15 minutes (BEME & KRASOVSKII, 1930).

The burrows are situated on the slopes or on the top of barkhans without any definite association with the bushy vegetation. Some preference is shown to the western slopes (ORLOV, 1957) and the bordering plots of the semi-overgrown and overgrown sands.

Dipus sagitta lives, as a rule, in the moist layer of sand, and because of that, at places where it is at a lower level (the tops of the barkhans) the burrows are correspondingly deep. The main tunnel is straight and slants into the sand at an angle ranging from 40 to 60°. At the depth of 15–40 cm it turns and sometimes branches and ends into a chamber (Fig. 13). The number of chambers in the burrows varies from 3 to 5. Generally the burrow is 60–150 cm deep, 2.5 m long and 70 cm deep. The animals usually live singly in burrows, rarely in pairs.

The movements of *Dipus sagitta* have been studied in detail as compared to those of other species because it is easier to observe their tracks on the barkhan sandso. The definite impressions of each finger can be identified and movements of every animal can be studied. KAZANTZEVA & FENJUK (1937) found the presence of individual ranges of *Dipus sagitta*. It has some characteristic features as compared to other rodents. Each of them consists of the burrow, feeding sites (Fig. 14) and paths for exploration (ORLOV, 1957). The animal constantly feeds in a particular range. Its size is 2–3 hectares, sometimes larger and depends chiefly on the availability of food and, perhaps, on the population density of the jerboas. In the home range there are several burrows, or burrow plots used at different times. Besides the main plot, the jerboas regularly explore the

506

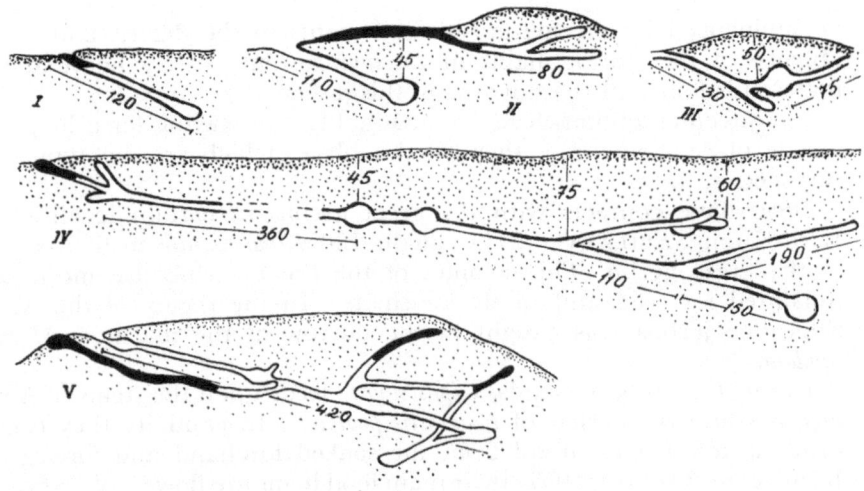

Fig. 13. Burrows of *D. sagitta*. I. temporary burrow; II, III. summer permanent burrows of an adult male; IV. summer permanent burrow of an adult female; V. winter burrow (after KAZANTZEVA & FENJUK, 1937) (Distances are shown in cm).

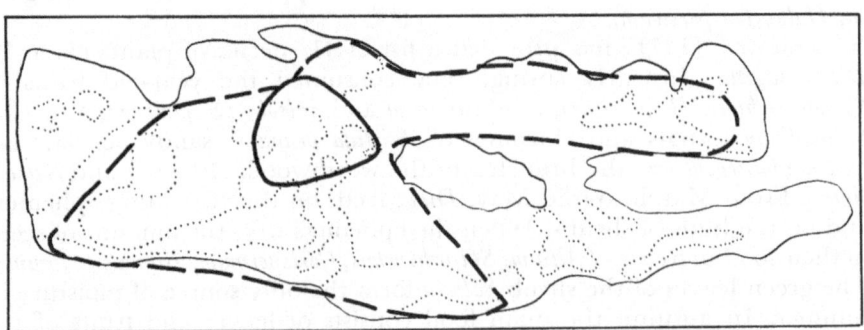

Fig. 14. The structure of an individual territory of *D. sagitta* (after ORLOV, 1957). The continuous thick line in the centre indicates the 'main' territory and the broken thick line indicates the feeding territory.

surrounding region which ranges from 12 to 15 hectares (upto 800 m across). The exploration areas of the young ones are very limited.

One more characteristic feature of the use of the territory by *Dipus sagitta* is the periodical change of the individual plots for the summer, though there are known cases of continuous use of a territory for about 1.5–2.5 months, practically on one place. Many individuals, however, during the period of activity, change their localities 3–4 times. During this the jerboas, sometimes like mice and *Microtus* (NAUMOV, 1951) may periodically return to the old places. These migrations to the new place

507

are influenced by the density of food plants in the deserts and by the absence of the storing habit in *D. sagitta*. Only rarely hoarded food material is found in the burrows of this jerboa.

The juvenile animals lead a nomadic life, not staying for a long time at one place. Sometimes they inhabit places which are not typical for this species.

Dipus have contacts with the great gerbils in certain localities. In Kizil-Kum, SABILAEV (1971) caught 130 *Dipus sagitta* in 650 colonies of the great gerbil. The colonies of the great gerbils are most likely inhabited for food and rarely for shelter. In the region of the Alaiski ridge one jerboa was caught in the burrow of the marmot, *Marmota caudata*.

In the Tersko-Kum sands, *Dipus sagitta* feeds upon the stems of *Elymus giganteus* and the earlets of *Eragrostis poaloides*. In captivity they eagerly drink water, licking it off from the soaked forehand and forelegs. In Kizil-Kum (ORLOV, 1957) their main food items are flowers of *Calligonum*, *Euphorbia heirolepis*, *Artemisia ereocarpa*, fruits of *Atriplex*, *Tournefortia sogdiana*, stems of *Corispermum papillosum* and the flowers of *Acanthophyllum borszewi*, and fruits of *Peganum harmala* and *Haloxylon persicum*. Among insects, the jerboas consume larvae of leaf-hoppers *(Psyllidae)* and galls on *Haloxylon persicum*.

SABILAEV (1971) has provided a list of 41 species of plants eaten by *Dipus sagitta*. In early spring, they consumed the year-old fruits of *Salsola richteri*, *H. persicum*, *Anabasis salsa* and *Salsola (S. paulseni, S. lanata, S. sogdiana)*, roots and rhizomes of *Aristida pennata;* sandy acacia, *Ammodendron argenteum*, the branches of the wormwood, *Artemisia* and *Salsola rigida*. From March to mid May, *Dipus* feeds on the vegetating ephemers and on the buds of shrubs. When the ephemers dry, the amount of seeds in their food increases *(Allium, Spirorhynchus, Corispermum, Bromus tectorum)*. The green leaves of the shrub *Salsola* form the only source of moisture in summer. In autumn the main food consists of leaves and fruits of the shrubs, especially frost-bitten stems of grey wormwood (sometimes destroying 75% of the vegetative parts of plants and also the perennial *Salsola*). In search of food the 'digs' of *Spermophilopsis leptodactylus* are usually visited.

In the Volgo-Uralski sands the list of the food plants includes 23 species. In the spring, their (KAZANTZEVA & FENJUK, 1937) main food comprises of the seeds of *Agriophyllum arenarium* and *Salsola kali* which are dug out of the sand by the animals. In summer the role of *Elymus giganteus* increases greatly from the beginning of June till the end of August. From the beginning of September the animals eat the seeds of *Agriophyllum*, *Corispermum intermedium* and *Kalidium caspicus*. The young stems of *P. harmala* and *Elymus* are also eagerly eaten. Insects are the constant, but not main, component of their food (KOLPAKOVA, 1932).

The time of emergence of the jerboas out of the burrows directly

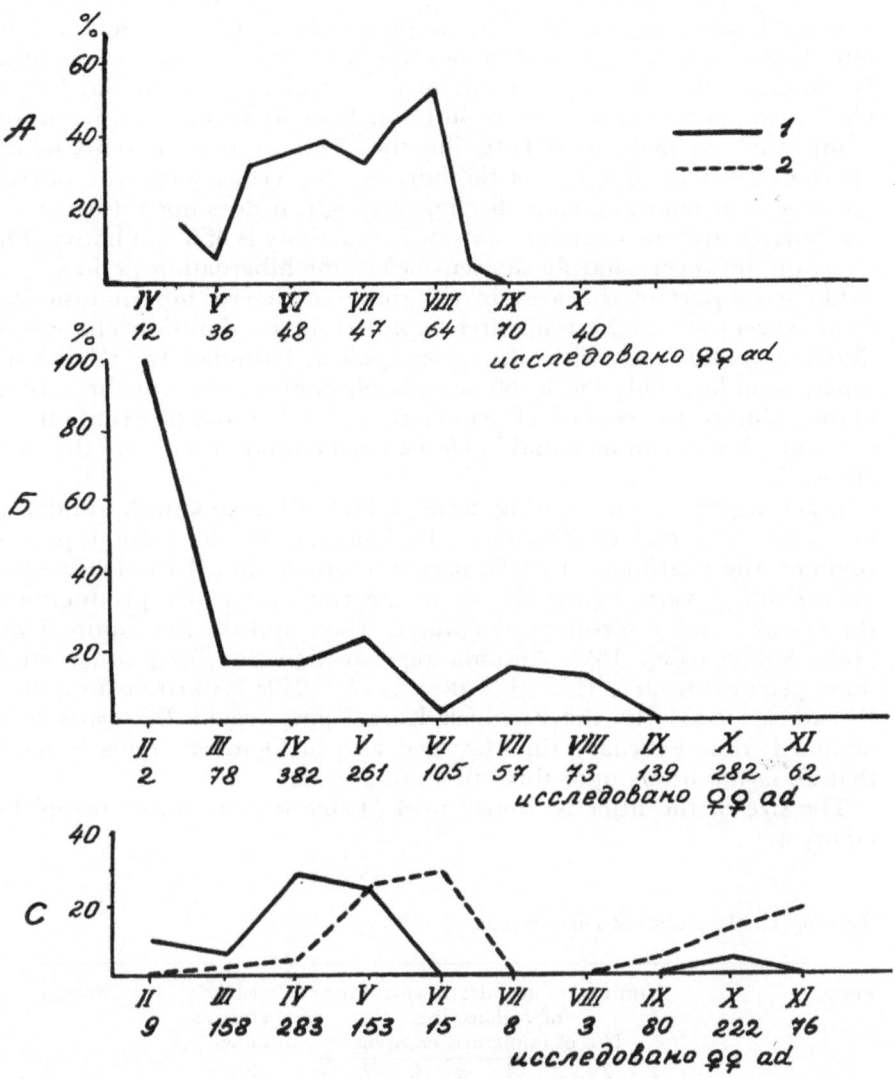

Fig. 15. Reproduction in *D. sagitta.* A. in the Volgo-Uralian sands (after KAZANTZEVA & FENJUK, 1937); B. in north-western Kizil-Kum (after SABILAEV, 1971b); C. in northern Turkmenia (our data). 1. percentage of pregnant females. 2. percentage of lactating females.

depends on the time of sunset. In the evening the first animal appears soon after the sunset in summer (July) and 25–40 minutes after the sunset in August, 42–60 minutes after the sunset in October, and 1–1.5 hours after sunset in November (KAZANTZEVA & FENJUK, 1937). Rain, wind and low temperature do not influence the activity of these rodents but their emergence time is somewhat delayed. In contrast to other Dipodidae, *Dipus* is very tolerant to low temperatures. In the Volgo-Ural sands, the temperature at dusk touches —8 °C and the maximum temperature at night as —17 °C but their activity pattern remains unaffected. After coming out of the burrow, the jerboa plugs the burrow opening with sand and starts feeding. At night it does not return to the rest burrow and the duration of its surface activity is of 9–9.5 hours. This duration, however, sharply shortens before the hibernation period.

On most part of the area in the southern deserts hibernation lasts from November till March, rarely upto April. Its duration changes in different years and with the geographical latitude. In Kizil-Kum hibernation lasts only for 36–60 days. Snow does not prevent the activity of the animals. In years of soft winter *D. sagitta* does not hibernate in the south at all, and can be found in December-February even over the snow cover.

Dipus sagitta begins mating from March (Tersko-kumsk sands) or even from the end of February (Turkmenia). In the second part of summer, the yearlings also take part in reproduction. In north-western Kizil-Kum (Sabilaev, 1971), there are two cycles of reproduction – the more intensive spring cycle (March–May) and the low summer one (July–August (Fig. 15)). Autumn reproduction has been observed at some places (Orlov, 1957; Kamnev et al., 1959; Nurgeldiev, 1960; Petrova, 1967). In the Zaunguz Kara-Kum, gravid *D. sagitta* were obtained from February till May and also in October. Thus it seems that *D. sagitta* litters upto three times in a year.

The size of the litter is from 2 to 8 (Table 8), the mean being 3–4 embryos.

Table 8. The fecundity of *Dipus sagitta*.

Place	Number of females having litters of various sizes No. of implanted embryos								Mean embryo number per gravid female	Source
	1	2	3	4	5	6	7	8		
Volga-Ural sands	—	17	50	10	3	—	—	—	3.0 ± 0.08	Kazantzeva & Fenjuk, 1937
Aral Kara-Kum	2	2	6	8	2	—	—	—	3.6 ± 0.08	Our data
North-west Kizil-Kum	1	16	49	63	36	23	9	4	4.2 ± 0.09	Orlov, 1957; Sabilaev, 1967; Our data
Zaunguz Kara-Kum	—	4	27	44	27	20	1	4	4.4 ± 0.08	Our data

510

Though it was earlier believed that moulting in *D. sagitta* takes place twice a year (BEME & KRASOVSKII, 1930) it now appears from observations made in Kizil-Kum that, as in other jerboas, the period of shedding of hair in *D. sagitta* is usually extended from the middle of May to the end of October. About 55–65% of the animals shed their hair in July–August.

THICK-TAILED THREE-TOED JERBOA

*Scirtopoda telum** Lichtenstein, 1823

S. telum is distributed in the southern feather-grass steppes, semi-deserts and north deserts, in the sands of the lower reaches of the Don (isolated population) between the Volga and Don river, in the region along north Caspic and north Aral Sea upto the Zaisan lake and in the Alacol depression.

The northern border of its range is spread from Kamyshin town on the Volga across the Inderborsk sands on the Ural to the Turgai valley (the region of several lakes), and to the southern border of Ulutau. Its range includes the area from the north Balkhash lake region to the Irtish river in the district of Semipalatinsk and in the south, southern Usturt, south-western Kizil-Kum (isolated population). It also occurs in the region of Jusali-Kizil-Orda towns, and along the south Betpackdala and the Ghu river till it reaches the south Balkhash lake.

From the beginning of middle Pleistocene it has been found in the Crimean foothills, and in the north, in and around the towns of Kremenschug near the Dnieper river and Volsk on the Volga river.

It inhabits different habitats, from steppes to deserts. In the north-western part of its range it dwells in the deserts in the central region in gravelly soil; in the south in the clayey wormwood-*Salsola* deserts and in the east, in the (eastern Kazakhstan and Mongolia) clayey deserts.

It is also found deep in the pine forests on the sand dunes in north Kazakhstan and in the birch forests along the Don and Rostov district, and in the afforestation areas in soil-conservation blocks at Povolje. It was also found in marshy fields (Kalmikiya), in arrow-grass and other types of steppes (Kazakhstan), in clayey and sandy deserts, rocky foothills and gravelly deserts (Kazak upland, Balkash Lake, Mongolia) and also along the fringes of the saline patches in melon fields, kitchen-gardens and along the roads.

S. telum avoids loose sands, sand hills, areas of wet soils with seed and *Tamarix* covers, sandy plains with dense grass covers and sparsely vegetated takirs and saline patches. *Scirtopoda telum* dwells in regions with relatively

---

* ELLERMAN & MORRISON-SCOTT (1951) have revised the nomenclature to *Stylodipus telum* (Eds.).

compact soils. It is not correct to consider it to be typically psammophile, as some authors have mentioned. Its morphological features also characterise this. The brush on its hind feet is not arranged perpendicularly, as in the real psammophiles, but fits closely to the fingers and is composed of comparatively soft short hair. Besides that, the distal end of middle metatarsal bone is much longer for *Scirtopoda telum* than in the other jerboas which are psammophile in habit *(Dipus, Paradipus)*. As it is known, the members of Dipodinae which inhabit soft quick-sands, lean while jumping on the all three fingers whereas in the species inhabiting the compact firm soils, the push is made only by one finger, and the two lateral ones play the role of shock absorbers (FOKIN, 1963). The abnormally long distal end of the middle metatarsal appendix also indicates a secondary adaptation for inhabiting firm, compact soils.

The occurrence of *Scirtopoda telum* in the sands in the north and south of the area, where other species like *A. sibirica* and *A. major* are also found, can be explained as follows. Firstly, the majority of the isolated populations of the jerboas in the sandy massifs of steppes and forest-steppes are the relics of the xerothermic period. Secondly, the habitat preference of *Scirtopoda telum* in certain localities in sands can be a result of increase in the density of the vegetation cover on clayey plots whereas most of the sandy region bear very sparse vegetation.

Scritopoda telum is thickly populated between the Don, the Ural, in Povolje, Predusturtje, and in the north Priaralje. In Aral Kara-Kum it is mostly found on the subsandy and loamy soils and sometimes on gravel plains and also along the sand dunes with vegetation cover of the grey worm-wood, *Carex physodes* and *Haloxylon aphyllum* (Fig. 8). On the sandy and subsandy soils it is mostly associated with the colonies of the great gerbils (Fig. 16), and it forms an elementary focus of plague (MURTAZANOVA et al., 1964). In the years of high density there are upto 12–20 specimens per hectare. On the firm subsands where also there is a large number of burrows of great gerbils, 7–8 specimen of *Scirtopoda telum* are found per hectare (max. 13).

The clayey plains having wormwood – *Anabasis* association and rare plants like *Haloxylon aphyllum* are inhabited by *Scirtopoda* more evenly. They are found both near the burrows of the great gerbils and outside (5–6 mean on 1 hectare). On the takirs grown with *Salsola* and *Anabasis salsa*, the density of this species is low (2–3 on 1 hectare).

During the periods of low density, however, *Scritopoda telum* stays only on clayey and loamy plots, over which *Anabasis salsa* grows abundantly. Its mortality increases due to the epizootics of plague in the colonies of gerbils and due to the high activity of predators. It is well known that the jerboas are highly sensitive to this disease (SHMUTER et al., 1957; MURTAZANOVA et al., 1964). The recovery of their numbers is usually preceded by the abundance of *Anabasis salsa* which plays an important role in their survival (Fig. 17).

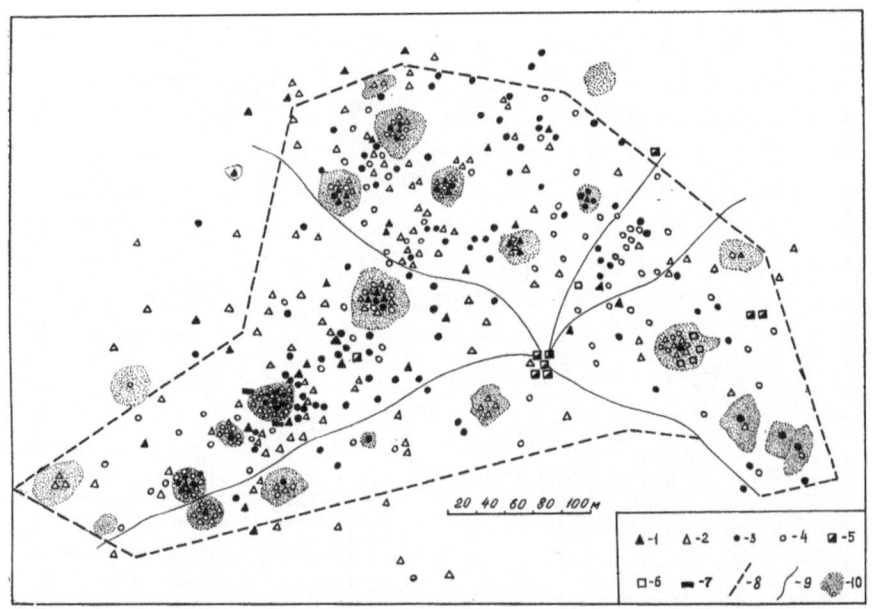

Fig. 16. Occurrence of *S. telum* and other Dipodid rodents in areas of preponderance of the great gerbil in Aral Kara-Kum during summer of 1964–65. 1. *S. telum* – adult males; 2. *S. telum* – adult females; 3, 4. *S. telum* – sub-adults; 5. *A. sibirica* – sub-adult males; 6. *Alactagulus pygmaeus* – subadult males and females; 7. *A. elater* – adult females; 8. the border of the study region; 9. roads; 10. colonies of great gerbils.

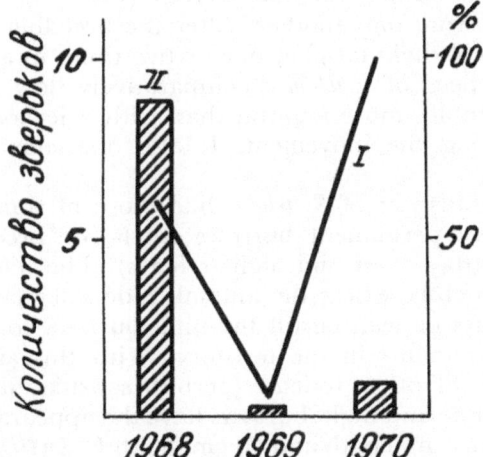

Fig. 17. Population dynamics of *S. telum* in Aral Kara-Kum. I. number of animals observed at night per km of route through an *Anabasis salsa* community; II. percentage of great gerbil burrows inhabited by *S. telum*.

513

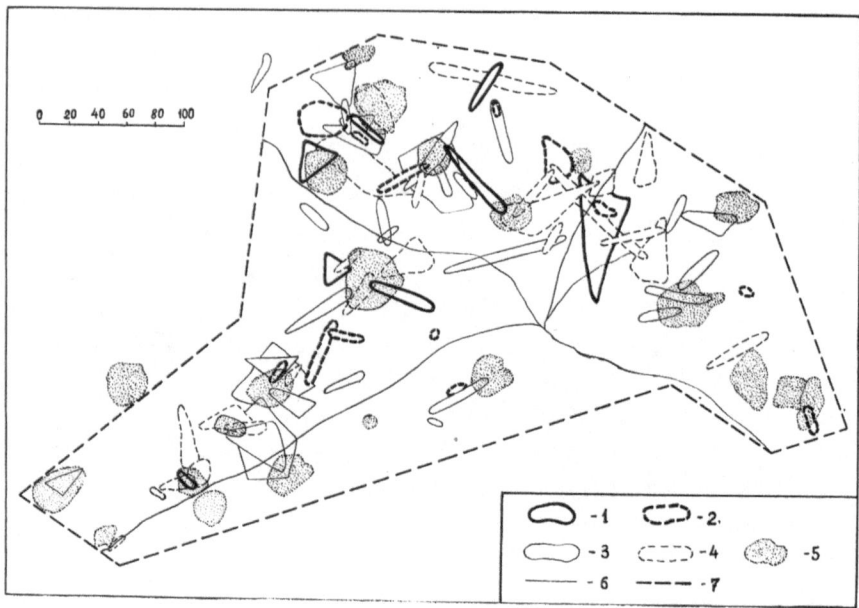

Fig. 18. Individual territories of *S. telum* in Aral Kara-Kum. 1. adult males; 2. adult females; 3. sub-adult males; 4. subadult females; 5. colonies of great gerbils; 6. roads; 7. the boundary of the study area. (Scale denotes distances in metres).

The individual ranges of the adult animals in summer are from 20 to 45 m in diameter which do not overlap (Fig. 18) and are situated 50–100 m apart from one another. After the breeding season *Scirtopoda telum* changes the individual plots one or two times in a month.

The running speed of *S. telum* is comparatively slow. In its behaviour it, perhaps, resembles more a gerbil than with a jerboa as it often uses the forelegs during the movement. It also does not jump over long distances.

The summer burrows of *S. telum*, like those of some other jerboas, are of two types; permanent burrows (nests, or breeding ones) and temporary ones (day's rest and night refuges). The temporary burrows are simple in structure where the animals hide safeguarding themselves from their enemies or waiting till the unfavourable conditions are over during their movements in the territory. With this aim, *S. telum* also uses the burrows of other rodents (gerbils, sousliks, jerboas). Its own temporary, day time and night burrows have the appearance of a straight, blind alley of upto 2 m length and 40 cm depth (LARINA, 1938). According to our observations near the Aral sea the inlet is usually very narrow. The day's resting burrows are usually plugged with soil, but those used during the night, as a rule, remain open. It is reported that the soil

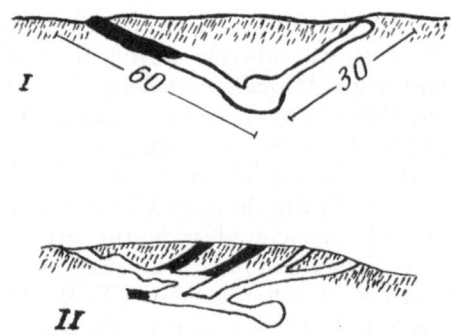

Fig. 19. Summer burrows of *S. telum* in the Volga river sand. I. the burrow of a female (after Larina, 1938); II. the burrow of a male, dug on November 4 (after FENJUK, 1928).

plugs play an important role in thermoregulation (MOKROUSOV, 1960). It must be noted that *S. telum*, in contrast to many other jerboas, do not always close the temporary burrows with plugs. But the permanent burrows are plugged by the animals not only at day time from inside, but at night also, particularly when the female leaves the litter while feeding. According to MOKROUSOV the temporary burrows usually have no chambers and nests in Pricaspi. A heap of sand is usually found near the burrow opening which is excavated out by the rodents.

The breeding and summer permanent burrows are very variable in structure and are composed of 1. an initial passage, 2. the main outlet, at day time it is plugged with soil, 3. the 'bolt run' or the emergency exit ending 3–5 cm below surface (Fig. 19) (it is used by the animal in case of a danger, when the enemy intrudes into the burrow), and 4. nest or day's rest chambers (LARINA, 1938). With prolonged use, the burrow becomes complicated and supplementary bolt runs and chambers appear (upto 3–4). The most-used outlet sometimes change and if so, the old one is securely closed with soil.

The initial passage of *S. telum* burrow takes a typical turn often at right angles and ends blindly into a chamber. The chamber is often extended into one or more passages. The length of permanent burrows of *S. telum* varies from 65 to 270 cm (mean 168), the depth from 20 to 70 cm (mean 39; average of 14 burrows, LARINA, 1938). In the littering chambers the ball-shaped nest is made of small fragments of rhizomes of *Poa* sedge.

The temperature in the nests of *S. telum* fluctuates at lesser depths from 22 to 24.5 °C in the beginning of August and from 18 to 23.5 °C at the end of the month (FENJUK, 1929).

The burrows in which the rodent hibernates are not very complicated, but extend deep into the earth, upto 120 cm. They are always closed with one or several soil plugs.

515

The associations of *S. telum* have been better studied than of other jerboas. In the Ilmenskaya subzone of the north-western Pricaspi they share their burrows with *Meriones meridianus, M. tamariscinus* and *Citellus pygmaeus*, and sometimes with *Allactaga elater* and *A. major, Alactagulus pygmaeus, Mus musculus, Cricetulus migratorius, Microtus socialis* and *M. arvalis*. It was observed that the social interactions, intraspecific as well as interspecific, increase with the shortage of forage. We have also observed that it is more closely associated with the great gerbil in the north Priaralje as it is found in their burrows in larger numbers as compared to that in the burrows of other rodents (Table 1). At many localities in the northern desert, *S. telum* is a predominant rodent species in the arid biogeocenosis.

There is scanty information about the feeding habits of *S. telum* in nature. FENJUK (1928) found in the stomachs of 11 animals from the Volga river valley, vegetative parts of wheat, the bulbs of *Tulipa biflora, Ornithogalum*, the rhizomes of *Poa bulbosa* and the lichens. In the Aral Sea territory we observed them feeding upon the green vegetative parts and seeds of *Anabasis salsa, Alyssum dasycarpum*, wormwood, the rhizomes of *Poa bulbosa* and the tulip bulbs (*Tulipa schrenki* and *T. biflora*).

In experiments conducted in burrows (FENJUK, 1929) *S. telum* were found to feed upon 19 out of 50 species of plants which wereo ffered (especially *Gagea bulitera, Agropyrum repens, Setaria viridis, Atriplex tatarica*). Many plants are eaten by the animals in certain periods only, the seasonal feed preferences are most marked during periods of drying and fruiting of plants like *Setaria, Agropyrum, Atriplex* etc. They shell out even the little seeds from the earlets.

Though *S. telum*, like all other Dipodidae, is active at night, the time of its coming out to the surface is very variable and to a great extent depends on the age and the physiological state of the animal. The young *S. telum* in the autumn, come out even at the beginning of dusk and sometimes also appear during the daytime. Direct observations made on the number of animals on constant routes indicate a bimodal activity pattern of *S. telum* in summer and autumn; the first peak appears between 10 and 12 o'clock at night and the second, the largest one, nearly at 3 o'clock in the morning (Fig. 20).

Before entering into hibernation, beginning with the second part of summer, the animals considerably fatten. There are typical large deposits not only of internal but also of subcutaneous fat. The diameter of the fat deposits in the tail increases upto 8–15 mm by September. The dates of hibernation depend on the fattening of the whole population and of the individuals. Because of this, the first to hibernate are the adult males, then the females and lastly the young ones. The animals enter their burrows for hibernation one by one.

The accumulation of large deposits of fat before hibernation is more marked in *S. telum* than in other species. The winter hibernation ends

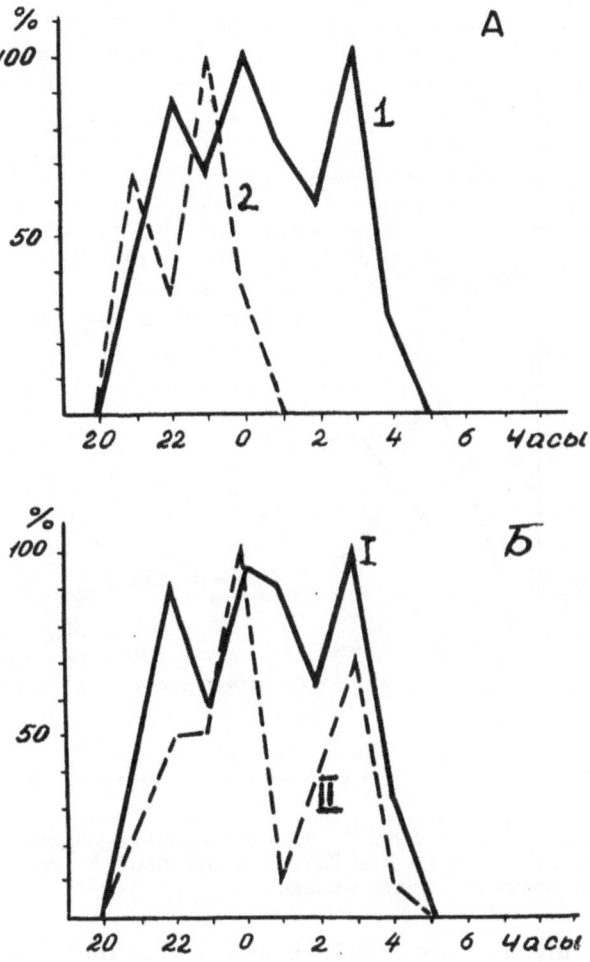

Fig. 20. Daily activity of Dipodidae (based on the number of animals observed on a permanent transect in the Aral Sea region). A. in summer (July 6, 1970); B. in autumn (September 21, 1970). 1 and I indicate *Alactagulus pygmaeus;* 2 and II indicate *S. telum.*

in the middle of March and the first to come out are the males. It coincides with the beginning of sprouting of vegetation and varies from 1 to 1.5 months in different geographical regions and in years. Hibernation of *S. telum* takes place 2–3 weeks earlier than in many other associated jerboas *(A. pygmaeus, A. elater, D. sagitta).* Usually it is during September or in the first half of October.

On the basis of the collection of lactating females and young ones from the beginning of June to early July and of pregnant females in the be-

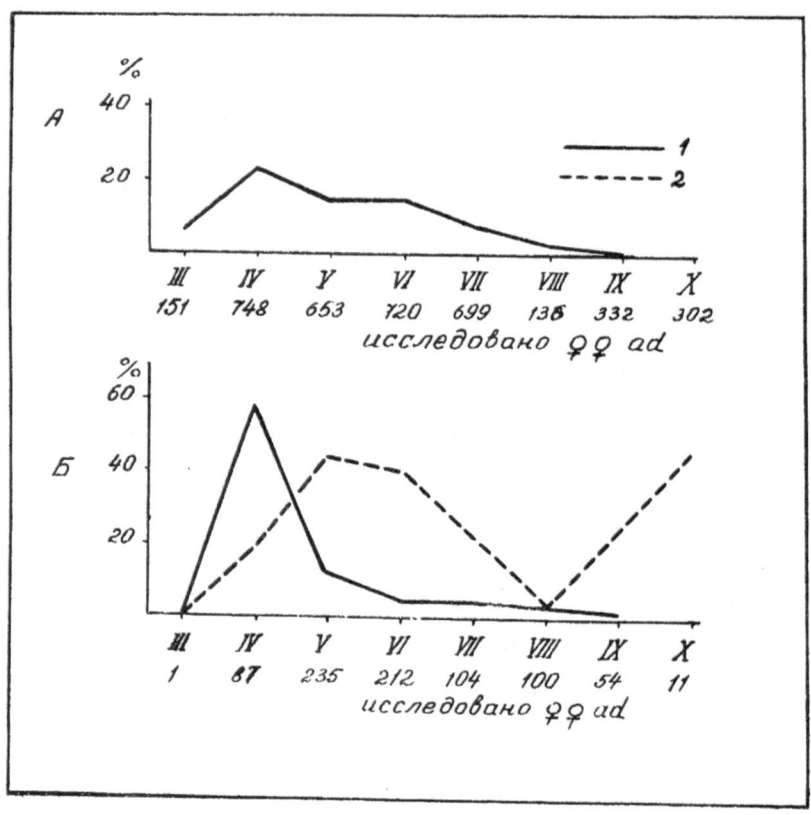

Fig. 21. Reproduction in *S. telum*. A. in the north-western Caspian Sea region (after MOKROUSOV, 1957); B. in the Aral Kara-Kum (our data). 1. percentage of pregnant females; 2. percentage of lactating females.

ginning of July, FENJUK (1928) suggested, that there are two litters of *S. telum* in a year, in the Low Volga region. He mentioned that *S. telum* has only one stretched period of reproduction during a year. LARINA (1938) wrote about a second autumn reproduction in Kalmikiya, as in the July– August period she obtained males with scrotal testes. MOKROUSOV (1957), however, described only one stretched breeding season from March to August and found that a female litters only once during a year (Fig. 21A). In Aral Kara-Kum the reproduction of *S. telum* is stretched from April to August (Fig. 21B). But, in our opinion without detailed investigations, the possibility of a second breeding cannot be ruled out.

The number of embryos of *S. telum* varies from 2 to 8 (Table 9), but, for majority of the animals, it is 3, 4 or 5 (mean 4) in Povolje, according to the data of MOKROUSOV, and 5.2 ± 0.19 in the north Aral sea region (Our data).

Table 9. The fecundity of *S. telum.*

Place	The number of females having litters of various sizes No. of implanted embryos								Mean embryo per gravid female	Source
	1	2	3	4	5	6	7	8		
Lower Volga	—	—	—	—	—	—	—	—	4.1	Mokrousov, 1957
Aral Kara-Kum	—	5	5	43	46	39	16	10	5.2 ± 0.19	Our data

A vast majority of *S. telum*, according to our data, perish during the second year of their birth. Only 5–7% attain an age of over two years. In the north populations the duration of life is apparently longer, like other species of the jerboas.

TURKMEN JERBOA

Jaculus turcmenicus Vinogradov & Bondar, 1949

This comparatively recently described species happens to be a common inhabitant of the Southern deserts of north-western Kizil-Kum and Kara-Kum. Its habitat adjoins that of *A. bobrinskii* and is situated along both sides of the modern valley of the river Amu-Darya. Only at places near the middle portion of this river *J. turcmenicus* does not penetrate into its right bank.

J. turcmenicus appears to have a similar evolutionary history as *A. bobrinskii*. Although it has been captured on the borders of clayey and sandy deserts, its characteristic morphological features (e.g. the hair on the fingers of its hind legs) would suggest that it belongs to the sandy terrains. Its recent finds on clayey plots in many parts of Turkmenia (Fenjuk *et al.*, 1957) have been rather intriguing. Stalmakova (1957) and Sapozhenikov (1963) have expressed the opinion *J. turcmenicus* depends on sandy soils for its food, but this has not been confirmed by more recent studies.

Like other jerboas, this species also avoids living on bare takir tract where no vegetable food is available. On the same *takirs*, however, where *Salsola* cover is present, *J. turcmenicus* is commonly found in abundance. It usually occurs on the periphery of sandy terrains and on plots with very thin vegetation covers. In these respects it differs from a typical inhabitant of the clayey deserts – *A. pygmaeus*. The food of *J. turcmenicus* is not particularly vegetarian in nature (Babaev & Ataev, 1962; Sabilaev, 1970b). In our several encounters with this animal we have never seen it in the sands and only once have we found foot prints of this

519

species crossing the relatively low sandy ridges and going from one takir to another.

Thus *J. turcmenicus* dwells on the takir-like clayey plots of the complex clayey-sandy desert. In north-western Kizil-Kum there are the takir-like alluvial plains of the ancient river-beds and the delta of Akhcha-Darya and of the southern part of the hill of Bell-Tau, overgrown with *Anabasis salsa* and *Salsola rigida* (0.1–0.3 specimen per km of route).

In north Turkmenia the species has been found on clayey-saline plains with covers of *Anabasis aphylla*, *Haloxylon aphyllum* and other *Salsola* (0.5 specimen per km) and on takirs with *Anabasis salsa* and *Haloxylon*.

Observations made by BABAEV & ATAEV (1962) in southern Turkmenia and by us in north-western Kizil-Kum indicate that *J. turcmenicus* come out for surface activity about 1.5–2.0 hours later than species like *A. elater* and *A. pygmaeus*.

J. turcmenicus feeds on the juicy (and in early spring and autumn on the dry) stems of *Anabasis salsa* and *Salsola rigida*. In July–August the green parts and the fruits of these plants are eaten in identical proportions.

STALMAKOVA (1957) working in northern Kara-Kum obtained, on May 22, a lactating female with 3 young, 6–7 weeks old and 5 young nearly 3 weeks old. FOKIN obtained gravid females on March 16, April 18 and October 7 and 16. SABILAEV (1970a) found one gravid female in Kizil-Kum in the middle of July which bore signs of earlier pregnancies. In 1971, we could obtain, in north-western Kizil-Kum, gravid females on July 6 and 17 and on July 1. Two adult (1 year old) females with scar marks of three previous pregnancies were obtained. Thus, reproductive characteristics of *J. turcmenicus* seem to be similar to those of the majority of the jerboas of Middle Asia. There are 3 litters in a year – in spring, summer and autumn – the yearling also taking part in the autumn reproduction.

Little is known about the litter size of *J. turcmenicus*. Observations made by our colleagues (LOBACHEV & SHENBROT, 1973) on 13 gravid females, obtained from north-western Kizil-Kum, point to a mean embryo number of 4.0 $\pm$ 0.27 (min. 3, max. 6).

LICHTENSTEIN'S JERBOA

Eremodipus lichtensteini Vinogradov, 1927

It dwells on the sandy deserts of Middle Asia and Kazakhstan, from the north-eastern Aral regions upto the south Balkhash Lake (Sari-Ishikotrau). It is common in Kizil-Kum and south Kara-Kum. In the north of this area isolated populations occur in the vicinity of moving sand masses and near the wells in grazing lands, and in areas with covers of *Haloxylon persicum* and *Acacia* sp. More to the south, in Kizil-Kum, for instance, this species is more numerous and is found in compact sandy massifs. It is characteristic of this species to dwell on the complex sandy deserts, with alternating clayey (takir-like) and saline patches and usually with ruderal vegetation. It has been reported to prefer dune areas (BONDAR & ZERNOVOV, 1960, SABILAEV, 1967) although opinions differ on this point (OSADCHAJ, 1959). In western Kizil-Kum SIROECHKOVSKII (1954) observed this animal at the highest level of sand mass. In Turkmenia there are more *E. lichtensteini* in the semi-stabilised dunes near the wells and especially on the large hillocks near bushes (STALMAKOVA, 1955; NURGELDIEV, 1950). With the exception of certain areas in north-western Kizil-Kum, this species is found in fewer number than a similar species, *Dipus sagitta*.

The structure of burrows of *E. lichtensteini* seems to be like that of *Dipus*, although the former's burrows are less deep and have more outlets, especially the temporary (protective) ones, since the animals usually dig them in the hillocks, near bushes where they hide when apprehending danger. There are 2–4 reserve outlets in the burrow which are closed with soil plugs. It feeds on the bulbs, seeds and green parts of plants.

Occasionally, *E. lichtensteini* has contacts with the great gerbil, but such occasions are rare because of the former's small numbers. Only in north-western Kizil-Kum (SABILAEV, 1967a), as many as 43 *E. lichtensteini* were captured from 650 colonies of the gerbils. In southern Pribalkhashje all *E. lichtensteini* captured were obtained from rodent colonies near the foothills of the southwestern ridges of sands bordering on the takirs (ZHURAVLEVA, 1958). It would seem that their occurrence there has some connection with the specific changes of soil and vegetation brought about by digging activity of the great gerbil in the area.

In Turkmenia (STALMAKOVA, 1955; SAPOZHENIKOV, 1963) and in Kizil-Kum (ORLOV, 1957; SABILAEV, 1967; BREER *et al.*, 1971), gravid females were obtained in April and May. Only once in Karshinskaya steppe (PETROVA, 1967) a gravid female with two placental scars was obtained in October, but it is possible that one of the scars represented the previous year's pregnancy. Thus, only one litter in a year is definitely known for *E. lichtensteini*. The number of embryos in this species varies from 2 to 8 (SABILAEV, 1967) and is usually 4–6, i.e. somewhat more than in other jerboas. In southern Kizil-Kum (BREER *et al.*, 1971) the mean number of embryos per gravid female was 4.7 in April and 6.5 in May.

COMB-TOED JERBOA

Paradipus ctenodactylus Vinogradov, 1929

Its range is limited to certain portions of the southern deserts, the southwestern Aral region and the deserts of Kizil-Kum and southern Kara-Kum. In the north it does not go upto the Syr-Darya river for about 100–150 km. To the south it goes upto Repetec, in the north-eastern outskirts of the Bucharski oasis.

It is a stenotopic species and is a typical psammophile particularly of the barkhans, covered with shrub vegetation like *Salsola richteri*, *Haloxylon persicum*, *Tamarix* and *Calligonum*. Isolated settlements of this species occur on the crests of stabilised dunes.

The species attains significantly high population density in restricted areas and in most parts of its range, population often comprises only a few dozen groups.

The burrows of *P. ctenodactylus* are similar in situation and structure to the burrows of *D. sagitta*. These are very simple with no reserve outlets. *P. ctenodactylus* do not, however, have any protective temporary burrows (ORLOV, 1957).

This species prefers to have its burrows for day-to-day use in wet, compact sand. On stabilised dunes, on the crest of the barkhans or on their west slopes, the burrows do not usually go below 15–20 cm and the choice of the burrowing site is not apparently dependent on the nature of vegetation. In contrast to other jerboas this species does not close its burrows, situated on the slopes of the barkhans with soil plugs in spring and autumn. These burrows serve the animals only for a day, while burrows dug in the sandy plains are used for longer periods of 6–12 days (in June–August the period of occupation may be from 3 to 30 days). During summer the inhabited burrows undergo some change. In a few cases repeated occupation of the old burrow has been observed.

The animals begin to dig their hibernating burrows from the middle of August on the slopes, protected from the wind. In excavated burrows, two chambers were found at depths of 25 and 100 cm, the lower one having a bedding of wool, paper and wadding. The depth of the hibernating burrows often reach upto 4–5 m and that of the nest chamber upto 3 m.

It is a very active, cautious and fearful animal. It can jump upto 3 m in length, leaving characteristic marks on the soil. It sometimes hops faster than other Dipodidae leaving pug marks alternatively, rather than simultaneously. It is good in climbing bushes. Although *P. ctenodactylus* reportedly live for a long time at one place (STALMAKOVA, 1945) and have their own burrows and individual plots, they are also known to move over great distances in search of food and promising territory. During such sojourns they usually follow the same route. Often during the night the animal moves from 7–8 to 10–11 km.

As a result, there is often considerable overlapping of individual occupation of particular plots. Such unusual use of territory has apparently, importance in its search for food. Besides that, during summer many individuals shift from their individual plots a number of times.

Contacts of *P. ctenodactylus* with other species are limited because of the unusual structure of its burrows. Its contacts with *D. sagitta* in the common temporary burrows and with *Meriones meridianus* have only been observed.

The similarity in the food of *D. sagitta* and *P. ctenodactylus* have been observed by several workers. Both species eat young stems, flowers and seeds, the last derived from bushes but not from grasses. The animals climb up the bushes, jump at them or pick up the feed materials from the ground. The broken branches are pulled away to an open place and eaten on the ground.

The main food of *P. ctenodactylus* in north-western Kizil-Kum comprise the generative organs of a large number of plants beginning with buds and ending with seeds and fruits. According to ORLOV (1957) there is considerable individual variability in respect of choice of food, and also there are clear seasonal and locational differences. Under different conditions the following plant parts have been identified as its food: the flowers of *Calligonum*, ovaries of *Euphorbia heirolepis*, flowers of *Artemisia ereocarpa*, ovaries and seeds of *Atriplex*, fruits of *Tournefortia sogdiana*, ovaries and seeds of *Corispermum papillosum*, flowers of *Acanthophyllum borszewi*, fruits of *Peganum harmala* and fruits of *Haloxylon persicum*.

P. ctenodactylus appear on the surface at daytime only when disturbed. In winter they go into hibernation when the temperature reaches 16–18° (in south Kara-Kum in December–January). It is interesting to note that their arousal is associated not only with an increase in temperature, but also with a sharp cooling of the soil layer. Active *P. ctenodactylus* have been observed during frosts at 20–25 °C (STALMAKOVA, 1954).

Only the spring reproduction of this species in Kara-Kum is known (STALMAKOVA, 1954; KAMNEV *et al.*, 1959) while in Kizil-Kum (ORLOV, 1957; GULIEVSKAJA, 1963) its reproductive season has been marked as April–May. SABILAEV (1969c) suggests the presence of two cycles of reproduction in Kizil-Kum – the spring and the summer ones, basing his opinion on the finding of gravid females in April and in July. There are no data about the age groups taking part in the second cycle of reproduction, but because of their presence in the second part of summer, we may suggest that yearlings also take part in it.

In Central Kara-Kum, north-eastern Kizil-Kum and in Zannguz Kara-Kum single females with 2 embryos have been captured. In Bepetec (south-eastern Kara-Kum) STALMAKOVA obtained a female with 3 embryos. ORLOV (1957) found in the western Kizil-Kum, animals with 3, 5 and 6 embryos; and SABILAEV (1969c) at the same places found 3 embryos in each of 3 females.

Gerbils

PERSIAN GERBIL

Meriones persicus Blanford, 1876

The Persian gerbil inhabits mountain steppes and semi-deserts of north-eastern Turkey, Iran, Afghanistan and Turkmenia (in the south-west on the Badchiz, Copet-Dag, Bolshiye and Maliey Balchans extending to the west upto the Krasnovodskiy plateau) and in the southern Za-Caucasus region upto Yerevan in the west. It lives in the dry subtropical zones of mountainous and foothill semi-deserts and deserts and mostly resides in the cavities of rocks and deposits at heights of from 1400 to 3000 m above sea level (POGOSJAN, 1949; NURGELDIEV, 1950; ALEK-PEROV, 1966). It has been found in Badhys in the same places in the rocky deserts where the great gerbil also lives, although *M. persicus* is known to prefer more rocky terrains like mountain gorges (HEPTNER, 1956).

The burrows of *M. persicus* are rather simple in structure and have only a few entrances. They are built not only in the rocky deposits, but also along stone fences and in buildings and ruins. Usually the burrows are located at some distance from each other, but in places there may be

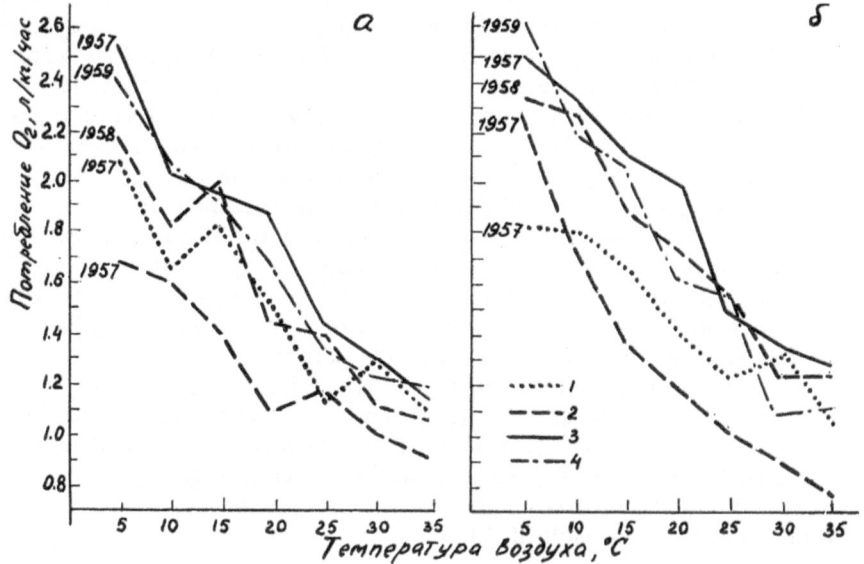

Fig. 22. Seasonal fluctuations in oxygen use by (a) Persian and (b) middle-Asian gerbils (after Kalabukhov, 1969). 1. March; 2. June/July; 3. November; 4. January. (ordinate: consumption in litres/kg/hr. abscissa: air temperature, °C).

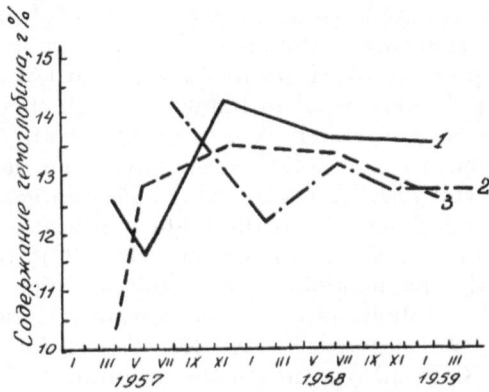

Fig. 23. Seasonal changes in haemoglobin concentration in the blood of great gerbil: 1. middle Asian gerbil; 2. Vinogradov's gerbil; 3. Persian gerbil.

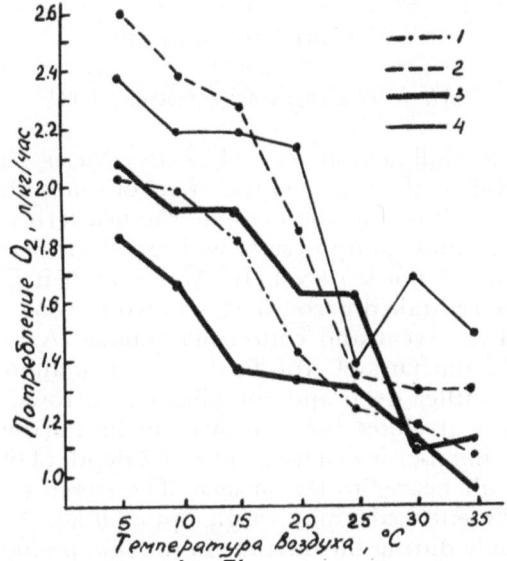

Fig. 24. Seasonal fluctuations in oxygen use by Vinogradov's gerbil (after KALA-BUKHOV, 1969): 1. February; 2. April; 3. July; 4. November. (ordinate: O_2 consumed in litres/kg/hr; abscissa: air temperature, °C).

groups of burrows present. The length of the tunnel is usually not more than 175 cm; there are large nest-chambers and chambers for the storage of food (ALEKPEROV, 1966). The gerbils eat the majority of plants growing near their burrows, preferring cereals, while they also eat insects and other small animals. They store food for the winter in the chambers of their burrows and under rocks.

Usually two litters are born in the warmer parts of the year (in spring and in autumn), resulting in the birth of upto 7 young ones. Sometimes these animals reproduce in winter in Azerbaidjan (ALEKPEROV, 1966). They are principally nocturnal in habit, although they may be surface active during the day in winter (NURGELDIEV, 1950).

Seasonal changes in metabolic indices have been determined for the Middle Asian gerbil *(M. blackleri)* and the Persian gerbil *(M. persicus)*. In both species metabolic rate is the highest in autumn and in winter, and the lowest in spring and in summer (Fig. 22); there are marked fluctuations in the haemoglobin concentration (Fig. 23) in the blood and in the basal metabolic rate. A similar picture is also seen in gerbils of Vinogradov *(M. vinogradovi)* (Fig. 24). The critical temperature for *M. persicus* is 35 °C in all seasons. In the autumn of 1959, however, the critical temperature was found to be 25 °C. Its oxygen use has been found to be maximum in spring and autumn and minimum in summer (KALABUKHOV, 1969).

VINOGRADOV'S GERBIL

Meriones vinogradovi Heptner, 1931

It inhabits the foothill semi-deserts of north-eastern Turkey, Palestine, Iranian Azerbaidjan, the mid (central) part of southern ZaCaucasus in Nakhitchivanian ASSR and the extreme southeastern part of Armenia. It inhabits loamy and sandy deserts with good grass covers at heights of 700–800 m above the sea-level. In Azerbaidjan it lives in the same places where the red-tailed gerbil does – in wormwood – *Salsola* deserts, on deposits and in cereal and cotton plantations (ALEKPEROV, 1966). It is found in the outskirts of crop fields also. The animal lives in small groups building rather deep and complicated burrows. During the hot period of the day it closes the entrances to its burrows with earthen plugs. The nest chamber is believed to be at a depth of 60 cm. The store-chambers are built nearer to the surface. The stored food in a burrow, consisting of different seeds, may weigh upto 3.5 kg. The Vinogradov's gerbil is active only during the daytime in the cold period of the year and only at night in the middle of summer. In one burrow two or more families can sometimes live. Reproductive activity takes place in spring and in autumn; the number of young ones in a litter varies usually from 6 to 8 (HEPTNER, 1940).

TAMARISK OR GREBENCHIKOVA GERBIL

Meriones tamariscinus Pallas, 1773

It inhabits deserts and semideserts of pre-Caucasus, Povolgie, Kazakhstan,

Uzbekistan, Tajikistan, north-western China and south-western Mongolia.

In the north-western Pri-Caspian region it is found in the south upto Grosny and Machatashkala, in the west upto Pricumsk and Atchiculac and in the north upto the Black Jar. It is quite numerous in the Prevolgan sandy terrains and in the Ilmensky district of the Astrachanian region. Between the Volga and the Ural it inhabits sandy terrains (upto 49 °N). Along the river Ural it is found upto 50 °N. In the east its range extends upto the Turgay river (Tusumian desert). It occurs throughout the central part of Betpackdala and in the regions of northern pre-Balchashie, lake Alakol and the Zaysan depression.

The Grebenchikova gerbil inhabits Usturt, the valley of Amu-Darya, the eastern PriAralian region and the valley of Syr-Darya (upto the mountains); it is common in the Bucharan and the Samarkandian oases, in the valley of Zeravshan and in the foothills of Tian-Shan and Pamiro-Alay, including the Fergana valley and the western part of the depression of Issyk-Kul. It has penetrated deeply into deserts along dry river-beds. It inhabits north Djungaria and the valley of Black Irtish in China upto Mongolia to the south of Mongolian Altai (Dulamzeren, 1970).

M. tamariscinus is most numerous in the grass and bush covered sands of river valleys and in irrigated lands. It is attracted to regions with a rather high level of sub-soil water and does not avoid salt-affected grounds. On the coast of the Caspian sea it is the first among the rodents to inhabit dried-up areas of the shore from where the sea has just retreated. It inhabits houses and commercial buildings. It is found in the mountains upto a height of 2000 m (western parts of the Chatkalian mountain range). It prefers bushy areas and saline marshes which are avoided by other species. In the valley of Amu-Darya its highest density of population is reached in the wet river-forest thickets. Its intensive penetration into the oases will be indicated by the fact that while only 0.5 to 1 individual per m could be trapped in the desert as many as 8 to 10 per m could be captured in the oases (Rejmov, 1971).

Usually these animals do not form settlements with a high density of population. Only in the most favourable areas, especially if such areas are of a limited size, the density of these animals may reach 20 to 30 per hectare (Afanasjev *et al.*, 1953).

M. tamariscinus can be found almost everywhere in the agricultural zone in northern and western Kizil-Kum, and there they live only along the dry river-beds in the vallies of Cuvan-Darya and Jana-Darya, particularly in association with the bushes of *Tamarix* (Kim, 1960). In the plains of Tajikistan it has been found on the virgin soils bearing irrigated crops; it inhabits banks of dykes, ancient deposits, crop fields of *Medicago*, and river-forest overgrowths, but does not avoid barkhan dunes. It is found in the sandy gravelly and rocky deserts, often in the vicinity of the burrows of *Citellus relictus*. In Fergana it inhabits saline-marshes while preferring hillocky areas with rich wormwood and *Salsola*

527

covers (DAVIDOV, 1964). In the southern part of Middle Tian-Shan their favourite haunts are the saline sandy regions, overgrown with *Eurotia ceratoides*, other species of *Eurotia*, *Juncus* and *Caragana*. The highest density of these animals is usually found along dykes and gorges, at the sites of old, dilapidated buildings and in abandoned cattle enclosures (SABILAEV & OSTROVSKII, 1967). In Mujun-Kum it inhabits the river valleys and the neighbouring regions of gray-ground plains, but they are most numerous in the dry river-beds, along old dykes in bushy overgrowths and in areas having wormwood-cereal-grass associations (KHRUSCELEVSKIJ *et al.*, 1963). In southern pre-Balchashie it inhabits oases covered with overgrowths of *Tamarix* and grass species and in fields of *Lasiagrostis splendens* in sandy terrains and in inter-barkhan areas where the sub-soil water table is high. These animals are also quite numerous in meadows in the valleys of Ili and Karatal and in the grain plantations and in settlements (ISMAGILOV, 1961). It usually lives in solitary burrows which are simpler than those of other gerbils. The burrows are usually built under the roots of trees, bushes of *Salsola* or turfs of *Lasiagrostis splendens* and other cereals. There are usually 2 to 4 entrances, the length of the main tunnel being usually 4 to 5 m, sometimes reaching 6 m or even more. There are large numbers of holes in the surrounding surface of the burrow, the majority of which are filled with soil. The nest chambers are at depths of 50 cm in the summer burrows and at 150 to 250 cm in the winter ones. The diameter of the nest-chamber is 20 cm. The nest is built of cut stems of cereal plants, together with feathers of birds and hair of mammals (usually on sheep pastures). Sometimes there are a number of small chambers filled with small stores of food. The summer burrows are not very deep. Spread among the actual nesting burrows there are many shallow burrows or simple holes on the ground with two or even one tunnel, which serve as refuges from danger during feeding time. The paths leading from the nest-burrows to these emergency holes are not easily visible.

The density of population of *M. tamariscinus* may reach 10 to 20 animals, and in some small pockets even 40 to 60 animals per hectare. The belief that these animals lead a solitary life needs substantiation. There are evidences of movements of these animals over a distance of 1.5 km. They are mostly active at night, but in autumn they are found on the surface during day time also (RALL, 1941).

The most important element of their feed comprise the green parts of plants. In the northern parts of their habitat their stomach contents have revealed a mixture of food items with green plant parts predominating. In spring this forage (i.e. green parts of plants) predominated in 32 per cent of the stomachs the contents of which were examined (PAVLOV, 1962). The green food consists mainly of the underground parts of plants containing a good deal of moisture. The principal plants that *M. tamariscinus* are known to eat in pre-Caucasus are *Kochia*, *Chenopodium*,

Atriplex, Salsola and *Alyssum* species and also some cultivated cereal crops, grasses and a number of vegetable and garden plants (e.g. water-melons, sunflowers, *Medicago* etc.).

The stomachs of these animals have also been found to contain the remains of Acrididae, beetles and of smaller rodents. In the Volgo-Uralian deserts they eat *Elymus*, wormwood, *Calligonum* and other species. The stores of food in their burrows may weigh as much as 4.5 kg and consist of the green parts of weeds (PAVLOV, 1962).

In Turkmenia they feed mainly on the fruits and branches of bushes, getting these from heights of 1.5 to 2 m. The stores of food may sometimes weight as much as 23–30 kg, a large part of which is not used at all (NURGELDIEV, 1969; FEDJANINA, 1968).

M. tamariscinus found in Kazakhstan, also consume forage with a good deal of water in it. The main foods are fruits and stems of *Elymus giganteus*, leaves of *Mulgeidum tataricum*, fruits and stems of *Tamarix*, *Atriplex tatarica*, *Salsola* and others, fruits of *Calligonum*, *Eleagnus*, *Nitraria*, *Rosa*, *Cannabina* and others. They readily eat cultivated plants and their fruits (AFANASJEV *et al.*, 1953).

The fecundity of *M. tamariscinus* is somewhat lower than that of *M. meridianus* which is reflected in the stability of its population. Although its reproduction takes place all the year round yet winter reproduction is mostly restricted to the most favourable years. The usual period of reproduction covers 200–240 days in the north-western Pricaspian and 130–200 days in the Volgo-Uralian deserts.

M. tamariscinus commonly mate from the middle to the end of March. After a gestation period of 20 days the first litters appear in the beginning of April while littering continues upto October. The maximum number of pregnant females can be seen in May and August and the minimum number in June. Some of the young females (about 3%) become sexually active in the year of their birth. The litter size varies from 1 to 8, the mean being 4 to 5. One female may have upto 3 litters during one season (RALL, 1941; AFANASJEV *et al.*, 1953).

M. tamariscinus damages cultivated plants in many regions. It also damages irrigation systems and pastures by its digging activity. It renders considerable damage to houses, commercial buildings and storehouses. It plays some role in the natural spread of plague, leptospirosis and paratyphe micro-organisms and takes part in the development of tularemia epizootics.

MINOR ASIAN GERBIL

Meriones blackleri Thomas, 1903

This gerbil lives in the deserts of Syria, Israel, Jordan, Turkey and West Iran. In the U.S.S.R. it lives in the Transcaucasus region in the north upto the river Cura and in the west upto the line connecting Yerevan-Tbilisi. It inhabits foothill wormwood deserts having loamy and loessed soils. In the mountains it can be found upto a height of 1500 m where it avoids the rocky areas usually occupied by *Meriones persicus*. It lives in wormwood *Salsola* deserts and in plantations of cereal crops and cotton in Azerbaidjan (POGOSJAN, 1949; ALEKPEROV, 1966).

Its burrows are similar to those of *M. tamariscinus* but the depth of the burrows is never more than 60 cm in Armenia. Some entrances are closed with soil plugs. In summer these animals are active mainly at night and in autumn and winter during the day time. They are not active in winter and if their stores of food are sufficient they may stay underground for 2 months at a stretch. On the plains of Azerbaidjan (Jebrailan region) it is active at night, while in the foothills it is found foraging in the twilight. In the foothills region, *M. persicus* are found to be active at night (ALEKPEROV, 1966).

They mainly eat the seeds of plants, but in spring they also eat the green parts of plants. They consume most of the plant species growing in their habitat, wild as well as cultivated, preferring seeds to the green parts. They frequently eat insects also. Their food stores are quite large and often include seeds of cereal crops like wheat. In Azerbaidjan, however, these animals do not, often, have stores of food for winter (ALEKPEROV, 1966).

It reproduces two times a year and there are 2 to 7 young ones in a litter in Armenia (POGOSJAN, 1949). In Azerbaidjan the reproduction continues in winter also, but in the middle of summer and in winter pregnant females can be found very rarely (ALEKPEROV, 1966).

MONGOLIAN GERBIL OR CLAWED GERBIL

Meriones unguiculatus Milne-Edwards, 1968

It inhabits the sandy steppes of middle, southern and north-eastern Mongolia, avoiding the mountainous regions. In the north it penetrates the Tuvinian ASSR and the southern part of Tchitinskaya district. During the forties of this century two isolated populations of this gerbil were found in the lower parts of Orhona and Urda, separated from its main habitat by more than 300 km (BANNIKOV, 1954). In recent times the gerbil has colonised the intermediate territories (NEKIPELOV, 1959; DULAMZEREN, 1970). The animals had reached the northern limits of its range presumably by being transported over there (HAMAGANOV, 1954).

In the south, the Mongolian gerbil inhabits the sandy deserts of north and north-eastern China. In the Tuvinian ASSR it lives mainly on the arable lands and deposits in the mountainous wormwood-cereal steppes, and at lower altitudes in the desert steppes and old deposits overgrown with *Caragana microphylla* (LAVRINENKO & TARASOV, 1967).

It is not found in steppe of mixed grasslands and in rocky and marshy areas in Baikalie. It is found around human settlements and more commonly in wormwood steppes (6 animals per hectare), saline areas (upto 15 animals per hectare) and in fields and pastures (upto 90 animals per hectare). They also inhabit dry sandy steppes, overgrown with *Caragana microphylla* and white wormwood and can be found in steppes with light loamy soils with a mixed cover of cereal plants and grass species of different types. It occurs there in fields of crops such as buckwheat, millet and wheat and is quite numerous on artificial earthen embankments along railway tracks and metalled and unmetalled highways and on the banks of irrigation systems. It digs burrows on rubbish heaps and in earthen buildings (NEKIPELOV, 1962). It avoids virgin soils (LEONTJEV, 1954). In the southern part of Mongolia this gerbil has no particular preference for sandy grounds. It is found to occur there on more compact soil. In the north east it has been found in steppes with cereal plants on sandy hillocks with *Caragana* cover and in saline patches and in salt-marshes carrying *Salsola*. They are usually most numerous on rubbish heaps on the outskirts of vegetable gardens (LIPAEV, 1967). Their most usual haunts are the areas of cereal-*Salsola*-*Caragana* association in the semi-deserts and sand dunes overgrown with *Nitraria*. They can easily adapt themselves to rocky habitats as well as to relatively humid areas with cover of *Iris tennifolia* and *Lycium* sp. So, from the south to the north these animals increasingly tend to be attracted to light dry soils. In the same direction the degree of sinantropism increases, and it can be seen most vividly in the northern settlements of these animals. In east Mongolia 36 per cent of the rodents found in homesteads are gerbils (KUCHERUK, 1946).

The number of clawed gerbils in an area vary considerably. Upto 32 burrows (with 760 entrances) per hectare have been recorded in Zabaikalie. The average density is lower at certain times (LEONTJEV, 1954). The data of LAVRINENKO & TARASOV (1967) for the Tuvinian ASSR reporting 70–176 animals per hectare, appear to be too much on the high side. The average number of burrow entrances per hectare was 42 in the Mongolian cereal – *Caragana* desert, while there are 504 such entrances per hectare in dune areas and 2800 in the culverts along the highways (Observations made in May 1944; BANNIKOV, 1954). In Zabaikalie in similar culverts there were 740–1420 entrances, while 2500 entrances were found on the deposits and only 30 to 32 entrances in virgin soil. The great fluctuations in the number of these openings indicated the instability of the resident population of this gerbil.

The clawed gerbils eat seeds and green parts of many plants. In summer they prefer to eat green parts and in winter the seeds and fruits. The composition of plant species, consumed by these animals, is found to be

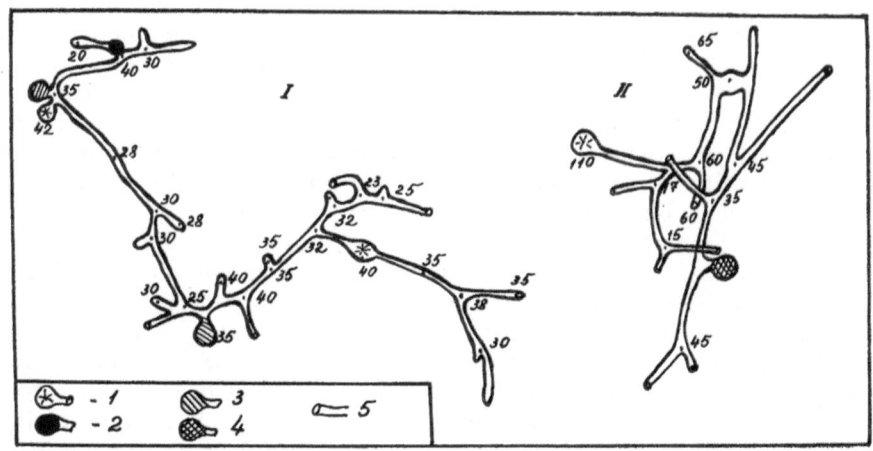

Fig. 25. Burrows of clawed gerbil. I. summer-burrow; II. winter-burrow (after
LEONTJEV, 1954). 1. dwelling nest; 2. deserted nest; 3. empty chamber; 4. food stores;
5. entrances (arabic numerals indicate depth in cm).

different for different regions. In the agricultural fields in Zabaikalie and
Tuva mostly the seeds of crop plants are eaten. In fallowlands their food
comprise cereals, wormwoods and plants of Chenopodiacea, Compositae,
Leguminosae and other families. In the food stores inside their burrows
seeds of buckwheat (upto 20.5 kg) wheat, millet, oats, *Setaria*, *Chenopodium*,
Caragana (upto 1.4 kg) and branches of wormwood have been found.
Oats are preferred to wheat as a feed (LAVRINENKO & TARASOV, 1967).
Usually the stores contain only one type of food. Storing of food begins
in the middle of August and continues upto the middle of autumn and
sometimes even upto winter. All animals of this species take part in
the storing; very often adult animals are seen collecting food items with
the help of their juvenile offspring. All members of a family group spend
the winter together in the same burrow (FETISOV & MOSKOVSKIKH,
1948).

The most important items of summer food are the green parts of hemp,
Chenopodium, buckwheat, *Atriplex*, *Artaphaxis scoparia* and other plants
(LEONTJEV, 1954).

The nesting burrow of the clawed gerbil is not very complicated in
structure. Its length is usually 5 to 6 m (sometimes upto 14 m) and
there are usually 5 to 10 entrances. The nest chambers are usually
situated at depths of 40–45 cm in the summer burrows and at dephts of
110–150 cm in the winter burrows (Fig. 25). The majority of the tunnels
are not very deep. The size of a nest-chamber is usually 15 to 40 × 13
to 20 cm. The nest is built of leaves and shoots of plants along with
some hair and feathers and sometimes of rubbish. The nesting burrow is

532

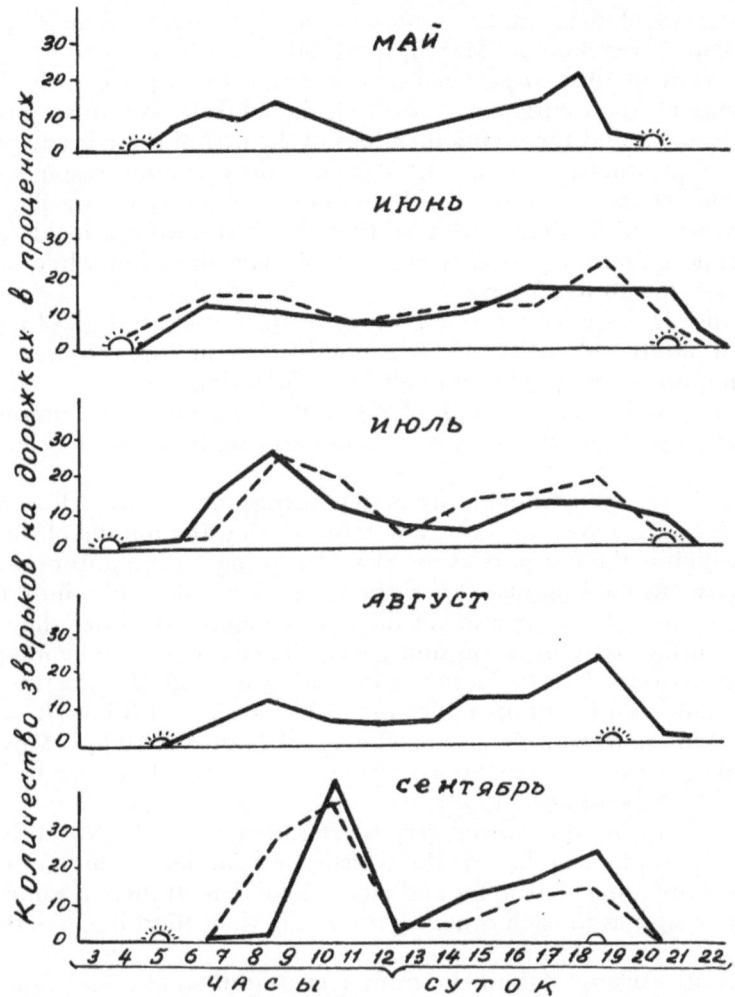

Fig. 26. Daily activity of Clawed gerbil (after LEONTJEV, 1954). Sequence of months shown from the top to the bottom: May, June, July, August, September. (ordinate: number of animals seen on the surface; abscissa: hours of the day).

surrounded by temporary burrows in a radius of 10 to 20 m. There are usually 1 to 3 entrances to each temporary burrow and sometimes there are small chambers where the remains of food may be found. The length of the temporary burrow is not more than 2 to 4 m (LEONTJEV, 1954). The clawed gerbil often uses the burrows of *Microtus brandti*.

In favourable years reproductive activity in Zabaikalie begins in February–March and ends in September or somewhat later. In February,

90 per cent of the females examined were pregnant, 13 to 27 per cent were found pregnant in March, 38 to 50 per cent in April–May, 30 to 33 per cent in June, 0 per cent in July and 0 to 33 per cent in August–September. Apparently, in some year, the adult females may have upto 3 litters each, and the young members of the first of these litters also take part in reproductive activity. In Mongolia the breeding season is shorter. The females seem to have two litters per year there, with a few having a third one (in September). The litter size varies from 2 to 11 (average 6.4). The gestation period is of 29 to 30 days duration while lactation continues for 20 to 25 days.

Young *M. unguiculatus* weigh 3 g at birth, 7 g at 10 days of age, 23 g when a month old, 40 to 41 g at 2 months, 46 g at 3 months and 48–60 g at 5 months of age. The animals born in spring attain a body weight of 60–70 g and a body length of 110–120 cm by autumn. Summer born animals also attain similar body weight and length by spring (LEONTJEV, 1954).

In summer these animals are active during the day as well as at night (Fig. 26). They are, however, not active during the middle of the day in July which is the hot part of the year. In spring and in autumn they are not active in cold nights. When the weather is cold and windy they do not leave their burrows and eat their stored food. In winter they appear on the surface only in warm sunny days. They are seen to be most active during periods of reproduction and storing of food (LEONTJEV, 1962). Their daily run in summer is of the order of 1.2 to 1.8 km (BANNIKOV, 1954). One marked animal moved over a distance of 50 km. Movements over distances of 15 to 20 km from their home burrows have been recorded (NEKIPELOV, 1959).

In early spring the clawed gerbils are rather silent. In November their whistles can often be heard; the whistles are similar to those of *Ochotona daurica*, only somewhat softer and more melodious. If there is some danger they give signals to each other by thumping their hind legs like the great gerbil.

In south-eastern Zabaikalie from 1 to 3 adult males and from 2 to 7 females have been caught from the same burrow in summer. So the total number of animals living in a burrow varied from 3 to 14. They spend the winter also in such groups consisting usually of parents and their latest litter. Observations indicated that the animals may visit some neighbouring nest-burrows, which are sometimes found to be inter-connected. So, as in the case of the great gerbils, the clawed gerbils also do not live in small family groups but in somewhat larger groups which may consist of some adult males, females, subadult and young ones of different generations. But a 'stranger' to the group is persecuted. When bred in captivity, these gerbils often bite to death any unknown animals introduced into their cage (LEONTJEV, 1954).

In the Altaian Gobi the Mongolian gerbils live in semi-deserts with cereal and *Salsola* plants and in the overgrowth of *Lycium*. They are common in the ravines. In the Zaaltaian Gobi they are found in the oases and on sand dunes overgrown with *Nitraria sibirica*. They are, however, most numerous in the vallies of rivers, near crop fields, around human settlements and in rubbish heaps (TARASOV, 1958).

The gerbil has been observed in a flat, barren steppe to the south of the Tannu-Ola range. In north-western Mongolia they avoid sandy barkhans but are quite numerous on flat *Salsola* plains (50 inhabited burrows have been recorded per hectare, NEKIPELOV, 1959).

In south-eastern Zabaikalie it feeds mainly on *Alhagi kirghisorum* and *Corispermum duriuscula*, which grow on the abandoned range of *Microtus brandti*. As the vegetation begins to change with the season, they begin to eat wild cereals which, however, do not satisfy their needs. The stores of food here consist partly of shoots of wormwood and *Suaeda*, and partly of *Eragrostis pilosa*.

The Mongolian gerbil, which was not found in the steppes of Zabaikalie upto 1939, seems to have lived here since long. The species evidently spread from its original habitat (the banks of brackish-water lakes) to the overgrowths of weeds on the fields and near human settlements, where a considerable number of them were found in 1952, 1953 and 1958.

According to marking data on clawed gerbils obtained in south-eastern Zabaikalie, they are a very mobile species, so much so that the population of different burrows mix intensively in summer and in autumn. This interesting information could be obtained after a relatively low number of animals were captured repeatedly, revealing as much as 10 to 32 per cent fluctuations in the population of adult males, 18 to 52 per cent in that of adult females and 22 to 28 per cent in the number of the young. A similar situation was found to hold true in respect of the percentage of animals captured for the first time in any particular burrow during March–August, 1958. This varied from 72 to 89 per cent of the total catch. In 43 per cent of the burrows from 1 to 6 animals could be captured during 120 days, in 44 per cent burrows from 7 to 16, and in 13 per cent burrows from 17 to 22. The largest number of animals was caught in the largest burrows or in those which were situated on the path of migration.

The potential longevity of these gerbils is approximately (or a little more than) 2 years. A marked adult female was caught for the second time after nearly a year i.e. after she had spent her second winter. However, the mean life span of this species may not be of more than 3 to 4 months. This would mean that an entire population is replaced within a year (LEONTJEV, 1962).

The movements of marked gerbils have not been found to be directed and never exceeding the limits of their range spread over 700 m. The use of the same burrow for the maximum period was found to be as follows: for an adult female – from May to August, for an adult male – from March to June and for a young female – for 3.5 months. However, long movements of these gerbils along roads, sometimes covering 40 km (village Solovievka-Borsa) are on record (Leontjev, 1962).

Midday gerbil

Meriones meridianus Pallas, 1773

It inhabits the sandy deserts of Mongolia (mainly central and southern regions), of north-western and central China, north-western Afghanistan and north-eastern Iran. In the U.S.S.R. it is found in north-eastern Caucasus (left banks of Terek to the east of Grosny town, the Astrachanian region, the eastern part of Volgogradian region; in the west it occurs upto the villages Termit, Pestchanoe and Beack Jar. In the east it inhabits the banks of the Volga and the sands on the right bank of the Ural river upto the village Calmykovo in the north), and all sandy deserts of Kazakhstan and Middle Asia, in the north upto the sandy terrain of Jety-Conur on the banks of the river Sarysu, the deserts of the Balkash region and the Alakolian depression. Isolated populations have been found in Zacaucases (Vedinian region of Armenia) on the left bank of the Don (between Volgograd and village Stepnoe), in the Zaisanian depression and in Tuvinian ASSR.

Among all gerbils, this species is the most desert-adapted, inhabiting sandy terrains of different degrees of stabilisation. In the Balkhash region it lives in semi-stabilised dunes covered with bushes and semi-bushy psammophytes. It also inhabits grass plots between barkhans and the slightly vegetated slopes of such sand masses. It also inhabits the sand dunes in loamy steppes. On the northern bank of the Balkhash it inhabits a narrow strip of sandy tract (Ismagilov, 1961; Burdelov & Leontjeva, 1956). In the Mujun-Kum desert it prefers the undulating and semi-stabilised dunes and hills and the surrounding plains. In other regions it is found only in sandy terrains (Khruscelevskij et al., 1963). In Kazakhstan it inhabits different types of deserts but in the north it prefers to live in lightly vegetated terrains. The highest density of population is found in desert regions with bushy covers. In the lower reaches of Amu-Darya the main habitats of *M. meridianus* are the rocky and grass covered deserts. A large number of these animals live in slightly hillocky alluvial land-forms on the outskirts of oases, where intensive cattle grazing has led to destruction of the pasture lands. In such areas, 49 per cent of the rodents captured have been found to be *M. meridianus* while in the oases they account for only 6 per cent of rodents (Rejmov, 1971). In Turkmenia it inhabits all of Kara-Kum, penetrating to the loessed deserts of Kara-Bill and Badhis. It prefers the upper parts of slopes of fixed barkhans, more rarely inhabiting the inter-barkhan areas, and in semi-stabilised areas they may live in the inter-dune regions (Nurgeldiev, 1969). In north-western Pricaspian, the range of *M. meridianus* is limited to the western border of the semi-desert zone. Since they are found only in the sandy regions, so there are large gaps between their inhabited areas. As there are more sandy terrains in the eastern part of this region, the number of gerbils is more there. These animals often live in steppe regions – on the slopes and saucer-like cavities of the dunes and along the roads. Very rarely they are found on the banks of lakes and dry river beds. The number of gerbils fluctuates here greatly, while in the Volgo-Uralian deserts their number is not only higher, but is much more stable also. The highest density of population is in the Pri-Volgian, Pri-Cumian and the Bajiganian deserts and in the Black lands where it fluctuates from 5 to 10 animals per hectare.

In the Tuvinian ASSR, *M. meridianus* prefers rocky areas and semi-stabilised dunes and more especially the regions covered with *Caragana* (LETOV *et al.*, 1963). In Armenia isolated populations of *M. meridianus* have been found on unstabilised sand dunes near village Gorovan (GAMBARJAN *et al.*, 1960; PAPANJAN, 1966).

In Mongolia *M. meridianus* mainly inhabits sandy areas with different degrees of vegetation cover (BANNIKOV, 1954). Its favourite habitats are semi-stabilised dunes with wormwood and *Agriophyllum* covers. In the Gobian Altai they have been found to inhabit sands with *Haloxylon* covers, while staying mainly in the overgrowths of *Caragana* and also on sand dunes with *Nitraria* covers. In the foothills of Ahy-Bogdojil it occurs along the beds of ravines overgrown with *Caragana* (TARASOV, 1958); it has also been found in houses and store houses. In the mountains in the U.S.S.R. it does not go higher than 600–800 m while in Mongolia it reaches 1600 m above the sea-level.

The burrows of these animals are usually situated under roots of plants growing in hillocks of sand. Its burrows are generally more complicated than those of the other gerbils except the great gerbil. Its complicated winter-burrows are 2 metres or more in depth (upto 4 m in the north); the length of the burrows is upto 4 m and there are a large number of outside openings and chambers for stores. The winter nests are usually situated at depths of 105 cm in the north-west of Caspian and at 200–250 cm in the Volgo-Uralian deserts (RALL, 1939; PAVLOV, 1962).

Although living in close association with *M. tamariscinus* in many regions, *M. meridianus* differs from the former species in its physiological peculiarities. *M. tamariscinus* does not exhibit any wide fluctuations in its use of oxygen and chemical thermoregulation. *M. meridianus* shows considerable variations in both these respects (Fig. 27). The increase in oxygen consumption in autumn over that in summer is of the order of 124–126 per cent for *M. tamariscinus* and 139–157 per cent for *M. meridianus*. In winter the level of metabolism increases in *M. tamariscinus* by only 6 to 13 per cent while in *M. meridianus* the increase is of the order of 76–87 per cent. The difference in the rate of fluctuations of metabolic

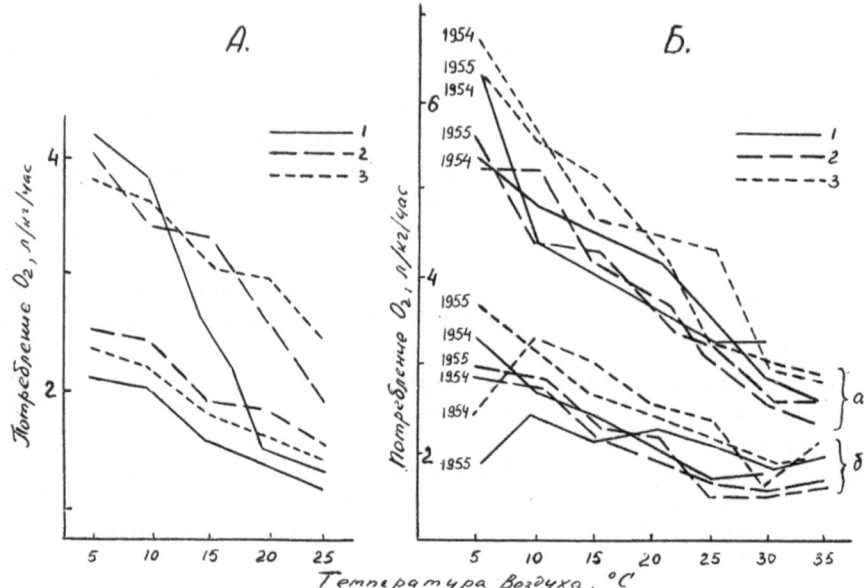

Fig. 27. Seasonal and yearly fluctuations in oxygen use by (a) *M. meridianus* and (b) *M. tamariscinus* on the right (A) and left (B) bank of the Volga (after KALABUKHOV, 1969). *M. meridianus:* 1. summer; 2. autumn and the beginning of winter; 3. the end of winter. *M. tamariscinus:* 1. winter; 2. summer; 3. autumn. (ordinate: O$_2$ consumption in litres/kg/hr; abscissa: air temperature, °C).

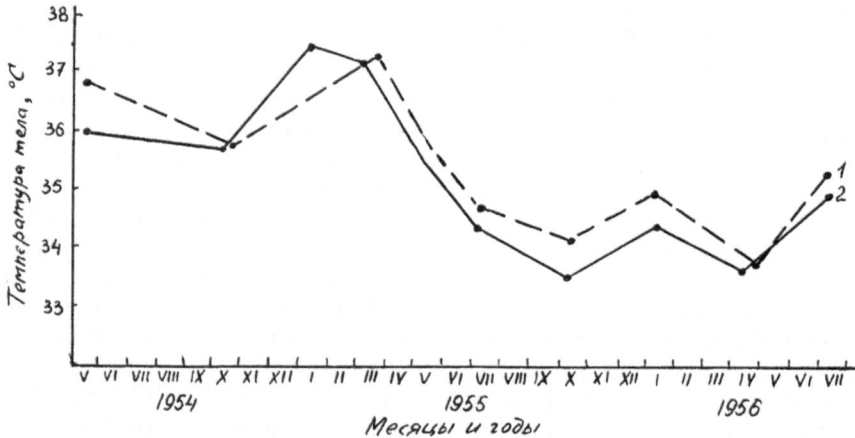

Fig. 28. Seasonal and yearly changes in body temperature of *M. tamariscinus* (after KALABUKHOV, 1969). 1. at air temperature 25 °C. 2. at air temperature 10 °C. (ordinate: body temperature, °C; abscissa: months and years).

538

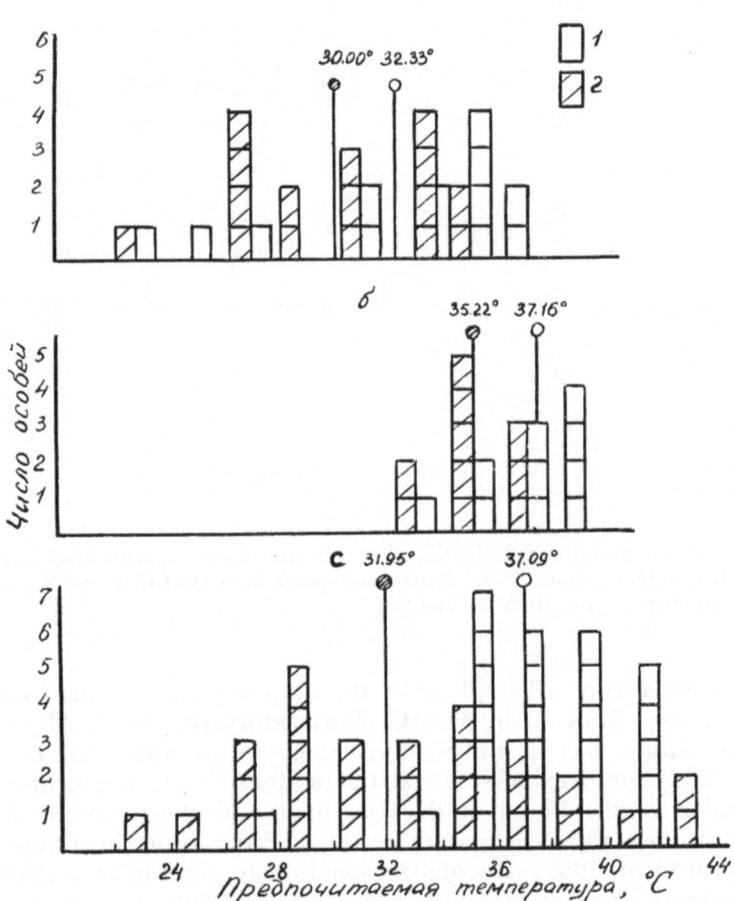

Fig. 29. Seasonal and yearly fluctuations in preferred temperature by 1. *M. meridianus* and 2. *M. tamariscinus* (after KALABUKHOV, 1969). a. May 30 to July 15, 1955; b. October 16–24, 1955; c. January 11–27, 1956. (ordinate: number of animals examined; abscissa: preferred temperature, °C).

indices of these two species are probably related to the large body size of *M. tamariscinus* in comparison to that of *M. meridianus*.

The critical air temperature for *M. tamariscinus* in summer has been found to be in the range of 25 to 30 °C, overheating of the organism begins at 35 °C, and at 37.5 °C the regulation of body temperature breaks down completely (Fig. 28). The critical air temperature for *M. meridianus* has been found to be between 30 and 35 °C, but its thermo-regulatory processes also break down at 37.5 °C. Its body temperature remains stable at air temperatures ranging from 10 to 35 °C. Irregular increases in its

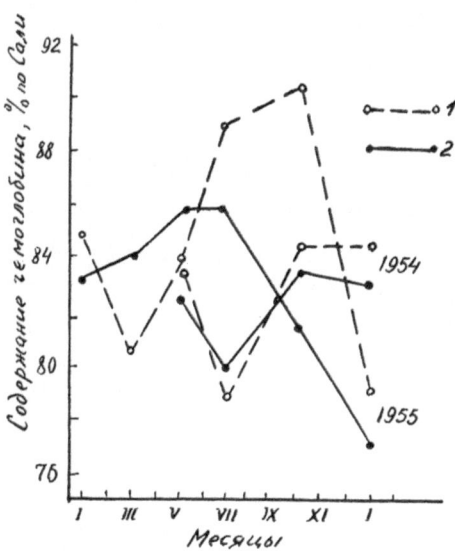

Fig. 30. Seasonal and yearly fluctuations in haemoglobin concentration in the blood of 1. *M. meridianus* and 2. *M. tamariscinus* (after KALABUKHOV, 1969). (ordinate: haemoglobin per cent; abscissa: months).

body temperature takes place as the air temperature falls from 10 to 5 °C, or rises from 35 to 37.5 °C (KALABUKHOV, 1969). *M. meridianus* almost always has a higher body temperature than *M. tamariscinus* (Fig. 29). Considerable seasonal fluctuations in the haemoglobin concentration of the blood of these animals have been noted (Fig. 30), although these differences were not of the same order in different years (MOKRIEVICH, 1957). Regular seasonal fluctuations in respect of several other physiological parameters (e.g. the body levels of ascorbic acid and tocopherols and the weight of liver and adrenals) have also been observed (Fig. 31). The degree of these fluctuations vary from year to year (depending upon the prevalent conditions). Marked differences in respect of these indices and of body weight in different populations (e.g. from the left and the right banks of the Volga) are very interesting. These indicate that two *M. meridianus* populations from different geographical regions have significant eco-physiological differences in respect of their level of metabolism, respiratory capacity of blood, indices of physical and chemical thermo-regulation, level of tocopherols and ascorbic acid in the tissues etc., all of which appear to be related to differences in the body size of the animals and have some adaptive significance (KALABUKHOV, 1969).

Eco-physiological differences between *M. meridianus* and *M. tamariscinus* are related not only to the differences in their body size but also to their

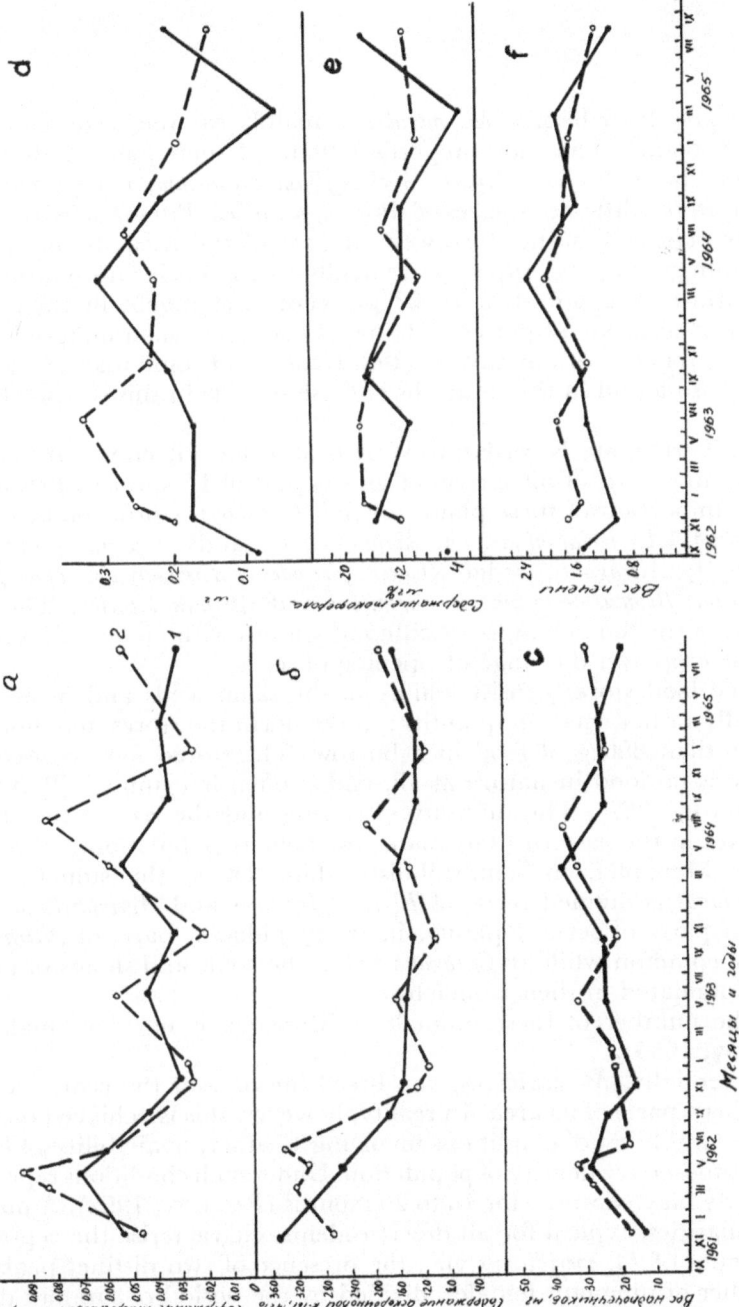

Fig. 31. Seasonal and yearly fluctuations in ascorbic acid status (a, b), adrenal weight (c), liver tocopherol status (d, e) and liver weight (f) in populations of *M. meridianus* from the right (1 – unbroken line) and left (2 – broken line) banks of the Volga (after KALABUKHOV, 1969). (ordinate: a. adrenal ascorbic acid content, mg; b. adrenal ascorbic acid concentration, mg per cent; c. adrenal weight, mg; d. tocopherol content in mg per liver; e. liver tocopherol concentration, mg per cent; f. liver weight, g; abscissa: months and years).

respective food habits. *M. meridianus* mainly eat the seeds and fruits of desert plants. The most important items of their food in the Caspian region are seeds of *Elymus, Agriophyllum arenarium, Corispermum,* sandy wormwood, different species of *Atriplex, Kalidium, Petrosimonia brachiata* and other *Salsola.* On the northwestern part of the area in spring and in autumn green parts of plants are avidly eaten. Their admixture (in small quantities) was found in 70–83 per cent of stomachs in these seasons; while in autumn 47 per cent stomachs contained such materials. Besides plants, remains of animals (beetles, locusts and sometimes even rodents) were also found in the stomachs of these rodents in this region (PAVLOV, 1962).

In Turkmenia Kara-Kum *M. meridianus* have been found to consume seeds and fruits of 52 species of grasses and of 16 species of shrubs. The most important of these plants are *Salsola richteri,* wormwoods, (*Artemisia santolina, A. terrae-albae, A. herba-albae),* sandy acacia (*Ammodendron conollyi),* *Astragalus,* sedge (*Carex physodes), Eremopyrum buonoparti, E. triticeum, Bromus tectorum, B. racemosus* and *Atriplex tatarica.* The composition of the food changes in different seasons with changes in the nature of the crop and the stage of ripening of seeds.

The food stores consist mainly of the same seeds and fruits and are usually rather small in quantity; in the north the stores may not contain more than 800 g of food in a burrow. The stored food is used only if there is no food in nature at all, and is often left unused (RALL, 1939; PAVLOV, 1962). The intensity of storing and the size of the stores are greater in the eastern (Tuvinian, Mongolian) populations.

In Mongolia, in semi-stabilised dune areas, the stomachs of *M. meridianus* contained seeds of *Elymus giganteus* and *Agriophyllum* and the green parts of several plants; in sandy hillocks, parts of *Nitraria* were more common while in *Caragana* fields, the seeds and shoots of this plant predominated in their stomachs.

The number of these animals in Mongolia is not very high (NEKI-PELOV, 1959).

Potentially, *M. meridianus* can breed throughout the year, even in the northern parts of its area. In reality, however, this is achieved only under a favourable set of conditions involving weather, availability of food and a not-too-heavy density of population. Under such conditions reproductive activity may continue for 18 to 20 months (PAVLOV, 1962). A number of peculiarities, typical for all desert rodents, characterise the reproductive activities of *M. meridianus* viz., the presence of two distinct peaks in the number of pregnant females, divided by the period of summer drought; the dropping of a second litter by the adult females and participation of young females in reproductive activity (sometimes they manage to drop even two litters).

The summer interval in reproductive activity is very vivid in the southern part of the area as compared to that in the north. The duration of the reproductive season is also different for the two regions. For example, in north-western Pri-Caspian (right banks of lower Volga) during 1948–1955, it fluctuated from 203 to 289 days (average 242 days), while on the right bank in the Volgo-Uralian sands it fluctuated from 120 to 250 days (average 172 days). Interestingly, the intensity of reproduction is higher in the latter region (PAVLOV, 1962).

In north-western Pri-Caspian, according to data collected over 8 years (1948–1955), the percentage of pregnant females was 0.4 to 0.9 in winter (December–January), 7.7 to 8.8 in spring (February–March), 20.3 to 22 in April–May, 15.6 in June, and 21.7–30.1 in July–August (the time of ripening of the majority of seeds). This percentage began to fall in September–October (18.9–13.5) and reduced to 1.7 in November (PAVLOV, 1962).

In the Volgo-Uralian sands the first pregnant females were noted in the end of February and beginning of March while mass pregnancy was observed in the end of March. Reproduction continued for the whole summer, usually occupying a period of 6–8 months (RALL, 1939). Nearly all adult females take part in the spring reproduction, bringing 1 to 3 litters while only 19.5 per cent of the females (including less than 1 year old) take part in the autumn reproduction. One litter is dropped by 85.4 per cent of the adult females, 2 litters by 13.5 per cent and 3 litters by 6 per cent. There are 1 to 9 (average 4.6) young ones in a litter in the Volgo-Uralian sands.

The age of the animals was found to be rather homogeneous in spring – 100 per cent of the animals having spent the winter. 8 per cent of the young ones appeared in the third ten-day-period in May, 30 per cent in the second ten days of June and 8 per cent by the end of September (ROSSOLIMO, 1957). In the Volgo-Uralian sands the population of young animals grew from 1 per cent of the total population in April to 85 per cent in September (Fig. 32, RALL, 1939). So the replacement of the population is achieved quite fast in the Volgo-Uralian sands. This is true for the periphery of the area also (NAUMOV, 1945).

The duration of the reproductive season in Kizil-Kum in spring is of three months (from the beginning of March to the beginning of June). In the end of summer only a few females drop new litters. The size of a litter fluctuated from 1 to 9 (average 5.8). The first ovulations take place in the end of March here (for those females who have passed the winter). By the middle of June only 3 per cent of the females were non-lactating, 80 per cent had dropped one litter and 17 per cent two litters; by September 33 per cent of the females had dropped two litters each. Spermatogenesis in the males began in early spring and was very vigorous in April in all the males examined (the average size of a testicle – 14.1 mm). The majority of the males remained in such a condition upto the end of May, but by the middle of June spermatogenesis had stopped in 80 per cent of the males. In young males, born in the same year, spermatogenesis

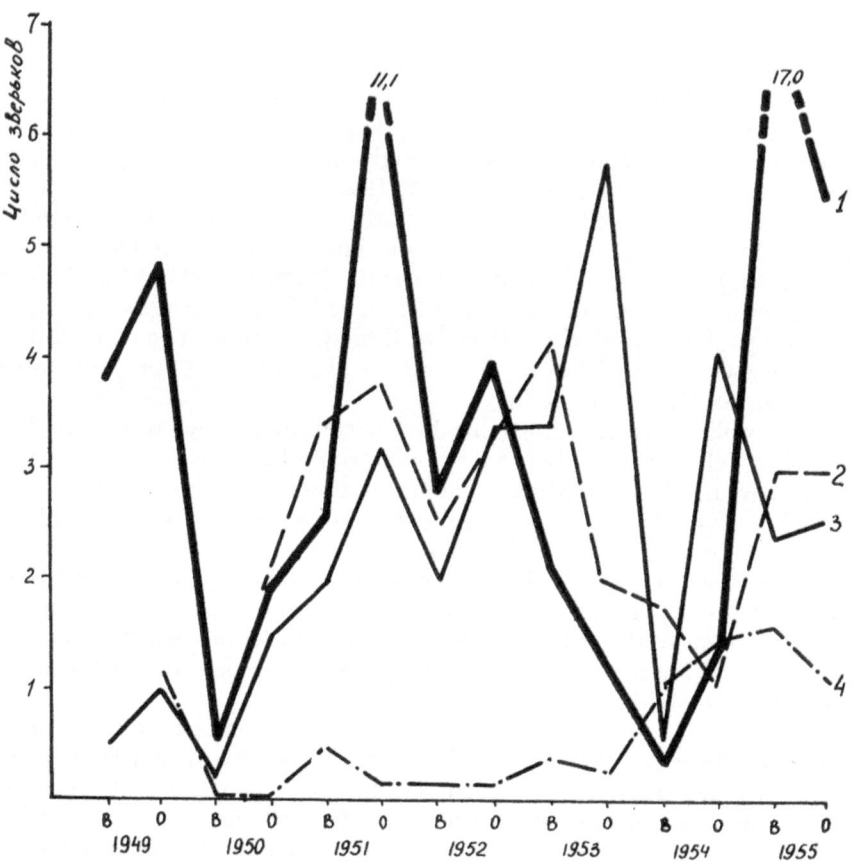

Fig. 32. Fluctuations in the number of *M. meridianus* in different populations in the north-western Caspian region (after PAVLOV, 1962). 1. Kumo-Manitchian; 2. Primorian; 3. Ilmenian; 4. Privolgian sands; (ordinate: number of animals; abscissa: years; B – spring; O – autumn.)

had begun in the end of June, but their sexual glands had degenerated by autumn. So, apparently, the testicles of all the males were in rest in autumn (ROSSOLIMO, 1957).

In Turkmenia (Kara-Kum) an intensive reproductive activity of *M. meridianus* takes place in March–June with a summer interval in July–August. A less intensive autumn reproduction takes place in September–October. The size of a litter fluctuated from 4 to 12 young ones (average 6.4). The duration of pregnancy was of 28 days (from 22 to 30), there being, usually, 2 to 3 litter per female. Shedding of hair is usually seen in April and in October.

Pregnant females of *M. meridianus* could be found in all the months, but in different years there were more or less prolonged intervals in

544

reproductive activity. Every year reproductive activity slackens or may even stop altogether in July–August and sometimes in the winter months. Sexual activity is apparently dependent upon the weather, and more particularly on the precipitation which is reflected by the growth of fodder plants. The average number of the young in a litter also fluctuated in different years and months (from 3.0 to 6.0 embryos in a litter). The reproductive activity in different populations was found to be quite different.

In summer, and especially in times of heat and drought, *M. meridianus* is active mainly at night; in spring and autumn it is active also in the morning and in the evening; and in winter, if it is warm enough, it comes on the surface during the day time.

In North-Caucasus, Kazakhstan and Middle Asia *M. meridianus* plays a very important role as the host of plague and some other dangerous infections. They render considerable damage to afforestation sites and nurseries of trees.

LIBYAN RED–TAILED GERBIL

Meriones libycus erythrourus Gray, 1842

It inhabits the deserts and semi-deserts of the Near East, Asia Minor, Middle Asia and the north-western part of China. The northern border of its range touches the lower Emba river, northern Usturt, lower Tchegan, St. Dongus, lake Tchubartenis, the desert of Arys-Kum (to the east of Jusaly), the central parts of Betpack-Dala (47 °N), the eastern foothills of the Tchu-Ilinian mountains, the southern foot-hills of the Jungarian and the northern foothills of the Zailinian Ala-taue along the valley of Ili river. They inhabit the east Caspian territory of Usturt, in the large balchans and in Kopet-Dag, Badhys and Kara-bill, in the northern, southern and eastern Aral region, in Mujun-Kum and near the Balkhas lake. They are also found in the foothills of Pamiro-Alay and Tian-Shan (upto Dushanbe in the north and Gulajb town in the east). Isolated populations have been encountered at Za-Caucasus at the eastern part of the Apsheronian peninsula, the Cura-Araksian low-land upto Talish, Shemacha; the eastern part of the Shiraksian steppes, the central part of Kizil-Kum (Tamdy) and the northern and southern parts of the Izzyk-Kul depression.

They inhabit foothill deserts with ephemeral vegetation covers. The soil in their habitat is usually loessed or loess-sandy and the terrain may include stabilised dunes, dry valleys and gorges. They are quite common in the unflooded areas of river-valleys in the desert zone. They also inhabit oases, gardens, and fields of cotton and *Medicago* and occur on the banks or dykes, along roads, on waste lands, in houses and other buildings and especially in mud-houses.

545

In western Kizil-Kum these animals are found on loamy and clay-loam deserts having *Salsola*-wormwood-ephemeral associations (RUDEN-TCHIK, 1959). In the southern Balkhash lake region it often inhabits fields of *Lasiagrostis splendens* and in overgrowths of reed near wells and basins.

It inhabits stabilised dunes with covers of *Ceratocarpus turkestanicus* and bushes of *Salsola laricifolia* and wormwood. It is also found on gravelly soil in the foothills (ISMAGILOV, 1961). In the eastern Priaralie it inhabits slopes of loamy and clayey soils (GINTLIS, 1959). In the northern Priaralie it inhabits dry cerealgrass valleys and stretches of sandy massifs with *Haloxylon aphyllum*. The distribution of the red-tailed gerbil is closely connected with that of the great gerbil whose burrows it often uses (SHILOV, 1953). The optimal ecological conditions for the red-tailed gerbil, in contrast to those for the great gerbil, are found more in the south. In the agricultural regions of Tajikistan, Uzbekistan, Turkmenia and Azerbaidjan the majority of the red-tailed gerbils live on the out-skirts of fallow land, on soil deposits of various ages, in fields and along roads. These animals are equally numerous in houses, commercial buildings, cattle yards, store houses, etc. (DAVIDOV, 1962, 1964; ALEK-PEROV, 1966).

The burrows of the red-tailed gerbil, although less complicated in structure than those of the great gerbil, are still more complicated than the burrows of other gerbil species. Usually their colonies occupy terri-tories covering several dozens of square metres and each has 10–20 entrances (sometimes even more than 50–60) situated quite near to each other. The burrows are usually two-storied with store-chambers situated in the upper part and nest chambers at depths of 1 to 1.5 m (RUDENTCHIK, 1959; ISMAGILOV, 1961). Besides nest-burrows there are protective, simple burrows with 2 to 3 entrances. If the number of animals is very high in any particular colony different burrow systems merge into rather large 'townships'. In winter, 25 to 30 red-tailed gerbils were found to occupy a burrow (in summer this group was found divided into 2 to 3 sub-groups each of which lived in neighbouring burrows (NURGELDIEV, 1960, 1969).

The radius of individual territory of red-tailed gerbils in Azerbaidjan has been found to depend on the density of population, ranging from 50 m if the density is of the order of a few dozens of animals per hectare to 100–120 m where the density is of the order of 10 animals per hectare (BAKEEV & KADAZKIJ, 1959). The mobility of the red-tailed gerbil is rather high and so they often change their burrows. ISMAGILOV (1961) dug out 35 gerbils on a single territory during the period 7 to 11 May, and later on, during the period 30 May to 13 June, he found 19 more animals in that territory. Seasonal migration of these animals from one station to another connected with changes in the vegetation and other conditions have been observed (DAVIDOV, 1964; ALEKPEROV, 1966).

The daily activity pattern of the red-tailed gerbil is more or less similar to that of other species of the genus *Meriones*. They are, however, somewhat more active not only in the cold part of the year, but in the beginning of summer and in autumn also. The daily activity patterns of populations, occupying different niches, differ considerably. For example, in Pri-Caspian part of Azerbaidjan during the whole of the year these gerbils are active only at night and in twilight, while those inhabiting the central regions of the Republic are active mainly during the day-light hours for most part of the year (BAKEEV & KADAZKIJ 1959). Red-tailed gerbils, living in the burrows of great gerbils often adopt a diurnal activity pattern, in common with the habit of the latter animals.

The red-tailed gerbils feed mainly on seeds. Even in summer 80 per cent of their food are seeds and fruits (GLADKINA & POLJAKOV, 1957). In spring green juicy forage, rich in vitamins, become important items of their dietary. In winter, after dry summers, they often feed on the underground parts of plants. They mostly consume plants like cereals, beans and also *Salsola* and sedges. They not only eat the seeds and fruits of cultivated plants but also store these for winter. They are also known to store the *Pistacia* (Anacardiaceae).

The stored feed materials weigh upto 10 kg. Storing is not very usual for populations of red-tailed gerbils inhabiting Azerbaidjan, but is very usual in Middle Asia where also the green parts of plants are important food items for this species.

Reproductive activity may continue for the whole year, but is usually restricted to 7 to 9 months – from the middle of February–March to the end of September–October. In the dry hot periods, sexual activity is somewhat less intensive, but it rarely stops. In spring-summer they usually have 2–3 litters and in autumn one more litter is added, a number of young females taking part in reproduction during autumn, some of which, in Azerbaidjan, even, produce two litters. The average number of young ones in a litter is 5–6.

The red-tailed gerbils are very tolerant to lack of water in their food. In the Kashka-Darjinian region these animals reproduced when their food contained only 20 per cent water, a level at which reproductive activity of the great gerbil is known to completely cease (GLADIKINA & POLJAKOV, 1957). In the northern Pri-Aralian and Pri-Caspian regions, the reproductive activity has been found to continue without any interruption from April to October (SHILOVA, 1956; GINTLIS, 1959). In Azerbaidjan and in the Kraznovodian peninsula the season of reproduction was even longer, occupying the whole of the winter (VASILIJEV et al., 1963; ALEKPEROV et al., 1967). The average number of embryos per pregnant female was 6.5 in the northern Pri-aralian region, 5.5 in Kashka-Darinian region and 3.5 in Turkmenia and Azerbaidjan.

Red-tailed gerbils in captivity produced 5 to 6 litters in a year with 2 to 8 young ones in every litter (average 5.3). The majority of litters were born during February–June; there were no litters in July–August and November–December and only 3 litters (out of a total of 29) were born in September. The duration of pregnancy in red-tailed gerbils ranged from 23 to 35 days (average 29 days). The young ones grow very quickly. By the age of 10 days nearly the whole of their body is covered with short fur, aural passages open, and eyes open by the age of 15–17 days. In a month, the young animals are quite independent of their mothers (VOLOGIN, 1967).

The intensity of reproduction fluctuates greatly with differences in conditions of existence. This is one of the causes of the observed high fluctuations in the number of red-tailed gerbils, such fluctuations being relatively higher than in other species of gerbils. The mortality of these animals rises considerably after dry summers and in cold snowy winters.

In Turkmenia the population of these animals increases markedly every 7–8 years. In certain years the population density may reach 150 to 200 animals per hectare. The greatest increase in population was marked in 1953; when the zone of high population density in western Turkmenia occupied 10 million hectares (NURGELDIEV, 1969). The direct cause of such a rise in population density was a favourable combination of weather conditions. The winter of 1950–1951 was severe and snowy and the summer of 1951 was very dry and hot, so reproductive activity was restricted to the middle of spring. Although sexual activity was resumed, in autumn the number of animals was rather small by that time. During the warm and wet winter of 1951–1952 reproductive activity of the red-tailed gerbil did not stop, and this became most intense in the spring of 1952, which was very wet – there having been 17.5 times more precipitation in that year than in 1951. Plant growth was very active in the spring of 1952 and continued in the summer also. The autumn and winter of 1952–1953 were also rather favourable, although the intensity of reproductive activity was not higher in those seasons than in the previous year. The dry, cold and prolonged spring of 1953 created a critical situation for the population. Reproductive activity became less intense in April and stopped altogether in May; this was accompanied by mass migration of these rodents. In April large scale invasion by the red-tailed gerbil of settlements and even of towns was observed and the animals were found to occupy unusual niches like takirs. The animals accumulated along railway lines and their bodies could be seen on the highways and on unmetalled roads. All usable stations were inhabited by the maximum possible number. The migrations and the development of epizootics caused not only an end to all reproductive activity but also resulted in a quick fall in the number of the animals, which led to a very serious reduction in the population size by the spring of 1954. In Turkmenia the zone of high density occupied territories from the southern shores of Korabogas-gol Bay of Caspian Sea, upto the northern parts of the Meshedy-Messerianian plain in the south; the eastern border of the territory touched the Usboy dry river-bed; in areas situated to the east of Usboy, where there had been no rise in the number of these animals in 1954 there had neither been any migration nor a reduction in population size (FENJUK, RADCHENKO & ZHERNOVOV, 1957). The red-tailed gerbil is a natural carrier of plague, skin leishmanioses and tickets (pincers) reflexive-typhae. It is not only a spreader of the causative agents, but in some places it serves as an important host of the organisms. In Tajikistan it destroys about 20 per cent of the plant cover, hinders soil conservation, harms plantings of young fruit trees and destroys agricultural crops.

AFGHAN GERBIL OR GERBIL OF ZARUDNY

Meriones crassus zarudnyi Heptner, 1937

It is a rare species, found in the deserts of Iran, Afghanistan and in the southern part of Turkmen SSR, the eastern parts of Badhys, south of Karabill and to the north – upto the town of Tachta-Bazar (HEPTNER, VOJCEKHOVSKII & NIKITIN, 1958). In Badhys it has been found in regions of jagged relief (to the east of Cushca town). They have been caught in the southern part of the slope of a deep gorge, densely over-grown with *Bromus*. The slope was full of burrows of this gerbil. In the mornings and in the evenings these animals were found foraging (HEPT-NER, 1956). The colonies of this gerbil in Turkmenia are situated on the lower parts of slopes of hills and on the depressions between the hills where the soil is condensed loamy (loessed) and cut with ravines. It inhabits areas with rich vegetation consisting of wormwood and *Poa bulbosa*, *Bromus* and other grasses. It prefers regions where the height of the grass is between 5 and 40 cm and the cover is of the order of 75–80 per cent. It is frequently found in the vicinity of the burrow – colonies of the great and the red-tailed gerbils. More rarely it establishes its own colonies on the ploughed up sides of unpaved highways, usually with a small number of animals. In winter and in the beginning of spring it eats fruits and seeds of grasses, then begins eating parts of ephemeral plants and by the end of summer returns to feeding upon flowers, seeds and fruits of grasses. In the beginning of autumn green plant parts again become very important items of its food and then it starts eating again fruits and seeds. The daily activity pattern of this animal is very similar to that of the red-tailed gerbil (NURGELDIEV, 1969).

GREAT GERBIL

Rhombomys opimus Lichtenstein, 1823

The great gerbil inhabits the desert and semi-desert zones of Kazakhstan,

Middle Asia, north-western China, South Mongolia and in the south-western parts of Iran and Afghanistan. Its morphophysiological and ecological peculiarities (thick winter fur, diurnal activity, well-developed instinct of storing food for winter, feeding mainly on the succulent bushes of salsola) show its connection with the Asiatic sandy and loessed deserts. It appears to be one of the oldest species of the region as indicated by the dimensions and the arrangements of its burrows which are situated on recent and ancient alluvial deposits in the southern parts of the Asiatic deserts.

Its range extends to the north of Guriev, includes the drainage-basin of the river Emba and areas to the north of Chelkar town upto the middle stream of the river Sarysu. It inhabits the desert of Mujun-Kum and areas further on, along the south bank of the lake Balkhash, to the south, round the mountains of Tian Shan. On the territory of the U.S.S.R. small colonies of this animal penetrate through the Jungaric gates to the Alaku depression and to regions along the river Ili almost upto halfway through its course. In China, great gerbils inhabit the deserts of the northwest, penetrating to the Zaaltayskaya Gobi (Mongolia) and to the inner Mongolia. In Afghanistan the great gerbil has been found only in the north, where large groups of its colonies are situated along the valleys of the Gurmatch mountain in the west to the Hulm mountain in the east. The southern most colonies are found about 5 kilometres to the south of the Pule-Humry mountain (AKIEV, 1968). The Iranian colonies of this gerbil in the Deshte-Kevir and the Deshte-Lut deserts are associated with the Turkmen colonies. 700 kilometres from these habitats, there is an isolated population of this gerbil in Iranian Azerbaidjan (LAY, 1967; MISONNE, 1959).

During the last twenty years, the range of the great gerbil has extended to the north, presumably being influenced by a higher rainfall in the deserts. The increased amounts of precipitation received during the last ten years have probably increased the reproductive rate of these animals in their principal habitat. The first gerbil colonies in the north were noted by the local inhabitants in the middle of the forties of our century. The animals have been regularly observed since the last thirty years.

In 1953 great gerbils were found on the right bank of the river Amba, about 30–40 kilometres to the east of Guriev, and in the north, near the Makat station. Considerably large, and probably old, colonies were found on the bend of the river Amba along its left bank, and further to the east, to the south of the Chelkar station on the Irgis river. Its range extends to the south-east, turning round the Tshelcar-Tenegis depression. The majority of the places mentioned above are situated in the desert zone (SHILOV, 1953).

Further observations have revealed some new colonies of great gerbils in the valley of the river Amba, in the lower reaches of Irgis and Turgay, 30–70 kilometres to the north from the border of the desert zone (Fig. 33).

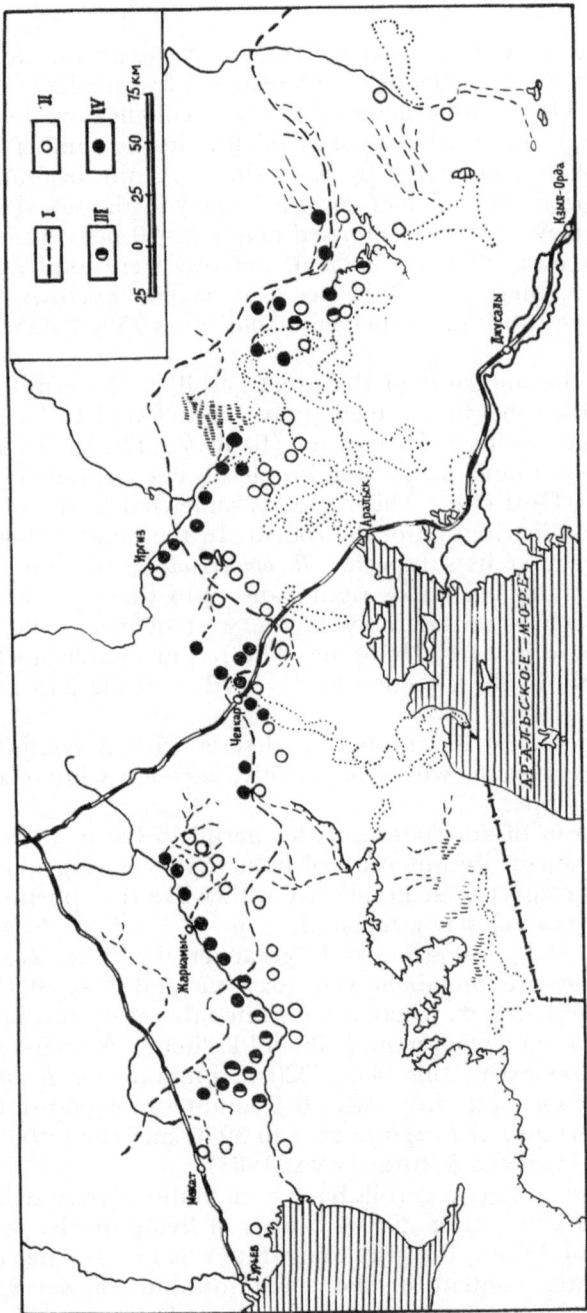

Fig. 33. Colonies of the great gerbil in the northern Aral region (after VARSHAVSKIJ *et al.,* 1969). I. northern border of desert zone; II. northern colonies observed in 1953; III. same as II, but not observed by SHILOV; IV. new colonies which appeared during 1958–1961. (Thatched portion on the right side of the Fig. is the Aral Sea)

The establishment of these colonies took place along the sandy massif and valleys of the rivers, the northern reaches of which have desert type of vegetation. The establishment of the new colonies has been proved by direct observations. Within a period of 10 years the number of burrows in the new colonies increased by 5–12 times. At the beginning of the observation period, the colonies consisted solely of the so-called 'recent' burrows (see below). These comprised about 40–60 per cent of the total number of burrows. 'Old' or 'ancient' burrows were totally absent. In the old colonies, there were 3–15 per cent 'recent' burrows, 49–62 per cent 'middle-age' and 22–48 per cent 'old' ones (VARSHAVSKIJ *et al.*, 1969).

Remains of the ancestors of the great gerbil (fossil gerbil-genus *Plio-rhombomys* – with roots in the teeth) have been found in later Pliocene deposits in south-western Turkmenia (Badhys). The genus *Rhombomys* has already been found in the middle pleistocene deposits in the town of Inder on the Ural river. This region is somewhat to the north of the present range of distribution of the rodent. In the middle Holocene, the great gerbil – rather its subspecies, *R. opimus obolenskii* Mal., inhabited the lower Volga region, on its right bank upto the river Terek in the south-west, which is considerably to the west of its present range. It disappeared there in early Holocene. The recent establishments of this gerbil are mostly in the direction of the borders of the past range of its distribution.

The great gerbil is distributed in deserts with a continental and relatively severe climate with cold, windy, snowless winters and partly in subtropical steppes.

The mechanism of adaptation of this gerbil to life in the deserts includes a reduction in the intensity of its activities during unfavourable periods. In this respect these gerbils are rather like the ephemeral plants with which they are closely associated.

The physiological peculiarities of the great gerbil indicate its adaptation to arid conditions. Its metabolic rate (oxygen intake) at 20 °C is of the order of 2100 ml/kg/hour, whereas for white laboratory rats of the same weight this rate is somewhat more (2460 ml/kg/hour). A similar difference in pulse rate also exists, this being 220–230/minute for *R. opimus* and 340–440/minute for white rats. After 8–9 minutes of exposure to the sun the body temperature of *R. opimus* rises to 39 °C and the respiration rate to 150/minutes (SLONIM & SHEGLOVA, 1963).

The body tissues of great gerbils have been found to resist dehydration, losing 0.5–2 per cent water during 5 days of living on dry food, while under similar conditions, the gray and white rats lose 1–3 per cent body water. The water content in the small intestine, however, remains constant in *R. opimus* during water restriction but it considerably decreases in the faeces (SHEGLOVA, 1962; SLONIM & SHEGLOVA, 1963).

Feeding on ephemers, succulent *Salsola* and wormwood, which possess

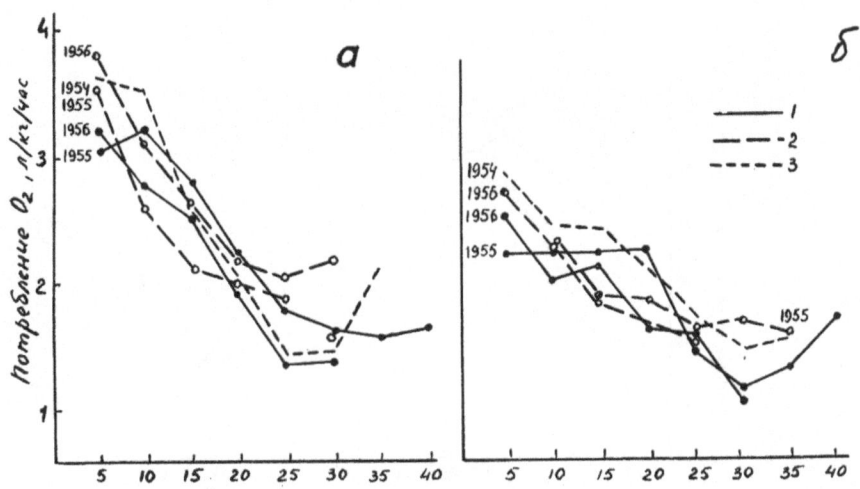

Fig. 34. Seasonal and yearly fluctuations in oxygen consumption by (a) red-tailed and (b) great gerbils (after KALABUKHOV, 1969). 1. February, 2. June, 3. October/ November. (ordinate: O_2 used in litres/kg/hr; abscissa: air temperature, °C).

large amounts of water, the great gerbil has quite a high rate of water turnover and it does not depend on its metabolic water to any great extent, unlike the real desert animals (SOKOLOV & SKURAT, 1962; SCHMIDT-NIELSEN, 1964). Its water balance is maintained by the water contained in the food, and by its economic use of the available water and its adaptive behaviour. The most important factor is that it stays inside its burrow for most of the time. The burrows are very complicated having a more or less constant micro-climate (ILJINSKAJA & KUZIN, 1965). The digging of burrows is an absolute reflex. The great gerbil and the young animals start digging irrespective of whether they were brought up alone or by their mothers (SLONIM & SHEGLOVA, 1963).

The great gerbil softens the influence of fluctuations of temperature, humidity and precipitation by changing its daily activity pattern according to the season. The metabolic rate of these gerbils regularly changes during different seasons. The curves of chemical thermo-regulation (changes in the use of oxygen connected with differences of temperature of the surroundings) are similar for the great and the red-tailed gerbil, although in the former species, the amplitude of fluctuation of the curve is relatively larger and it lies on a higher level (Fig. 34). Both of these species show seasonal fluctuations and differences in metabolic rate in various years depending on the availability of food and on the weather (KALABUKHOV, 1969).

Definite seasonal and year to year differences are also noticeable in the temperature of the ground that gerbils prefer (the ground temperatures

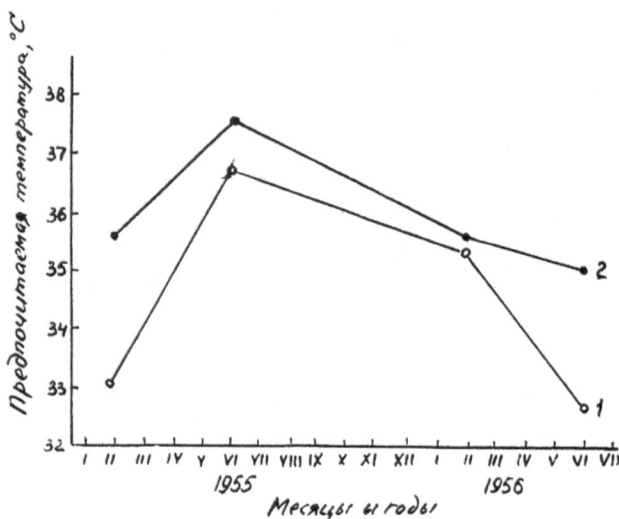

Fig. 35. Seasonal and yearly fluctuations in preferred temperature for (1) red-tailed and (2) great gerbils (after KALABUKHOV, 1969). (ordinate: soil temperature, °C; abscissa: months and years).

have been determined with the thermo-gradient device of Herter-Kalabukhov). These differences are lower for the great gerbil than those for the red-tailed gerbil (Fig. 35). In the great gerbil, the haemoglobin concentration ranges from 11.7 to 12.9 per cent and in the red-tailed gerbil, from 11.5 to 13.5 per cent. The pattern of thermoregulation in the red-tailed gerbil, which is nocturnal and is more active in the colder part of the year, differs from that in *R. opimus*, the latter species apparently possessing some adaptive, thermo-regulatory attributes. It now appears that the great gerbils, with their diurnal activity pattern, their long, deep and complicated burrows with a constant microclimate, and with their short period of stay on the surface during the cold period of the year, manage to maintain their heat balance mainly with the help of adaptive behaviour (KALABUKHOV, 1969). The numbers of different types of leucocytes in the blood of the great gerbil have been found to be higher in comparison with those of the red-tailed gerbil.

It is now generally believed that the most important factor influencing the distribution of the great gerbil is the condition of the soil. The ideal set of soil conditions allows the digging of complicated systems of underground burrows with the necessary microclimate. Apart from these, the micro- and the mesorelief of the surface and the composition of the plant covers, providing nourishment for the animals are also important considerations.

The great gerbil does not seem to possess any outstanding digging talents and, as a rule, it tries to avoid hard ground, in which it is very difficult for them to dig. *Citellus* and some Dipodidae are known to be able to dig not only in loamy soil but also in gypsum beds. However, gophers and Dipodidae usually make individual and simple burrows which are often abandoned later on while the great gerbil makes large and complex refuges.

The great gerbils work in groups while digging burrows, and improve upon the original structures making them more and more complicated, through hard work of several generations. Because of this, these complex burrows of great gerbils are justifiably called 'colonies' by many authors.

Our observations made over a period of nearly 20 years on the same burrow-colonies in the northern Aral region showed that the number, position, size, number of entrances and other external signs of the burrows did not change essentially, or changed to different degrees in some colonies of these animals, testifying to the stability of the colony complex. It can be explained only by assuming that these gerbils regularly repair their burrows, keeping them in a good condition. The outward look and the plan of the subsidiary burrows were the most stable; these burrows were used regularly although periodically and not very intensively (so called 'recent' burrows). During prolonged and severe reductions in gerbil numbers, the entrances to these burrows become filled with sand and these are no longer visible from the surface. But as soon as the animal number catches up, these burrows are restored at the same places, with their familiar appearances. The main burrows, which are used by the animals more or less regularly, change their outward appearances and even position. Sometimes, however, the centre of the burrow system is shifted from one place to another but the entire colony remains at one place.

There are convincing evidences, based on data obtained by the radio-carbon (C^{14}) method, that the main burrows are frequently used by many generations of these animals over many hundreds and even thousands of years (DINESMAN, 1968).

The burrows of great gerbils, as a rule, are situated in places where the ground stratum has the so-called 'non-washing' type of water regime. In this situation there is a well-aired impervious stratum which has a stable dryness during the whole year between the strata of wetting and capillary borders of the subsoil waters, having more or less a stable hydro-thermal regime with a high stability of humidity of air and without intense fluctuations of temperature. All these characterize the so-called 'zone of aeration'. The majority of the tunnels and pantry-chambers are situated in the stratum of wetting, while the innermost nest-chambers are apparently situated on the border of this stratum with the first tier of impervious stratum (LEONTJEVA, 1966).

LEONTJEVA (1966) has observed that the association of the great gerbil 'with deserts can be explained not by the food specialisation but by the mechanical properties of the ground, which create quite an even microclimate in the burrow, with a stable temperature and good aeration during all seasons of the year'. She considers that such factors depend upon 1. the depth of subsoil water, 2. the mechanical structure of the soil, which influences the height of the capillary border of the subsoil waters, water capacity and ventilation, and 3. the relief.

The most favourable conditions for the gerbils are those of subsandy grounds of younger and, especially, ancient alluvium. The more complex and profusely used burrows are found in such soil. These are very comfortable during the winter, summer, drought spells and the reproductive period. If the particle size of soil is small in the higher stratum, the possibilities of building burrow-colonies there and of the duration of their occupation are limited.

The characteristics of burrow-colonies of great gerbils have been elaborated by many authors, the majority of whom consider that their structure is determined by the qualities of the ground (NAUMOV, 1954; NAUMOV & KULIK, 1955; VARSHAVSKIJ & SHILOV, 1956; ROTSHILD, 1957; DUBJANSKIJ, 1962, 1963; KASATKIN et al., 1963; BALABAS et al., 1965; GVOZDEVA, 1965; LEONTJEVA, 1966). The systems of classification, so far suggested, are rather complicated and distinguish six or more types of burrows according to different signs, showing no difference between 'types', 'age of development' and 'variants' of burrows. However, all authorities consider that there exists three main types of colonies, the structures of which depend upon the qualities of the ground: 1. complicated and deep, many-staged burrow structure with firm walls and roof of nest-chamber, tunnels and food-chambers. The winter nest-chambers are situated 1.5 to 2.5 m below the surface. The microclimate in these chambers is very stable. These burrows are inhabited both in hot and cold periods of the year. They are very compact but may be rather large. They are usually situated in the depressions and are often surrounded with a 'wreath' of large trees of *Haloxylon aphyllum* (Fig. 36). Sometimes two or even three families of gerbils live in one burrow (a female with young ones being considered a family). There is usually quite a large number of different species of vertebrate and invertebrate animals living together as 'room-mates' and 'lodgers'. These burrows are usually situated in the subsandy ground with a variable mineral structure, which has an alluvial (partly proluvial) origin. 2. Not very deep and not so complicated burrow structures, usually of two stages; the majority of the tunnels and chambers are situated near the surface. The tunnels are not so compact, long and superficial, and the entrances are situated far from each other; there is no concentration of entrances in the centre of the colony. These burrows are much less firm, the microclimate (hydrothermal regime) is not very stable and therefore, not favourable for the animals. These can rarely serve as nest-burrows (only when the number of animals increase). These burrows are situated in

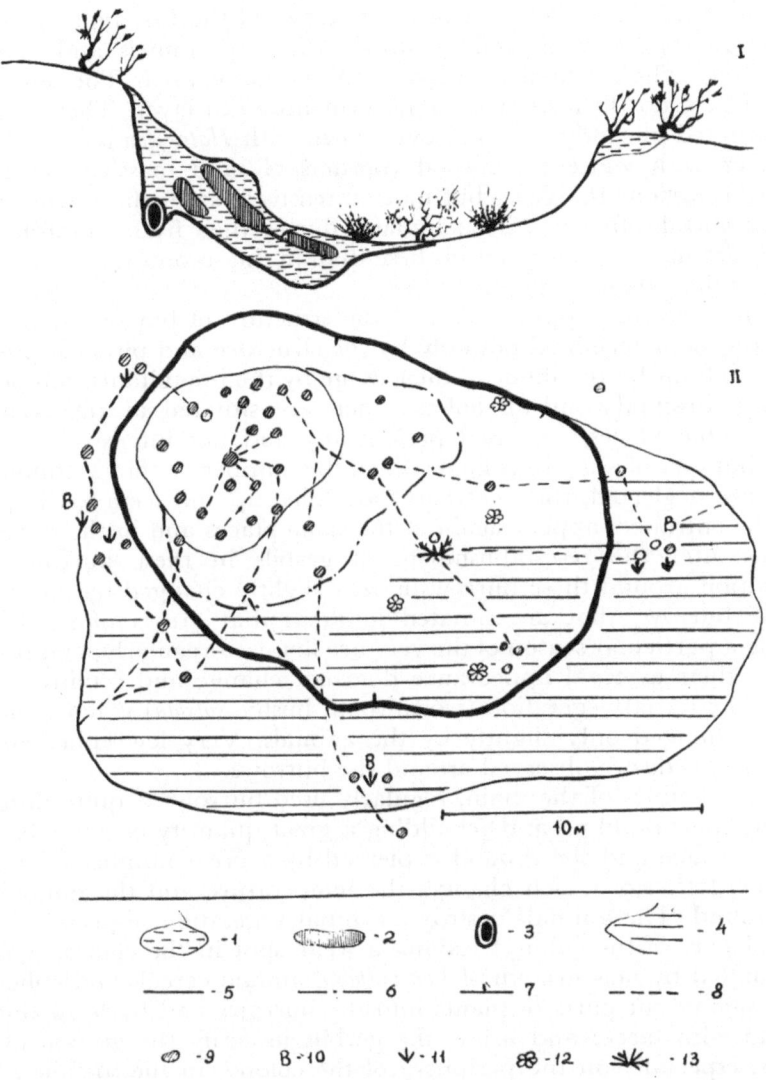

Fig. 36. The structure of a burrow-colony of great gerbils in ancient alluvium (after NAUMOV *et al.*, 1972). I. crosssectional view; II. view from above. 1. area of the colony; 2. food-chambers; 3. winter nest; 4. foraging zone; 5. the centre of the colony; 6. the peripheral zone; 7. depression; 8. paths; 9. exits; 10. adjacent burrows (new colonies); 11. *Haloxylon;* 12. *Eurotia;* 13. *Calligonum.*

the clayey and loamy grounds, with a superficial layer which is cemented with different salts and so it is very hard to dig in it. These are used by the animals mainly in the warm period of the year, especially during early establishment. The number and type of room-mates and lodgers

557

are much less in comparison to the burrows of the first type. 3. Rather deep, but not complicated burrows with long tunnels and scattered entrances. These usually occupy a rather large area. The superficial tunnels are not firm as there is no cemented salt-layer. These burrows are situated in sandy grounds, overgrown with *Haloxylon persicum* bushes together with white wormwood (borders of sand massifs). These are used throughout the year, but more intensively so in the warm period. Young gerbils often stay in such places during their fresh establishments. There are more room-mates in these than in the second type of burrows, but less than those in the first type.

So the external appearance and the structure of burrow-colonies are found to be determined not only by the structure and physical property of the soil but by the mode of their usage by the inhabitants. Simple and mainly subsidiary outside-holes, which are situated at relatively un-comfortable places, are used periodically and not intensively, usually if the number of animals is high. When the number of animals diminishes these are neglected, their entrances are filled up and seem to disappear. But the entrances appear again at the same places and in the same con-ditions after the re-establishment of gerbils in these burrows. The vegetation around these burrows is very slightly changed by the gerbils.

The burrows that are situated in areas that are comfortable only during a particular season of the year are not used regularly, and because of this their external appearance does not change and retains a 'new' look. The initial vegetation (wormwood, bushy *Salsola*) is not destroyed and is changed only slightly by the animals. Very few ephemers and inedible weeds are observed around the burrows.

The dynamics of the main, regularly used burrows is quite different. During their building and rebuilding a great quantity of soil is brought to the surface and the ground is pierced by a great number of tunnels, the ventilation of which changes the temperature and the humidity of the ground. The animals destroy all initial vegetation (especially in the central part of the colony) leaving a weak spot in the centre, which is surrounded by new growth of *Poa bulbosa*, annual cereals and ephemers. By dragging cut parts of plants into the burrows and by fertilizing the ground with faeces and urine, the gerbils intensify the growth of new plants, especially on the periphery of the colony, in the so-called 'con-sumed zone'*. Such a colony with its bare spot, surrounded by a rather rich green vegetation of the 'consumed zone' strongly differs from the rest of the desert and can be easily identified even from a height of some hundreds of metres.

On the sandy and sub-sandy grounds the continuous work of the gerbils over the main nest-colonies is accompanied by dispersal of the soil brought up from the deeper layers of the ground, and this leads to

* An area with all plants consumed.

the appearance of depressions which in time become larger and deeper. Thus, if the soil stratum is fully occupied by the main tunnels, the centre of the colony is sometimes shifted to the slopes of the burrow systems. Because of this the depressions become wider and sometimes ellipsoidal in shape.

The depth and the size of the depression may indicate its age. The largest 'ancient' depressions are found in the most stable settlements of the great gerbils. In the Aral Kara-Kum region the size of such depressions ranged between 12 and 35 m in width and from 1.2 to 2.1 m in depth. In the unstable settlements on the southern border of the Bolskiye Barsuky desert (Big Badgers) these depressions are often upto 80 metres wide, but are never deeper than 1 to 1.5 m. The majority of the depressions have been found to be 15 to 25 m in diameter; very often shallow, saucer-like depressions with no burrows have been found which were probably occupied in the past. Because of the shallow soil layer the existence of the burrows in such places over a prolonged period is impossible and so the once 'utilised' places are soon abandoned by the animals. Nearly 2 percent of the burrows in Big Badgers have been found to be such abandoned ones, while the old stable burrows accounted for 43 percent and burrows in 'new places' 55 percent of the total. There were only 0.5 percent abandoned burrows, 62.5 percent old ones and 37 percent burrows in 'new places' in the valleys of the north-western Aral Sea territory.

The burrows which are used regularly over a prolonged period very rarely appear in dense grounds. More often these are built in loamy alluvium along gulleys of erosion or in proluvium along slopes and precipices. These are sometimes a few metres in depth. These burrows, with their layer of cementing salts (sulphates and carbonates), are not easy to excavate and remain in the central part of the colony, composing a little hill ('dome') in which new tunnels and chambers for stores of food are dug, although the roof is not solid and often collapses. The size of the colony area without plants and the size of the 'dome' indicate the age of the colony. The size and the depth of the depression indicate the intensity of its usage. Studies have revealed that the main complex of nest-burrows and 'outside holes' (burrow openings) are built simultaneously, but their patterns change considerably with time. The extensively used burrows maintain their initial external appearance while the main nest-burrows pass through different stages of development. These may be used very little (in the light grounds – on sandy and sub-sandy soils) or may not be used at all (in the loamy and clayey grounds). But even in the latter situations the burrow needs a very long period of time to finish the whole cycle of development and to disappear, during which period the burrow must be regularly used and rebuilt. The periodical fluctuations in the number of resident animals and the consequent weakening of the digging activity prolongs the existence of the colony.

Many workers have subdivided the colonies into groups based upon their age: 'young', 'middle-aged' and 'ancient' (SHEKHANOV, 1952; NAUMOV, 1964; NAUMOV & KULIK, 1955; VARSHAVSKIJ *et al.*, 1952; ROTSHILD, 1957). The 'age' of a colony is very useful as one of its ecological characteristics, showing not only the real age of the settlement, but its stability as well.

The number of great gerbils in any area, as already mentioned, is limited by the relief of the area and the structure of the ground. These factors are important not only for digging, but these also determine the composition of vegetation. The most favourable complex of these factors can be seen in the regions of the modern and ancient alluvial deposits. There are complex landscapes where great areas of flood plains and marshy terraces are combined with alluvium, alternating with the remains of ancient river-beds and takirs or flat areas with loamy and sub-sandy soils having *Salsola*-wormwood complex of vegetation. All these combine with ridges or isles of sandy hillocks of different vegetation covers. Such complex, alluvial subsandy tracts serve ideally for the existence of the most stable settlements of great gerbils where they find the most favourable conditions for wintering, but such areas are also intensively used in the warm period of the year. Successful wintering is secured by the most deep, complex and firm burrows with large chambers for the storage of food while survival during summer is ensured by the availability of the rich and variable vegetation of these places where *Haloxylon aphyllum*,

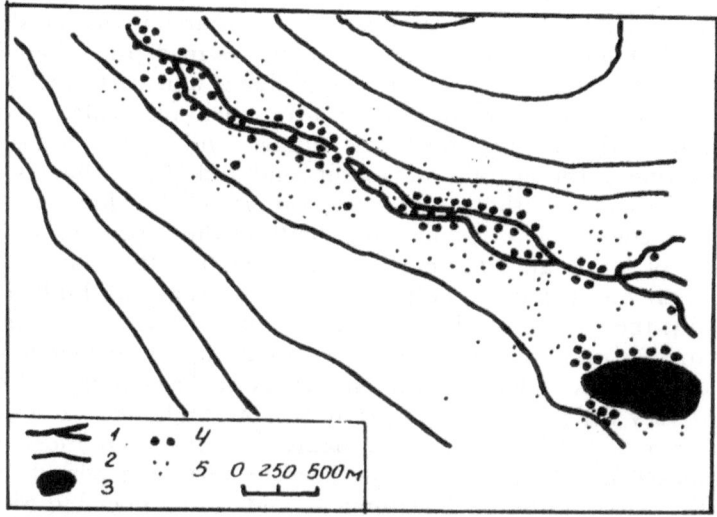

Fig. 37. Colony of the great gerbil in a dry valley (north-western Aral sea territory (after NAUMOV *et al.*, 1972). 1. talweg of a dry river-bed; 2. 4 m contours; 3. takir; 4. burrows inhabited in winter; 5. burrows inhabited only in the summer.

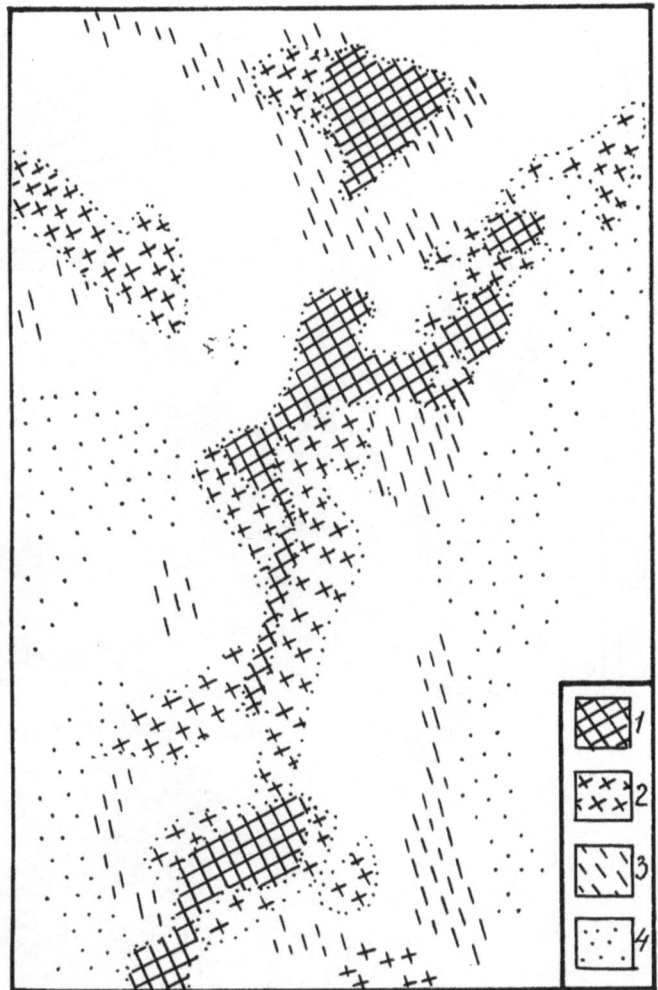

Fig. 38. Colony of the great gerbil in an ancient valley in the Aral-KaraKum (after NAUMOV *et al.*, 1972). 1. colonies of high density (more than 5 burrows per km of a route); 2. colonies of medium density (2.5–5 burrow per km); 3. colonies of low density (1–2.5 burrow per km); 4. colonies of very low density (less than 1 burrow per km).

white wormwood *Poa bulbosa, Crucifera* and *Alyssum* grow abundantly.

Sand hills are a somewhat less favourable habitat for these animals, but even then such habitats are often used, especially if the hills are covered with *Haloxylon,* white wormwood, *Eurotia* and different ephemers and ephemeroids. However, the possibilities of building permanent burrows in these locations are meagre as the conditions for wintering

561

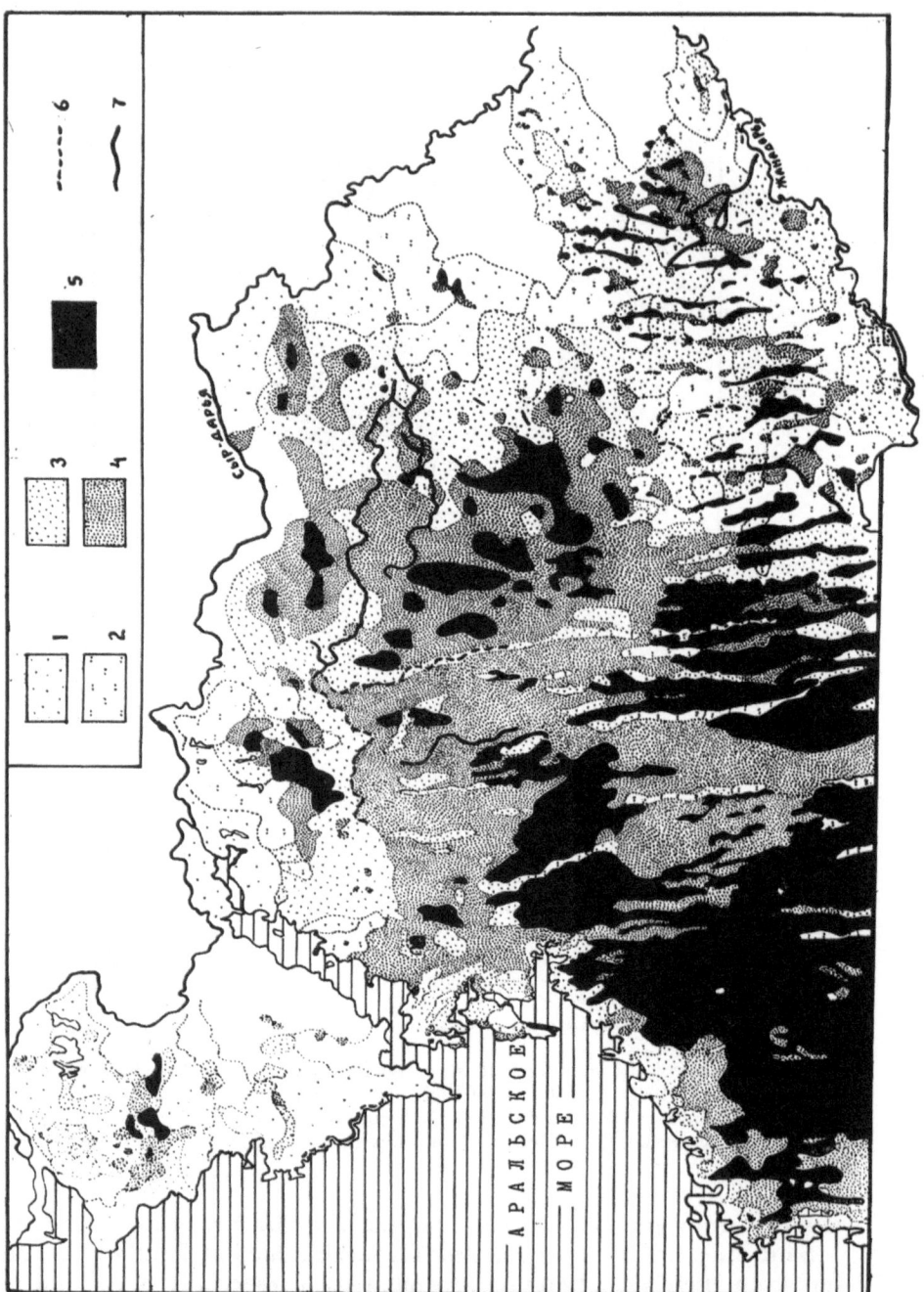

Fig. 39. Distribution of the colonies of gerbils in northern Kizil-Kum (after NAUMOV *et al.*, 1972). 1. less than 0.33 colonies per km, 2. 0.4–1 colony per km, 3. 1.1–2.5, 4. 2.6–5, 5. more than 5 colonies per km, 6. chain of colonies with 5–10 burrows per km, 7. chain of colonies with more than 10 burrows per km. (The thatched portion at the bottom is the Aral sea; the Syr Darya is shown on the left of the Fig.).

there are not so stable. The conditions become better in years of low precipitation and scanty vegetation. However, some stable colonies are found in such habitats and the activity in those colonies is maximum during summer (NAUMOV *et al.*, 1972). In the eroded valleys and gorges of north Priaraliye and in the littoral saline marshes of the Caspian and Aral Seas, where the soil depth is adequate and is capable of holding permanent colonies, deep and complex burrows of the great gerbil, which can serve for wintering, have been found (Fig. 37). At other places fragile burrows with superficial tunnels and chambers are built. These are intensively used in the warm period of the year, especially when young animals grow up and start establishing themselves. Because of this factor the number of animals living in the loamy and saline regions fluctuates rather strongly (in the same way, as on the dunes) not only from year to year but also during different seasons of the same year.

The most stable colonies of the great gerbil are found mainly in ancient valleys of older alluvial elements (Fig. 38). Such colonies are present in the lower and central parts of the Aral-Turgayan strait. The most stable and active natural foci of plague are connected with these settlements. Similar colonies are also found between the Syr-Darya and the Sarysu rivers (Kizil-Kum, Fig. 39), as well as in the valley of the river Emba, along the ancient valleys of north Predusturtie, and to the south – in Turkmenia, Zaungus Kara-Kum and along the ancient valley of Usboy. Again, at these very places the most stable and active natural foci of plague exist. The occurrence of very large colonies of gerbils in the valleys of Mugab and Zaravshan and in Badhys appears to be related to the ancient alluvial deposits in those areas.

Sub-montane deserts (proluvial), which are often composed of sub-sandy deposits, are also intensively used by the great gerbils. Such colonies are found in northern and north-western Usturt and Mangish-lack peninsula, on the foothills of Copet-Dag and on the Balchans of southern Usturt. But these settlements are smaller and are not as stable as the ones thriving on ancient alluvium in the valleys. Only the settlements of great gerbils in depressions of lake Balkhash are not stable.

The great gerbil colonies in China and Mongolia are strictly confined to sandy grounds. The most usual places inhabited by these animals there are sandy islets with covers of *Tamarix* and *Haloxylon* and the thin strata of sands on dried lake-beds overgrown with large *Salsola*, *Kalidium*, *Suaeda* and *Reaumuria* (BANNIKOV, 1954). In the Zaaltaian Gobi these animals inhabit sandy terrains which have luxuriant covers of *Haloxylon*, *Tamarix* and poplar (NEKIPELOV, 1959).

The most stable colonies of great gerbils usually consist of small, scattered populations which occupy different parts of the complex alluvial landscape. Each of the smaller population groups occupy rather large areas (hundreds and thousands of hectares) and, hence, these may be called 'large-block' settlements (NAUMOV, 1964, 1967; NAUMOV &

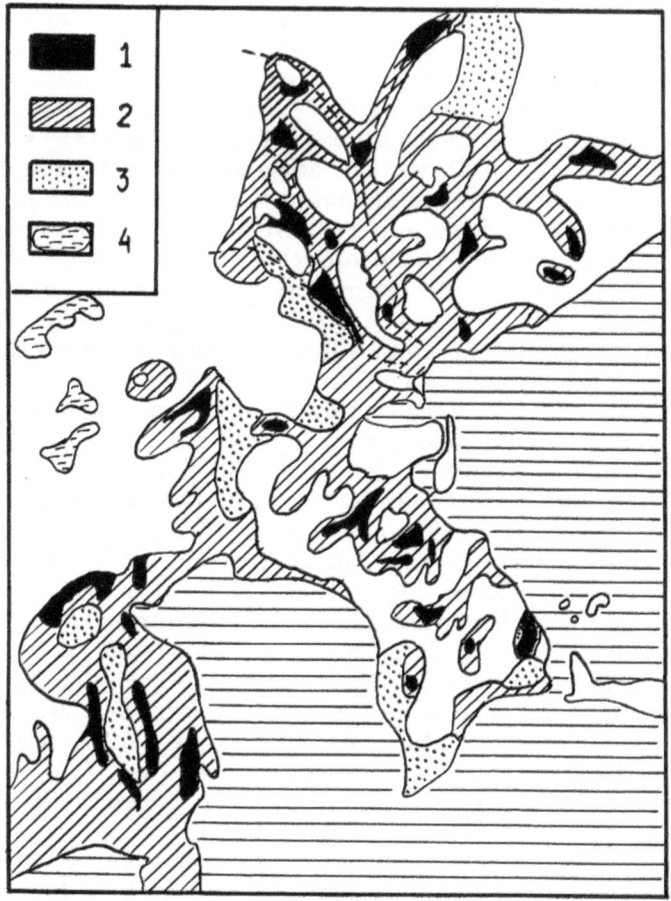

Fig. 40. Chain of colonies of great gerbils in a periodically water-logged area in the northern Aral region (after NAUMOV *et al.*, 1972). Colonies of (a) high; (2) medium; (3) low density; and (4) takirs.

LOBACHEV, 1965). These serve as the base or 'kernel' or 'nucleus' of the settlements.

Such 'nuclei', as a rule, are surrounded by thinly populated and rather un-stable settlements on plain, clayey-sandy and sub-sandy regions, in cellular and semi-cellular sands and on saline marshes also. These un-stable settlements occupy rather large areas, much greater than the 'nuclei', but the density of burrows in these colonies is much lower and the majority of the burrows are 'young' and 'middle-aged'. The populations in these colonies fluctuate considerably and do not attain high densities. These habitations are found in the complexes of valleys (slopes)

564

and on eroded gorges in the northern part (Fig. 40), and also on the foothills in the southern part of the area.

In the southern part of the northern zone and in the southern zone of deserts of the moderate zone, great gerbils are found in the majority of the biotopes and usually in 'continuous' stretches. In these vast ranges of great gerbils, stable 'nucleus' colonies and peripheral unstable colonies are easily distinguishable. The latter usually occupy the space between two 'nucleus' colonies, inter-connecting them over a very large area (NAUMOV & KASATKIN, 1963).

Most probably, the success of the great gerbil in occupying a wide variety of desert habitats lies in its capacity to feed on a wide spectrum of desert plants (LOBACHEV & KCHAMDAMOVA, 1972). It does not, however, consume any animal food. Their main foods are desert succulents (bushy *Salsola*), wormwoods, annual cereals *(Poa bulbosa)* and so on, desert and sandy sedges, weeds – such as *Atriplex, Chenopodium, Crucifera* – ephemers, and many other plants of sandy and saline soils *(Kochia, Eurotia* etc.).

Gerbils usually do not feed upon narrow-leaved cereals *(Stipa, Agropyron desertorum,* etc.) as well as on such psammophytes as, *Calligonum, Astragalus,* etc. The animals eat the leaves, shoots, bark, roots, rhizomes and bulbs and also flowers, seeds and fruits.

In the northern desert zone the main food of the great gerbil is *Haloxylon aphyllum*. In localities where this plant grows in abundance, the winter stores and the summer diet of the animals mostly consist of it. In regions where *Haloxylon aphyllum* is absent, its place is taken over by white wormwood, *Anabasis aphylla*. In the gorges and dry valleys some other plants *(Eurotia ceratoides, Kochia* etc.) are also included in their diet. Spring-summer ephemers like *Alyssum, Sisimbrium* and *Poa bulbosa* form part of the food of the great gerbils only seasonally.

In the southern zone *Haloxylon* is still an important food of the gerbils but there the role of ephemers and ephemeroids increases. 115 species of desert plants are known to be consumed by great gerbils in Turkmenia, 68 species (59 percent) of which are eaten in spring, 72 species (62.6 percent) in summer, 63 species (54.8 per cent) in autumn and 50 species (43.5 percent) in winter.

Ephemers and ephemeroids, which are rich in water content are the most preferred items of food during summer and these constitute 80–85 percent of the total food intake. Shoots of *Haloxylon, Salsola richteri* and other bushes are also eaten. A large number of vegetatively propagating plants such as *Euphorbia cherolepis, Aristida pennata, Cousinia bipinnata, Tournefortia sogdiana, Heliotropium arguzioides, Atriplex dimorphostegia, Senesio subdentatus,* onion *(Allium sabulosum)* and wormwoods are also available to them. In summer and autumn, the great gerbil digs onions and rhizomes of *Poa, Carex physodes, Ceratocarpus,* etc. 95 per cent of its autumn food is composed of underground parts of plants, ephemers and ephemeroids. Stores of *Carex*, *Ceratocarpus* and *Bromus*, gathered since spring, may sometimes weigh as much as 64 kg. Shoots of *Haloxylon*, *S. richteri* and other bushes are also added to the stores (NURGELDIEV, 1969; LOBA-CHEV & KCHAMDAMOVA, 1972). In Mongolia, *R. opimus* is know to eat *Haloxylon*, *Zygophyllum*, poplar leaves and shoots, reed, *Calligonum*, and bushy *Salsola* (BANNIKOV, 1954).

The habit of storing food is an important adaptation mechanism in these animals, aimed at ensuring an even level of nutrition. The storing activity begins in spring, comes to a halt during the dry summer period, and is at its maximum during autumn. Spring stores are used during the summer and the autumn stores are the basic (in the south) or the only (in the north) source of food in winter.

The spring stores mostly consist of stems, flowers and seeds of annual cereals *(Bromus, Rheum)*, ephemers, *(Alyssum, Sisimbrium)*, bulbs of *Poa bulbosa* (sometimes weighing upto several kg), and flowers of white worm-wood. Branches of *Haloxylon aphyllum*, wormwood, *Salsola richteri*, and the bushes, *Eurotia* and *Salsola* (*Anabasis* and others) are stored in autumn. The stored items in the same chamber are usually found to be the same always. The intensity of storing and the size of the store depend upon the climate of the year. Sometimes the gerbils may not store food even if there is an abundance of vegetation around and this is usually followed by mild snowless winters. But in other years they store food very inten-sively even if the vegetation is rather scanty and this is often followed by severely snowy winters. The Turkmen shepherds say that 'these animals know what type of winter will come and whether they will need their stores of food or not' (NURGELDIEV, 1952).

As a rule the food stores are kept in special chambers in the burrow system. But in the north zone, and sometimes in the south, in loamy and clayey grounds where it is difficult to dig large chambers, some part of the stores is left in stacks on the surface near the entrances. The size of these stacks may be 2 to 3 m in diameter and upto 1 m in height. The local people believe that the size of these stacks grow before snowy winters.

In the north-east Caspian Sea region the stores in the chambers usually weigh from 1.5 to 4.5 kg, besides that piled in stacks on the surface, measuring upto 50 cm in height and 8 kg in weight (VANSULIN, 1962). The majority of the stacks consist of semi-bushy plants – *Salsola*, *Anabasis ammodendron*, *Anabasis salsa*, gray wormwood, *Eurotia ceratoides*, *Kochia prostrata*, *Atriplex*, *Ceratocarpus*, etc.

The size of the food-stores is larger in the northern colonies as com-pared to those in the southern zone (SHEKHANOV, 1952). The stores of food found within one burrow in BetpackDala were found to weigh 35–40 kg, (ISMAGILOV, 1961) while they exceeded 8–10 kg in Turk-menia (NURGELDIEV, 1969).

The digging, feeding and storing activities of great gerbils, and their contribution to the fertility status of the soil with their urine, faeces and

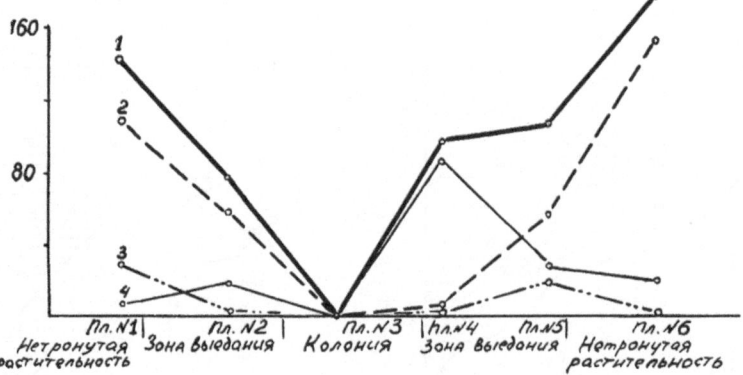

Fig. 41. Consumption of vegetation by the great gerbil in a colony (after NAUMOV, 1963). 1. total amount of vegetation, irrespective of species; 2. wormwood, *A. terrae-alba;* 3. annual cereal crops; 4. annuals with large flowers. (abscissa: N_1 & N_6 area unaffected by rodents; N_2, N_4 & N_5 foraging zone; N_3 colony proper).

remains of food influence the vegetation cover of the areas. The activities of this rodent usually bring about a depletion of the vegetation cover, although, to some extent, they also enrich and improve the fodder quality of the pastures. In the central parts of the colonies nearly all plants disappear, and only a few inedible species of *Salsola* remains unaffected. In the principal feeding areas of these animals (Fig. 41), wormwood becomes rare but due to fertilization of the ground brought about by the animals, a large number of ephemeral plant species like *Crucifera*, *Eurotia ceratoides*, *Kochia* and *Poa bulbosa*, which are very valuable fodder plants, appear. In some areas of great gerbil infestation, however, white wormwood is found to grow and thrive quite well.

In areas where the populations of great gerbil have been maintained at very low levels through control measures over a period of 4 to 6 years, profound qualitative changes have been found to occur in the pastures mostly because of the overgrowth of ruderal vegetation. One year after the initiation of control operations, the vegetation cover reached 30–40 percent instead of the 10–15 percent as found on areas having inhabited colonies. Great gerbils cause considerable damage to *Haloxylon aphyllum*. But the favourable moisture conditions in the depressions where the colonies are situated and where the soil is fertilized with the rodents' urine, faeces and decaying food stores, lead to vigorous growth of *Haloxylon aphyllum*. This is clearly indicated by the occurrence of the largest and the best developed trees of this species on the periphery of the depressions.

The great gerbils are most active at dawn. They emerge out of the burrows 20–40 minutes before sunrise during summer (between 4.30 A.M. and 5.00 A.M.). Two peak activity periods have been recorded in June

567

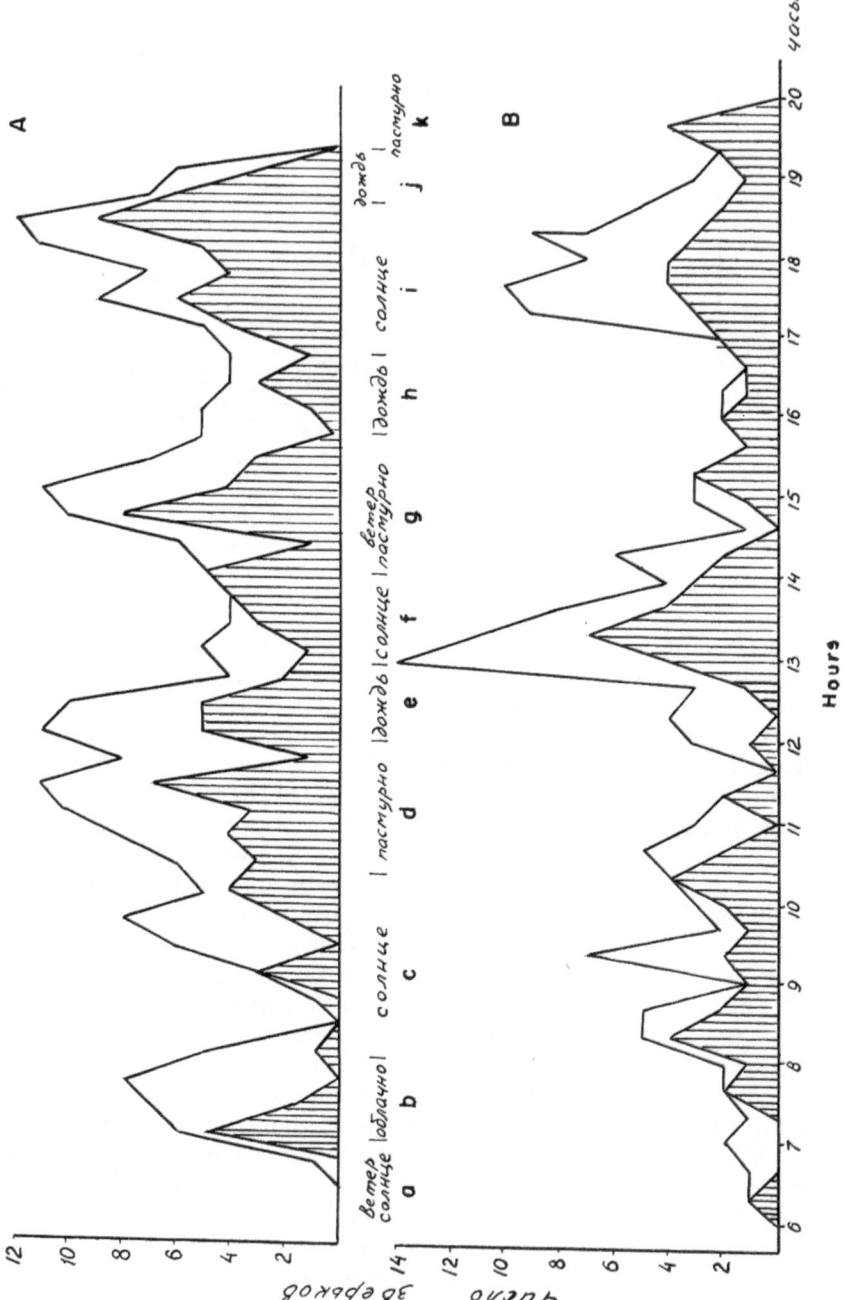

Fig. 42. Daily activity of great gerbil in the northern Aral region. A. in different weather conditions; B. at different hours; the upper curve in both Figs. shows the number of animals observed on the surface and the thatched one shows the number of animals seen feeding (after NAUMOV, 1971). a. windy and sunny; b. cloudy; c. sunny; d. cloudy; e. rainy; f. sunny; g. windy and cloudy; h. rainy; i. sunny; j. rainy; k. cloudy.

– one from 6 to 10 hours in the morning and the other from 18 to 22 hours in the evening. Their activity pattern also varies with variations in the air temperature in the range 22 ° to 37 °C (LOBACHEV & PASKHINA, 1972). In winter, the two peak activity phases merge together and the animals remain continuously active on the surface. In spring and autumn, again there are two activity peaks at different hours (Fig. 42). In cold weather, especially after snowfalls, these animals emerge out of their burrows only at 9 to 10 hours in the morning. The gerbils cease their surface activity at dusk, at 18 to 19 hours in spring and autumn, and at 20 to 21 hours in summer. The gerbils are most active in the morning and in the evening in summer, less active during the mid-day (13 to 15 hours) and least active at mid-day on very hot days. When the animals are on the surface they eat and store food for most of the time while only a small proportion of the time is spent in digging and in other kinds of activity.

The weather greatly influences the activity pattern of these animals. High ambient temperature and intensive radiation regimes suppress their activity. Warm and cloudy weather induces these animals to be highly active even during the hot periods of the day. Rain and strong winds (in all seasons) suppress their activity.

That the great gerbils are also active by night had remained unknown in the past although many scientists had, reportedly, heard the voices of these animals at night (KAMBULIN, 1941). Observations made in Aral Kara-Kum showed that the sub-adults of this species leave the burrows of their parents and start new establishments at night as it is presumably safer for them to do so under cover of darkness (ROTSHILD, 1955; MARIN & ROTSHILD, 1965).

The settled animals do not move far from the entrances of their main burrows while foraging on the surface. They move with slow jumps around the colony, spending most of the time in looking for food, and in eating. They also occasionally engage themselves in digging and in clearing their burrow entrances. They rise on their hind legs and look around quite often. When faced with danger they lean upon their tails and emit repeated alarm-calls, each of 85 msec in duration, 2 to 3 kc/s in frequency and having a period of 320 to 340 m/sec (LISICINA & NIKOLSKY, 1967). At the same time they make a thumping noise with their feet. These alarm-calls are heard not only by the gerbils which are on the surface, but also by those which are in their burrows. So all animals belonging to the colony pop out of their burrow openings in a state of excitement and panic.

The great gerbils are precocious, highly fecund rodents, having a potentiality to reproduce throughout the year. The intensity of reproduction regularly falls from spring to autumn. Some of the young animals are able to reproduce within a year of their birth (at the age of 3–4 months), delivering two litters. Extensive field data obtained by

569

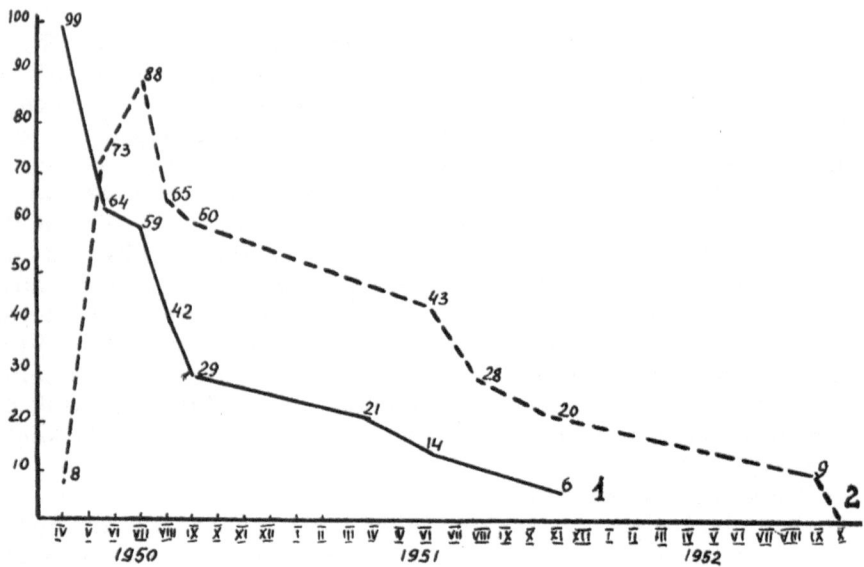

Fig. 43. Life span and mortality rate of the great gerbil in the Balkhash lake region (after BURDELOV, 1958). 1. animals born in 1949 or earlier; 2. animals born in 1950. (ordinate: percentage of animals surviving; abscissa: months and years).

Soviet zoologists indicate that even in the northern parts of the desert, in some years, these gerbils reproduce late in the autumn and in the winter months. In the southern parts of the desert this phenomenon is more pronouncedly marked (MOKROUSOV *et al.*, 1965). Compared to other species of gerbils, however, the great gerbil is relatively less fecund.

The age of sexual maturity for some animals from the early litters (February to March) is 3–4 months. Animals of later litters and the majority of the animals from the early litters reach sexual maturity only at the age of 10 months. The prolongation of this period chiefly depends upon the conditions in which the young ones grow, especially upon the state of the available vegetation. Young great gerbils grow vigorously for the first 2 to 3 months after their birth, then there is a decline in their growth rate and their body weights tend to fluctuate rather widely. The fluctuations may range upto 100 gm (sometimes even more). The highest body weight in males in pri-Balchashie has been recorded as 430 g and that in females, 300 g (BURDELOV, 1958). According to observations made on marked animals in pri-Balchashie, 4 to 5 percent of the young die during every ten-day-period of their growth (Table 10). The mortality rate reaches 10 percent during every ten-day-period when new colonies are being established. The mortality rate is, however, low in winter, being 1.5 percent during every ten-day-period in this season

570

Table 10. Sex ratio of marked adult great gerbils in pri-Balchashie (BURDELOV, 1958).

Period of observation	Adults		Percent females	Old ones		Percent females
	males	females		males	females	
March-April 1951	15	15	50	3	17	85
June 1951	14	20	58.8	1	11	92
July 1951	13	12	48	—	2	100
September 1952	7	16	70	4	6	60
Total	49	63		8	36	
Mean % females			56.25			81.81

and the rate again rises to 3 to 4 percent in summer (Fig. 43). There is a sex-difference in the mortality rates of adult animals.

The maximum life span of gerbils in this region is not more than 3–4 years for females and 2–3 years for males.

The litter size in this species varies from 1 to 14, usually from 4 to 7 (average 5 to 6). The number of embryos is usually greater than the number of young ones born. Resorption of foetuses usually takes place in dry years when food is scarcely available.

The duration of pregnancy is from 23 to 32 days. In captivity a female can produce 6 litters in 6 months (AKOPJAN & KRIVONOSOV, 1965). In the northern parts of this gerbil's range, in populations that survive the winter, there are always 2 to 3 litters during the April–September period. The first two litters appear in quick succession of each other as the females conceive soon after the delivery and even on the day of parturition. The first two litters are borne by all adult females, while a third and possibly a fourth litter are borne by only 15 to 30 percent of the females that survive. A very small proportion of the females born during a year sometimes deliver upto four litters within the same year.

In Kizil-Kum, the duration of the spring-summer breeding activity of the great gerbils never exceeds 3 months (from the end of March to the beginning of June). In July–August the second peak of reproduction is observed although it is not so well-expressed. Mainly young females (born in the same year) take part in this breeding spell. In normal years 100 percent of females take part in reproductive activity and more than half of them usually produce two litters. In the summer 2 of 3 percent of the young females take part in reproductive activity and the litter size varies from 1 to 14. The average number of embryos per female at first littering is 6.7 and in the second, 6.2. The litter size decreases from 7.3 in April to 5.5 in the end of May (ROSSOLIMO, 1957).

On the Krasnovodsk peninsula (Turkmenia) the season of reproduction of red-tailed and great gerbils coincide with the period of vegetative growth of plants. The maximal number of pregnant female great gerbils is observed while the majority of the plants bloom, and that of red-tailed gerbils – while the seeds ripen. In good years, the great gerbil reproduces during 11 months while the red-tailed gerbil for 12 months. If the year is dry reproduction begins in March and is terminated by June. During such a situation 40 percent of adult females of the great gerbil take part in reproduction, while 50 percent red-tailed gerbils breed (VASILIJEV et al., 1963). The average size of a litter fluctuates in different seasons from 4.5 to 6.7 embryos per pregnant female of the great gerbil with the maximum in April. The corresponding figures for the red-tailed gerbils are 4.1 to 6.4, with two peaks during May and October. Minimum reproductive activity in the two species is observed in August and January (VASILIJEV et al., 1963). Fluctuations of number are rather large in both the species in the Krasnovodsk peninsula in Turkmenia. An inverse relationship between the intensity of reproduction (the number of embryos per 100 adult animals) and the density of the great gerbils (the average number of animals per ha) has been observed. Deviations from this general trend in 1957 and 1958 may be ascribed to unfavourable weather (cold spring followed by dry summer).

The litter size of great gerbils in western Kara-Kum (Turkmenia) fluctuates from 1 to 14, the average number being 6.4. It changes from 3.7 to 4 in July, August and September to 6.8 to 7.1 in February, March and April. In the autumn-winter months the litter size is 4 to 5 (KAMNEV et al., 1959).

The fluctuations in the intensity of reproduction shown in Table 10 indicate that there are considerable differences between some populations of the great gerbil. Reproductive activity starts in January in the south – in the Turkmenian (western) Kara-Kum, in February in the north-eastern Caspian region, and in March in the southern Balkhash territory. In the southern zone of deserts, in some years, the autumn-winter period of reproduction continues upto the spring-summer period and pregnant females are found in every month of the year. In the north zone at least a short interval occurs between two reproductive phases.

The period of intensive reproduction in the west zone (north-eastern Caspian region, Turkmenian Kara-Kum) is during February–May while in the east (Balkhash lake region) this period is restricted more often to April to June.

The fluctuations in the intensity of reproduction (Table 11) also varies in different populations. The fecundity of the great gerbils in the Mangish-lack peninsula can be trebled while in the north-eastern Caspian region, in the Turkmenian Kara-Kum and in the low-land Darialyk-Takir such fluctuations in fecundity rarely occur. The smallest fluctuations are observed in the Ilian depression and in Aral Kara-Kum, where the number of great gerbils is very stable (MOKROUSOV et al., 1965). The intensity of reproduction and the indices of fecundity are strictly de-

Table 11. Indices of reproductive activity of great gerbils in different regions based upon the data of several years (MOKROUSOV *et al.*, 1965).

Regions	Number of adult females examined	Sum of monthly percentage of pregnant females (over a year)	Average number of embryos per female	Intensity of reproduction*	Period of intensive reproduction**
North-east Caspian region	35921	162	6.0	972	February–March
Mangishlack peninsula	23999	128	5.8	742	April–May
Usturt	14189	160	5.6	896	April–July
Aral Kara-Kum	41452	185	6.0	1110	April–July
Darilyk-takir	15294	154	6.1	939	April–June
Inter Ili-Caratal river region	24166	117	5.8	680	April–June
Western Kara-Kum	36206	180	6.2	1118	February–April
Lowland Kara-Kum	44509	163	5.9	962	February–April
South-eastern Kara-Kum	50322	196	4.5	882	February–April

* The intensity of reproduction is the product of the sum of monthly percentages of pregnant females (over a year) and the average number of embryos per female.
** The period of intensive reproduction is the period when more than 20 percent of the females are pregnant.

pendent on the sex and the age structure of a population. According to observations made on marked animals the maximum life span of the great gerbils is of more than 3 years, but only 1 to 3 animals out of a thousand usually reach to that age. Trapped animals from the north-western Aral region also show fluctuations in the rate of fecundity in the population (Fig. 44).

The sex ratio in the newborn in nature is nearly 1:1. The sex ratio in adult animals also is nearly 1:1 when the pooled yearly collection is considered. But it changes with the age of the population as a large number of the males die at a younger age as compared to the females. As a result, the older (more than 2 years of age) groups consist only of females.

Mating takes place in these animals usually just after the birth of the young ones. The newborns weigh 4.5 to 5 g. During the first 12–13 days they survive solely on their mothers' milk and at the age of 22 days they turn to vegetation. The female carries its young ones over to a new place and makes a new nest there if there is some danger or if the old nest

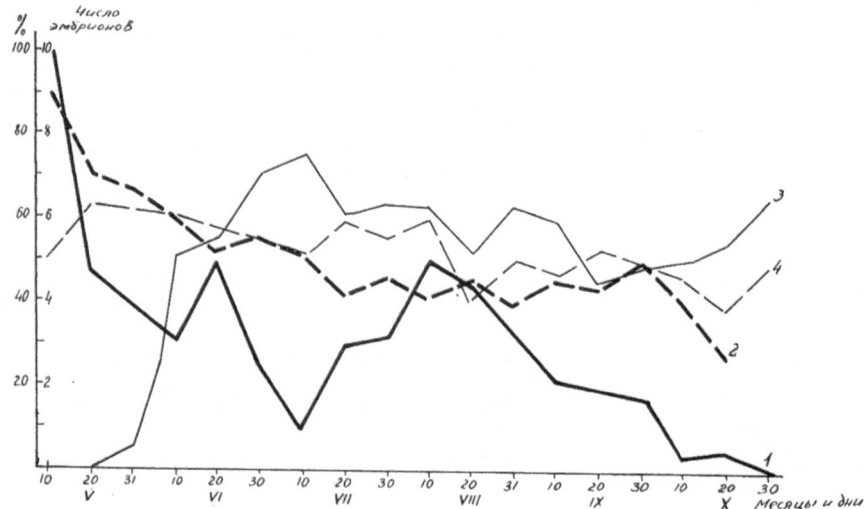

Fig. 44. Population structure and reproductive activity in the great gerbil in the north-western Aral sea region (after NAUMOV *et al.*, 1972). 1. percentage of pregnant females; 2. the number of embryos per female; 3. percentage of young ones. 4. percentage of adult females; (ordinate: percentage and number of animals; abscissa: days and months).

becomes uncomfortable. The male also takes part in the bringing up of the young ones, in the building of new nest, and in the carrying of the young ones. The old males are more careful towards their progeny but they do not distinguish between their own offspring and those of others, treating all the young ones in the same way. The attitude of animals of all ages towards the young ones is one of tolerance (AKOPJAN & KRI-VONOSOV, 1965).

The social structure of the population of the great gerbils is very well organised and at the same time very dynamic. The basis of it is the constant use of burrow – colonies, regulated with the help of different kinds of signals. Accoustic, visual and olfactory signals are the most important in the deserts. The acoustic signal of the great gerbils is rather complex in nature. The animals not only use their voices, but also stamp (thump) their feet. The perception of the thumping noise is possibly associated with the structure of the middle ear, the capsule *(bullae ossae)* of which is filled with a spongy matter which enables the animal to separate biologically important signals passing through the ground from the unimportant ones. The voice of the gerbils as well as their signals usually have a small range and are perceived by the neighbouring animals who react to the alarm calls.

Olfactory signals are communicated with the help of odorous glands. The members of every group mark each other and so all of them have

574

the same 'group smell'. In this way animals of 'their' and 'other' groups are distinguished. Smell marks are left on the paths and the walks in the colony, and the latter are marked (mainly by the males) by raising some special 'gerbil hillocks'* on which an animal often leaves a smell mark, by rubbing its abdomen against the mound. The occurrence of such gerbil-hillocks also indicate that the colony is occupied.

It is known that the great gerbils use several burrows simultaneously (STALMAKOVA, 1942; SIDOROVA, 1959; NAUMOV & LOBACHEV, 1965; LEONTJEVA, 1966). The number of such burrows may vary from 1 to 12 depending on the season and the size of the group.

The primary group is a family, consisting, in spring, of a male and a female and later their young ones also. Studies made on marked great gerbils indicate that in certain cases a couple may form a permanent pair for their life time. But due to the high rate of mortality of males in this species, such long-term pairing is barely achieved. The male members of the pairs are, therefore, being continuously replaced, resulting in the formation of more complex and mobile groups. We call these dynamic social groups 'populational parcells' (NAUMOV, 1967). These consist of individuals from different families, and hence, there are a few adult males, a larger number of adult females and several young ones. All females inhabit their own burrows and feed their young ones. Sometimes several females live in one and the same burrow, but in different parts of it. The males maintain some sort of a communication system because they periodically visit different females. New social relations in the populational parcell are developed within the ranges of the animals. The parcell may be small consisting of 1 or 2 males and 2 to 4 females along with young ones or these may be rather complex comprising the total population of more than a dozen colonies (LOBACHEV, 1973). Our observations indicate that the large and complex parcells appear in the most favourable places where the number of animals is quite stable, while the smaller parcells, sometimes even separate families, usually live in unfavourable localities, are not stable at all, and very often they break up and disappear.

Social relations among the great gerbils are not very simple. Different females, even though living in one and the same colony, do not tolerate each other and are rather aggressive towards the young ones of other females. A male coming to a new female, may at first be treated rather agressively, but such an unfriendly attitude of the female usually lasts for a short period only. Even if there are no 'friendly' relations between the members of one parcell they are never agressive to each other. But an animal coming from some other group faces a different treatment. If a member of a different grouping is introduced into the colony fierce fights occur and often they kill each other. Inhospitable treatment is

* Heaps made of sand mixed with faeces or urine and used for signalling.

575

meted out to a new-comer only on the territory of the nest-burrows while the neighbours are freely allowed to visit the peripheral territories. The delineation of the territory between different animals, families and parcells of great gerbils is chiefly done for the possession of the nest burrows and food resources and for the use of space.

The movements of the great gerbils are also associated with the above-mentioned three factors. The primary aim of their movements is to find food. The animals move along well-known paths in the vicinity of their family seat and foraging territory, or in individual territory, and these movements never extend beyond a hundred metres. The second type of movement is associated with changing from one burrow with its fodder territory to a new burrow. Such migrations are always preceded by long (hundreds of metres) exploratory excursions, far from the already exploited feeding range. These two types of movements account for 85 percent of the total migrations of the great gerbils with mostly the adult animals taking part in these movements as is evidenced by the observations made on marked rodents in the northern pri-Aral region (NAUMOV & LOBACHEV, 1965). The third type of movement is undertaken by the sub-adult rodents when shifting to the unoccupied burrows in the neighbouring colonies. This third type of movement, which is distinctly different from the first two types as exploratory investigations, play practically no role in the choice of the direction of shift. The young ones usually follow the paths which have scent markings left by the adults which have already migrated. The movements of the young ones are particularly well marked in the continuous settlements along the dry valleys and gorges in the northern pri-Aral region, or along the sandy beds in Kara-Kum and Kizil-Kum. The presence, in this species, of the 'reflex of following the scent' left by migratory animals emphasises the 'nomadic way' of life of these rodents. Wild animals of diverse species usually take to the paths of the least danger and the greatest convenience, usually on the strength of their olfactory perceptibilities.

In the course of their migrations, the great gerbils follow the paths used by human beings and the motor roads. We have been able to follow their migrations upto 10 km and even more. These observations made it possible for us to explain why the majority of the colonies of the great gerbils are situated near to motor roads, and paths. However, not all animals of this species follow the nomadic way of establishing new settlements at intervals, although the seemingly sedentary populations also are known to shift over distances of a few kilometres. This type of shifts over short-distances, however, appears to take place in only ten percent of the cases when these animals move. Sometimes the younger generation migrates to much greater distances than the adult ones. Migrations over distances of 7, 12 and 18 km have been recorded (SHEKHANOV, 1952; NAUMOV et al., 1972; LOBACHEV, 1961).

The mobility of the great gerbils changes considerably in different

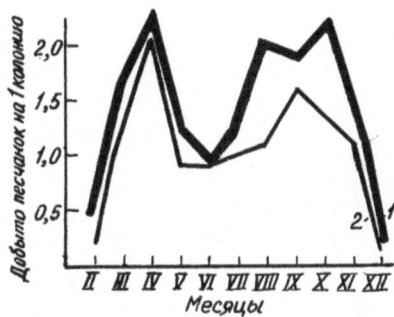

Fig. 45. Monthly variations, indicating mobility, in the number of great gerbils trapped from the same colonies in the Balkhash lake region during 1957–1961 (after BURDELOV *et al.*, 1964). 1. 25 colonies along the road; 2. 25 colonies away from the road. (ordinate: number of animals trapped per colony; abscissa: months).

seasons. It is the least during the winter, increases in spring, the same level being maintained during the summer period (June–August) and rises again in autumn and upto the beginning of winter. During the first phase of increased mobility the majority of the settling animals are adult which have survived the winter and are now looking for new partners and new burrows for purposes of reproduction. During this period almost 100 percent of the trapped animals are found to be males. The second phase of increased mobility is associated with the movements of the younger animals which disintegrate themselves from these old colonies and start their own establishments in order to form new groups for wintering (Fig. 45).

While movements over short distances may take place during any part of the year, the movements of the young ones in search of new abodes usually take place in the middle of summer (before the dry period) and towards the end of summer (prior to the period of storing food and of preparing for spending the winter). The mobility becomes greater if the number of gerbils increases. When the density of gerbils was rather high, there were 78.5 percent young animals out of a total of 335 gerbils caught far from their own burrows (MARIN, 1959).

Short migrations of the young animals unite smaller groupings (population parcells) into a larger population, securing communication between animals in it. However, the migrations of the adult males and females from one colony to another are also important. So both types of migrations play important roles in the population dynamics of the great gerbils.

Several zoologists have observed that fluctuations in the number of the great gerbils are related to the amount of precipitation received during the cold period of the year, which in turn determines the development of vegetation and, hence, the quality and the size of the stores of

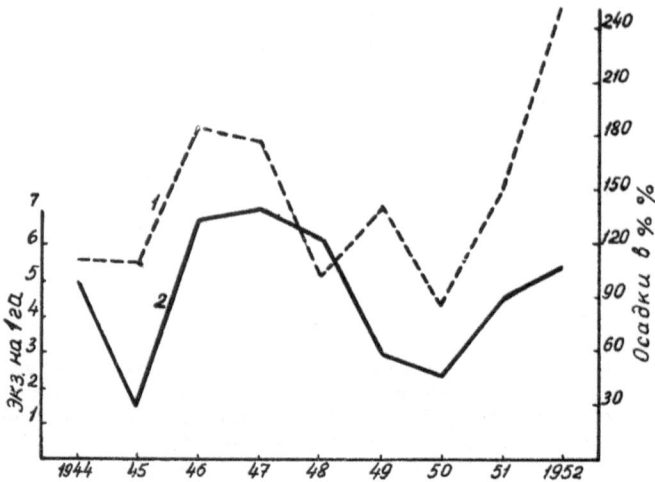

Fig. 46. Relationship between the amount of winter rainfall and the density of great gerbil population (after Burdelov, 1958). 1. precipitation received during the years 1944–1952 expressed as percentage of precipitation received in 1947–48; 2. the average number of animals per hectare during the following summer. (ordinate: left, number of animals per hectare; right, percentage of rainfall; abscissa: years).

food. There are reliable data showing correlation of the amount of precipitation and the number of animals (Fig. 46, Shekhanov, 1952). It is, however, a point of debate whether the influences of precipitation can be seen in the same year (Burdelov, 1958) or a year later (Buliginskaja, 1954; Poljakov, 1954).

In reality the picture is much more complicated because the populations of the great gerbils are, in a way, influenced by the 'extreme' conditions* of the desert by virtue of their adaptive mechanisms. There are many factors which influence the animals' activity patterns and thereby cause considerable cyclic fluctuations in the number of the great gerbil. Each cycle may have a duration of 5 to 7 years (Naumov et al., 1972).

The population dynamics of a species depends not only on the relations of the animals with their surroundings, but also on the relations between animals within a population and within the species as a whole. So population dynamics is a phenomenon which develops not only in time but in space as well (Naumov, 1965). It is based on the social hierarchy of the species as an integrated population of a higher range. The interactions between animals of different ranks is also a factor influencing population dynamics (Naumov, 1971). The results of the interactions

* We do not think it proper to call arid conditions 'extreme' for an animal which has developed in the desert during the course of evolution.

578

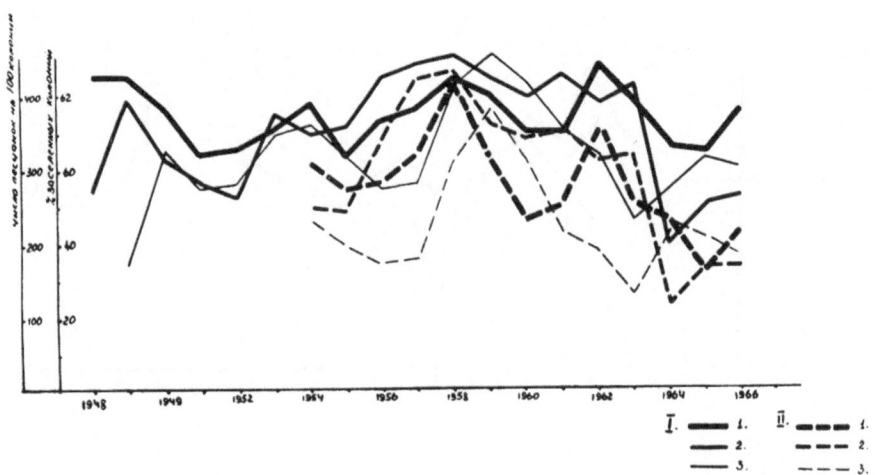

Fig. 47. Fluctuations in the number of great gerbils in three geographical populations in the Aral region (after NAUMOV *et al.*, 1972). Continuous lines (I) percentage of inhabited burrows; dotted lines (II) number of animals per 100 burrows; I_1 and II_1. Aral Kara-Kum; I_2 and II_2. desert, north of Aral sea; I_3 and II_3. northern Kizil-Kum. (ordinate: outer   the number of great gerbils per 100 burrows; inner – percentage of the colonies inhabited; abscissa: years).

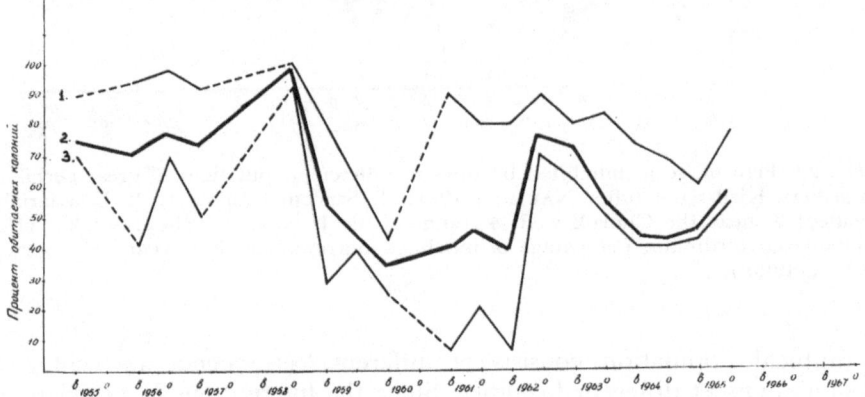

Fig. 48. Fluctuations in the number (based on the proportion of inhabited burrows) of great gerbils in colonies of eastern Kara-Kum, Aral (after NAUMOV *et al.*, 1972). 1. high; 2. medium; 3. low density; (ordinate: percentage of inhabited burrows; abscissa: years. B – spring; O – autumn).

are the migration of animals and especially of the young ones, which connect populations of different ranks together.

Changes in the population of animals occupying a large geographical region in central Kazakhstan have been studied (Fig. 47). This geo-

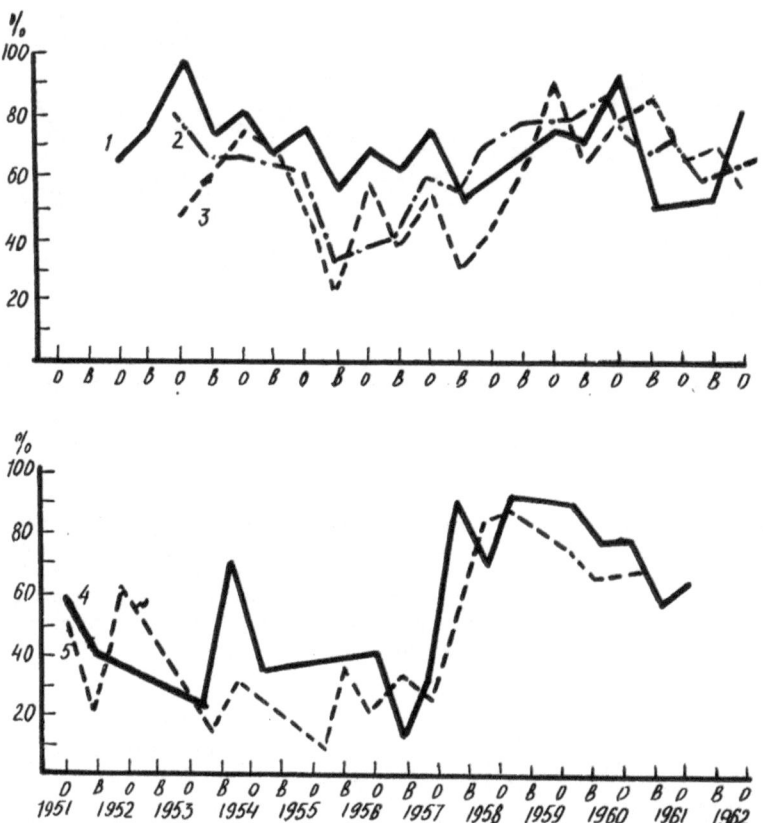

Fig. 49. Percentage of inhabited burrows in different populations of great gerbils in northern Kizil-Kum (after NAUMOV, 1971). 1. Sarybulakian sands; 2. Eskedarialyk valley; 3. near the Churuk well; 4. sands of the Janadarian valley; 5. takirs along Janadaria. (ordinate: percentage of inhabited burrows; abscissa: years. B – spring; O – autumn).

graphical population consists of different 'settlements' or ecological populations at different localities. Since the fluctuations in gerbil numbers do not occur at the same time in all ecological situations, the ecological populations are also called 'independent populations' (BEKLE-MISHEV, 1962). The exchange of animals from one population to another occurs only at their peripheries and in years when gerbil numbers increase to a large extent.

In the various ecological populations within a geographical population the fluctuations in numbers are not synchronised. In the colonies of eastern Kara-Kum the main difference was in the amplitude of fluctuations (Fig. 48), while in Kizil-Kum the difference was in time also

580

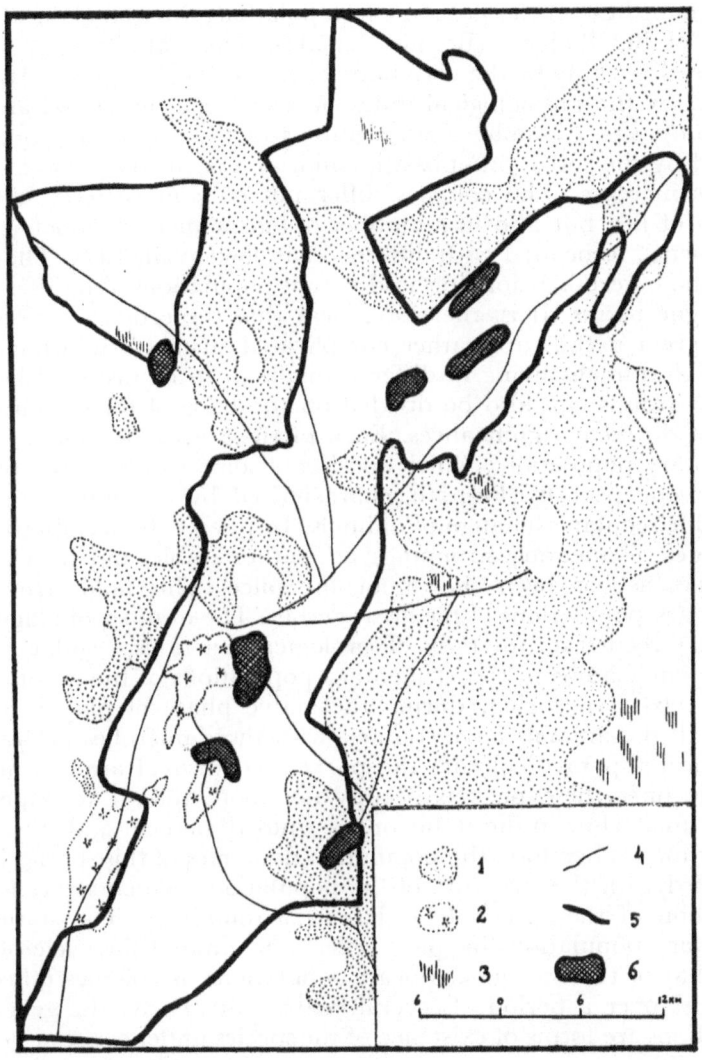

Fig. 50. Habitat types colonized by great gerbils in Aral Kara-Kum (after NAUMOV, 1971). 1. hummocky sands; 2. takirs; 3. saline flats; 4. roads and paths; 5. borders of main colonies; 6. pockets of survival during periods of low population density.

(Fig. 49). The exchange of animals between different ecological populations is a regular feature and plays an important role in their population dynamics.

The stability of the colonies is closely related to their structure. The colonies located in the north-western Aral Sea region are much more

581

stable than those of the permanent populations situated along the sandy terrain of Big Badgers (Bolshie Barsuki). The stability of a colony is influenced not only by the variability of conditions in a mosaic landscape, but also by the physiological differences between individual animals in the colony. For example, some members of the group may remain un-affected by the spread of epizootics and, as a result the numbers of such animals increase. The colonies differ not only in their structure and rhythm of life, but also in a number of biochemical parameters of their constituents, associated with differences in food availability. The sensitiv-ity of these gerbils belonging to different populations also differ towards the plague toxins. It is, therefore, clear that the ecological populations of the great gerbils are rather complicated structure in which several groups of animals, living in different niches, can be distinguished. Each of these groups can also be divided on the basis of their occupation of areas of different significances. Each of these areas is inhabited by an 'elementary population', which consists of some (sometimes – of a large number of) 'parcell' (Fig. 50), characterised by a common rhythm of life and a regular exchange of animals. Inspite of the possibility of such exchanges, elementary groupings are rather stable and have a varied genotype, as has been shown for the mice. This holds true for the elementary populations of the great gerbils. The number of plague-prone animals screened on the basis of serological reactions, regularly changes in different seasons in the elementary populations of the gerbils (LOBA-CHEV, 1964). In spring, during the active phase of plague microbes, nearly all the animals have contacts with those microbes *(P. pestis)*.

In the temporary colonies the pockets of survival are absent, in the unstable ones these pockets occupy not more than 1 to 3 percent of the territory while in the stable ones – upto 10 percent and even more of the territory. Therefore, the population dynamics of this species is closely connected with the structure of the populations, which serves for better adaptation of the species to different surroundings. The status of the elementary populations in space reflects the annual life cycle of gerbils, while that of the ecological (local) populations or colonies reflects their life cycles over a period of several years. And, lastly, the geographical populations are forms of existence of the species under concrete geograph-ical conditions.

The decline in the rank order of a population of these animals may be ascribed to their preference for particular land forms as habitat. But spatial differentiation is not a passive reaction of the animals to differences in habitat conditions. Different colonies of the gerbils develop, get complicated and become more stable not only due to the favourable conditions offered by their surroundings, but also to the changes, favour-able to their own well-being, which they themselves introduce into the habitat. Such adaptation of the territory to the needs of the gerbils can be most vividly seen in areas where their digging activity is easily pursued.

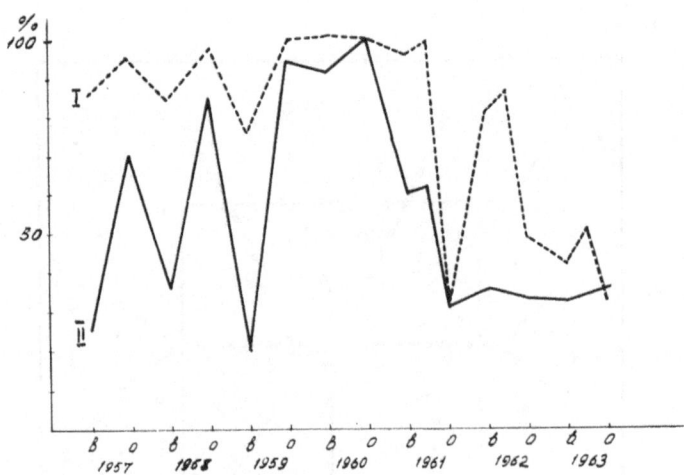

Fig. 51. Percentage of inhabited burrows of great gerbils in (I) survival pockets and (II) in surrounding populations in Otarbaytube, Aral Kara-Kum (after NAUMOV, 1965). (ordinate: percentage of inhabited burrows: abscissa: years. B – spring; O – autumn).

The above considerations would tend to explain the different types of changes in the population structure observed in different and sometimes even in neighbouring colonies.

Increases and decreases in the number of the animals are displayed not so much through gross changes in the density of the population as in the persistent presence of the animals at places most favourable to them and their appearance and disappearance from the rest of the biome (Fig. 51). The most stable populations are those in which the survival rates have remained more or less uniform over a prolonged period. In the territories where such uniform survival rates are observed usually have summer conditions fit for reproduction. In such tracts seasonal colonies appear every year like a circle of country-houses around a large town. In these colonies a new generation grows up; some of the young ones immigrate far from their native settlements, the rest secure a high density of population to ensure survival in winter (Fig. 52). Knowledge of all of the mechanisms of the dynamics of its population is essential for understanding the role of the great gerbils in desert biocenosis as well as for undertaking control measures against these carriers of a terrible infection and vermins of desert pastures.

The great gerbil may be called a background-species for sandy deserts due to its very large number, extensive digging habit and intensive digging activity.

The burrows of these gerbils serve as permanent and temporary refuges for a great number of vertebrates and invertebrates. The gerbil

583

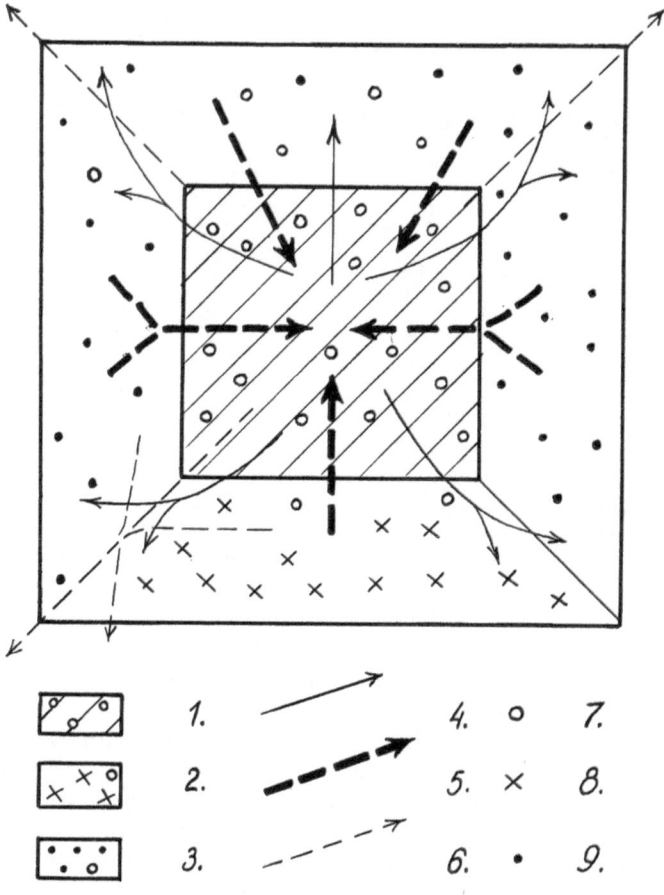

Fig. 52. Colonies and survival pockets of great gerbils (after NAUMOV, 1971). 1. region of loamy alluvium; 2. hummocky sands; 3. loamy region with wormwood – *Salsola* cover; 4. spring emigration of gerbils out from places of wintering; 5. autumn invasions of young ones; 6. distant emigration of young ones; 7. deep and permanent burrows; 8. less frequented burrows; 9. shallow, temporary burrows.

burrows are the only shelters for the other animals against the vagaries of the desert climate. The presence of such refuges enable mesophyles and even some hygrophyle animals to inhabit the desert. At a depth of 50 cm the microclimate in the burrows is rather stable, with a high relative humidity of the air (80 to 95 percent). In the Guriev district, the temperature in the burrow fluctuates around 22.0 °C (at a depth of 50 cm) and only around 18.5 °C at a depth of 150 cm, while on the surface the temperature touches 54 °C (SHIRANOVICH *et al.*, 1965). The temperature and humidity of the air at a depth of 50 cm differ from those on the ground. The difference in temperature was of the order of 5 to 7 °C in June, 1 to 2 °C in September and 0.5 to 1 °C in October in pri-Balchasie (ILJINSKAJA, 1963). In Mujun-Kum the temperature in the tunnels and food-chambers of the burrows at a depth of 20 cm was 10 to 16 °C in the middle of April, 23 °C at a depth of 50 cm in the middle of May to the beginning of June, 25 to 29 °C at 20 and 50 cm depths in the end of June to the beginning of August and 10 to 12 °C in the middle of October. During the same periods the temperature in the nest-chambers at depths of 150 to 250 cm was 8 °C in April, 20 to 22 °C in July–August and 15 to 17 °C in the middle of October. The relative humidity in the same periods fluctuated in the superficial tunnels from 80 to 100 percent in April–May, 50 to 80 percent in July–August, and 40 to 60 percent in October. In summer and in autumn the daily fluctuations of humidity varied from 40 to 60 percent. In the deep tunnels and nest-chambers the relative humidity was rather stable (80 to 85 percent) in all the seasons with almost no daily fluctuations, whereas the humidity at the ground level was 2 to 8 percent (ILJINSKAJA & KUZIN, 1965).

In regions where the great gerbils are relatively abundant the number of their burrow openings exceed the number of burrow openings of other rodents as observed in a study area covering 242 hectares (Table 12).

Very often the burrows of other rodents are situated around those of the great gerbils. *Scirtopoda telum* and *Meriones meridianus* live in the colonies of the great gerbils, and, in the northern part of the area, *Microtus socialis* are also found in association. A great number of subfossil skulls of *Lagurus luteus* were found in the colonies of the great gerbils and this showed that the former animals were once closely associated with the latter. About a hundred years ago *Lagurus luteus* vanished in large numbers in its range (LOBACHEV, 1961).

The richness and variety of invertebrate fauna found in the burrows of the great gerbils near Ashkhabad has been shown by VLASOV (1937). He counted not less than 210 species of insects and spider-like animals. DUBININ (1946–1954) studied the biocenosis of the burrows of the great

Table 12. Relative number of burrow-openings of rodents.

Type of habitat	Number of entrances		Total
	great gerbils	other rodents	
Beds of sub-sandy alluvium	275	18	293
Flat clay regions	133	22	155
Subaeral sands and sub-sands	163	0	163
Average	199	16	215

and the red-tailed gerbils and of *Citellus fulvus, Ellobius talpinus* and other animals in Kazakhstan and found 219 species of invertebrates in them, while only 38 species were found on the surface. DUBININ showed that the richness of invertebrate population in a burrow depends on the micro-climate of the burrow and on the length of time that the owner of the burrow spends in it. 9 species of mosquitoes out of the 12 species occurring in Kazakhstan have been found in the burrows of rodents only and especially in the burrows of the great gerbils (SHAKIRZJANOVA, 1954).

Of the invertebrate fauna found in the burrows, the most important are the fleas and pincers which transfer many communicable diseases of man and his domestic animals. The following fleas have been identified from the burrows of the great gerbils: *Pulex irritans, Echidnophaga oschanini, Xenopsylla conformis, X gerbilli, X. hirpites, X. nuttali, X. skrjabini, Synosternus pallidus, S. longispinus, Coptopsylla bairamaliensis, C. lamellifer, Oropsylla* sp., *Rostropsylla daca, Ceratophyllus tersus, C. aralis, C. laeviceps, C. turcmenicus, C. tesquorum, C. trispinus, C. mocrzeckii, Paradoxopsyllus tuschkan, Leptopsylla taschenbergii, Ctenophtalmus dolichus, Rhadinopsylla cedestis, Neopsylla setosa, Stenoponia conspecta* and *S. vlasovi.* Pincers which inhabit the burrows are *Haemophysalis numidiana, Rhipicephalus schulzei, R. pumilio, Hyalomma asiaticum, Ornithodorus tartakowskii* and Gamasidae pincers. Lice and other ectoparasites are also found on the body of the great gerbils and in their burrows.

Many terrestrial vertebrates use the burrows of the great gerbils quite intensively. According to an account, 61 species of animals have been known to visit the burrows of the great gerbils in pri-Balchasie; of these there were 24 species of mammals, 12 of birds, 10 of reptiles and 15 species of large insects (LESNJAK, 1959). In the sands of Saryishikotrau, a total of 3043 mammals of 15 species were trapped in 4 colonies of *Rhombomys* during 8 years (1958–1965). These included 3 species of gerbils, 2 species of Dipodidae, *Mus musculus, Ellobius talpinus, Cricetulus migratorius, Lepus tolai,* 3 species of Insectivora, and 3 others. Further observations indicated that 9 more species had also visited the colonies (*Spermophilopsis leptodactylus, Jaculus lichtensteini, Apodemus sylvaticus,* fox, *Vulpes corsac, Canis lupus,* badger and steppe cat). A number of birds of different species have also often been found near the entrances to the burrows, e.g. *Oenanthe isabellina* and *O. deserti* and *Athene noctua.* On the surface of the colonies *Passer domesticus, Alauda* and *Melanocorypha, Podoces panderi, Falco tinnunculus* and *Buteo rufinus* have been seen. Nearly all species of reptiles are closely associated with the burrows of the great gerbils. Only *Phrynocephalus mystaceus* and *P. interscapularis* have not been found to be so closely associated. On the surface of the colonies the green toad *(Bufo viridis)* have also been found (SHEKHANOV, 1952; KRILOVA, 1967). In Mujun-Kum the colonies of the great gerbils have been found to harbour the largest number of outsiders, followed by those of *M. meridianus* (9 percent), the red-tailed gerbils and *Citellus fulvus* (each

4 percent) (KHRUSCELEVSKIJ & MUKHAMEDJAROVA, 1963). In the northern Aral desert 15 species of rodents as well as 3 species of Insectivora, 3 species of small mammals, 3 species of reptiles and a green toad use the burrows of the great gerbils (SHEKHANOV, 1952; KRILOVA et al., 1957). A great number of animals have also been found to live in the burrows of the great gerbil in Kizil-Kum. But for the gerbil burrows the survival of the green toad, *Agama sanguinolenta* and of *Eremias lineolata*, *E. media* and the steppe tortoise *(Testudo horsfieldi)* would have been impossible. 50 to 90 percent of these animals live in the gerbil colonies. When the number of gerbils decline and the entrances to the burrows are plugged, the number of non-gerbil occupiers of the burrows and especially that of the reptiles also declines (KRIVOSHEEV, 1958). In the burrows of *Rhombomys* some ducks *(Tadorna rutila* and *Tadorna tadorna)* readily nest. Some pole cats, *Mustela eversmanni* and foxes *Vulpes corsac* make their burrows in the colonies of the gerbils. 51 species of vertebrates and 155 species of invertebrates live in the colonies of the great gerbil in Turkmenia. 26 out of these 51 species are reptiles, some of which are permanently associated with the burrows and even lay their eggs in them (e.g. *Teratoscincus scincus, Phrynocephalus interscapularis, Gymnodactylus caspius, Grossobommon eversmanni* and *Testudo horsfieldi*). The green toad also live in the burrows. Some birds also use these as their nests, e.g. *Oenanthe lugens, Oenanthe isabellina, Oenanthe pleschanka,* and, sometimes, *Athene noctua bactriana* and *Upupa epops.*

Of the invertebrates found in the burrows there were 29 species of fleas, 20 species of Ixodidae pincers, 4 species of Gamasidae, cockroaches, beetles and Diptera (Mosquitoes) (NURGELDIEV, 1969). In the western and eastern (Turkmenian) Kara-Kum, 16 species of vertebrates were found in the burrows of the great gerbils. The most usual visitors are the red-tailed gerbils. Along the Kara-Kum canal, *Microtus afghanus* are often found in the colonies of the great gerbils (ATAEV, 1962), while in Karabill, Badhis and the Obrutchev steppes *M. afghanus* have common colonies with the great gerbils.

In Aral Kara-Kum 87 percent of the mammals, 20 percent of the birds and 81 percent of the reptiles use the burrows of the great gerbils but the intensity and mode of use of the burrows vary. Of the regular burrow-mates (which permanently live and reproduce in the burrows of the great gerbils) mention may be made of: *M. libycus, M. meridianus, Allactaga elater, Scirtopoda telum, Cricetulus migratorius, C. fulvus, C. pygmaeus, Mustella eversmanni, M. nivalis, Vormela peregusna* and *Hemiechinus auritus.* Similarly, of the birds, the more important ones are: *Aquila heliaca, A. nipalensis, Buteo urlinus, Bubo bubo, Athene noctua* and *Corvus uficollis.* Among the reptiles the regular burrow users are: *Agama sanguinolenta, Testudo horsfieldi, Ancistrodon halus* and the green toad. The 'lodgers', which though live in the colonies frequently, yet have their own refuges, are – Mammals: *Meriones tamariscinus, Allocricetulus eversmanni, Allactaga*

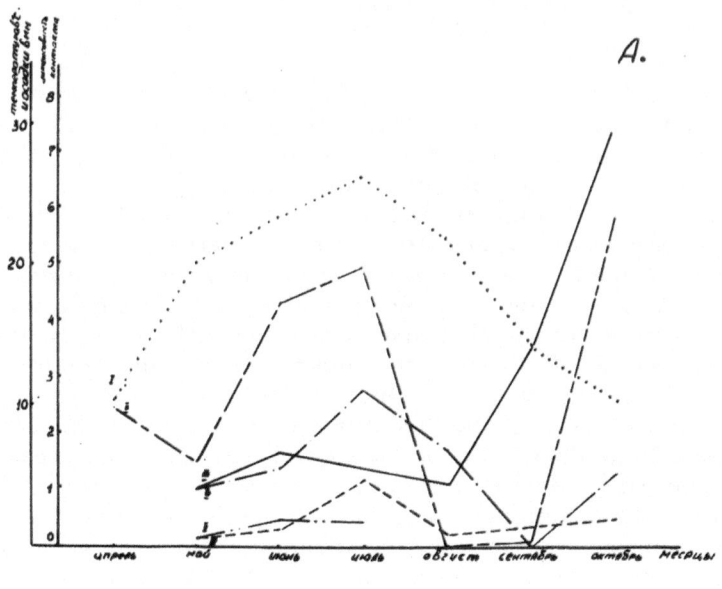

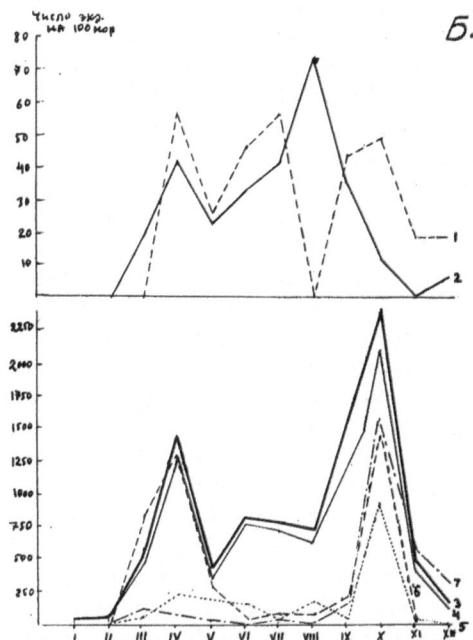

Fig. 53. The intensity of contacts of different species of small mammals with the burrows of great gerbils. A. in the Aral Kara-Kum (after NAUMOV *et al.*, 1972). I. mean monthly temperature; II. precipitation. The number of animals of different species seen entering the burrows of the great gerbil in 4 hours; III. the great gerbil; IV. *S. telum;* V. *M. meridianus;* VI. others. B. in the Balkhash region (after BONDAR, 1967). 1. insectivora; 2. carnivora; 3. all rodents; 4. *M. meridianus;* 5. red-tailed gerbil and *M. tamariscinus;* 6. Dipodidae; 7. house mice. (A. ordinate: outer, temperature, °C and rainfall, mm; inner, intensity of contacts; abscissa: months from April to October. B. ordinate: number of animal runs per 100 burrows; abscissa: months).

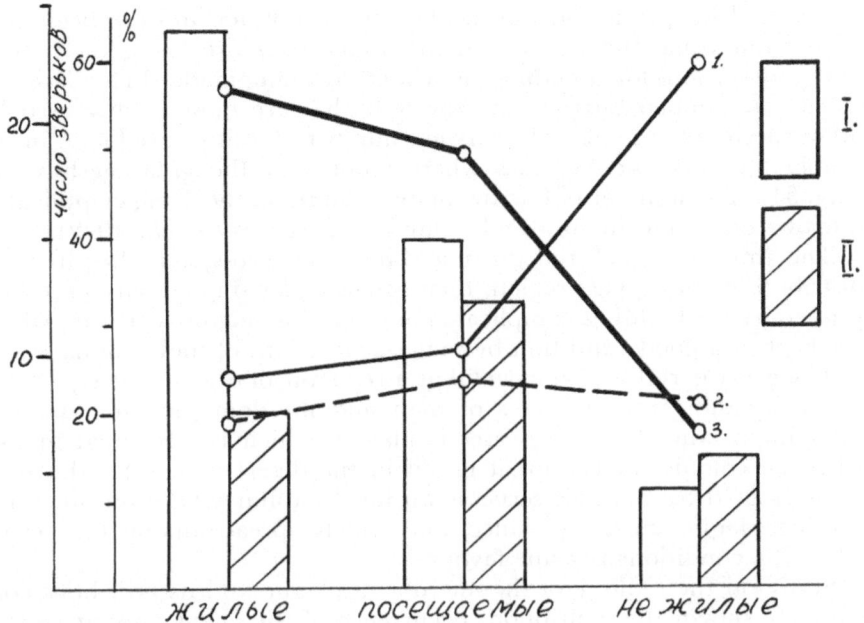

Fig. 54. Use of different types of burrows in Aral Kara-Kum in 1960–63 (after NAUMOV *et al.*, 1972). The number of trapped rodents during a period of 70 days: I. great gerbil; II. other species sharing the burrows. Percentage of: 1. Dipodidae; 2. *M. meridianus* and *M. libycus;* 3. other mammals. (ordinate: outer, number of animals; inner, percentage of animals; abscissa: inhabited, visited, uninhabited burrows).

severtzovi and *Microtus socialis;* Birds: *Tadorna tadorna, Casarca ferruginea, Merops superciliosus;* Reptiles: *Eremias velox, E. arguta* and *Taphrometopon lineolatum.* The 'guests' that often visit the colonies for short periods are – Mammals: *Alactagulus pygmaeus, Allactaga saltator, Allactaga major,* the house mouse, *Ellobius talpinus, Diplomesodon pulchellum;* Birds: *Calandrella cinerea, C. rufescens, Melanocorypha leucoptera, Oenanthe pleschanka, O. hispanica* and *O. deserti;* Reptiles: *Eremias media, E. lineolata, Teratoscincus scincus, Gymnodactylus* sp. and *Eryx miliaris.*

The 'visitors' which accidentally use the gerbil colonies are – Mammals: *Jaculus lichtensteini, Pygeretmus platyurus, Dipus sagitta, Lepus europaeus, Lepus tolai, Crocidura suaveolens;* Birds: *Streptopelia senegalensis, Pterocles alchata, P. orientalis, Surrhaptes paradoxus, Otis tarda, Chlamydotis undulata, Cuculus canorus, Caprimulgus europaeus, Corvus corone, Passer domesticus, Emberiza bruniceps, Lanius excubitor, L. collurio, Prunella sylvianana* and *Erithropygia galactotes;* Reptiles: *Eremias grammica* and *Phrynocephalus helioscopus.* In this category there are 69 species of vertebrates.

The mode of use of the burrows of the great gerbils by different types of animals differ in different seasons and years. The most intensive use

589

is observed in spring and during the end of summer and the beginning of autumn when the colonies of the young ones are being established and preparations for spending the winter are being made (Fig. 53a & b).

The abandoned burrows are the ones that are most readily used by other species of animals. The smallest number of visitors are found in the simple burrows and colonies where families of the great gerbils live (Fig. 54). The number of burrow-mates is larger in the deep, complicated burrows occurring in sub-sandy alluvium (NAUMOV et al., 1972).

The importance of the great gerbil to the eco-system lies in their distribution over a vast region, their intensive digging activity and their propensity for building complicated and durable burrow-colonies, which are kept in a good condition by many generations of these animals.

The great gerbil is also harmful as a reservoir of organisms responsible for many dangerous diseases of man and his domestic animals. The most important of such diseases is plague, which is associated mainly with the colonies of the great gerbil in the desert areas. The burrow-colonies of these animals serve as an ideal habitat where the microbes causing plague exist, reproduce, and widely spread among the gerbils when the conditions become favourable.

Study on the ecology of the microbe of plague and its host have conclusively shown the continuous occurrence of the agent only at certain 'primary *niduses*' of plague. If these *niduses* are systematically tackled, it becomes possible to improve the health conditions in a region without the general extermination of gerbils and other bearers of plague (NAUMOV et al., 1972).

Besides plague, the gerbils play some role in the maintainance and spread of the organisms responsible for some other diseases like Leishmaniasis *(Leishmania tropica)*, tickets (pincers) and spirochetosis in some parts of Asia. When the epizootic of tularemia breaks out among the Balkhash depression (in the valley of the river Ili) the great gerbils are also found to be involved in the epizootic.

REFERENCES

AFANASJEV, A. V., BAZHANOV, V. S., KORELOV, M. N., SLUDSKY, A. A. & STRAUTMAN, E. I. 1953. Zveri Kazakhstana. Alma-Ata: 1–535.

AKOPJAN, M. M. & KRIVONOSOV, K. I. 1965. O razmnozenii i povedenii bolshykh peschanok. '*Materiali IV. nauchnoi conferencii po prirodnoi ochagovosti i profilactike chumy*. Alma-Ata: 14–15.

AKIEV, A. K. 1968. O bolshoi peschanke v severnom Afganistane. Sbornik '*Gryzuni i ikh ektoparasiti*'. Saratov: 422–424.

ALEKPEROV, H. M. 1966. Mlekopitajushie jugo-zapadnogo Azerbajdzhana. Baku.

ALEKPEROV, H. M., EIGELIS, U. K., POLTAVCEV, N. N. & AHVARDOV, N. U. 1967. Dinamica razmnozhenija krasnokhvostoi peschanki *(Meriones erythrourus)* v Azerbajdzhanskoi SSR. *Izv. AN Azerb. SSR, ser. biol.*, N 1: 70–76.

ALIEVA, S. B. 1965. Materiali po faune i ecologii gryzunov Nakhichevanskoi ASSR. *Trudi Instituta Zoologii Acad. Nauk Azerbaidzhanskoi SSR*, 25.

ANDRUSHKO, A. M. 1939. Dejatelnost gryzunov na sukhih pastbishakh Srednei Azii. *Izd. LGU.*

ARGIROPULO, A. I. 1939. K rasprostraneniju i ecologii nekotorych mlekopitaushich Armenii. *Zoolog. sbornik*, I. Erevan.

ATAEV, C. 1962. Novie mestonachozdenia afganskoi polevki v jugo-Vostochnoi Turkmenii. *Izvestia Acad. Nauk Turkm. SSR. Biologia*, 2.

ATTALLAH, S. I. & HARRISON, D. L. 1968. On the conspecificity of *Allactaga euphratica* Thomas, 1880 and *A. williamsi* Thomas, 1897 with the complete list of subspecies. *Mammalia*, 32(4): 628–638.

BABAEV, C. & ATAEV, C. 1962. O nachozhdenii *Jaculus turcmenicus* v juzhpoi Turkmenii. *Izvestia Acad. Nauk Turkm. SSR. Biologia*, 3.

BAKEEV, N. N. & KADAZKIJ, N. G. 1959. Sutochnaja aktivnost krasnokhvostoi peschanki Azerbaidzhanskoi SSR. *X Soveshanie po parazitologicheskim problemam i prirodnoi ochagovosti*, 2.

BALABAS, N. G., KUZIN, N. P. & TROFIMENKO, I. P. 1965. O stroenii nor bolshikh peschanok v severo-vostochnikh Mujunkumakh. *'Materiali IV nauchnoi conferencii po prirodnoi ochavosti i profilactice chumy'*, Alma-Ata.

BANNIKOV, A. G. 1947. Materialy k posnanju nekotorych mlekopitaushich Mongolii. I – Tushkanchiki. *Bull. Mosc. Obshestva Ispitatelei Prirody*, 52, 4.

BANNIKOV, A. G. 1954. Mlekopitajushie Mongoljskoy narodnoy respubliki. Moscow.

BARABASH-NIKIPHOROV, I. I. 1957. Zveri jugo-vostochnoi chasti chernosemnogo centra SSSR. Voronezh.

BEKLEMISHEV, V. N. 1962. Prostranstvennaja i funktionalnaja structura populatii. *Bull. MOIP. Biol.*, 65, 2.

BEME, L. B. & KRASOVSKII, D. B. 1930. Materiali k posnaniu ecologii nogaiskogo tushkanchika. *Ezhegodnic Zool. Museum Acad. Nauk SSSR. XXXI*, 5.

BERMAN, D. I. 1962. Cardiocranius paradoxus-novyi vid v faune SSSR. *Bull. Mosc. Obshestva Ispitatelei Prirody:* 67, 5.

BOGORODSKI, U. V. 1967. O biologii tushkanchika-pryguna v zapadnom Zabajcalje. *Zool. J. ILVI*, 4: 632–633.

BONDAR, E. P. 1956. Materiali po mlekopitajushim pustyni Betpackdala i jugo-zapadnoi chasti Kasachskogo nagorja. *Trudi Sredneasiatskogo protivochumnogo Instituta*, 4, Alma-Ata.

BONDAR, E. P. & ZERNOVOV, I. A. 1960. Ecologo-faunisticheskij ocherk gryzunov Zapadnoj Turkmenii. *'Voprosi prirodnoj ochagovosti i epizootologii chumy v Turkmenii'*. Ashkhabad.

BREER, V. D., LOBIZOVA, V. P. & RACHININA, N. A. 1971. Materiali po rasprostraneniju i biologii nekotorych vidov tushkanchikov Central Kizil-Kum. *Mater. VII conf. protivochumnykh uchrezhdenii Srednei Asii i Kazakstana*. Alma Ata.

BULIGINSKAJA, M. A. 1954. Prognozi chislennosti bolshoi peschanki. *III Ecologicheskaja conferentia. Tezisi dokl.* III, Kiev.

BURDELOV, A. S. 1958. Prodolzhitelnost zhizni bolshikh peschanok i vozrastnoj sostav ich populatii. *Trudi Sredne-Aziatskogo n.-i. protivochumnogo Inst.* Alma-Ata.

BURDELOV, A. S., BONDAR, E. P. & ZHURAVLEVA, V. I. 1964. Podvizhnost bolshikh peschanok i Ikh epizootologicheskoe znachenie v uslovijah severnoi pustyni. *Zool. J.* XLIII, 1.

BURDELOV, A. S. & LEONTJEVA, M. N. 1956. Peschanki severnogo Pribalkhashja. *Trudi Sredne-Aziatskogo n.-i. protivochumnogo inst.*, 3, Alma-Ata.

CHABAEVA, G. M. 1968. K ecologii tushkanchika-pryguna jugozapadnogo Zabaikalja. *Uchenye zapiski Burjatskogo pedagogicheskogo Instituta*, 31: 23–24.

DAVIDOV, G. S. 1962. K rasprostraneniju peschanok v Tadzikistane. *Trudi Inst. Zoologii i parazitologii*, 22: 58–69.

DAVIDOV, G. S. 1964. Gryzuny severnogo Tadzhikistana. Dushanbe.

DINESMAN, L. G. 1968. Izuchenie istorii biogeocenozov po noram zhivotnikh. Moscow. Nauka.

DUBININ, V. B. 1946. Obitateli nor mlekopitajutshikh. Juzno-Kazakhstanskoj oblasti i ich znachenie dlja cheloveka. *Izv. AN Kazakh. SSR, ser. parazitologii*, 4.

DUBININ, V. B. 1954. K voprosu o faune i ecologii mlekopitajushikh Khavastskogo raiona Tashkentskoi oblasti Uzbekskoi SSR. *Trudi Inst. zool. i parazitol. AN UzbSSR*, 3.

DUBJANSKIJ, M. A. 1962. Zvisimost stroenija kolonii bolshykh peschanok ot pochvenno-gruntovykh uslovii. *Bull. MOIP, Biologia*, 67, 4.

DUBJANSKIJ, M. A. 1963. Tipy poselenii bolshoi peschanki i ikh epizootologicheskoe znachenie v Priaralskikh Karakumakh. *Zool. J.* 42, 1.

DULAMZEREN, S. 1970. Mongol orni chachten amtan todorchojloch bichuch (mongolian). Ulanbator: 1–238.

EISENBERG, J. 1967. A comparative study of rodent ethology with emphasis on evolution of social behavior. *I. Proc. Un. Stat. Nat. Mus.* v. 122.

ELIZARIEVA, M. B. 1949. Nakhozhdenie Salpingotus crassicauda v predelakh SSSR. *Doklady Acad. Nauk SSSR*, 16, 3: 495–498.

ELLERMAN, J. R. & MORRISON-SCOTT, T. C. S. 1951. Checklist of Palearctic and Indian mammals. B.M.N.H., London.

FEDJANINA, T. F. 1966. Materiali po chislennosti i ecologii grebentshikovoj peschanki. *Sb. 'Vrednyi gryzuni Kirgizii'.*

FENJUK, B. K. 1928. K biologii tushkanchikov. *Materiali k poznaniu fauny Nizhnego Povolgja*, 2: 1–40.

FENJUK, B. K. 1929. Eshe o biologii tushkanchicov i o merach borby s nimi. *Materiali k posnaniu fauny Nizhnego Povolgia*, 3: 1–53. Saratov.

FENJUK, B. K. & KAMNEV, P. I. 1957. Zametki o faune mlekopitaushich Mangyshlaka i Usturta. Sbornik: *'Gryzuni u borba s nimi'.* 5. Saratov.

FENJUK, B. K., RADCHENKO, A. G. & ZHERNOVOV, I. V. 1957. Massovoe uvelichenie chislennosti krasnokhvostoi peschanki v zapadnoi Turkmenii v 1953 g. *'Voprosi ecologii'*, 2, Kiev.

FETISOV, A. S. & MOSKOVSKIKH, A. A. 1948. Kogtistaja peschanka v Zabajkalje. *Trudi Irkutskogo universiteta, ser. biol.*, 3, 4.

FLINT, V. E. 1970. Materiali po ecologii tushkanchika pryguna. Sbornik *'Fauna i ecologia grysunov'.* 9. Moscow.

FOKANOV, V. A. 1954. Mlekopitajushie juzhnoi chasti doliny reki Ural. *Trudi Zoolog. Instituta*, 16. Leningrad.

FOKIN, I. M. 1963. Osobennosti bega tushkanchikov. *Bull. Mosc. Obshestva Ispitatelei Prirody. Biologia*, 63, 5.

FOKIN, I. M. 1969. K ecologii i rasprostraneniju tushkanchika Bobrinskogo v Kara-Kumakh. Sbornik *'Voprosi ecologii i biocenologii'*, 9. Leningradskii University.

FORMOZOV, A. N. 1929. Skotoboi, ego znachenie dlja stepnoi fauni i borby s vrediteljami. *'Priroda'*, 11.

GAMBARJAN, P. P., PAPANJAN, S. B. & MARTIROSJAN, B. A. 1960. Material po biologii poludennoi peschanki Meriones meridianus v Armjanskoi SSR. *Bull. Moscow Obshestva Ispytatelei Pryrody, Biologia* 65, 6: 17–22.

GINTLIS, R. V. 1959. Osnovnye cherti ecologii krasnokhvostoi peschanki vostochnogo Prikaspija v predelakh Gurjevskoj oblasti. *'Nauchnaja conf. po prirodnoi ochagovosti i epidemiologii chumy'.* Gurjev.

GLADKINA, T. S. & POLJAKOV, I. J. 1957. Kriterii prognoza chislennosti krasnokhvostoi peschanki v Zakavkazje i Srednej Azii Sb. *'Voprosi ecologii'*, 2, Kiev: 132–140.

GULIEVSKAJA, N. S. 1963. Grebnepalyi tushkanchik v severo-vostochnych Kizil-Kumakh, *Zool. J.* 42, 7: 1110–1111.

GVOZDEVA, L. P. 1965. Rastitelnost poselenii bolshikh peschanok v Mujunkumakh. *'Materiali IV nauchn. conf. po prirodnoi ochagovosti i profil. chumy'*. Alma-Ata.

HAMAGANOV, S. A. 1954. K biologii kogtistoi peschanki v rajone Torejskikh ozer. *Izvestija Irkutskogo gos. n.-i. protivochumnogo inst. Sibiri i D. Vostoka.* 12: 150–155.

HEPTNER, V. G. 1940. Fauna peschanok Irana i zoogeograficheskie osobennosti malo-azijsko-irano-afganskikh stran. *'Novie memuari MOIP'*, 20: 1–72.

HEPTNER, V. G. 1956. Fauna pozvonochnikh Badkhiza (juzhnii Turkmenistan). Ashkhabad: 1–334.

HEPTNER, V. G., VOJCEKHOVSKII, D. P. & NIKITIN, V. P. 1958. Zametki o peschankakh (Gerbillidae, Glires), XV Novie dannie o Meriones zarudnii Heptner. *Trudi Inst. Zool. i parazitol. Acad. Nauk Turkm. SSR*, III: 141–147.

ILJINSKAJA, V. L. 1963. K izucheniju temperaturnogo rezhima v norakh bolshikh peschanok. *'Materiali nauchn. conf. po prirodnoi ochagovosti i profilactike chumy'*, Alma-Ata.

ILJINSKAJA, V. L. & KUZIN, I. P. 1965. O vlazhnosti vozdukha i temperature v norakh bolshikh peschanok v Mujunkumakh. *'Materiali IV nauchn. conf. po prirodnoi ochagovosti i profilactike chumy'*. Alma-Ata.

ISHUNIN, G. L. & PAVLENKO, T. A. 1966. Materiali po ecologii zhivotmych pastbish Kizil-Kumov. *Sbornik: 'Pozvonochnie zhivotnie Srednei Azii'*. Tashkent.

ISMAGILOV, M. I. 1948. K ecologii zaica rusaka, malogo tushkanchika i serogo hom-jachka na ostrove Barsa-Kelmes. *Izvestia Acad. Nauk Kazakhskoi SSR. Zoologia*, 8.

ISMAGILOV, M. I. 1961. Ecologia landshaftnikh gryzunov Betpak-Dali i juzhnogo Pribalkhashja. Alma-Ata.

KALABUKHOV, N. I. 1956. Spjachka zhivotnych. Isd. 3. Kharkov.

KALABUKHOV, N. I. 1969. Periodicheskie (sezonnie i godichnie) izmenenija v organizme gryzunov, ikh prichini i posledstvija Leningrad, *'Nauka'*: 1–248.

KAMBULIN, E. A. 1941. Materiali po ecologii bolshoi peschanki v Kazakhstane. *Sb. 'Gryzuni i borba s nimi'*, 1, Saratov.

KAMNEV, P. I., SKVORTZOV, G. N. & GURJEVA, I. M. 1959. Zametki po ecologii nekoto-rikh vidov gryzunov zapadnoi chasti centralnikh Karakumov. *Sb. 'Gryzuni i borba s nimi'*, VI, Saratov.

KAPITONOV, V. I. 1972. Nabludenia po ecologii pjatipalogo Karlikovogo tushkanchika v Kazakhstane. *Izvestia Acad. Nauk Kazakhskoi SSR. Biologia*, 5: 38–43.

KASATKIN, B. M., LEONTJEVA, M. N. & TOMILOVA, T. P. 1963. K ocenke razlichnikh tipov kolonii bolshikh peschanok i ikh raspredeleniju v sploshnikh poselenijakh Ilijskoi chasti juzhnogo Pribalkhashja. *'Materiali nauchn. conf. po prirodn. ochagovosti i profilactike chumy'*, Alma-Ata.

KAZANTZEVA, J. M. & FENJUK, B. K. 1937. K ecologii mokhnonogogo tushkanchika (Dipus sagitta). *Uchenye zapiski Saratovskogo Universiteta. Biologia*, I, 1.

KHRUSCELEVSKIJ, V. P., CHERNONOG, N. F., SHAREC, A. S. & DMITRJUK, G. Y. 1963. Landshaftnye osobennosti raspredelenia i kolebanii chislennosti peschanok v Mujun-Kumakh. *Mater. nauchnoi conf. po prirodnoi ochagovosti chumy*. Alma-Ata.

KHRUSCELEVSKIJ, V. P. & MUKHAMEDJAROVA, N. A. 1963. Materiali po podviznosti gryzunov v Mujunkumach. *'Materiali nauchnoi conf. po prirodnoi ochagovosti i profilactike chumy'*, Alma-Ata: 235–238.

KIM, T. A. 1960. Materiali po ecologii tamariskovoi peschanki v pustyne Kizil-Kum. *Zool. J.* 39, 5, 759–763.

KOLESNIKOV, I. I. 1934. Vrednie gryzuny kauchuconosa tau-sagyz.

KOLESNIKOV, I. I. 1953. Mlekopitajushie. *Fauna Uzbekskoi SSR*, III, 5. Gryzuni. Tash-kent.

KOLOSOV, A. M. 1939. Fauna mlekopitajushkh Altaja i smezhnikh oblastey Mongolii v svjazi s nekotorimi problemami zoogeografii. *Zool. J.* 18, 2.

KOLPAKOVA, S. A. 1932. K kharakteristike uslovii mestoobitanija gryzunov novogo Ushtagana, bivshego Azgirskogo raiona (Zap. Kazakhstan). *Vestnik medic. epide-miologii i parazitologii*, 2, 1, Saratov.

KONDRASHKIN, G. A. 1959. Probuzhdenie ot zimnei spjachki zemljanych zajchikov raznogo pola. '*Gryzuni i borba s nimi*', 6. Saratov.

KONDRASHKIN, G. A. & JEDIKINA, V. S. 1957. Ocherk ecologii zemljanogo zaichica delty Volgi. '*Gryzuni i borba s nimi*', 5. Saratov.

KRILOVA, K. T. 1957. O rationalizatii metoda uchota effectivnosti borbi s suslikami i bolshimi peschankami. *Sb.* '*Gryzuni i borba s nimi*, V, Saratov.

KRILOVA, K. T., VARSHAVSKIJ, S. N., SHILOVA, E. S., SHILOV, M. N. & PODLESSKII, G. U. 1957. Lanshaftno-geograficheskie osobennnosti mezhvidovogo kontakta v poseleniakh bolshikh peschanok v Priaralje. *Materiali k soveshaniju po voprosam zoogeografii sushi.* Lvov.

KRIVOSHEEV, V. G. 1958. Materiali po ecologo-geograficheskoi kharakteristike fauni nazemnikh pozvonorhnikh severnikh Kizil-Kumov. *Uchenie zapiski Moskovskogo pedagogicheskogo Inst. im. Lenina*, Moscow, 124, 7.

KUCHERUK, V. V. 1946. Gryzuni-obitateli zhilish v vostochnoj Mongolii. *Zool. J.* 24, 2.

LARINA, N. I. 1938. Zametki po ecologii melkich Dipodidae Kalmyzkich stepei. *Uchenye zapiski Saratovskogo Universiteta. Biologia*, 1(14).

LAVRINENKO, A. E. & TARASOV, N. S. 1967. Opit istreblenija mongolskich peschanok otravlennoy primankoy v Tuve. '*Izvestija Ircutskogo protivochumnogo Inst*'. 27, 397–399.

LAY, D. M. 1967. A study of the Mammals of Iran. *Fieldiana Zoology*, v. 54.

LEONTJEV, A. N. 1954. K ecologii kogtistoi peschanki v BM ASSR. *Izvestija Irkutskogo protyvochumnogo Inst.* 12: 137–149.

LEONTJEV, A. N. 1962. K sutochnoi aktivnosti mongolskoi peschanki i polevki Brandta. *Izv. Irkutskogo g s. n.-i. protivochumnogo Inst. Sibiri i D. Vostoka*, 16: 78–84.

LEONTJEV, A. N. 1962. K izucheniju populjacii mongolskikh peschanok metodom mechenija. *Izv. Irkutskogo gos. n.-i. protivochumnogo Inst. Sibiri i D. Vostoka*, 24: 296–302.

LEONTJEV, A. N. & HAMAGANOV, S. A. 1957. Otravlennie primanki v borbe s mongolskoi peschankoi. *Izv. Irkutskogo gos. n.-i. protivochumnogo Inst. Sibiri i D. Vostoka*, 16.

LEONTJEVA, M. N. 1966. Gruntovie vodi i bolshaja peschanka. Autoreferat candidatskoi dissertacii, Gorkii.

LESNJAK, A. P. 1959. Contakt boljshoy peschanki s drugimi zhivotnimi v chumnom ochage Pribalhashja. '*X sovetch. po parasitol. problemam u prirodnoochogovim boleznjam*', Moscow.

LETOV, G. S., EMELJANOV, N. D., LETOVA, G. I. 1963. Materiali po rasprostraneniju i ecologii peschanok v Tuve i prilezhashei chasti Mongolii. *Izv. Irkutskogo gos. n.-i. protivochumnogo Inst. Sibiri i D. Vostoka*, 25.

LIPAEV, V. I. 1967. Materiali po stationarnomu razmesheniju gryzunov v stepjakh severo-vostochnoi Mongolii. *Izv. Irkutskogo Inst. Sibiri i D. Vostoka*, 27, 168–174.

LISICINA, T. J., NIKOLSKY, A. A. 1967. Preduprezdajushiy ob opastnosti signal gryzunov otkritih prostranstv. '*Conf. molodikh uchenyh MGU*'. Moscow.

LOBACHEV, V. S. 1961. Materiali po biologii kanjuka kurgannica v juzhnom Kazakh-stane. *Nauchnye doklady vyshey shkoly. Biologicheskie Nauki*, 1: 37–43.

LOBACHEV, V. S. 1964. Epizootologicheskoe znachenie podviznosti i kontaktnih svjazey bolshikh peschanok v elementarnih ochagah chumy. '*I godichn. nauchno-otchetnay conferencija biol. faculteta MGU*'.

LOBACHEV, V. S. 1971. Nahodka karlicovogo tuchkanchika (*Salpingotus crassicauda*, Dipodidae) v severnom Priaralje. *Zool. J.* 50, 2: 305–306.

LOBACHEV, V. S. 1973. O dalnich migracijach bolshikh peschanok i osobennostjach ich izuchenia. *Vestnik Moscow University*, 5: 29–34.

LOBACHEV, V. S., KCHAMDAMOVA, T. U. 1972. Pitanie bolshoi peschanki. *Bull. Moscov. Obshestva Ispitatelei Prirody. Biologia*, XXVII, 5: 40–54.

LOBACHEV, V. S. & PASKHINA, N. M. 1972. Povedenie i aktivnost bolshoi peschanki. *Sbornik 'Povedenie zhivotnykh.' I Vsesojusnoe soveshanie evolutionnich i ecolog. aspectov povedenia zhivotnich.* Moscow.

LOBACHEV, V. S., SHENBROT, G. I. 1973. Sravnitelnyi analis metodov ucheta chis-lennosti tushkanchikov (Dipodidae). *Bull. MOIP. Biologia*, XXVIII, 2: 47–57.

LOBIZOVA, V. P. 1971. K izucheniju tushkhanchikov juzhnogo Uzbekistana. *Mater. VII confer. protivochumnykh uchrezhdenii Srednei Azii, i Kazakhstana.* Alma-Ata.

MARIN, S. N. 1959. O znachenii podvizhnosti bolshikh peschanok dlja sokhranenia chumnoi infekcii v ochage. *X sovechanie pa parasitol. problemam. I.* Moscow-Leningrad.

MARIN, S. N. & ROTSHILD, E. V. 1965. O nochnoy activnosti bolshih peschanok. *Bull. MOIP,* 65, 5.

MISONNE, X. 1959. Analyse zoogéographique des manifères de l'Iran. *Mem. Inst. Royal des sciences naturelles de Belgique.* Bruxelles.

MOKRIEVICH, N. A. 1957. Sezonnye izmenenija nekotorikh ecologo-fisiologicheskih osobennostey poludennykh i grebenchikovykh peschanok v Volgo-Uralskikh peskakh. *'Gryzuni i borba s nimi'.* Saratov, 5: 29–49.

MOKROUSOV, N. J. 1957. Periodica zhiznedejatelnosti i razmnozhenia emuranchica (Scirtopoda telum) v severo-zapadnom Prikaspii. *Sbornik: 'Gryzuni i borba s nimi',* 5. Saratov.

MOKROUSOV, N. J. 1960. Landshaphtnaja priurochennost poselenii emuranchika. *Materiali confer. po voprosam zoogeographii sushi.* Alma-Ata.

MOKROUSOV, N. J., JACOVLEV, M. G., BONDAR, E. P., NAIDEN, P. E. & SAMARIN, E. G. 1965. Sravnitelnaja kharakteristica rasmnozhennia bolshoy peschanki v raslichnih chastjakhee areala. *'Materiali IV. confer. po prirodn. ochagovosti, chumy.'* Alma-Ata.

MURTAZANOVA, E. S., LOBACHEV, V. S. & SAVCHENKOV, J. I. 1964. Epizootologichescoe znachenie tushkanchikov v Priaraljskih Karakumah. *Bull. Moscow Obshestva Ispitatelei Pryrody. Biol.,* 5.

NAUMOV, N. P. 1945. Dynamica tschislennosti i evoluzia. *J. Obshei Biologii.*

NAUMOV, N. P. 1951. Novii metod izuchenia ecologii melkih lesnih gryzunov. *'Fauna i ecologia gryzunov',* 4, MGU.

NAUMOV, N. P. 1954. Tipy poselenii gryzunov i ich epizootologicheskoe znachenie. *Zool. J.* XXXIII, 2: 268–289.

NAUMOV, N. P. 1963. Ecologia zhivotnikh. Moscow.

NAUMOV, N. P. 1964. Microstructura i ustoichivost prirodnikh ochagov boleznei. *Zool. J.* 43, 3: 322–333.

NAUMOV, N. P. 1965. Prostranstvennie osobennosti i mekhanizmi dinamiki chislennosti nazemnikh pozvonochnikh. *J. obshei biologii,* 26, 6.

NAUMOV, N. P. 1967. Structura populatii i dinamika chislennosti nazemnikh pozvonochnikh. *Zool. J.* 46, 10: 1470–1486.

NAUMOV, N. P. 1971. Prostranstvennaja structura vida mlekopitaushich. *Zool. J.* 50, 7: 965–980.

NAUMOV, N. P. & KASATKIN, B. M. 1963. Organizacionno-metodicheskii ukasania po istrebleniju bolshoy peschanki s celju ozdorovlenia territorii ot chumi. Isd-vo MGU.

NAUMOV, N. P. & KULIK 1955. O kostjah mlekopitajushih, sobranih na kolonijah bolshykh peschanok. *'Voprosi kraevov obshey eksperimentalnoy parasitologii i medzoologii'.* 9.

NAUMOV, N. P. & LOBACHEV, V. S. 1965. Structura poselenii i podvizhnost bolshikh peschanok. *Mater. IV nauchnoi confer. po prirodnoi ochagovosti chumy.* Alma-Ata.

NAUMOV, N. P., LOBACHEV, V. S., DMITRIEV, P. P. & SMIRIN, V. M. 1972. Prirodnii ochag chumy v Priaralskih Karakumah. Moscow University.

NEKIPELOV, N. V. 1935. Materiali po ecologii gryzunov v okrestnostjah ozera Barun-Torey. *Izvestia Irkutskogo protyvochumnogo Insti.*

NEKIPELOV, N. V. 1940. Novie dannye po biologii Allactaga saltator. *Zool. J.,* XIX, 2.

NEKIPELOV, N. V. 1959. Osnovnie osobennosti chumnykh ochagov MNR. *'X sovetshanie po parasitol. problemam i prirodnoochagovim bolesnjam.'* Tesisi dokladov. Moscow.

NEKIPELOV, N. V. 1962. Raspredelenie mlekopitajushich po biotopam jugo-vostochnogo Zabaikalja. *Izvestia Irkutskogo protivochumnogo Instituta,* 24.

NERONOV, I. M., USACHEV, G. P. & JACOVLEV, E. P. 1964. K rasprostraneniju i ecologii tushkanchika Severtzova v jugo-zapadnom Tadzhikistane. *Istvestia Akad. Nauk. Tadzhik. SSR. Biologia,* 2(16): 105–106.

NURGELDIEV, O. N. 1950. O geographicheskom rasprostranenii i ecologii krasnokhvostoy peschanki v Turkmenii. '*Izv. Turkmenscoy philial AN SSSR*', 6: 30–39.

NURGELDIEV, O. N. 1952. Vlijanie gryzunov na kustarnikovii i drevestnii porodi na trasse glavnogo Turkmenscogo kanala. '*Isvestia Acad. Nauk Turkmenskoi SSR*', 4: 11–18.

NURGELDIEV, O. N. 1960. Materiali po faune i ecologii mlekopitajushich trassy Karakum Kanala. Ashkhabad.

NURGELDIEV, O. N. 1969. Ecologia mlekopitajushikh ravninnoi Turkmenii. Ahkhabad.

OGNEV, S. I. 1948. Zveri SSSR i prilezhashich stran, V, I. Acad. Nauk SSSR.

ORLOV, O. J. 1957. Materialy po biologii tushkanchikov v zapadnych Kizil-Kumach. *Sbornik nauchnych studencheskih rabot.* Moscow University: 21–27.

OSADCHAJ, N. P. 1959. O pitanii necotorych vidov pustynnykh gryzunov. *Uchenye Zapiski Moscow University*, 189.

PAPANJAN, S. B. 1966. K ecologii poludennoy peschanki v Armjanskoy SSR. *Biol. J. Armenii*, 19, 5.

PARASKIV, K. P. 1960. Novye dannye o karlikovom tushkanchike. *Trudi Instituta zoologii Acad. Nauk Kazakhskoi SSR.* 12: 100–110.

PAVLENKO, T. A. & HUBAIDULINA, S. T. 1970. Pozvonochnie khrebta Nara-Tau. Ashkhabad.

PAVLOV, A. N. 1962. Osnovnii cherty ecologii peschanok severozapadnogo Prikaspija. Voronezh: 1–21.

PETROVA, A. A. 1967. Gryzuni. *Sbornik 'Ecologia, mery ochrani i racionalnoe ispolzovanie pozvonochnych zhivotnich Karshinskoi stepi.*' Tashkent.

POGOSJAN, A. R. 1949. Ecologija i biologija peschanok v Armjanskoy SSR. '*Zool. sbornik Inst. Zool. i fitopatol*'. *AN Armjanskoy SSR*, 4: 99–125.

POGOSJAN, A. R. 1955. Materiali po ecologii tushkanchikov rasprostranennych v Armjanskoy SSR. *Nauchnye trudi Erevanskogo University. Biologia*, 5: 153–165.

POLJAKOV, I. J. 1954. Sistema meroprijatii po borbe s bolshoi i krasnokhvostoi peschankami – vrediteljami pastbisch Srednei Asii. *Sbornik 'Voprosi ulutschenia kormovoi basy v stepnoi, polupustynnoi i pustynnoi zonakh USSR*'. Acad. Nauk SSSR. Moscow–Leningrad.

RALL, J. M. 1939. Teplovie uslovia v norach peschanich gryzunov i metodika ich izuchenia. *Zool. J.* 18, 1.

RALL, J. M. 1941. Ocherk ecologii grebenchukovoi peschanki. *Sbornik 'Gryzuni i borba s nimi*', 1. Alma-Ata.

REJMOV, R. 1971. K morphologii i ecologii Meriones tamariscinus oasisa nisoviev Amu-Darji. *Vestnik Kara-Kalpacskogo filiala Acad. Nauk Uzbekskoi SSR*, 1.

ROSSOLIMO, O. L. 1957. Osobennosti rasmnozhenia bolshoi i poludennoi peschanok. *Uchenye zapiski Moscow pedagogicheskogo instituta Potemkina*, 65, 6. Moscow.

ROTSHILD, E. V. 1955. Nochnaj activnost bolshykh peschanok. '*Priroda*', N 7: 119.

ROTSHILD, E. V. 1957. O metodakh ucheta chislennosti boljshikh peschanok. '*Gryzuni i borba s nimi*'. Saratov.

RUDENTCHIK, J. V. 1959. K rasprostraneniju i ecologii krasnokhvostoy peschanki v zapadnikh Kizil-Kumakh. '*Trudi sredneaziatskogo protivochumnogo Inst.*', 5, 207–212.

SABILAEV, A. S. 1965. Blochi tushkanchikov Karacalpakii. *Materiali IV nauchnoi conferencii po prirodnoi ochagovosti chumy.* Alma-Ata.

SABILAEV, A. S. 1967. Ecologia tushkanchikov Karacalpakii i ich rol v epizootologii chumy. *Autorepherat candidat. dissertation.* Tashkent.

SABILAEV, A. S. 1967a. Materiali po ecologii tushkanchica likhtensteina v severozapadnych Kizil-Kumakh. *Uzbekskii biologicheskii J.* 3: 55–58.

SABILAEV, A. S. 1968. Materialy po rasprostraneniu i ecologii bolshogo tushkanchica na Karacalpakskom Usturte. *Acad. Nauk Uzbekskoi SSR. Uzbekskii biologicheskii J.* 1: 57–59.

SABILAEV, A. S. 1969a. Ecologia tushkanchica Severtzova (*Allactaga severtzovi*) v Karacalpakii. *Zool. J.*, XLVIII, 6: 902–910.

SABILAEV, A. S. 1969b. Rasprostranenie emuranchica na territorii Karacalpakskogo Usturta i Kizil-Kumov. *Uzbekskii Biol. J.* 1: 49–51.

SABILAEV, A. S. 1969c. Materiali po rasprostraneniu i ecologii *Paradipus ctenodactylus* v severo-zapadnych Kizil-Kumakh. *Sbornik 'Ecologia i biologia zhivotnych Uzbekistana'* Tashkent.

SABILAEV, A. S. 1970a. Mesto tushkanchikov (Rodentia, Dipodidae) v prirodnom ochage chumy na Usturte i v severo-zapadnych Kizil-Kumach. *Zool. J.* XLIX, 6: 916–919.

SABILAEV, A. S. 1970b. Ob ecologii i rasprostranenii v Karacalpakii *Allactaga saltator* i *Jaculus turcmenicus. Vestnik Karacalpakskogo phyliala Acad. Nauk Uzbekskoi SSR,* 4: 38–43.

SABILAEV, A. S. 1971. Ecologia mochnonogogo tushkanchika v severozapadnijch Kizil-Kumakh. *Zool. J.* L, 10: 1553–1563.

SABILAEV, A. S. 1971a. K ecologii malogo tushkanchica (Allactaga elater, Dipodidae) na Usturte i v severo-zapadnych Kizil-Kumakh. *Nauchnye doclady vyshei shkholy. Biol. Nauki,* 2: 15–21.

SABILAEV, A. S. 1971b. Ob ecologii tarbaganchica na Usturte i v severo-zapadnych Kizil-Kumakh. *Bull. Moscovskogo obshestva ispitatelei prirody. Biologia,* 4: 16–20.

SABILAEV, A. S. & OSTROVSKII, I. B. 1967. O roli tushkanchikov v epizootologii chumy na territorii Karakalpakii. *'Trudy VI conferencii po prirodnoi ochagovosti bolesnei.'* Dushanbe.

SAMSONOV, B. P. 1953. Tipy kostno-myshechnich system i gryzunov, vedushich razlichnyi obraz zhizni. *Autorepherat candidat. dissertation.* Moscow University.

SAPOZHENIKOV, J. F. 1963. Osobennosti rasprostranenia pustynnych mlekopitajushich v peschanoi pustyne Karakum. *Thesis III Vsesojuznogo soveshania po zoogeographia sushi.* Tashkent.

SARZINSKI, V. A. 1963. Zoologicheskie issledovanija v chumnom ochage Gornogo. Altaja. *Izvestija Irkutskogo protivochumnogo Instit.* 25.

SCHMIDT-NIELSEN, K. 1964. Desert animals: Physiological problems of Heat and Water. Oxford-London.

SHAKIRZJANOVA, M. S. 1954. Norovie moskiti Kazakstana i ikh rol v peredache visceralnogo lejshmanioza v Kzil-Ordinskoj oblasti. *'Prirodnaja ochagovost zaraznich bolezney v Kazakstane'.* Alma-Ata.

SHEGLOVA, A. T. 1962. Bolshay peschanka kak predstavitel zhiznennoy formi pustini. *'Voprosi ecologii',* VI, Kiev.

SHEKHANOV, M. V. 1952. Biologia bolshoy peschanki v Severnom Priaralje. *Autoref. cand. diss.* Moscow.

SHILOV, M. N. 1953. Severnaja granica areala boljshoy peschanki v uslovijah severnogo Priaralja. *'Bull. MOIP',* Biol. 53, 4.

SHILOVA, E. S. 1953. O rasprostranenii i stacionarnom razmeshenii krasnokhvostoy peschanki v severnom Priaralje. *'Bull. MOIP', Biol.,* 58, 2: 3–7.

SHILOVA, E. S. 1956. O nekotorkh osobennostjah razmnozhenia krasnokhvostoy peschanki v severnom Priaralje. *'Trudi Sredneaziatskogo n.-i. protivochumnogo Inst.'* 3. Alma-Ata.

SHIRANOVICH, P. I., MOLODOVSKY, A. V., OSOLINKER, B. S., DEREVJANCHENKO, B. E. & SAMARIN, E. G. 1965. O microclimate nor boljshoy peschanki (*Rhombomys opimus* Licht.). *Zool. J.,* 44, 8.

SHMUTER, M. F., FEDOROVA, T. V. & STOLGENOVA, H. A. 1957. Vospriimchivost tushkanchikov (emuranchika i tarbaganchica) k chume v zavisimosti ot sezonov goda. *'Nauch. conf. po prirodnoy ochagovosti osobo opasnih zabolevanii'.* Tezisi dokl., Saratov.

SHNITNIKOV, V. N. 1936. Mlekopitajushie Semirechja. Moscow-Leningrad.

SHUBIN, I. T. & ISMAGILOV, M. I. 1969. Ecologija karlikovogo tushkanchika *(Salpingotus crassicauda)* v Zajsanskoy kotlovine. *Zool. J.* 48, 11: 1722–1726.

SIDOROVA, G. A. 1959. Necotorii cherti biologii bolshoy peschanki v ochagah zoonoznogo kozhnogo leishmanioza v Bucharskoy oblasti Uzbekskoy SSR. *'X sovesht. po parazitol. problemam i prirodnoochagovim boleznjam',* AN SSSR, 2.

SILVERSTOV, V. B., LOBACHEV, V. S. & SHILOV, M. N. 1969. Novii dannii o rasprostranenii i biologii priaraljskogo tolstohvostogo tushkanchika. '*Bull. MOIP*', *Biol.*, 84, 3: 113–131.

SIROECHKOVSKII, E. E. 1954. O razmeshenii nekotorikh tushkanchikov v peschanoi pustine i metodi ucheta ikh chislennosti. *Zool. J.* 33: 6.

SKVORTZOV, G. N. 1955. Ob uslovjah zimney spjachki zemljanogo zaychika v Turkmenii. '*Gryzuni i borba s nimi*', 4, Saratov 39–51.

SLONIM, A. D. & SHEGLOVA, A. I. 1963. Fisiologicheskii osobennosti mlekopitajushih pustin Sredney Azii. '*Prirodnie uslovij. zhvotnovodstvo i kormovaj baza pustin*', Ashchabad.

SOKOLOV, V. E. & SKURAT, L. N. 1962. Ecologicheskie i physiologicheskie prisposoblenia bolshikh peschanok k besvodnym usloviam pustyni. *Sbornik 'Voprosi ecologii'*, 6, Moscow.

STALMAKOVA, V. A. 1942. O lesokhozjaistvennom znachenii bolshoi peschanki v vostochnikh Kara-Kumakh. *Trudi Turkmenskogo philiala Acad. Nauk SSSR.* 2.

STALMAKOVA, V. A. 1945. K ecologii grebnepalogo tushkanchika v Kara-Kumakh. *Izvestia Turkmenskogo philiala Acad. Nauk SSSR*, 3–4.

STALMAKOVA, V. A. 1954. Gryzuni Karakumov ikh ecologia i khosjaistvennoe znachenie. *Sbornik 'Pustyni SSSR i ich osvoenie'*, 2. *Acad. Nauk SSSR*. Moscow.

STALMAKOVA, V. A. 1955. Mlekopitajushie Repetecskogo peschanopustynnogo zapovednica i prilezhashikh rajonov Kara-Kumskoi pustyni. *Trudy Repetecskoi peschanopustynnoi stancii*, 3.

STALMAKOVA, V. A. 1957. O nachozdenii turkmenskogo tushkanchika (*Jaculus turcmenicus*) v severnych Kara-Kumach u nekotorykh ego ecologo-morphologicheskikh osobennostjakh. *Zool. J.* XXXVI, 2.

TARASOV, P. P. 1958. Gryzuni jugo-vostochnoi chasti Mongoljskogo Altaja i prilezasthey Gobi. '*Izvestija Irkutskogo n.-i. protivochumnogo Inst. Sibiri i Dalnego Vostoka*, 19.

TCHUGUNOV, J. D. 1962. Materiali po mlekopitajustshim Gobiyskogo Altaja. *Bull. MOIP. Biol.*, 67, 6.

TEMBOTOV, A. K. 1972. Geographia mlekopitajushikh severnogo Kavkasa. Nalchik.

TOKTOSUNOV, A. 1958. Gryzuni Kirgizii. Frunze.

TRUKHACHEV, N. N. 1965. O nachodke karlikovogo tushkanchika v Uznom Prybalkhashije. *Zool. J.* 44, 9: 1428–1429.

VANSULIN, S. A. 1962. Nekotorii kharakternii cherty ecologii bolshih peschanok severo-vostochnogo Prikaspija. '*Voprosi ecologii*', 4, Kiev.

VARSHAVSKIJ, S. N. & SHILOV, M. N. 1956. Ecologo-geographicheskii osobennosti rasprostranenija i territorialnogo raspredelenja bolshoy peschanki v Serevnom Priaralje. '*Trudi Sredne-Aziatskogo n.-i. protivochumnogo Inst.*', 3.

VARSHAVSKIJ, S. N., SHILOV, M. N. & GARBUSOV, V. K. 1962. Landshaftnii osobennosti strukturi poselenji bolshikh peschanok v severnom Priaralje i ikh svjaz s rasseleniem vida. '*Voprosi ecologii*', 6.

VARSHAVSKIJ, S. N., SHILOV, M. N., GARBUSOV, V. K., MARIN, S. N. & PONOMAREV, H. A. 1969. Sovremennoe rasselenie bolshoy peschanki v Severnom Priaralje i ego epizootologicheskoe znachenie *Zool. J.* 48, I: 126–134.

VASILIJEV, S. V. EFIMOVA, V. I. & ZARKHIDZE, V. A. 1963. Osobennosti razmnozhenija bolshoy i krasnokhvostoy peschanok na Krasnovodskom poluostrove. '*Trudi VIZR*', 18.

VINOGRADOV, B. S. 1937. Fauna SSSR. Mlekopitajuschie, 3. Moscow-Leningrad.

VLASOV, J. P. 1937. Nora kak svoeobrazniy biotop v okrestnostjah Ashkhabada. '*Problemi parazitologii i fauni Turkmenii*'. Moscow-Leningrad.

VOLOGIN, N. I. 1967. O potenciale plodovitosti krasnohvostoy peschanki Turkmenii. '*Gryzuni i ikh ektoparaziti*', Saratov: 44–48.

VORONTZOV, N. N., ORLOV, O. J. & SMIRNOV, V. M. 1969. Biologija i rasprostranenie karlicovih tushkanchikov (*Salpingotus crassicauda*) v Zaysanskoy kotlovine. '*Mlekopitajushie*', Novosibirsk.

ZHURAVLEVA, V. I. 1958. O nachodke tushkanchica Lichtensteina (*E. lichtensteini*) v juzhnom Pribalkhashje. *Trudi Sredneaziatskogo protivochumnogo Instituta*, 4: 269-270. Alma-Ata.

AUTHOR INDEX*

Numbers in roman type refer to the page on which the reference is cited. Numbers in italics refer to the page on which the reference is listed.

* Our grateful thanks are due to Mrs ANJANA GHOSH and Miss SHUBHRA MANI for their help in the preparation of the three indices of this book. – Eds.

599

600

602

603

604

605

606

GENUS AND SPECIES INDEX

607

Mus musculus 60–62, 64–66, 68, 69, 71, 72, 74, 81, 172, 340, 344, 360, 371, 378a, 378b, 516, 586
Mus musculus bactrianus 82, 85, 86, 88, 89, 94, 108, 398
Mus musculus gentilis 20–22, 31, 32, 35
Mus platythrix sadhu 85–89, 92–94, 107, 109, 397, 398, 408
Mustella eversmanii 587
Mustella nivalis 587
Myocricetodon cherifiense 374
Myomimus 339, 342, 343, 349, 370, 378a
Myomimus judaicus 339, 342, 349, 352
Myomimus maritsensis 378a
Myomimus personatus 342, 349, 378a
Myomimus roachi 339, 342, 349, 353, 356, 371

Nannocricetus 340, 345, 350, 352, 369
Nanophyton erinaceum 468, 469, 473, 495, 499
Nanorrhops 51
Neopsylla setosa 504, 586
Neospirocerca rajasthanensis 458, 459
Neotoma 155, 169, 193, 241, 413, 417, 430, 459
Neotoma albigula 269, 306, 308, 312, 313, 423
Neotoma fuscipes 155, 167, 169, 191
Neotoma lepida 191, 192, 217, 269, 306, 308, 312, 313, 418, 422
Neotoma micropus 167
Nesokia bacheri 332
Nesokia indica 61, 62, 64–66, 68–74, 368, 372
Nesokia indica indica 82, 85, 89, 109, 398
Nitraria 529, 531, 537, 542
Nitraria schoberi 469
Nitraria sibirica 535
Notomys 193, 196, 197, 221, 384, 385, 389, 394, 395
Notomys alexis 167, 168, 391, 392
Notomys cervinus 167, 391
Notomys mitchelli 218

Occitanomys 378a
Occitanomys anomalus 378a
Ochotona daurica 534
Ochotona roylei 74
Ochotona rufescens 74
Oedaleonotus enigma 291
Oenanthe deserti 586, 589
Oenanthe hispanica 589
Oenanthe isabellina 586, 587
Oenanthe leucopyga 335

Oenanthe lugens 587
Oenanthe plesehanka 587, 589
Onychognathus tristrami 332
Onychomys 192, 196, 219, 241, 254, 255, 262, 264, 314, 384
Onychomys leucogaster 306, 308, 313–315
Onychomys torridus 191, 226, 306, 308, 313–316
Ophtalmopsylla volgensis 504
Opuntia 217, 418, 422
Opuntia engelmanni 423
Ornithodorus tartakowskii 586
Ornithogalum 516
Oropsylla 586
Oryzomys 159
Oryzomys longicaudatus 162, 165
Otis tarda 589

Pachyuromys 179, 180, 181, 220
Pachyuromys duprassi 191
Pachyuromys steatomys 404
Paleocricetus 365
Panicum 39
Panicum turgidum 89
Paradipus 193, 512
Paradipus ctenodactylus 466, 505, 521–523
Paradoxopsyllus tuschkan 586
Paraethomys 378b
Paraethomys anomalus 378b
Paraethomys filfilae 373, 378b
Parallactaga 339, 343, 350, 352, 357, 369
Paraphiomys 374
Paraphiomys simonsi 372
Parapodemus jordanicus 340, 354, 356, 369
Paraspiculuris, see Aspiculuris
Paraxerus 193
Parotomys 186
Passer domesticus 586, 589
Pasturella pestis 582
Pedetes 193
Peganum harmala 474, 508, 523
Pelomys 378a
Pelomys europlus 378a
Pentaphylloides dryadanthoides 74
Perognathus 155, 172, 193, 194, 196, 210–212, 215, 216, 218, 221, 254, 265, 317, 431
Perognathus amplus 244, 246, 306, 308, 316
Perognathus baileyi 244, 246, 251, 306, 308, 316, 413, 423
Perognathus californicus 191, 192, 210, 215, 216
Perognathus fallax 167, 191, 423
Perognathus fasciatus 413
Perognathus flavescens 413

SUBJECT INDEX

617

621

623